MEDICAL RADIOLOGY

Diagnostic Imaging and Radiation Oncology

Gastrointestinal Cancer

Radiation Therapy

Contributors

V. J. Belcastro · P. P. Carbone · B. J. Cummings · L. J. A. DiDio
R. R. Dobelbower, Jr. · D. E. Dosoretz · R. A. DuBrow · R. J. Fadell
F. B. Gelder · P. J. Goldblatt · M. Haghbin · E. J. Hinson · J. M. Howard
M. J. Katin · K. Kim · L. Komarnicky · J. Korinek · B. Levin
J. H. Meerwaldt · W. M. Mendenhall · R. R. Million · M. Mohiuddin
V. B. Pinto · G. Ramirez · B. Sischy · St. M. Wagner

Edited by

Ralph R. Dobelbower, Jr.

Foreword by

Luther W. Brady and Hans-Peter Heilmann

With 76 Figures

Springer-Verlag Berlin Heidelberg New York
London Paris Tokyo Hong Kong

Ralph R. Dobelbower, Jr., M.D., Ph.D., F.A.C.R.
Professor and Chairman
Department of Radiation Therapy
Professor of Neurological Surgery
(Radiation Therapy)
Medical College of Ohio
3000 Arlington Avenue
C.S. 10008
Toledo, OH 43699
USA

MEDICAL RADIOLOGY · Diagnostic Imaging and Radiation Oncology
Continuation of
Handbuch der medizinischen Radiologie
Encyclopedia of Medical Radiology

ISBN 3-540-50505-9 Springer-Verlag Berlin Heidelberg New York
ISBN 0-387-50505-9 Springer-Verlag New York Berlin Heidelberg

Library of Congress Cataloging-in-Publication Data. Gastrointestinal cancer : radiation therapy / contributors, V.J. Belcastro ... [et al.] ; edited by Ralph R. Dobelbower, Jr. ; foreword by Luther W. Brady and Hans-Peter Heilmann. p. cm. - (Medical radiology) Includes bibliographical references.
ISBN 0-387-50505-9 (U.S. : alk. paper)
1. Digestive organs - Cancer. 2. Digestive organs - Cancer - Radiotherapy. I. Belcastro, V.J. (Vincent J.) II. Dobelbower, Ralph R. III. Series.
[DNLM: 1. Digestive System Neoplasms - radiotherapy. WI 149 G2568] RC280.D5G374 1990 616.99'4330642 -dc20
DNLM/DLC 89-21870 CIP

Typesetting, printing and bookbinding: Appl, Wemding
2122/3130-543210 - Printed on acid-free paper

This book is dedicated to my wife, Mary Louise Dobelbower, and our seven children, without whose patience, love, dedication, and sacrifice this work would not have come to fruition.

Ralph R. Dobelbower, Jr., 1989

"The answers to my questions are not found in any book"

RALPH R. DOBELBOWER, SR.,
in his ninth decade 1981

List of Contributors

VINCENT J. BELCASTRO, M. D.
Belcastro and Carrasquillo
Surgical Associates
Cape Coral Hospital
708 Del Prado Blvd.
Cape Coral, FL 33990
USA

PAUL P. CARBONE, M. D., Professor
University of Wisconsin
Department of Human Oncology
Wisconsin Clinical Cancer Center
600 Highland Avenue
Madison, WI 53792
USA

B. J. CUMMINGS, M. B., Ch. B., FRCPC,
F. R. C. R., F. R. A. C. R.
Radiation Oncologist, Professor
Department of Radiology
University of Toronto
The Princess Margaret Hospital
500 Sherbourne Street
Toronto, Ontario, M4X 1K9
Canada

LIBERATO J. A. DIDIO, M. D., D. Sc., Ph. D.,
Professor of Anatomy
Department of Anatomy
Medical College of Ohio
3000 Arlington Avenue
C. S. 10008
Toledo, OH 43699
USA

RALPH R. DOBELBOWER, JR., M. D., Ph. D.,
F. A. C. R., Professor and Chairman
Department of Radiation Therapy
Professor of Neurological Surgery
(Radiation Therapy)
Medical College of Ohio
3000 Arlington Avenue
C. S. 10008
Toledo, OH 43699
USA

DANIEL E. DOSORETZ, M. D.
Radiation Therapy Regional Center
Radiation Therapy Associates
3680 Broadway
Fort Myers, FL 33901
USA

RONELLE A. DUBROW, M. D.
Associate Professor
Division of Diagnostic Imaging
and Department of Diagnostic Radiology
The University of Texas System Cancer
Center
M. D. Anderson Hospital and Tumor
Institute
1515 Holcombe Blvd.
Houston, TX 77030
USA

RONALD J. FADELL, M. D.
Department of Radiology
Medical College of Ohio
3000 Arlington Avenue
C. S. 10008
Toledo, OH 43699
USA

FRANK B. GELDER, Ph. D.
Associate Professor of Surgery, Pathology,
and Microbiology/Immunology
Director of Clinical Immunology
Louisiana State University Medical Center
1501 Kings Highway
Shreveport, LA 71130
USA

PETER J. GOLDBLATT, M. D.
Professor and Chairman
Department of Pathology
Dean of Graduate School
Medical College of Ohio
3000 Arlington Avenue
C. S. 10008
Toledo, OH 43699
USA

Mahroo Haghbin, M.D.
Clinical Assistant Professor of Radiation Oncology
School of Medicine and Dentistry
University of Rochester Medical Center
Rochester, NY 14620
USA

E. Josephine Hinson, R.T.T.
Research Assistant, Daisy Marquis Jones Radiation Oncology Center
Highland Hospital
1000 South Avenue
Rochester, NY 14620
USA

John M. Howard, M.D.
Professor of Surgery
Department of Surgery
Medical College of Ohio
3000 Arlington Avenue
C.S. 10008
Toledo, OH 43699
USA

Michael J. Katin, M.D.
Radiation Therapy Regional Center
Radiation Therapy Associates
3680 Broadway
Fort Myers, FL 33901
USA

Kitai Kim, M.D.
Professor of Pathology
Department of Pathology
Medical College of Ohio
3000 Arlington Avenue
C.S. 10008
Toledo, OH 43699
USA

Lydia Komarnicky, M.D.
Department of Radiation Therapy and Nuclear Medicine
Thomas Jefferson University Hospital
Eleventh and Walnut Street
Philadelphia, PA 19107
USA

Josef Korinek, M.D., Ph.D.
Section of Gastrointestinal Oncology and Digestive Diseases, Division of Medicine
The University of Texas
M.D. Anderson Cancer Center, Houston
P.O. Box 78, 1515 Holcombe Blvd.
Houston, TX 77030
USA

Bernard Levin, M.D.
Section of Gastrointestinal Oncology and Digestive Diseases
Division of Medicine
The University of Texas
M.D. Anderson Cancer Center, Houston
P.O. Box 78, 1515 Holcombe Blvd.
Houston, TX 77030
USA

J.H. Meerwaldt, M.D., Ph.D.
The Dr. Daniel den Hoed Cancer Center and Rotterdam Radio-Therapeutic Institute
Department of Radiotherapy
P.O. Box 5201
3008 AE Rotterdam
The Netherlands

William M. Mendenhall, M.D.
Associate Professor of Radiation Therapy
Department of Radiation Therapy
University of Florida College of Medicine
Box J-385, J. Hillis Miller Health Center
Gainesville, FL 32610-0385
USA

Rodney R. Million, M.D.
Professor of Radiation Therapy
Department of Radiation Therapy
University of Florida College of Medicine
Box J-385, J. Hillis Miller Health Center
Gainesville, FL 32610-0385
USA

Mohammed Mohiuddin, M.D.
Professor of Radiation Therapy and Nuclear Medicine
Department of Radiation Therapy and Nuclear Medicine
Thomas Jefferson University Hospital
Eleventh and Walnut Street
Philadelphia, PA 19107
USA

VALERIAN B. PINTO, Doctoral Candidate
Department of Microbiology/Immunology
Louisiana State University Medical Center
1501 Kings Highway
Shreveport, LA 71130
USA

GUILLERMO RAMIREZ, M.D.
Professor of Oncology
University of Wisconsin
Department of Human Oncology
Wisconsin Clinical Cancer Center
600 Highland Avenue
Madison, WI 53792
USA

BEN SISCHY, M.D.
Director
Clinical Professor of Radiation Oncology
School of Medicine and Dentistry
University of Rochester Medical Center
Daisy Marquis Jones Radiation Oncology Center
Highland Hospital
1000 South Avenue
Rochester, NY 14620
USA

STEVEN M. WAGNER, M.D.
Clinical Associate Professor of Medicine
Department of Medicine
Medical College of Ohio
3000 Arlington Avenue
C.S. 10008
Toledo, OH 43699
USA

Foreword

Primary malignant tumors of the gastrointestinal tract account for 23% of all invasive cancers and 24% of all the deaths in 1988. In the United States, the American Cancer Society estimates that there will be 227 500 new cases of gastrointestinal malignancy diagnosed, with 122 350 deaths. This includes 9800 new cases of carcinomas of the esophagus, 24 800 carcinomas of the stomach, 2500 new cases of malignant tumors of the small intestine, 105 000 new cases of tumors of the large intestine, and 42 000 new cases of carcinoma of the rectum, 14 000 new cases involving the liver and biliary passages, 27 000 new cases of carcinoma of the pancreas, and 2400 other unspecified digestive tract malignancies. This combined incidence is second only to the number of new cases of lung cancer.

Bleeding from the gastrointestinal tract represents one of the earliest findings with regard to malignancy of the gastrointestinal tract. In general, personal or family history of colon and rectal cancer, personal or family history of polyps in the colon or rectum, or inflammatory bowel disease are high risk factors with regard to tumors in these locations. Dietary behavior is also important in that a diet high in fat and/or low in fiber content may be a significant causative factor.

In the assessment of patients with lesions involving the colon and rectum, a rectal examination should be performed every year after the age of 40, and if stool specimens reveal the presence of blood then a proctosigmoidoscopy should be carried out in order to precisely define the cause of the rectal bleeding. As the sites of most colorectal cancers appear to be shifting higher in the colon, the longer flexible instruments that are now available are required for adequate exploration of the entire colon, which is not possible with rigid proctoscopes. If abnormalities are found, then more extensive studies such as colonoscopy, barium enema, etc., should be carried out.

In primary malignant tumors of the esophagus, the onset of symptoms is insidious in character and unfortunately delays diagnosis. The same is true for primary malignant tumors involving the stomach, small intestine, liver and biliary passages, and pancreas. In these instances, there is a pressing need for appropriate diagnostic studies that would allow earlier diagnosis. Any insidious symptom relative to esophagitis, dysphagia, abdom inal pain, discomfort, or distress should be investigated fully in order to define the cause of the difficulty.

In 1989, the management of malignant tumors involving the gastrointestinal tract has progressed to where combined integrated multimodal programs of management are appropriate for essentially all tumor sites. Surgery plays an important role not only from the standpoint of diagnosis but in many instances from the standpoint of surgical resection. Oftentimes postoperative radiation therapy is appropriate in the period following surgical resection. Multidrug systemic chemotherapy is being studied more actively now in the treatment not only of advanced disease but also in those patients at high risk or local persistence or recurrence of their disease process.

Various techniques for radiation therapy are finding important applications in malignant tumors of the gastrointestinal tract including not only intraluminal placement of radioactive sources in order to boost the radiation dosage beyond that achieved by external beam radiation therapy but also intraoperative radiation therapy for malignant tumors of the stomach, pancreas, biliary tree, and retroperitoneal lymph nodes. Interstitial implantation techniques using radioactive sources are appropriate for certain tumor sites such as the pancreas.

With the growing number of primary malignant tumors of the digestive tract, more effort is now toward multimodal management than in the past, with gradual progressive improvement in terms of long-term survival without disease.

The book put forth by Ralph R. Dobelbower, Jr. et al. addresses these questions in-depth and makes recommendations as to appropriate techniques for early diagnosis, tumor assessment, and staging as well as treatment procedures.

LUTHER W. BRADY
Philadelphia

HANS-PETER HEILMANN
Hamburg

Preface

During my residency training at the Thomas Jefferson University Hospital, I became interested in the treatment of deep-seated malignancies, particulary cancer of the pancreas, when the Department of Radiation Therapy and Nuclear Medicine acquired a 45 million Volt Brown Bovari betatron which was capable of generating electron beams with energies up to 45 meV, as well as 45 meVp photons. This eventually led to development of a general interest in gastrointestinal cancers that was fueled by interaction with other gastrointestinal oncologists at that institution, particularly Dr. Gerald Marks, the noted colorectal surgical oncologist. For the last nine years I have had the pleasure of pursuing my interests in gastrointestinal malignancies at the Medical College of Ohio, often in collaboration with John M. Howard, M. D., and Hollis W. Merrick III, M. D., both of whom harbor abiding interests in gastrointestinal tract neoplasia. Dr. Howard was instrumental in founding the SACAAPACA group (Scientists and Clinicians Allied Against Pancreatic Cancer) at the MCO. This group has provided not only a forum for discussion of problems related to pancreatic cancer research, but also a vehicle for prospective study in these regards. For the last six years I have served with John Earle, M. D. as Co-Chairman of the Radiation Therapy Committee of the Eastern Cooperative Oncology Group, and as the Radiation Therapy Co-Chairman of the Gastrointestinal Disease Committee of that group. This has served to introduce me to a whole new spectrum of oncologists with special interests in gastrointestinal tumors. From these experiences I have selected a panel of experts that I consider particularly well qualified to review recent developments in the field of gastrointestinal cancer. This panel comprises the group of chapter authors of this book. My purpose in organizing and editing the work is to present what my colleagues and I feel represents the state of the art in the understanding, diagnosis and treatment of gastrointestinal cancer.

This book is designed to provide a comprehensive review suitable for the practicing radiation oncologist, medical oncologist, and surgical oncologist, as well as residents in training in the oncologic specialties. The general surgeon, the general internist and the gastroenterologist will also find the work of interest as a reference. The intent is to provide a general overview rather than just the experience of selected individuals or institutions. Rare tumors are discussed as well as garden variety neoplasms. The list of references for each chapter is up-to-date, and is meant to direct the reader's attention to keystone articles as well as classic pieces of oncologic investigation. Much of the work is organized by disease site, however, major chapters are devoted to radiographic diagnosis, tumor markers, pathology, chemotherapy, and follow-up of patients treated for gastrointestinal malignancy.

This book should aid the oncologist in understanding the basic nature of gastrointestinal oncology problems, as well as assist him in the diagnosis and treatment (especially by radiation therapy) of gastrointestinal malignancies.

Chapter 1 is an overview of the gastrointestinal cancer problem. The etiology of gastrointestinal tumors is discussed here only briefly, but is addressed in some detail in many of the disease site oriented chapters.

In Chap. 2, Drs. Gelder and Pinto discuss the applications of tumor markers to the diagnosis and treatment of gastrointestinal malignancies.

Chapter 3, by Drs. Goldblatt and Kim, is essentially an atlas of the pathology of tumors of the gastrointestinal tract. Distinctive gross and microscopic features of each neoplasm are mentioned, and several illustrative photographs emphasize selected tumors and aspects of histopathology.

Chapter 4, by Dr. DuBrow, gives an overview of the radiologic approaches to diagnosis of gastrointestinal cancer, punctuated by many illustrative radiographs.

Chapter 5, written by Drs. Mendenhall and Million, presents the natural history, epidemiology, staging, prognostic factors and treatment for neoplasms of the esophagus with an emphasis on the radiation treatment techniques as well as the results. This general format is followed in Chaps. 5 through 11.

In Chap. 6, Drs. Katin, Belcastro, and Dosoretz present an excellent synopsis of the management of gastric cancer.

Chapter 7 is devoted to an extensive discussion of innovations in the diagnosis and management of pancreatic cancer over the course of the last decade. The authors of this chapter are myself, Drs. Wagner, Fadell, Howard, and DiDio.

Dr. Meerwaldt discusses biliary tumors in Chap. 8, and Drs. Mohiuddin and Komarnicky discuss colon cancer in detail in Chap. 9.

Chapter 10 is devoted to recent developments in radiation therapy of rectal cancer presented by Dr. Cummings.

In Chap. 11, the treatment of anal cancer is outlined in some detail. This chapter is authored by Drs. Haghbin and Sischy, in collaboration with E. Josephine Hinson.

The role of the medical oncologist in the management of gastrointestinal cancers is presented in comprehensive form by Drs. Ramirez and Carbone in Chap. 12.

Drs. Korinek and Levin address issues pertinent to the follow-up of patients with gastrointestinal cancer in a final Chap. 13.

It is hoped that this book will serve as a timely update on the subject of management of patients with gastrointestinal malignancies as part of the Springer-Verlag Diagnostic Imaging and Radiation Oncology series. To the best of our knowledge, no other volume presents a cohesive compilation of these many facets of the gastrointestinal cancer problem.

Acknowledgements. Many individuals made significant contributions to this work. In addition to the authors of the various chapters of the book, many other individuals have made significant contributions. I am indebted to medical mentor, Simon Kramer, M. D., for inspiration and encouragement. Special thanks are due Mark Calcamuggio for his assistance in editing portions of the manuscript. I wish to thank many of my past and present associates, including Rajender K. Ahuja, M. D., Munther I. Ajlouni, M. D., Andrew J. Milligan, Ph. D., Donald G. Bronn, M. D., Ph. D., Andre A. Konski, M. D., and Farideh R. Bagne, Ph. D., J. D. for many hours conducting the clinical and academic affairs of the Department of Radiation Therapy at the Medical College of Ohio when I was absent from the clinic or administrative offices while in the process of preparation of this book. I am grateful to Robin France, Beth Comte, and Sandy Price for secretarial assistance in the preparation of the manuscript.

RALPH R. DOBELBOWER, JR.

Contents

1 Introduction

Ralph R. Dobelbower, Jr.

Cancer is the second leading cause of death in the United States today. It is estimated by the American Cancer Society that nearly one million new cancer cases will be diagnosed in the United States in the year 1989. This figure excludes carcinomas in situ and nonmelanoma skin cancers. If these highly curable cancers were included, the number of new cancer cases would be increased by over 50%. The total number of cancer deaths in the United States during calendar year 1989 is estimated to be approximately half the number of new cancer cases diagnosed (roughly one-half million).

The gastrointestinal (GI) tract accounts for more new cancer cases than any other body system, approximately a quarter of a million new cases annually. And roughly half that number of patients are expected to die of GI cancers each year. The overall importance of the GI cancer problem cannot be overstated.

Some of the cancers arising in the GI tract are amenable to the usual forms of antineoplastic therapy (surgery, radiation therapy, and chemotherapy) while others (cancers of the biliary tree and the pancreas, for example) are steadfastly refractory to conventional treatment. In such cases, new and effective means of therapy must be identified.

During the calendar year 1989, it is estimated that 10100 Americans will develop esophageal cancer and that 9400 patients will die of this disease. Clearly the results of current therapy leave a great deal to be desired. It is estimated that 20000 Americans will develop stomach cancer and that 14000 will die of that disease. Here again there is considerable room for improvement of clinical results.

Tumors of the small intestine are the least common of the GI malignancies. It is estimated that 2700 Americans will develop this disease in 1989 and that 800 will die therefrom. On the other hand, the large bowel is one of the most common sites of neoplasia in the human body in both sexes with 107000 patients expected to develop the disease and 54000 expected to die therefrom in 1989. Forty-four thousand Americans are expected to develop cancer of the rectum in 1989, and it is expected that 7900 will die of that disease.

Tumors of the liver and biliary ducts will account for 14500 new cases and 11400 deaths during the year, while pancreatic cancer will account for 27000 new cases and 25000 deaths. These figures, when taken in toto, indicate the magnitude of the GI cancer problem.

Unlike lung cancer, the etiology of most of the tumors of the GI tract is enigmatic. The marked geographic variations in incidence of GI cancers, especially esophageal and gastric neoplasms, suggests a high degree of dependence on external and geographic factors; however, these tantalizing bits of epidemiological information (discussed in detail elsewhere in this work) have been insufficient to lead us to the direct causes of GI cancer in most cases. The decreasing incidence of gastric cancer is a case in point, as is the changing pattern of incidence of colorectal cancer.

In recent decades, it was stated that 50% of colorectal cancers were within the reach of the examining finger and 75% within reach of the sigmoidoscope. This is no longer true with recent findings showing a higher incidence of right-sided colon cancer.

This decade has seen an explosion in techniques of genetic engineering which, hopefully, will provide new insight into tumorigensis in general and, into the development of neoplasms of the gastrointestinal tract in particular.

Recognition of the acquired immune deficiency (AIDS) during this decade is having major implications for our understanding of the etiology of certain GI tumors, particularly anal cancers and lymphomas. Much remains to be learned in these regards.

The development of DNA flow cytometry also has major implications for the diagnosis, treatment,

---

Ralph R. Dobelbower, Jr., M.D., Ph.D., F.A.C.R., Professor and Chairman, Department of Radiation Therapy, Professor of Neurological Surgery (Radiation Therapy), Medical College of Ohio, 3000 Arlington Avenue, C.S. 10008, Toledo, OH 43699, USA and follow-up of GI malignancies; however, the surface of the body of knowledge in this regard has just been scratched and there is a great deal more to learn.

The development of new imaging techniques such as magnetic resonance imaging (MRI), endoscopic ultrasound and, especially, computerized body section radiography (CT) has revolutionized the diagnosis of deep-seated malignancies. The CT scan enables the radiation oncologist to vigorously tailor the dose profile to the tumor profile in many GI cancer situations. This, along with the use of radiation therapy simulators, which have now proliferated to every corner of the world, and the development and wide-spread use of high-energy photon and electron beams, enables the radiation oncology team to effectively treat tumors that could not be treated in prior decades. Furthermore, tailoring the dose profile carefully to the disease profile with the aid of these new technologies permits higher local radiation tumor doses with lower doses (and thus fewer complications) in normal tissues. Many pilot studies and clinical trials will be necessary to accurately define the role of the implementation of these new technologies.

Surgical and anesthesia techniques have changed considerably in recent decades. The risk of anesthesia per se is now almost nil. More knowledge of patients' nutritional needs, careful preoperative care, and meticulous postoperative care with antibiotics and other drugs permit bold surgeons to develop new techniques for dealing with GI malignancies. Combinations of surgical, radiotherapeutic, and medical antineoplastic abilities to enhance local tumor clearance, decrease the incidence of metastasis, and improve the patients' survival are being studied on every front. Not only are these studies being conducted at individual institutions, but now more widely by cooperative group (and intergroup) studies. Such efforts can only increase our knowledge base when it comes to dealing with the problems of GI cancer.

The recent resurgence of intraoperative radiation therapy, interstitial radiation therapy, and intraluminal radiation therapy techniques is being directed primarily to GI neoplasia. The tumor most often irradiated with intraoperative radiation therapy has been cancer of the pancreas. Although the survival results have been far from spectacular, the palliative benefit in terms of pain relief has been marked.

In spite of new technologies (MRI, CT, DNA flow cytometry, genetic engineering, high-energy electron, photon and particle beams) and the resurgence and reapplication of old technologies (intraoperative radiotherapy, intraluminal radiotherapy, interstitial implants), some tumors simply remain refractory and elude effective treatment – tumors such as cancers of the biliary system and the pancreas. Yet, the last decade has seen some major improvements in antineoplastic therapy in the GI tract. The changing pattern of treatment for anal cancer is an example. It is now becoming clear that the highly morbid but previously standard treatment, abdominoperioneal resection, may not be necessary for the majority of patients with this disease when they are treated with moderate dose radiation therapy along with radiation sensitizers and followed closely.

Hopefully, other tumor sites will yield to new applications of old technology and the use of new technology, including radioprotectors, radiosensitizers, and biologic response modifiers in combined modality treatment programs during the ensuing decade. The subsequent chapters in this work provide a comprehensive review of the state of the art in GI cancer and also identify problem areas where additional clinical and basic research is needed to advance the cause against GI cancer.

2 Tumorigenesis and Tumor Markers

FRANK B. GELDER and VALERIAN B. PINTO

CONTENTS

2.1 Introduction

Until recently, the bulk of basic information about cancer came from studies which defined phenotypic differences between malignant and normal cells and tissues. The information gained from these studies indicates that cancer cells share certain characteristics. However, these comparative studies failed to define clear-out cancer cell specific alterations or explain the mechanisms responsible for both the induction and the aberrant behavior of tumor cells. Recent technological advances, especially at the molecular and genetic level, promise to complement substantially and extend these earlier comparative studies. Although several areas of investigation have contributed to the rapidly expanding knowledge defining cancer biology, two avenues were of central importance: (a) the biological and biochemical characterization of DNA and RNA tumor viruses leading to techniques in somatic cell hybridization, chromosomal transfer, and DNA transfection, and (b) the identification and utilization of bacterial restriction endonuclease enzymes leading to the molecular cloning and detailed analyses of discrete DNA segments.

All cancers, as a group, are composed of cells that have lost the mechanisms controlling the critical balance of cell number in a normal organ system. These changes result in a proliferative increase in cells unresponsive to cell growth regulatory mechanisms. In the past few years, some of the answers to questions about cell regulation have been answered. Cancer genes (oncogenes) have been discovered in the chromosomes of some tumor cells and are thought to represent the driving force behind initiating and directing uncontrolled growth in many cancers. The activation of these genes appears to be associated with the conversion of normal cells into cancerous cells. Once activated, these genes continually direct cells to behave in an abnormal pattern. The abnormal behavior of cancer cells is characterized by several distinctive features. The most obvious is uncontrolled growth. A second characteristic is invasiveness. Cancer cells fail to respect the territorial boundaries that confine normal cells to particular tissues. They detach and establish residence at distant locations (metastasis). This change is defined in vitro as a loss of contact inhibition and anchorage dependency. The loss of both of these cell growth regulation characteristics may simply represent the loss or alteration of membrane receptors responsible for turning off cell division following contact with a neighboring cell.

The anchorage-independence of transformed baby hamster kidney (BHK) cells is lost after fusion

FRANK B. GELDER, Ph.D., Associate Professor of Surgery, Pathology, and Microbiology/Immunology, Director of Clinical Immunology, Louisiana State University Medical Center, 1501 Kings Highway, Shreveport, LA 71130, USA

VALERIAN B. PINTO, Doctoral Candidate, Department of Microbiology/Immunology, Louisiana State University Medical Center, 1501 Kings Highway, Shreveport, LA 71130, USA with normal human fibroblasts and is reexpressed when one copy of human chromosome number 1 is lost (STOLER and BOUCK 1985). Direct evidence for the involvement of chromosome 1 in human malignancies comes from the work of Benedict and colleagues (BENEDICT et al. 1984). They showed that the ability of the suppressed hybrids produced by the fusion of HT1080 (human fibrosarcoma line) and normal human fibroblasts to form tumors in nude mice, was associated with the loss of both copies of chromosome 1 and one copy of chromosome 4. Thus, it appears that a locus on chromosome 1 governs the expression of anchorage independence. This in turn is directly related to tumorigenicity (MARSHALL and DAVE 1978).

Several biochemical changes have been identified in malignant cell populations when compared to their normal cell counterparts. Deficiencies in complex glycolipids and glycoproteins are frequently demonstrated along with concomitant increases in simple gangliosides (G_{m1}, G_{m3}). Presumably, some of these biochemical alterations result in the expression of characteristics consistent with those defining cancer, including uncontrolled growth, invasion, detachment, residence in distant structures, altered cell morphology, reduced serum requirement, increased saturation density, extended or indefinite life span, alteration of karyotype from diploid to aneuploid, alteration of cell surface antigens, expression of fetal antigens and the synthesis of ectopic proteins, production of proteinases (e.g., plasminogen), impairment of terminal differentiation, and tumorigenicity in a susceptible host.

What follows is a discussion of tumorigenesis in general and is not exclusively related to tumors of the gastrointestinal (GI) tract. Illustrative examples from other body systems are employed when appropriate.

2.2 Molecular Basis of Cancer

Although the mechanism of cellular transformation is not well-defined, certain facts are known. First, experimental, clinical, and epidemiologic data support the hypothesis that transformation is a multistep event. Secondly, transformation is frequently associated with an alteration of the genome. Thirdly, a genomic alteration may express its transforming potential by either recessive allele expression, which requires the inactivation of the normal homologous chromosome, or oncogene activation, which acts dominantly.

2.2.1 Recessive Allele Expression

A number of childhood tumors fit the "two-mutation hypothesis" proposed by KNUDSON (1971) for the initiation of cancer by recessive allele expression. According to this hypothesis, two mutations are required for cell transformation. The first mutation can be either somatic or germinal but the second must always be somatic. When the first mutation occurs in a germinal cell, the tumor is hereditary, and the gene carrier may develop 0, 1, or multiple tumors as seen in inherited retinoblastomas (KNUDSON et al. 1975), neuroblastomas, and Wilms' tumors (KNUDSON and STRONG 1972). When the first mutation occurs in somatic cells, as seen in sporadic cases of retinoblastoma and neuroblastoma, the tumor is nonhereditary and single. In both cases, the development of tumors requires inactivation of the normal allele on the homologous chromosome.

An hereditary form of the disease arises as a germinal mutation of the locus [Rb-1 locus in retinoblastomas (KNUDSON et al. 1976), 11 p 13 locus in Wilms' tumors (KOUFOS et al. 1984), and loss of genes on chromosome 22 (SEIZINGER et al. 1986)], and is inherited by an individual who then carries a mutation in all of his somatic or germinal cells. Any subsequent event in such predisposed cells which result in homozygosity for the mutant allele will also result in a tumor clone. Possible chromosomal mechanisms that lead to homozygosity of the mutant allele include: (a) mitotic nondisjunction with loss of the wild-type chromosome resulting in homozygosity at all loci on the chromosome, and (b) mitotic nondisjunction with reduplication of the mutant chromosome resulting in homozygosity at all loci on the chromosome. Other possible mechanisms include mitotic recombination, gene conversion, deletion, and point mutation (CAVENEE et al. 1983). Genetic and molecular studies demonstrate that mitotic nondisjunction and mitotic recombination are involved in the loss of the homologous wild-type chromosome 13 resulting in the development of retinoblastoma. Somatic deletion and duplication of genes on chromosome 11 and loss of genes on chromosome 22 result in Wilms' tumor and acoustic neuroma, respectively. Recent work has also revealed homozygous deletions in band 14 on chromosome 3 of oat cell carcinoma (WHANG-PENG et al. 1982) and in chromosome 11 of hepatoblastoma and rhabdosarcoma (KOUFOS et al. 1985). Thus, it seems very likely that similar homozygous genetic defects will be found that lead to an overt predisposition to a number of familial carcinomas, including renal carcinomas (PATHAK et al. 1982) and perhaps some of the GI cancers that manifest a familial predisposition.

2.2.2 Oncogenes and Oncogene Activation

Oncogenes are highly conserved cellular genes, probably involved in the process of growth and differentiation, in which alteration to either their coding potential or their expression is an essential step in neoplastic transformation. Viral oncogenes are known as v-*onc*s and their cellular progenitors are termed c-*onc*s. The term proto-oncogene is used to denote either the cellular progenitors of retroviruses v-*onc*s or the cellular genes whose damage gives rise to the active oncogenes in tumor DNA.

Employing genetic engineering techniques, several laboratories have demonstrated that the activation of oncogenes, under certain circumstances, results in the development of cancer. Two distinct mechanisms have been described for oncogene activation. In the first mechanism, oncogenes are introduced into the cell by retroviruses following infection and reverse transcription of RNA and DNA. These retrovirus-associated oncogenes are of cellular origin. They have been picked up from animal cells and carried (transduced) by these viruses, possibly providing the virus with an increase in survival potential. In the second mechanism, oncogene expression is associated with a mutagenic event. Nonviral carcinogens, including various forms of radiation and a wide range of chemical agents, can induce cellular transformation. In contrast to the insertion of genetic codes by viruses into mammalian gene sequences, carcinogens act by damaging DNA, thus potentiating gene mutation. Nonlethal mutations at specific sites may result in cancer. Gene transfer experiments provided direct evidence of the role of altered DNA in cell transformation. In these experiments, DNA was extracted from methylcholanthrene-induced mouse fibrosarcoma cells and transferred to cultures of normal mouse fibroblasts. Shortly thereafter, several colonies of transformed cells appeared. The information governing the initial methylcholanthrene-induced tumor cell had been transferred to a normal cell by DNA molecules. Subsequent studies demonstrated that the transfer of DNA obtained from malignant cells to normal cells consistently results in transformation. Using serial gene transfer, discrete segments of human DNA (oncogenes) were identified which carried the information required to transfer normal cells into cancer cells. Employing cloned oncogenes as probes in hybridization experiments, it was determined that similar regions in normal human DNA existed. Further characterization of these hybridizing regions from normal DNA demonstrated and pinpointed minor regions of difference. Therefore, it became clear that the oncogene, capable of transforming normal cells, was a slightly altered version of a normal gene (proto-oncogene). However, unlike the oncogene, the proto-oncogene lacked transforming capabilities.

Two distinct sets of normal proto-oncogenes and their respective activated oncogenic counterparts have been described based on their mode of activation. One set was discovered during gene transfer experiments and comprises oncogenes that arise from the mutation of proto-oncogenes following insult by numerous environmental factors. The other set is composed of genes that are carried by and activated by retroviruses. These data support the hypothesis that the same proto-oncogene can be activated by two independent mechanisms: (a) carcinogen induced mutation and (b) retrovirus acquisition.

The well-studied mechanisms of transformations mediated by carcinogens and/or retroviruses that lead to oncogene activations are: (a) chromosome translocation (qualitative), (b) oncogene amplification (quantitative), and (c) promoter/enhancer insertion.

2.2.2.1 Chromosome Translocation

The process of transformation mediated by chromosomal translocation is best illustrated by Burkitt's lymphoma and murine plasmacytomas. Transformation involves translocation of genetic material in which known oncogenes have been located to a new site. In Burkitt's lymphoma, the *myc* gene (v-*myc:* myeloblastosis virus in chickens), normally located on chromosome 8, relocates in the vicinity of one of the three immunoglobulin loci on chromosome 16, 2, or 22. In chronic myeloid leukemia, the *abl* gene (v-*abl:* Abelsons leukemia virus in mouse) in chromosome 8 translocates to chromosome 22 and in mouse plasmacytomas; the *myc* gene translocates to the vicinity of the immunoglobulin gene loci on chromosome 12 or 6 from chromosome 15 (ROWLEY 1984).

The mechanism of transformation mediated by chromosome translocation has yet to be elucidated. Although enhanced transcription of the translocated *myc* gene is proposed as a mechanism of continued stimulation for cellular multiplication, it is by no means the rule in either Burkitt's lymphoma or mouse plasmacytoma (KLEIN and KLEIN 1985). Neither is mutation nor a cis/trans mechanism solely responsible for activation of the translocated *myc* gene. However, the evidence to date clearly indicates that translocation of the *myc* or *abl* gene in cells appears to confer selective multiplication advantages to these cells.

2.2.2.2 *Oncogene Amplification*

Oncogene amplification has been observed in a number of cytological preparations of malignant tumors and occurs mainly in the form of tandem expansions of particular genes or in the form of supernumerary chromosomes. Examples of oncogene amplification include c-*myc* in breast carcinomas (KOZBOR and CROCE 1984) and neuroblastomas (SCHWAB et al. 1983) and amplification of c-*myc* and n-*myc* in small cell carcinoma of the lung (LITTLE et al. 1983). However, in no case is amplification of a particular oncogene consistently associated with a particular kind of tumor. Nor is gene amplification limited to malignant cells, since amplification of c-Ha-*ras* oncogene is observed in normal diploid human fibroblasts during their limited replicative life span in vitro (SRIVASTAVA et al. 1985). Therefore, although amplification of oncogenes may not be essential for transformation, amplification seems to confer some selective advantage to the cells both in vitro and in vivo.

2.2.2.3 *Promoter Insertion*

Certain retroviruses, without oncogenes, responsible for chronic leukemias with a long latent period after infection cause neoplasia by the "promoter insertion mechanism." When the provirus integrates in the immediate vicinity of a cellular oncogene, the strong promotor/enhancer sequences present in the long-terminal repeats (LTRs) of the provirus activate the cellular oncogene, resulting in constant overproduction of the oncogene product.

There is significant experimental proof for the promoter insertion mechanism. The avian leukemia virus integrates in the vicinity of the c-*myc* gene either downstream and in the same transcription orientation as c-*myc* or upstream of c-*myc* in either orientation (PAYNE et al. 1981, 1982; NEEL et al. 1981). Transformation by the promoter insertion mechanism also has been observed in T cell lymphomas of AKR mice induced by mink cell, focus-forming retroviruses (LI et al. 1984) and in some thymic lymphosarcomas in cats associated with feline leukemia virus (NEIL et al. 1984).

2.2.3 Human T Cell Leukemia Virus Induced Transformation

Human T cell leukemia viruses (HTLVs) do not contain onc genes and do not seem to activate a cellular *onc* gene by insertional activation because the proviral integration shows no common sites in cellular DNA of primary leukemic cells. Therefore, a trans-acting protein is suspected of being the transforming element in infected cells. The pX protein, a product of the *tat* gene, is a specific transcriptional activator of the LTR of the HTVL and is essential for viral replication (CHEN et al. 1985; FUJISAWA et al. 1985; FELBER et al. 1985; SODORSKI et al. 1984; ROSEN et al. 1985). In HTLV 1 and HTLV 2, the pX protein is thought to act on cellular regulatory sequences, thereby activating certain cellular *(onc?)* genes resulting in transformation. However, the pX protein of HTLV 3 is thought to be responsible for host cell (T4) mortality.

2.3 Oncogene Products

Oncogenes mediate their tumorigenic activity through the proteins they encode, which include growth factors, growth factor receptors, and intracellular signalling molecules (effector molecules). Transformation of cells mediated by oncogenes is the result of either uncontrolled synthesis of self-stimulatory growth factors, alteration in the growth factor receptor resulting in a mitogenic in stimulus in the absence of ligand interaction, or the uncontrolled synthesis of effector molecules (also the synthesis of abnormal molecules). These mechanisms of transformation are termed "autocrine stimulation," a term first used by SPORN and TODARO (1980) to describe proliferation in transformed cells via the uncontrolled synthesis of growth factor by the target cell (transformed cell) itself.

2.3.1 Growth Factors

Growth factors appear to be produced by the same cell on which they act. Platelet-derived growth factor (PDGF) is a potent mitogen for connective tissue cells and is synthesized not only by platelets but also by normal cells such as cytotrophoblasts

(GOUSTIN et al. 1985). This growth factor must be supplied exogenously to stimulate growth of connective cells. The PDGF/v-sis (v-sis: simian virus in woolly monkeys) dependent autocrine stimulation occurs in sarcomas and glioblastoma (EVA et al. 1982) and its expression in NIH-3T3 cells transforms the cells morphologically to cause anchorage independence (CLARKE et al. 1984).

Like PDGF, TGFα (transforming growth factor alpha) is a peptide that is structurally related to EGF (epidermal growth factor) and is synthesized only for a short period during embryogenesis (TWARDZIK et al. 1982). However, it also is synthesized by certain transformed cells such as human carcinomas and virally transformed rodent fibroblasts (TODARO et al. 1976, 1980). Similarly, the effects of bombesin-like peptides produced by small cell carcinomas of the lung are inhibited both in vitro and in vivo by antibodies against bombesin (CUTITA et al. 1985).

2.3.2 Receptor Alterations

Mechanisms such as structural modification of receptor proteins, over-expression, and possibly abnormal or ectopic expression may lead to the receptor molecule functioning as a transforming protein. Truncated forms of the EGF receptor are found in human tumors, but since they lack the transmembrane and cytoplasmic catalytic domains, over expression, as seen in glioblastomas and the majority of squamous cell carcinomas (LIBERMANN et al. 1985; OZANNE et al. 1985), may be responsible for transformation. In these tumors, the over expression of EGF receptors is a direct result of receptor gene amplification, thereby sensitizing these cells to low levels of growth factors resulting in transformation.

The c-*fms* (v-*onc:* McDonough sarcoma virus in chicken) gene appears to encode the receptor for CSF-1 (colony stimulating factor), the growth factor for mononuclear phagocytic cells. In this case, the cells are provided with a receptor for the growth factor they normally produce or c-fms-expressing cells must be exposed to CSF-1 in order to be transformed.

2.3.3 Nuclear Oncoproteins

Proteins encoded by viral oncogenes like v-*myc* (MC29 myelocytomatosis virus in chicken), v-*fos* (FBJ osteosarcoma virus in mouse), v-*ski* (SKV 770 virus in chicken), and v-*myb* (myeloblastosis virus in chicken) and their cellular counterparts (c-*myc,* c-*fos,* c-*ski,* and c-*myb*) are located in the nucleus (BISHOP 1985). These proteins, along with p53 (a nuclear oncoprotein complexed with the large T antigen in SV40 transformed cells), have short half-lives and, thus, may function as effectors to link events at the cell surface to the nucleus to regulate gene expression and DNA synthesis. Elevated levels of c-*myc,* p53, and c-*fos* expression, as judged by RNA levels, are found following stimulation of quiescent fibroblasts (KELLY et al. 1983; REICH and LEVINE 1984; KRUIJER et al. 1984; MULLER et al. 1984). Increased levels of c-*myc* expression are also seen in B-lymphoid cells as a result of specific chromosomal translocation. While there are some indications that c-*myc* expression is regulated in the cell cycle, as judged by the rise in c-*myc* RNA (up to 40-fold) following stimulation, there is at present no compelling evidence for cell cycle regulation of the other nuclear proteins.

2.3.4 GTP Binding Proteins

The ras oncogenes exist as a family of related genes.

n-*ras*	chromosome # 1
c-Ki-*ras*-1	chromosome # 6
c-Ha-*ras*-1	chromosome # 11
c-Ki-*ras*-2	chromosome # 12
c-Ha-*ras*-2	chromosome # 10

The oncogene encodes for a protein of M_r 21,000 that binds GTP and has GTPase activity. Single amino acid substitutions at sites 12, 13, 59, 61, and 63 in the ras protein can lead to transforming activity (FASANO et al. 1984). Mutations at three of these sites, amino acids 12, 13, and 61 (BOS et al. 1985; YUASA et al. 1983; TABIN et al. 1982), have been identified in human malignancies. All *ras* oncogenes found in human tumors are "activated" by single point mutations leading to amino acid substitutions that change the protein product, such as inactivating the GTPase activity. There is some evidence that ras protein interacts with the EGF receptor (KAMATA and FERAMISCO 1984) and ras protein in yeast appears to be part of the adenylate cyclase system (BROEK et al. 1985; TODA et al. 1985). Thus, there is circumstantial evidence that the p21 *ras* molecule interacts with growth factor receptors and preliminary evidence suggests that it is involved in events that occur up until late GI and commits the cell to S phase.

2.4 Mechanism of Transformation

Certain growth factors elicit their transforming potential by stimulating the hydrolysis of phosphatidylinositol 4,5-biphosphate (PIP2) to diacylglycerol (DG) and inositol 1,4,5-triphosphate (IP3). The hydrolysis of PIP2 represents a bifurcation point in that the two products act as second messengers to initiate two separate pathways (NISHIZUKA 1984; BERRIDGE 1984a+b; BERRIDGE and IRVINE 1984). The two signals are protein kinase C activators. The precise role of protein kinase C as the receptor and site of action of the potent tumor-promoting phorbol esters is not clearly understood. However, it appears to be important in controlling cellular proliferation (CASTANAGA et al. 1982; KIKKAWA et al. 1983; LEACH et al. 1983). It may act by switching on an Na/H exchanger to induce an increase in pH that appears to have a permissive effect on DNA synthesis.

The other limb of the signal pathway is controlled by IP3, which is responsible for intracellular calcium mobilization (BERRIDGE 1984a+b; BERRIDGE and IRVINE 1984). An increase in the intracellular level of calcium has been implicated in the control of growth of many cell types (BERRIDGE 1975; METCALFE et al. 1980). However, there is no direct evidence to show that an increase in intracellular calcium is necessary for DNA synthesis.

There are other growth factors, such as epidermal growth factor (EGF) and insulin, which can stimulate an increase in pH independent of the protein kinase C pathway (MOOLENAAR et al. 1983; HESKETH et al. 1985). Thus, although EGF does not appear to stimulate inositol lipid breakdown, it activates the same second messengers operating in the dual signal pathway.

Oncogenes also elicit their effect on cell morphology and adhesion through the action of a unique protein kinase that phosphorylates tyrosine. Although some normal cells contain tyrosine specific protein kinases, the concentrations of these enzymes in malignant cells are severalfold higher. It appears that the phosphorylation of tyrosine in certain functional molecules alters cell growth and morphology. One protein affected by this novel group of protein kinases is vinculin. Vinculin is one of the proteins responsible for cell adhesion, both in vivo and in vitro. It has been demonstrated that vinculin in transformed cells is phosphorylated at tyrosine residues to a much greater extent than in normal cells, resulting in a reduction in cell adhesion. In addition, evidence strongly suggests that proto-oncogene products are involved in the control of cell growth, division, and possibly embryonic development. Events which alter normal proto-oncogene expression release the cell from the internal regulatory controls, both contact and receptor mediated, responsible for confining normal cells within specific boundaries.

2.5 Clinical Markers of Cancer

Many biochemically and immunochemically identifiable markers have been studied to determine their usefulness as diagnostic adjuncts and prognostic indicators in various cancers. As a result of these studies, several measurable factors have been identified which differ in cancer patients compared to normal individuals. Although none of these markers are specific enough to be used alone as diagnostic probes, they clearly are of assistance in determining tumor burden and response to therapy in specific patient populations. These markers fall into a classification loosely termed tumor-associated antigens. Tumor-associated antigens can be divided into five categories based upon their cellular origin: (a) proteins of embryonic origin, (b) normal cellular proteins synthesized in excess during malignancy secondary to increases in cell number, synthetic rates, or both, (c) ectopic proteins, defined as proteins produced by cancers arising from cells not normally associated with their production, (d) reactive proteins, including those spuriously elevated in the presence of tumor, and (e) oncogene products.

2.5.1 Monoclonal Paraproteins

One of the first tumor-associated proteins to be identified was Bence Jones protein. The presence of this protein in the urine of patients is associated with "mollities ossium," now known as multiple myeloma. This protein, as well as intact monoclonal immunoglobulin molecules, is now easily identified and quantified using immunochemical procedures. The concentrations of these proteins accurately reflect the disease status of a given patient and can be employed to monitor therapeutic responses.

2.5.2 Prostatic Acid Phosphatase

The measurement of prostatic acid phosphatase is one of the most frequently employed clinical enzymatic tests for cancer. This assay has been employed for several decades to assist in the diagnosis of metastatic adenocarcinoma of the prostate. In patients with carcinoma of the prostate, the concentration of this enzyme often accurately reflects clinical status. However, in patients without clinical signs of prostatic cancer, an elevated value for this enzyme is not easily interpreted. First, there are other tissue sources for acid phosphatase and these different isoenzymes are not always easily distinguished. Second, there are other cancers which have been shown to have ectopic acid phosphatase production. Third, there are some benign patholigies associated with elevated acid phosphatase levels.

2.5.3 Carcinoembryonic Antigen

One of the most extensively studied tumor markers in man is the carcinoembryonic antigen (CEA). GOLD and co-workers (GOLD and FREEDMAN 1965; GOLD et al. 1968) first detected CEA in the sera of patients with colorectal carcinomas and later localized to the surface of the colon cancer cells. Over the last 15 years, assays have increased in sensitivity to where CEA is now detected in a variety of clinical states and in normal individuals. Elevated CEA levels are found in patients with different types of GI tract malignancy, most notably stomach, colorectal, and pancreatic. However, elevated CEA values are also found in other malignant and benign pathologies (Table 1).

It has become quite clear that CEA assays are most useful in the management of GI cancers. Two pitfalls in CEA measurements as a screening test for colon cancer are a high number of false positives and a lack of sensitivity in detecting small lesions. In general, CEA levels increase with advancing stages of cancer and therefore, the test is less likely to detect the early asymptomatic lesion. KOHLER et al. (1980) noted significantly longer survival times in patients with preoperative CEA values less than 20 ng/ml as compared to patients with CEA levels greater than 20 ng/ml. The CEA assay has been most effective in patients with documented colorectal carcinoma who are followed with serial serum determinations. The results of several studies suggest that a positive CEA in a patient with a previous negative postoperative CEA is a sensitive and reliable indicator of recurrence or continued tumor growth.

Table 1. Serum CEA in patients

Clinical status	Percent positive (> 2.5 ng/ml)
Primary malignancy	
Colorectal	78-83
Stomach	59-61
Pancreatic	80-92
Breast	47-68
Lung	66-77
Prostatic	40-67
Benign diseases	
Cirrhosis	45
Chronic lung disease	57
Ulcerative colitis	32
Pancreatitis	43

ALSABTI (1979) studied 49 patients with resectable colon cancers and noted a 3% recurrence rate at 15 months in the group which had normal postoperatice CEA levels compared to an 85% recurrence rate in a group which remained abnormal postoperatively. Other investigators (GOLD and FREEDMAN 1975; SUGARBAKER et al. 1976; SAVRIN et al. 1979; BEATTY et al. 1979; ARNAUD et al. 1980; LIM et al. 1980) have noted that up to 90% of patients with recurrent colon cancer have progressively elevated CEA levels which precede routine laboratory or physical evidence by 3-15 months.

The rate of increase in serum CEA levels also might be of use in predicting the extent of recurrence and overall survival. In a study by WOOD et al. (1980), two patterns of serum CEA increase were noted. Of those patients who had a rapid increase, i.e., up to 100 ng/ml within 6 months, 35% had only a local recurrence of cancer and none survived past 4 years. In contrast, of those exhibiting a slow increase, i.e., CEA values of less than 75 ng/ml for at least 12 months, 64% exhibited only local recurrences and 40% were alive after 4 years.

Serial serum CEA values also have been used to document patient response to radiation and chemotherapy in the adjuvant and palliative settings. AL-SARRAF et al. (1979) followed up 49 patients with colorectal carcinoma who had an elevated serum CEA prior to chemotherapy. Of the 18 patients who had a partial response to the drug regimen, serum CEA levels decreased in 89%, increased in one patient, and exhibited no change in one patient. The 31 patients with increasing CEA values had a significantly decreased mean survival time compared to those patients with decreasing CEA levels (20 weeks versus 51 weeks respectively). HERRERA et al. (1977) using a different chemotherapy protocol, noted a similar survival pattern based on serum CEA levels.

In addition to GI tract tumors, elevated serum CEA values have been documented in patients with cancers of the ovary, uterus, thyroid, breast, and lung (KHOO et al. 1979; LINDGREN et al. 1979; MADEDDU et al. 1980; SHOUSHA et al. 1979; STEWARD et al. 1974; WAALKES et al. 1980; VINCENT and CHU 1973; VINCENT et al. 1979). The association of abnormal serum CEA values with smoking also has been noted. The CEA levels return to normal, however, within 3 months of cessation of smoking (ALEXANDER et al. 1976). In patients with small cell carcinoma of the lung, up to 75% have serum CEA levels greater than 2.5 ng/ml and 48% have levels greater than 5 ng/ml (WAALKES et al. 1980). Similar findings occur in patients with breast cancer. For example, abnormal CEA levels occur in 27% of patients with disease localized to the breast and 79% of patients with metastatic breast cancer (STEWARD et al. 1974). SHOUSHA et al. (1979) found that breast cancer patients with CEA-negative tumors have significantly greater 5 to 10 year survival rates unrelated to stage, history, or therapy.

CEA is found in various body fluids and is of value in diagnosing metastatic disease. Patients with meningeal and brain metastases from breast or colon cancer have elevated CSF CEA levels unrelated to plasma CEA, CSF protein, or CSF erythrocyte numbers (BRESALIER and KARLIN 1979; SNITZER et al. 1975; YAP et al. 1980). Patients without clinical or laboratory evidence of meningeal metastasis have undetectable CSF CEA in spite of CEA serum values greater than 100 ng/ml, implying that CEA does not cross the blood-brain barrier. Elevated CEA values are also found in serous effusions in over 90% of the patients with a positive cytology (WHITESIDE and DEKKER 1979). One-half of the patients with negative cytology but an effusion CEA greater than 2.5 ng/ml had underlying malignancy. The CEA levels are more consistently elevated in patients with primary tumors of the GI tract, lung, or breast.

The most significant problem associated with elevated CEA is knowing what to do about it. It is difficult to justify administering chemotherapy in the absence of any other laboratory or physical evidence of recurrent disease. Without knowledge of disease location, radiotherapy is not an option. There has been a marked increase in "second and third look" operations bases on increasing CEA levels. When CEA measurements are used in conjunction with prompt reoperation following the identification of an increase in CEA level, cure rates of 68% have been recorded. These data suggest that the overall cure rate for colorectal carcinoma could be increased by employing this procedure. Recommended guidelines for monitoring CEA in patients with colorectal carcinoma are as follows:

1. A preoperative serum CEA value should be obtained to establish a baseline and assist in staging.
2. Following clinically definitive surgery, postoperative serum CEA measurements should be performed monthly for 3 months, then every 3–6 months following surgery for 5 years.
3. If CEA levels increase significantly from the patient's baseline established during the initial follow-up, repeat testing should be performed at 2-week intervals and evaluated in conjunction with other physical and laboratory data to determine if therapeutic intervention is required.

2.5.4 Alpha Fetoprotein

Alpha fetoprotein (AFP) is an alpha globulin synthesized by the fetal liver, yolk sac, and GI tract. It is the major serum protein of the human fetus during early development (GITLIN et al. 1972). Serum AFP levels fall rapidly after birth to 20 ng/ml or less in the normal adult (RUOSLAHTI and SEPPALA 1971; WALDMANN and MCINTIRE 1974). The association of AFP with malignancy was first demonstrated by ABELEV et al. (1963), who found elevated levels of this protein in mice with chemically induced transplantable hepatoma. TATARINOV (1966) demonstrated the clinical usefulness of AFP quantitation in sera from patients with primary hepatocellular carcinoma. The availability of this assay prompted several studies directed at detecting early disease, especially in areas where there is a high incidence of primary liver cancer. ISHII (1973) demonstrated elevated AFP levels in 93% of Chinese patients with hepatoma and MCINTIRE et al. (1972) demonstrated similar findings in Ugandan patients. Although the majority of patients with hepatocellular carcinoma have elevated AFP levels, a percentage of patients with this pathology do not produce this oncofetal protein. AFP elevations also have been demonstrated in other malignancies as well as several benign conditions. A number of patients, 27% with viral hepatitis and 15% of those with Laennec's cirrhosis, have AFP levels greater than 40 ng/ml. WALDMANN and MCINTIRE (1974) demonstrated elevated levels in sera from patients with various nonhepatic malignancies (Table 2). It is now clear that pathology resulting in liver injury followed by hepatocyte regeneration is accompanied by increases in serum AFP which subside following the regenerative process. In addition, cancer

Table 2. Serum AFP in patients

Clinical status	Percent positive (> 40 ng/ml)
Primary malignancy	
Testicular teratocarcinoma	75
Pancreatic carcinoma	23
Gastric carcinoma	18
Colon carcinoma	5
Bronchogenic carcinoma	7
Breast carcinoma	0

patients with liver metastases and obstruction tend to have higher AFP levels than similar nonobstructed patients. Following either intrahepatic or extrahepatic decompression employing a T tube or percutaneous needle drainage, AFP levels fall and on occasion approach normal limits.

2.5.5 Human Chorionic Gonadotropin

Under normal circumstances, human chorionic-gonadotropin (hCG) is secreted by syncytiotrophoblastic cells of the placenta and serves to maintain corpus luteum function. The hCG glycoprotein hormone contains alpha and beta subunits (VAITUKAITUS 1978). Immunologically and biochemically, the alpha subunit is almost identical to the alpha subunits of the pituitary hormones, luteinizing hormones (LHs), follicle-stimulating hormone (FSH), and thyroid-stimulating hormone (TSH). The beta subunit of hCG contains a unique carboxyl terminus consisting of approximately 30 amino acids not found in the other glycoprotein hormones. This distinction is associated with substantial biological and immunological specificity. As a result, immunoassays have come to rely on the use of monoclonal antibodies directed at this unique domain for the quantitation of hCG. Since the introduction of a specific RIA for beta hCG, a number of trophoblastic and nontrophoblastic tumors that secrete hCG have been identified (VAITUKAITUS 1978; BRAUNSTEIN et al. 1973) - see Table 3.

In addition to its use in diagnosing suspected ectopic pregnancy, hCG RIA is used as a marker for chorionic tumors (SCHWARTZ and DIPIETRO 1980; JONES 1975). In addition, incomplete involution of the uterus and vaginal bleeding following evacuation of a hydatidiform mole are suggestive of persistent trophoblastic disease. Serial beta hCG quantitation is the most reliable method for identifying those patients who require additional therapy. In a

Table 3. Serum beta hCG in patients

Clinical status	Percent positive
Primary malignancy	
Gonadal	
Adenocarcinoma, ovary	40
Testicular	
Embryonal	58
Seminoma	38
Choriocarcinoma	100
GI tract	
Stomach	22
Small bowel	13
Pancreas	33
Biliary tract	11
Liver	21
Colorectal	12
Miscellaneous	
Lung	10
Breast (stage III and IV)	21
Melanoma	9
Hodgkin's disease	2
Multiple myeloma	6
Chronic myelogenous leukemia	1

series of 81 patients following evacuation of a hydatidiform mole, 23.5% had elevated serum beta hCG levels after 8 weeks. Approximately one-half of this group subsequently developed choriocarcinoma (DELFS 1957).

The beta hCG assay accurately reflects regression of trophoblastic neoplasms and can be used to predict therapeutic responses. BERKOWITZ and GOLDSTEIN (1979) studied 51 patients with nonmetastatic gestational trophoblastic tumors who were treated with high-dose methotrexate and citrovorum factor rescue. Only three of these patients failed to demonstrate a ten fold decrease in beta hCG concentration within 18 days and required additional chemotherapy to obtain the desired clinical response. Choriocarcinomas developing after a full-term pregnancy occur less frequently and generally respond less favorably to therapy than the above group. In addition, the absolute serum concentration of pretreatment beta hCG is a useful prognostic indicator. Twenty-four hour urinary levels greater than 100,000 IU are associated with a 55% mortality as compared to a 25% mortality when the initial beta hCG is less than 100,000 IU for 24 h. The presence of hCG in the sera of some men with germ cell tumors was first noted by ZONDEK (1930). However, the clinical use of this marker did not attain its full potential until a sensitive radioimmunoassay was developed. With this assay, hCG is found in a significant number of patients with testicular germ cell tumors. As shown in Table 3, it also is identified in the sera of patients with a variety of neoplastic malignancies. The marker hCG has proven to be a sensitive and reliable marker for testicular tumors, and along with AFP has helped in staging and determining response to chemotherapy. The hCG level can be elevated in many GI neoplasms (Table 3).

2.5.6 Blood Group Antigens

Neoplastic transformation is accompanied by changes in both the glycoprotein and the glycolipid composition of the cell. These changes are frequently associated with the appearance of tumor-associated carbohydrate antigens. Incomplete synthesis, elongation, or abnormal attachment points for these sugar groups are associated with the appearance of neoglycoprotein and/or neoglycolipids. Although tumor-specific epitopes have not yet been identified, several new monoclonal antibodies detecting these carbohydrate tumor-associated antigens have been described. These new antibodies not only detect antigens which are present in sera and on cells from certain cancer patients but are also found in embryonic tissues and at low levels in serum and tissues from normal individuals. Because of their lack of cancer specificity these antibodies are used clinically to assist in the diagnostic workup and monitor therapy rather than for cancer screening. Table 4 lists these monoclonal antibodies and their cancer specificity.

The monoclonal antibody CA 19-9 identifies an epitope termed 19-9. This epitope is present on both glycoproteins and glycolipids and is increased in the sera of patients with gastrointestinal cancers (KOPROWSKI et al. 1980). The 19-9 epitope has been identified as a sialylated Le^a (Lewis blood group) derivative. Independently, two other monoclonal antibodies have been developed with the same specificity as CA 19-9. These are the CA 50 antibody of Holmgren (HOLMGREN et al. 1984) and the DU-PAN-2 antibody of METZGAR et al. (1984). Similarly, GOOI et al. (1981, 1983) developed a monoclonal antibody to the Le^x epitope. This sialylated form of Le^x blood group antigen has been used for serological testing for cancers of the pancreas and colon (CHIA et al. 1985).

Table 4. Monoclonal antibodies to tumor-associated blood group antigens

Antibody	Isotype	Cancer specificity
CA 19-9	IgG_1	Pancreatic and gastrointestinal
CA 50	IgM	Pancreatic and gastrointestinal
DU-PAN-2	IgM	Pancreatic
CO-51.4/CA 19-9	IgG_3	Gastrointestinal
CS Le^x	IgM	Breast and colon
F 36/22	IgG_3	Breast
MoV_2	IgM	Ovarian
CA 125	IgG_1	Ovarian

Another of the described cancer markers is termed CA 125. This monoclonal antibody recognizes an antigen common to most nonmucous ovarian carcinomas (KABAWAT et al. 1983a+b). In Several studies (BAST et al. 1983; BAST and KNAPP 1985; KLUG et al. 1984; RICOLLEAU et al. 1984; CANNEY et al. 1984; KIVINEN et al. 1986) CA 125 concentrations were elevated in 50%–96% of the patients with non-mucinous ovarian carcinoma. Serum CA 125 levels correlate with clinical status and are a good predictor of residual disease.

CA 125 antigen is also elevated in other carcinomas of the genitourinary system, including carcinoma of the fallopian tube (100%), endometrial carcinoma (78%), and endocervical carcinoma (83%).

2.5.7 Pancreatic Oncofetal Antigen

A new tumor marker for pancreatic cancer was described (BANWO et al. 1974), purified, and characterized (GELDER et al. 1978). The pancreatic oncofetal antigen (POA) was first identified in extracts of fetal pancreas and pancreatic cancer. HOBBS et al. (1980) reported a rocket immunoelectrophoresis assay for POA which had a 97% sensitivity and a 98% specificity for pancreatic cancer. Several laboratories have now independently developed assays for POA and have reported elevated values in 53%–98% of the patients with pancreatic cancer (BANWO et al. 1974; GELDER et al. 1978; HOBBS et al. 1980; SHIMANO et al. 1981, 1983; KAWA et al. 1983; NISHIDA et al. 1985). Present in small amounts in the sera of normal individuals, POA is increased in patients with different malignancies as well as in a few patients with benign conditions (Table 5). The highest frequency of increased concentrations and the highest absolute concentrations of POA are detected in sera from patients with pancreatic cancer. Because of the lack of absolute specificity for pancreatic cancer, assays for POA are not suited for generalized population screening but are useful in the differential diagnosis of pancreatic cancer. In addition, serial measurements of POA

Table 5. Serum POA in patients

Clinical status	Percent positive (> 14 units/ml)
Primary malignancy	
Pancreas	53-98
Biliary tract carcinoma	36
Hepatoma	10
Colorectal	2
Stomach	28
Breast	15
Lung (adenocarcinoma)	15
Lung (bronchogenic)	24
Benign diseases	
Benign biliary disease/cirrhosis	18
Benign lung disease	0
Pancreatitis	12
Benign breast disease	0
Inflammatory bowel disease	1

are valuable in detecting recurrences of pancreatic cancer (GELDER et al. 1983).

CEJKA and KITHIER (1985) employed an assay for POA to monitor patients with breast cancer. In their study 37 of 196 (19%) women with breast cancer had serum POA values greater than 15 units/ml. In this study, the highest incidence of increased POA concentrations was in patients with stage III and metastatic disease. They concluded that monitoring POA concentrations during therapy of breast cancer patients was useful, especially in patients with repeatedly negative results for CEA. In a recent study, we have measured POA in cytosol preparations of breast cancer tissue submitted for estrogen receptor studies. This study identified a subpopulation of breast cancer patients with elevated cytosol POA values. This study permitted us to identify the breast cancer patients population that could benefit from serum POA measurements throughout therapy.

2.5.8 Human Tumor-Associated Antigens: Concluding Remarks

The number of new cancer markers reported in the literature has grown considerably over the last few years, but the fact remains that only a few markers are actually used at present and none of them have reached the eminence enjoyed by the CEA. Like CEA, none of the new markers are tumor specific. Although the search for clinically useful tumor specific markers still goes on, it is unlikely that one will be found with the possible exception of virus associated antigens. With this in mind, four rules governing the use of tumor markers have emerged:

1. No tumor-associated antigen is tumor specific.
2. No tumor-associated antigen is present in all tumors of the same histological type.
3. The presence or absence of a tumor-associated antigen may enable tumors to be divided into subsets which represent different stages of differentiation and/or different cell origins. Different subsets may respond differently to different therapeutic modalities.
4. A tumor-associated antigen may be widely distributed in many tumors or limited in its distribution to only a few tumor types. The more restricted tumor-associated antigens are likely to be found on tumors of the same embryonic origin.

Although all of the tumor-associated antigens discussed in this text are selectively elevated in certain cancers and their measurement is both noninvasive (venipuncture) and inexpensive, they lack the specificity to be used as a screening test. Ideally, a cancer marker used for early diagnosis would have a high sensitivity and specificity, be present in easily accessible body fluids, and be simple and inexpensive to assay. The test sensitivity should be such as to allow cancer detection at an early enough stage to enhance the chances for cure, and specificity should be such as to discriminate cancer from benign diseases of the same organ system. If the shortcoming is a lack of specificity, then one must consider what to do with a positive test result. It seems unlikely that a test based on the detection of a tumor associated antigen of low specificity will be of practical use in cancer screening.

The tumor-associated antigens, discussed in this text, are used extensively to follow cancer treatment. In this application, certain markers have proven to be of significant value and are most frequently employed to monitor patients following resection for cure or other therapeutic intervention.

Acknowledgment. The authors wish to acknowledge Ms. Suzan Florentine for her help in the preparation of this manuscript.

References

Abelev GI, Perova SD, Khramkova NI, Postnikova ZA, Irlin IS (1963) Productional of embryonal α-globulin by transplantable mouse hepatomas. Transplantation 1: 174

Alexander JC Jr, Silverman NA, Chretien PB (1976) Effect of age and cigarette smoking on carcinoembryonic antigen levels. JAMA 235: 1975

Alsabti E (1979) Carcinoembryonic antigen (CEA) as a prognostic marker in colonic cancer. J Surg Onocl 12: 127
Al-Sarraf M, Baker L, Talley RW, Kithier K, Vaitkevicius VK (1979) The value of the serial carcinoembryonic antigen (CEA) in predicting response rate and survival of patients with gastrointestinal cancer treated with chemotherapy. Cancer 44: 1222
Arnaud JP, Koehl C, Adloff M (1980) Carcinoembryonic antigen (CEA) in diagnosis and prognosis of colorectal carcinoma. Dis Colon Rectum 23: 141
Banwo O, Versey J, Hobbs JR (1974) New oncofetal antigen for human pancreas. Lancet 1: 643-645
Bast RC Jr, Knapp RC (1985) Recent advances in the immunodiagnosis of epithelial ovarian carcinoma. In: Alberts DS, Surwit EA (eds) Ovarian cancer. Nijhoff, Boston, pp 23-35
Bast RC Jr, Klug TL, John ES, Jenison E, Niloff JM, Lazarus H, Berkowitz RS, et al. (1983) A radioimmunoassay using a monoclonal antibody to monitor the course of epithelial ovarian cancer. N Engl J Med 309: 883-887
Beatty JD, Romero C, Brown PW, Lawrence W, Terz JJ (1979) Clinical value of carcinoembryonic antigen. Arch Surg 114: 563
Benedict WF, Weissman BE, Mark C, Stanbridge EJ (1984) Tumorigenicity of human HT1080 fibrosarcoma x normal fibroblast hybrids: Chromosome dosage dependency. Cancer Rs 44: 3471-3479
Berkowitz RS, Goldstein DP (1979) Methotrexate with citrovorum factor rescue for nonmetastatic gestational trophoblastic neoplasms. Obstet Gynecol 54: 725
Berridge MF (1975) Control of cell division: a unifying hypothesis. J Cyclic Nucleotide Res 1: 305-320
Berridge MJ (1984a) Inositol trisphosphate and diacylglycerol as second messengers. Biochem J 220: 345-360
Berridge MJ (1984b) Oncogenes, inositol lipids and cellular proliferation. Biotechnology 2: 541-546
Berridge MJ, Irvine RF (1984) Inositol trisphosphate, a novel second messenger in cellular signal transduction. Nature 312: 315-321
Berridge MJ, Heslop JP, Irvine RF, Brown KD (1984) Inositol trisphosphate formation and calcium mobilization in Swiss 3T3 cells in response to platelet-derived growth factor. Biochem J 222: 195-201
Bishop JM (1985) Viral oncogenes. Cell 42: 23-38
Bos JL, Toksoz D, Marshall CJ, Verlaan de Vries M, Veeneman GH, van der Eb A, van Boom JH, et al. (1985) Aminoacid substitutions at codon 13 of the N-ras oncogene in human acute myeloid leukaemia. Nature 315: 726-730
Braunstein GD, Vaitukaitus JL, Carbone PO, Ross GT (1973) Ectopic production of human chorionic gonadotropin by neoplasms. Ann Intern Med 78: 39
Bresalier RS, Karlin DA (1979) Meningeal metastasis from rectal carcinoma with elevated cerebrospinal fluid carcinoembryonic antigen. Dis Colon Rectum 22: 216
Broek D, Samiy N, Fasano O, Fujiyama A, Tamanoi F, Northup J, Wigler M (1985) Differential activation of yeast adenylate cyclase by wild type and mutant RAS proteins. Cell 41: 763-769
Canney PA, Moore M, Wilkinson PM, James RD (1984) Ovarian cancer antigen CA 125, a prospective clinical assessment of its role as a tumor marker. Br J Cancer 50: 765-769
Castagna M, Takai Y, Kaibuchi K, Sano K, Kikkawa U, Nishizuka Y (1982) Direct activation of calcium-activated, phospholipid-dependent protein kinase by tumor-promoting phorbol esters. J Biol Chem 257: 7847-7851
Cavenee WK, Dryja TP, Phillips RA, Benedict WF, Yodbout R, Yallie BL, Murphree AL, et al. (1983) Expression of recessive alleles by chromosomal mechanism in retinoblastoma. Nature 305: 779-786
Cejka J, Kithier K (1985) Immunoassay for pancreatic oncofetal antigen. Clin Chem 31: 780-781
Chen ISY, Slamon DJ, Rosenblatt JD, Shah NP, Quar SG, Wachsan W (1985) The x gene is essential for HTLV replication. Science 229: 54-58
Chia D, Terasaki PI, Suyama N, Galton J, Hirota M, Katz D (1985) Use of monoclonal antibody to sialylated Le^a and sialylated Le^x for serologic tests of cancer. Cancer Res 45: 435-437
Clarke MF, Westin E, Schmidt D, Josephs SF, Ratner L, Wong-Staal F, Gallo RC, Reitz MS (1984) Transformation of NIH-3T3 cells by a human c-sis cDNA clone. Nature 308: 464-467
Cutita F, Carney DN, Mulshine J, Moody TW, Fedorko J, Fischler A, Minna JD (1985) Bombesin-like peptides can function as autocrine growth factors in human small-cell lung cancer. Nature 316: 823-826
Delfs E (1957) Quantitative chorionic gonadotropin. Prognostic value in hydatidiform mole and chorionepithelioma. Obstet. Gynecol 9: 1
DelVillano BC, Brennan S, Brock P, Bucher C, Liu V, McClure M, Rake B, et al. (1983) Radioimmunometric assay for a monoclonal antibody defined tumor marker, CA 19-9. Clin Chem 29: 549-552
Eva A, Robbins KC, Andersen PR, Srinivasan A, Tronick SR, Reddy EP, Elmore NW et al. (1982) Cellular genes analogous to retroviral onc genes are transcribed in human tumor cells. Nature 295: 116-119
Fasano O, Aldrich T, Tamanoi F, Taporowsky E, Furth M, Wigler M (1984) Analysis of the transforming potential of the human H-ras gene by random mutagenesis. Proc Natl Acad Sci USA 81: 4008-4012
Felber BK, Paskalis H, Kleinman-Ewing C, Wong-Staal F, Pavlakis GN (1985) The px protein of HTLV-I is a transcriptional activator of its long terminal repeats. Science 229: 672-679
Fujisawa JI, Seiki M, Kiyokawa T, Yoshida M (1985) Functional activation of the long terminal repeat of human T-cell leukemia virus type I by a trans-acting factor. J Biol Chem 82: 2277-2281
Gelder FB, Reese CJ, Moosa AR, Hall T, Hunter R (1978) Purification, partial characterization and clinical evaluation of a pancreatic oncofetal antigen. Cancer Res 38: 312-324
Gelder FB, Barr LH, Goldman LI (1983) Application of tumor markers in the immunodiagnosis of cancer. Radiobioassays, Vol II., (Askar FS, Ed), CRC Press, Inc 99-115
Gitlin D, Perricelli A, Gitlin GM (1972) Synthesis of α-fetoprotein by liver, yolksac, and gastrointestinal tract of the human conceptus. Cancer Res 32: 979
Gold P, Freedman SO (1965) Demonstration of tumor-specific antigens in human colonic carcinomata by immunologic tolerance and absorption techniques. J Exp Med 121: 439
Gold P, Freedman SO (1975) Tests for carcinoembryonic antigen. Role in diagnosis and management of cancer. JAMA 234: 190
Gold P, Gold M, Freedman SO (1968) Cellular location of carcinoembryonic antigen of the human digestive system. Cancer Res 28: 1331
Gooi HC, Feizi T, Kapadia A, Knowles BB, Solter D, Evans MJ (1981) Stage-specific embryonic antigen involves alpha 1-3 fucosylated type 2 blood group chains. Nature 292: 156-158

Gooi HC, Williams LK, Uemura K, Hounsell EF, McIlhinney RA, Feizi T (1983) A marker of human foetal endoderm defined by a monoclonal antibody involves Type 1 blood group chains. Mol Immunol 6: 607-613

Goustin AS, Betsholtz C, Pfeifer-Ohlsson S, Persson H, Rydnert J, Bywater M, Holmgren G, et al. (1985) Coexpression of the sis and myc proto-oncogenes in developing human placenta suggests autocrine control of trophoblast growth. Cell 41: 301-312

Herrera MA, Chum TM, Holyoke ED, Mittelman A (1977) Cea monitoring of palliative treatment for colorectal carcinoma. Am J Surg 185: 23

Hesketh TR, Moore JP, Morris JDH, Taylor MV, Rogers J, Smith GA, Metcalfe JC (1985) A common sequence of calcium and pH signals in the mitogenic stimulation of eukaryotic cells. Nature 313: 481-484

Hoobs JR, Knapp MI, Branfoot AC (1980) Pancreatic oncofetal antigen (POA): its frequency and localization in humans. Oncodev Biol Med 1: 37-48

Holmgren J, Lindholm L, Perssor B, Lagergard T, Nilsson O, Svennerholm L, Rudenstam RM et al. (1984) Detection by monoclonal antibody of carbohydrate antigen CA50 in serum of patients with carcinoma. Br Med J 288: 1479-1482

Ishii M (1973) Radioimmunoassay of alpha-fetoprotein. Gann Monogr 14: 89

Jones WB (1975) Treatment of chorionic tumors. Clin Obstet Gynecol 18: 247

Kabawat S, Bast RC Jr, Bhan AK, Welch WR, Knapp RC, Colvin RB (1983a) Tissue distribution of a coelomic epithelium related antigen recognized by the monoclonal antibody OC 125. Int J Gynecol Pathol 2: 275-285

Kabawat S, Bast RC, Welch WR, Knapp RC, Colvin RB (1983b) Immunopathologic characterization of a monoclonal antibody that recognizes common surface antigen of human ovarian tumors of serous, endometrioid, and clear cell types. Am J Clin Pathol 79: 98-104

Kamata T, Feramisco JR (1984) Epidermal growth factor stimulates guanine nucleotide binding activity and phosphorylation of ras oncogene proteins. Nature 310: 147-150

Kawa S, Homma T, Oguchi H, Nagata A, Furuta S, Usuda N, Nagata T, Fukui H (1983) Clinical application of the enzyme immunoassay of pancreatic oncofetal antigen. Ann NY Acad Sci 417: 400-409

Kelly K, Cochran BH, Stiles CD, Leder P (1983) Cell-specific regulation of the c-myc gene by lymphocyte mitogens and platelet-derived growth factors. Cell 35: 603-610

Khoo SK, Whitaker S, Jones I, MacKay E (1979) Predictive value of serial carcinoembryonic antigen in long-term follow-up of ovarian cancer. Cancer 43: 2471

Kikkawa U, Takai Y, Tanaky Y, Miyaka R, Nishizuka Y (1983) Protein kinase C as a possible receptor protein for tumor-promoting phorbol esters. J Biol Chem 258: 11442-11445

Kivinen S, Kuoppala T, Leppilampi M, Vuori J, Kaupilla A (1986) Tumor-associated antigen CA 125 before and during the treatment of ovarian carcinoma. Obstet Gynecol 67: 468-472

Klein G, Klein E (1985) Evolution of tumors and the impact of molecular oncology. Nature 315: 190-195

Klug TL, Bast RC Jr Niloff JM, Knapp RC, Zurawski VR Jr (1984) Monoclonal antibody immunoradiometric assay for an antigenic determinant (CA 125) associated with human epithelial ovarian carcinomas. Cancer Res 44: 1048-1053

Knudson AG (1971) Mutation and cancer: statistical study of retinoblastoma. Proc Natl Acad Sci USA 68: 820-823

Knudson AG, Strong LC (1972) Mutation and cancer: a model for Wilms' tumor of the kidney. JNCI 48: 313-324

Knudson AG, Hethcote HW, Brown WB (1975) Mutation and childhood cancer: a probabilistic model for the incidence of retinoblastoma. Proc Natl Acad Sci USA 72: 5116-5120

Knudson AG, Meadows AT, Nichols WW, Hill R (1976) Chromosomal deletion and retinoblastoma. N Engl J Med 295: 1120-1123

Kohler JP, Siminowitz D, Paloyan D (1980) Preoperative CEA level: a prognostic test in patients with colorectal carcinoma. Ann Surg 46: 449

Koprowski H, Herlyn M, Steplewski Z, Sears HF (1980) Specific antigen in serum of patients with colon carcinoma. Science 212: 53-55

Koufos A, Hansen MF, Lampkin BC, Workman ML, Copeland NG, Jenkins NA, Cavenee WK (1984) Loss of alleles at loci on human chromosome 11 during genesis of Wilms' tumor. Nature 309: 170-172

Koufos A, Hansen MF, Copeland NG, Jenkins NA, Lampkin BC, Cavernee WK (1985) Loss of heterozygosity in three embryonal tumors suggests a common pathogenic mechanism. Nature 316: 330-334

Kozbor D, Croce CM (1984) Amplification of the c-myc oncogene in one of five human breast carcinoma cell lines. Cancer Res 44: 438-441

Kruijer W, Cooper JA, Hunter T, Verma IM (1984) Platelet-derived growth factor induces rapid but transient expression of the c-fos gene and protein. Nature 312: 711-716

Kuusela P, Jalanko H, Roberts P, Sipponen P, Mecklin JP, Pitkanen R, Makela O (1984) Comparison of CA19-9 and carcinoembryonic antigen (CEA) level in serum of colorectal diseases. Br J Cancer 49: 135-139

Leach KL, James ML, Blumberg PM (1983) Characterization of a specific phorbol ester aporeceptor in mouse brain cytosol. Proc Natl Acad Sci USA 80: 4208-4212

Li Y, Holland CA, Hartley JW, Hopkins N (1984) Viral integration near c-myc in 10-20% of MCF 247-induced AKR lymphomas. Proc. Natl Acad Sci USA 81: 6808-6811

Libermann TA, Nasbaum MR, Razon N, Kris R, Lax I, Soreq H, Whittle N, et al. (1985) Amplification, enhanced expression and possible rearrangement of EGF receptor gene in primary human brain tumours of glial origin. Nature 313: 144-147

Lim CN-H, McPherson JA, McClelland AR, McCoy L, Koch M (1980) Value of serial CEA determinations in a surgical adjuvant trial of colorectal and gastric carcinoma. J Surg Oncol 14: 275

Lindgren J, Wahlstrom T, Seppala M (1979) Tissue CEA in premalignant epithelial lesions and epidermoid carcinoma of the uterine cervix: prognostic significance. Int J Cancer 23: 448

Little C, Nau MM, Carney DN, Gazdar AF, Minna JD (1983) Amplification and expression of the myc oncogene in human lung cancer cell lines. Nature 306: 194-196

Madeddu G, Langer M, Dettori G, Costanza C (1980) Role of serum carcinoembryonic antigen in preoperative diagnosis of cancer in patients with thyroid nodules. Cancer 45: 2607

Marshall CJ, Dave H (1978) Suppression of the transformed phenotype in somatic cell hybrids. J Cell Sci 33: 171-190

McIntire KR, Vogel CL, Princler GL, Patel IR (1972) Serum alpha-fetoprotein as a biochemical marker for hepatocellular carcinoma. Cancer Res 32: 1941

Metcalfe JC, Pozzan T, Smith GA, Hesketh TR (1980) A calcium hypothesis for the control of cell growth. Biochem Soc Symp 45: 1-26

Metzgar RS, Rodriguez N, Finn OJ, Lan MS, Daash VN, Fernsten PD, Meyers WC, et al. (1984) Detection of pancreatic cancer associated antigen (DU-PAN-2 antigen) in serum and ascites fluid of patients with adenocarcinoma. Proc Natl Acad Sci USA 81: 5242-5246

Moolenaar WH, Tsien RY, van der Saag PT, de Laat SW (1983) Na+/H+ exchange and cytoplasmic pH in the action of growth factors in human fibroblasts. Nature 304: 645-648

Muller R, Bravo R, Burkhardt J, Curran T (1984) Induction of c-fos gene and protein by growth factors precedes activation of c-myc. Nature 312: 716-720

Neel BG, Hayward WS, Robinson HL, Fang J, Astrin SM (1981) Avian leukosis virus-induced tumors have common proviral integration sites and synthesize discrete new RNAs: oncogenes by promoter insertion. Cell 23: 323-334

Neil JC, Hughes D, McFarlane R, Wilkie NM, Orions DE, Lees G, Jarret O (1984) Transduction and arrangement of the myc gene by feline leukemia virus in naturally occurring cell leukemias. Nature 308: 814-820

Nishida K, Sugiura M, Yoshikawa T, Kondo M (1985) Enzyme immunoassay of pancreatic oncofetal antigen (POA) as a marker of pancreatic cancer. Gut 26: 450-455

Nishizuka Y (1984) The role of protein kinase C in cell surface signal transduction and tumor promotion. Nature 308: 693-697

Ozanne B, Shum A, Richards CS, Cassells D, Grossman D, Trent J, Gusterson B, Hendler F (1985) Evidence for an increase of EGF receptors in epidermoid malignancies. In: Feramisco J, Ozanne B, Stiles CD (eds) Cancer cells. Cold Spring Harbor Laboratory Press, New York, pp 41-46

Pathak S, Strong LC, Ferrell RE, Trindade A (1982) Familial renal cell carcinoma with a 3; 11 chromosome translocation limited to tumor cells. Science 217: 939-941

Payne GS, Courtneidge SA, Crittenden LB, Fodly AM, Bishop JM, Varmus HE (1981) Analysis of avian leukosis virus DNA and RNA in bursal tumors: viral gene expression is not required for maintenance of the tumor state. Cell 23: 311-322

Payne GS, Bishop JM, Varmus HE (1982) Multiple arrangements of viral DNA and an activated host oncogene in bursal lymphoma. Nature 295: 209-216

Reich NC, Levine AJ (1984) Growth regulation of a cellular tumor antigen, p53, in nontransformed cells. Nature 308: 199-201

Ricolleau G, Chatal JF, Fumoleau P, Kremer M, Douillard JY, Curtet C (1984) Radioimmunoassay of CA 125 antigen in ovarian carcinomas: Advantages compared with CA 19-9 and CEA. Tumor Biol 5: 151-159

Ritts RE, DelVillano BC, Jo VLW, Herberman RB, Klug TL, Zurawski VR Jr (1984) Initial clinical evaluation of an immunometric assay for Ca 19-9 using the NCI serum bank. Int J Cancer 33: 339-345

Rosen CA, Sodorski JG, Kettman R, Burny A, Haseltin WA (1985) Trans-activation of the bovine leukemia virus long term repeat in BLV-infected cells. Science 227: 320-322

Rowley JD (1984) Biological implications of consistent chromosome rearrangements in leukemia and lymphoma. Cancer Res 44: 3159-3168

Ruoslahti E, Seppala M (1971) Studies of carcinofetal proteins. III. Development of a radioimmunoassay for α-fetoprotein in serum of healthy human adults. Int J Cancer 8: 374

Savrin RA, Cooperman M, Martin EW Jr (1979) Clinical application of carcinoembryonic antigen in patients with colorectal carcinoma. Dis Colon Rectum 22: 211

Schwab M, Alitalo K, Klempnauer KH, Varmus HE, Bishop JM, Gilbert F, Brodeur G, et al. (1983) Amplified DNA with limited homology to myc cellular oncogene is shared by human neuroblastoma cell line and a neuroblastoma tumor. Nature 305: 245-248

Schwartz RO, DiPietro DL (1980) B-hCG as a diagnostic aid for suspected ectopic pregnancy. Obstet Gynecol 56: 197

Seizinger BR, Martuza RL, Yusella JF (1986) Loss of genes on chromosome 22 in tumorigenesis of human acoustic neuroma. Nature 322: 644-647

Shimano T, Loor RM, Papsidero LD, Kuriyama M, Vincent RG, Nemoto T, Holyoke ED, et al. (1981) Isolation, characterization and clinical evaluation of a pancreatic cancer-associated antigen. Cancer 47: 1602-1613

Shimano T, Mori T, Kitada M, Maruyama H, Kosaki G (1983) Purification and characterization of a pancreatic cancer-associated antigen (PCAA) from normal colonic mucosa. Ann NY Acad Sci 417: 97-102

Shousha S, Lyssiotis T, Godfrey VM, Scheuer PJ (1979) Carcinoembryonic antigen in breast cancer tissue: a useful prognostic indicator. Br Med J 1: 777

Snitzer LS, McKinney EC, Tejada F, Sigel MM, Rosomott HL, Zubrod CG (1975) Cerebral metastases and carcinoembryonic antigen in CSF. N Engl J Med 293: 1101

Sodorski JG, Rosen CA, Haseltine WA (1984) Trans-acting transcriptional activation of the long terminal repeat of human lymphotropic viruses in infected cells. Science 225: 381-385

Sporn MB, Todaro CJ (1980) Autocrine secretion and malignant transformation of cells. N Engl J Med 303: 878

Srivastava A, Norris JS, Schmooker-Reis RJ, Goldstein S (1985) c-H-ras 1 proto-oncogene amplification and overexpression during the limited replicative life-span of normal human fibroblasts. J Biol Chem 260: 6404-6409

Steward AM, Nixon D, Zamcheck N, Aisenberg A (1974) Carcinoembryonic antigen in breast cancer patients: serum levels and disease progress. Cancer 33: 1246

Stoler A, Bouck N (1985) Identification of a single chromosome in the normal human genome essential for suppression of hamster cell transformation. Proc Natl Acad Sci USA 82: 570-574

Sugarbaker PH, Zamcheck N, Moore FD (1976) Assessment of serial carcinoembryonic antigen (CEA) assays in postoperative detection of recurrent colorectal cancer. Cancer 38: 2310

Tabin CJ, Bradley SM, Bargmann CI, Weinberg RA, Papageorge AG, Scolnick EM, Dhar R, Lowy DR, Chang EH (1982) Mechanism of activation of a human oncogene. Nature 300: 143-149

Tatarinov YS (1966) Content of embryo-specific alpha-globulin in fetal and neonatal sera from adult humans with primary carcinoma of the liver. Fed Proc [Suppl] 25: 344

Toda T, Uno I, Ishikawa T, Powers S, Kataoka T, Broek D, Broach J, et al. (1985) In yeast RAS proteins are controlling elements of the cyclic AMP pathway. Cell 40: 27-36

Todaro CJ, Delarco JE, Cohen S (1976) Transformation by murine and feline sarcoma viruses specifically blocks binding of epidermal growth factor to cells. Nature 264: 26

Todaro GJ, Fryling CM, Delarco JE (1980) Transforming growth factors produced by certain human tumor cells: polypeptides that interact with epidermal growth factor receptors. Proc. Natl Acad Sci USA 77: 5258

Twardzik DR, Ranchalis JE, Todaro GJ (1982) Mouse embryonic transforming growth factors related to those isolated from tumor cells. Cancer Res 42: 590-593

Vaitukaitus JL (1978) Tumors and human chorionic gonadtropin. In: Ruddon RW (ed) Biological markers of neoplasia: basics and applied aspects. Elsevier/North-Holland, New York, p 317

Vincent RG, Chu TM, Lane WW (1979) The value of carcinoembryonic antigen in patients with carcinoma of the lung. Cancer 44: 685

Vincent RJ, Chu TM (1973) Carcinoembryonic antigen in patients with carcinoma of the lung. J Thorac Cardiovasc Surg 66: 230

Waalkes TP, Abeloff MD, Woo KB, Ettinger DS, Ruddon RW, Aldenderfer P (1980) Carcinoembryonic antigen for monitoring patients with small cell carcinoma of the lung during treatment. Cancer Res 40: 4420

Waldmann TA, McIntire KR (1974) The use of a radioimmunoassay for alpha-fetoprotein in the diagnosis of malignancy. Cancer 34: 1510

Whang-Peng J, San-Kao CS, Lee CE, Bunn PA, Carney DN, Yazdor AF, Minna JD (1982) Specific chromosome defect associated with human small cell lung cancer: deletion 3p(14-23). Science 215: 181-182

Whiteside TL, Dekker A (1979) Diagnostic significance of carcinoembryonic antigen levels in serous effusions. Acta Cytol (Baltimore) 23: 443

Wong AJ, Ruppert JM, Eggleston J, Hamilton SR, Baylin SB, Vogelstein B (1986) Gene amplification of c-myc and N-myc in small cell carcinoma of the lung. Science 233: 461-464

Wood CB, Ratcliffe JG, Burt RW, Malcolm AJH, Blumgart LH (1980) The clinical significance of the pattern of elevated serum carcinoembryonic antigen (CEA) levels in recurrent colorectal cancer. Br J Surg 67: 46

Yap BS; Yap HY, Fritsche HA, Blumenschein G, Bodey GP (1980) CSF carcinoembryonic antigen in meningeal carcinomatosis from breast cancer. JAMA 244: 1601

Yuasa Y, Srivastava SK, Dunn CY, Rhim JS, Reddy EP, Aaronson SA (1983) Acquisition of transforming properties by alternative point mutations within c-has/bas human protooncogene. Nature 303: 775-779

Zondek B (1930) Versuch einer biologischen (hormonalen) Diagnostik beim malignen Hodentumor. Chirurg 2: 1072-1073

3 Pathology of Gastrointestinal Cancer

PETER J. GOLDBLATT and KITAI KIM

CONTENTS

PETER J. GOLDBLATT, M.D., Professor and Chairman, KITAI KIM, M.D., Professor of Pathology, Department of Pathology, Medical College of Ohio, 3000 Arlington Avenue, C.S. 10008 Toledo, OH 43699, USA

3.1 Introduction

An exhaustive discussion of the pathologic features of neoplasms of the alimentary tract would require a text twice the size of this entire volume. The organs of the alimentary tract are among the most frequent site of origin of malignancies in the human body (SILVERBERG 1985; ROSAI 1981). Thus, the diversity of malignant new growths found in this system is great, and virtually every type of neoplasm that has been identified has been described as arising from the alimentary tract, or involving it secondarily. For this reason, we will discuss only the most important malignancies below, and the lesser ones will be mentioned only cursorily for completeness. While there may be reasons to disagree with any classification, we have generally organized the discussion to follow that of the Histological Classification of Tumours of the World Health Organization (OOTA and SOBIN 1977; MORSON and SOBIN 1976; GIBSON and SOBIN 1978). In keeping with the remainder of this volume, we have left out discussion of tumors of the mouth, oropharynx, and salivary glands, but have included discussion of important tumors of the liver and biliary tree and those of the exocrine and endocrine pancreas.

As stated above, the alimentary tract is particularly prone to the development of malignancies. The reasons for this are not entirely clear, but in all probability invole its particular function of absorbing and metabolizing materials presented to it, including potentially carcinogenic substances, and the almost constant turnover (cell cycling) of its principal constituent cells. In this regard, it is not surprising that the epithelial components are the most frequent cells of origin of malignancies of the alimentary tract. Nonetheless, we also find significant involvement of this system by malignancies arising in lymphoid cells, and we are becoming increasingly aware of the major role of the alimentary tract in the immune system.

In general the discussion that follows will mention the distinctive gross and microscopic features of each neoplasm, some special features or techniques which may be helpful for precise identification or classification, important prognostic features, including some indication of the method of pathologic staging and grading, a brief exposition of the usual behavior, and finally, where possible, some indication of the pathogenesis of the lesion. The literature cited emphasizes recent contributions, especially those which have extensive references to the more fundamental literature. Some useful general references are also included, which can be consulted for further details.

3.2 Malignancies of the Esophagus

3.2.1 Epithelial Malignancies

Esophageal cancer represents approximately 1% of all malignant neoplasms (SCHOTTENFELD 1984) and 10% of gastrointestinal (GI) malignancies (ROBBINS et al. 1984). Well over 90% of esophageal malignancies are of epithelial origin and the vast majority of these are squamous cell (epidermoid) carcinomas. The latter represent 1.1% of all cancers and 7% of all GI cancers in the United States (FEIN et al. 1985; SAITO et al. 1985).

3.2.1.1 Squamous Cell Carcinoma

The distribution of this important cancer shows a marked variation worldwide. In areas of the Middle East and Far East, it is one of the most frequent cancers. In the United States, it is a male predominant disease and shows a propensity for the elderly. Blacks are more prone to develop the tumor than whites, but the male preponderance is not as marked in blacks (ROBBINS et al. 1984). Racial, sexual, and age relationships vary greatly on a regional basis, with a marked female preponderance being shown in the Transkai region of Iran (ROBBINS et al. 1984). Although multiple etiologies have been suggested, none has been unequivocally established. A relationship with alcohol ingestion and cigarette smoking has been suggested in the United States (ROBBINS et al. 1984). A relationship to benign stricture and achalasia has been suggested (ROBBINS et al. 1984).

Important gross features of the tumor (Fig. 1) include a tendency to be bulky, oval lesions with their long axis parallel to the long axis of the esophagus. They frequently show central ulceration and typically undermine the mucosa at the periphery. This tendency makes diagnosis more difficult (BRALOW 1982) since viable tumor is frequently covered by normal epithelium. *Biopsy* and *brushing* cytology are strongly recommended to improve diagnostic yield. The lack of an anatomic serosa in the esophagus, the rich lymphatic supply, and the propensity of these tumors to invade the muscular wall early dictate that metastatic spread principally to mediastinal lymph nodes occurs early in this disease. Transdiaphragm spread to abdominal lymph nodes is also frequent (AKIYAMA et al. 1981). The majority arise in the middle third of the esophagus with about 20% arising in the upper third, where radiation therapy is the treatment of choice. In the lower third there is an admixture of adenocarcinomas (see below).

Microscopically, these tumors are easily recognized as squamous in origin. They are graded histologically as well differentiated when they show keratin pearls (Fig. 3) and easily demonstrated intercellular bridges. They are usually graded as well, moderately, or poorly differentiated on the ease with which these features are demonstrated. In a large series (Earlam and Cunha-Melo 1980) 5-year survival was a dismal 4%, although cures by radiation therapy and/or surgery have been achieved. A variant with less aggressive behavior termed verrucous carcinoma is recognized (Rosai 1981).

3.2.1.2 Adenocarcinoma and Its Histologic Variants

Malignancies of glandular origin represent about 10% of tumors of the lower third of the esophagus, their principal site of origin, and about 3%–4% of all esophageal malignancies (Fein et al. 1985). They are thought to arise from two histogenetically different origins: submucosal esophageal glands and heterotopic glandular (columnar) mucosa (Barrett's esophagus) (Rosai 1981). The latter has been frequently debated, since these tumors tend to arise close to the cardia of the stomach and may involve the stomach by direct extension. The only unequivocal cases are those with squamous mucosa intervening between the tumor and the stomach. Those involving both sides of the cardioesophageal junction should be considered gastric in origin.

As stated, histologically the tumors form glands which resemble the esophageal submucosal glands or heterotopic gastric mucosa. Benign squamous differentiation (adenocanthoma) is sometimes present, but the behavior is similar to the common forms of adenocarcinoma. Papillary (finger-like) configurations are also seen in the common forms of esophageal adenocarcinoma. A variant with a characteristic cribriform pattern is designated *adenoid cystic carcinoma (cylindroma)* and is composed of cells of two types: duct-like cells and myoepithelial cells. These tumors are rare in the esophagus, but are similar to those seen arising in salivary glands. Two other variants should be noted: *muco epidermoid carcinoma,* which consists of three cell types (squamous cells, mucus-secreting cells, and intermediate cells), and *adenosquamous carcinoma,* in which both a malignant squamous and a malignant glandular cell population is present (Oota and Sobin 1977). Each of these again, is such a rare tumor that epidemiologic and prognostic factors are difficult to define. The risk of developing an adenocarcinoma of the esophagus in heterotopic glandular mucosa of the type described by Barrett has been debated, but this probably should be considered a major risk factor until proven otherwise (Cameron et al. 1985).

3.2.1.3 Undifferentiated and Unclassified Carcinoma

Tumors of presumed epithelial origin which show neither glandular nor squamous differentiation have been placed in this category. With the use of electron microscopy and immunocytochemistry fewer and fewer tumors should remain unclassifiable. Histochemical and immunocytochemical markers (Boland et al. 1982; Cooper and Giles 1984; Zalcberg and McKenzie 1985) such as the epithelial membrane antigen and low molecular weight cytokeratin are particularly useful in establishing the epithelial origin of a malignant neoplasm, while demonstration of bundles of intracellular intermediate filaments on the one hand or intracellular canaliculi on the other may serve to point to a squamous (former) or glandular (latter) origin. Well developed intercellular junctions including desmosomes seen in electron micrographs may also help to point to an epithelial origin. Thus only a few, usually quite aggressive, malignancies remain in this category after adequate pathologic assessment. *Oat cell carcinoma* presenting as a fungating mass in the esophagus composed of small cells like those of its pulmonary counterpart has been reported. These can be associated with ACTH or serotonin production (Oota and Sobin 1977).

3.2.2 Nonepithelial Malignancies

As noted above (Sect. 3.2.1) epithelial malignancies account for well over 90% of all esophageal neoplasms. Indeed, even benign mesenchymal neoplasms are relatively rare, though the leiomyoma is the most common of these, and its malignant counterpart (leiomyosarcoma) is the most frequent nonepithelial neoplasm.

3.2.2.1 Leiomyosarcoma

Though leiomyosarcoma is said to be the most frequent mesenchymal neoplasm of the esophagus

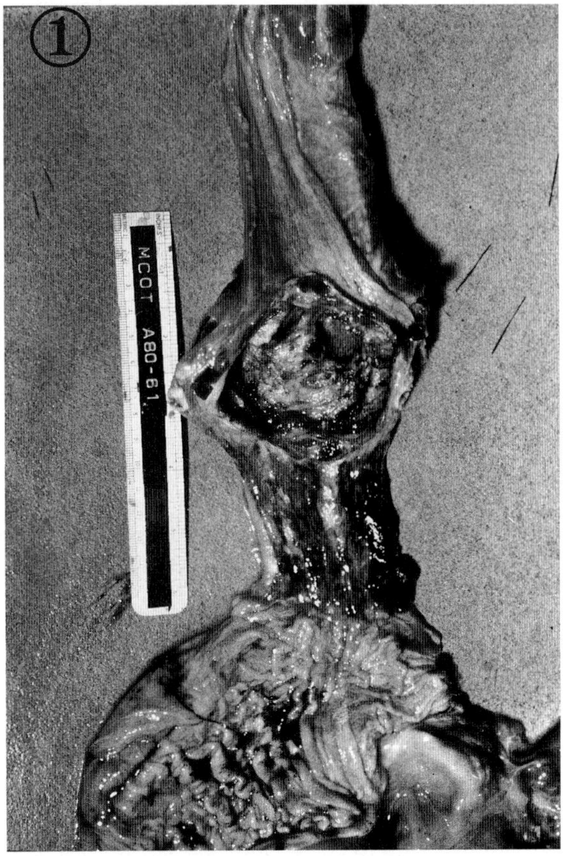
①
MCOT A80-61

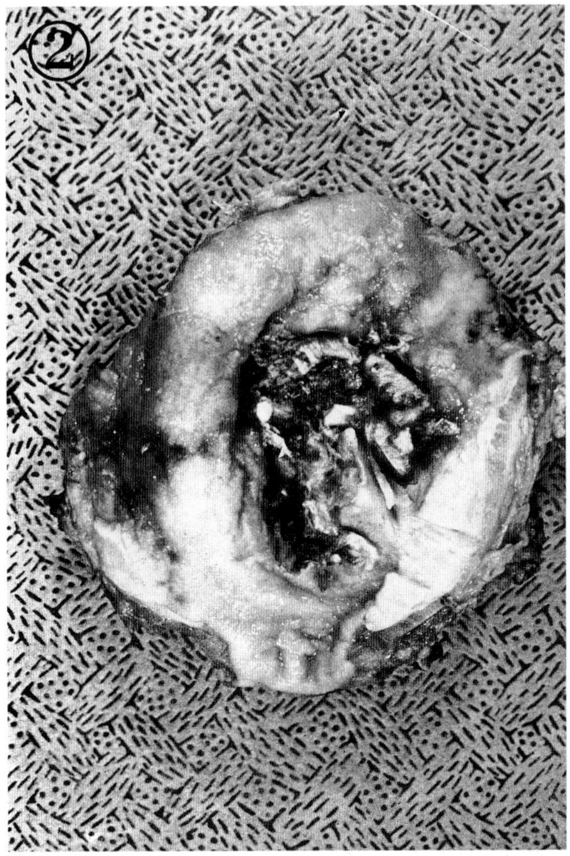
②

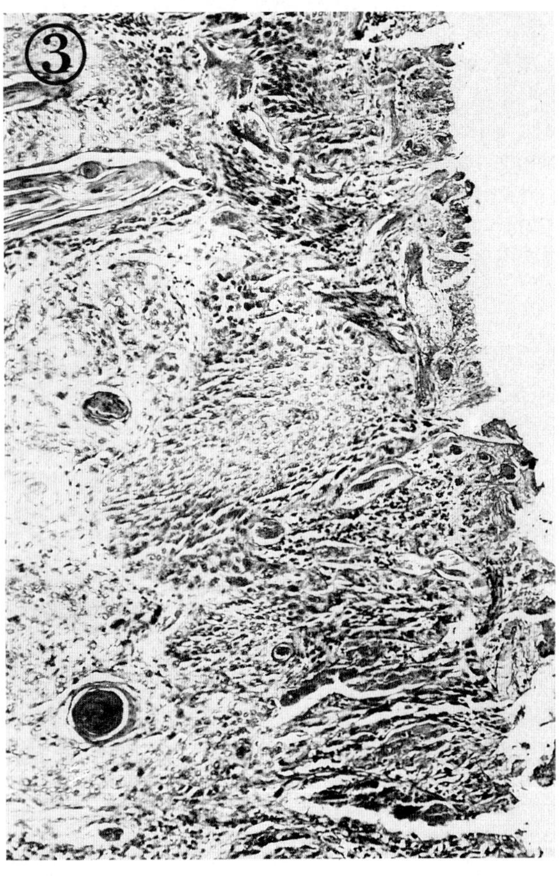
③

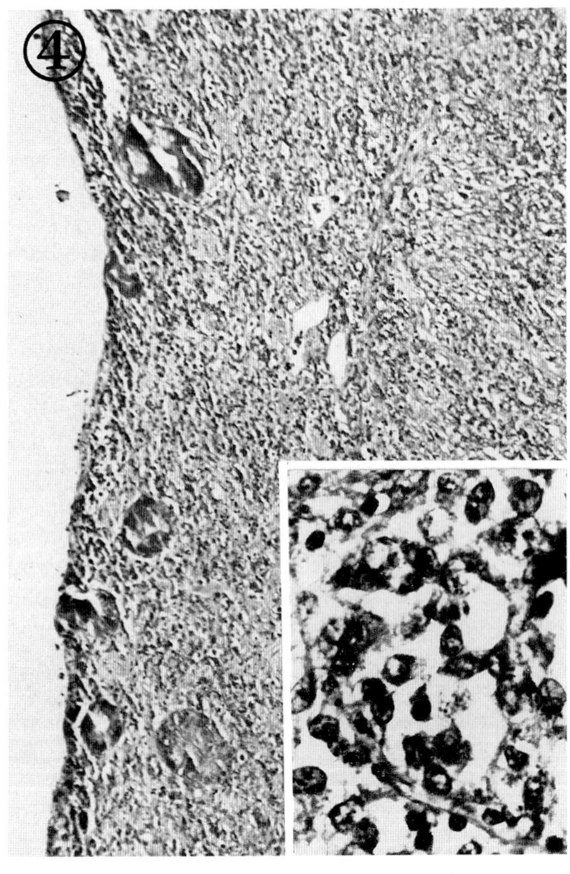
④

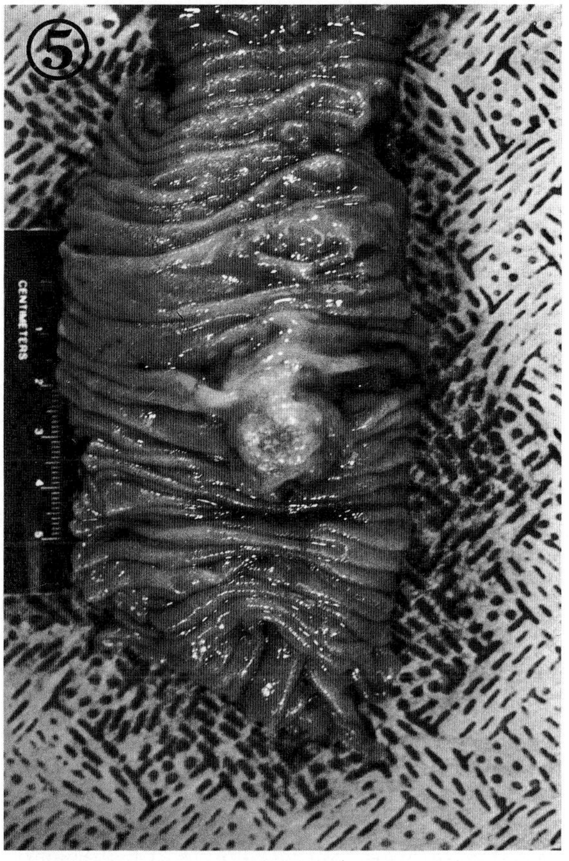

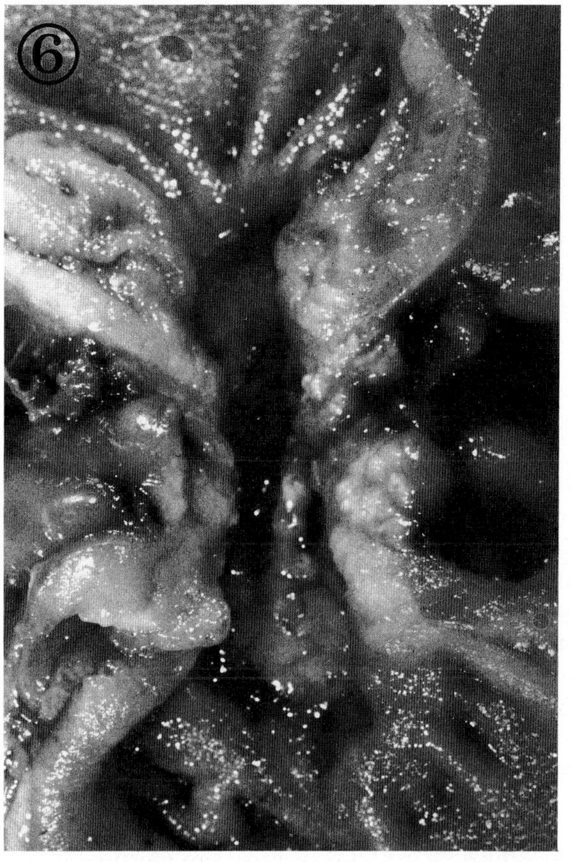

Fig. 1. Photograph of esophagus and stomach. The expansile mass is a squamous cell carcinoma of the middle third of the esophagus. This is the most common site for this tumor

Fig. 2. Photograph of stomach with primary large cell lymphoma. Note diffuse thickening of wall. This picture would be difficult to distinguish from diffuse involvement by carcinoma (linitis plastica)

Fig. 3. Photomicrograph of esophagus with moderately differentiated squamous cell carcinoma. Normal mucosa is undermined by the tumor at the *upper right*. A keratin pearl is seen at the *lower left*. ×20

Fig. 4. Photomicrograph of gastric mucosa with lymphomatous infiltration. The small round cells have pushed the glands apart. Higher magnification (*inset, right* ×300) shows the cells to have large vesicular nuclei, some of which show indentations. This is consistent with histiocytic lymphoma. ×20

Fig. 5. Photograph of duodenum. A mass is seen in the area of the ampulla. Note central ulceration. Microscopic study showed this to be a well differentiated adenocarcinoma (see Fig. 7)

Fig. 6. Photograph of carcinoma of sigmoid colon. Note circumferential constricting growth. This is characteristic of left-sided colonic lesions as contrasted with bulky, fungating tumors in the cecum

(ROSAI 1981), only 38 examples were listed in a 1968 review (ATHANASOULIS and ARAL 1968). They tend to be larger and softer than their benign counterpart and may show central necrosis and hemorrhage. The main histologic criterion of malignancy is the presence of a large number of mitoses.

3.2.2.2 Others

It should be recalled that the upper third of the esophagus contains skeletal muscle which extends to the middle third. Thus, rhabdomyosarcomas have been reported to arise from esophageal muscle (MING 1973), but they are extremely rare. The esophagus contains considerably less lymphoid tissue than does the remainder of the GI tract so that primary lymphoma is less common at this level than lower down. Nonetheless, primary lymphoma of various histologic types, including plasmacytoma presenting with dysphagia due to diffuse involvement, has been reported in this site (ROSAI 1981). A scattering of isolated cases of other mesenchymal malignancies can be cited (ROSAI 1981; MING 1973; OOTA and SOBIN 1977).

3.2.3 Miscellaneous Malignancies

Of the sundry tumors which may arise in the esophagus on occasion, two deserve particular mention: carcinosarcoma (spindle cell carcinoma) (ANTONIADES 1982b) and malignant melanoma (ANTONIADES 1982g).

The spindle cell carcinoma is probably a poorly differentiated squamous cell tumor. Also termed *pseudosarcoma* and *carcinosarcoma*, spindle cell carcinomas typically arise as large polypoid neoplasms in which the epithelial component is inconspicuous, at times, histologically. They may metastasize as only the epithelial component, though rarely the spindle cells predominate in metastases. Survival in up to 50% of resectable cases has been reported (ANTONIADES 1982b). A resemblance to so-called fibrous histiocytoma has been mentioned (ANTONIADES 1982b).

Unquestioned cases of primary malignant melanoma arising in the esophagus are documented (ANTONIADES 1982g). They are described as bulky and exophytic and arising mostly in the middle and lower third. Weight loss is a bad prognostic sign but the overall outlook is poor with rapid demise being the usual course.

Additionally, isolated instances of carcinoid of the esophagus are reported (ROSAI 1981). How they relate to the oat cell carcinoma mentioned above is not entirely clear. These tumors will be described more extensively below (see Sects. 3.4.2 and 3.5.2).

3.2.4 Secondary Malignancies Involving the Esophagus

As noted above (Sect. 3.2.1), the esophagus is richly invested with lymphatics so the potential for malignant neoplasms which spread via this route to involve it secondarily is great. However, the most frequent route of involvement by malignancies from other sites involves direct extension from either the lung or the stomach. The latter also may spread via lymphatics (FEIN et al. 1985; SAITO et al. 1985).

3.3 Malignancies of the Stomach

Just as squamous cell carcinoma is by far the commonest malignant tumor arising in the esophagus, adenocarcinoma of the stomach is the most significant malignant gastric neoplasm. Again, a marked variation has been noted worldwide, and while the incidence is declining in the United States, where it is still predominantly a male disease, it is still an extremely common tumor in Japan and the Scandinavian countries (BRALOW 1982; RYAN and LIVSTONE 1985). A second significant type of neoplasm is the primary gastric lymphoma (BROOKS and ENTERLINE 1983; SANDLER 1984; SKUDDER and SCHWARTZ 1985).

3.3.1 Epithelial Malignancies

Virtually all of the epithelial malignancies of the stomach arise from the mucous secreting gastric crypt cells. The prognosis correlates best with the gross appearance, with the polypoid tumors that project into the lumen having a better outlook in general than those which penetrate the wall. In an extensive series, LAUREN (1965) divided gastric carcinomas into two histologic patterns which he termed *intestinal* (the most frequent, better prognosis) and *diffuse* (worse prognosis). ROSAI (1981) has commented that this classification is difficult to apply, but offers another classification which is a similar mixture of gross, microscopic, and histogenetic features. Suffice it to say that the depth of penetration appears to be the best predictor of biologic behavior. Multiple lesions may also arise at the same time, but the prognosis is still good (90% survival) if detected early (NOGUCHI et al. 1985).

The relationship of malignancy, whether it is grossly ulcerating or not, to chronic peptic ulceration has been widely debated. Currently (BRALOW 1982), the association seems to be with the chronic gastritis which is almost always present in association with ulcers, rather than the ulceration per se. The association with achlorhydria and chronic atrophic gastritis has been noted for a long time, particularly the type associated with pernicious anemia (fundic). Whether this relates to specific antibodies either to parietal cells or to intrinsic factor has not been established.

3.3.1.1 Adenocarcinoma and Its Histologic Variants

As noted previously, the majority of primary gastric neoplasms are mucin producing adenocarcinomas. The mucin is most frequently an acid mucoprotein as opposed to the neutral secretion of normal gastric mucin producing cells. In Japan, where the incidence of gastric malignancy is particularly high, gastric adenocarcinomas are detected in the in situ stage through the aggressive use of cytologic screening (NOGUCHI et al. 1985). The prognosis is thus significantly improved over the rather dismal 5-year survival in most series in this country. Histologic patterns which are recognized in the WHO classification (OOTA and SOBIN 1977) include papillary, tubular, and acinar structures, as well as mucinous and a signet ring cell carcinoma (Fig. 9). The latter may have three distinct cell types (signet ring, goblet cells with acid mucin, and eosinophilic cells with neutral mucin), and is the tumor most frequently associated with metastasis to the ovary, termed Kruckenberg tumor. The adenocarcinomas may be well, moderately, or poorly differentiated. Their depth of penetration is extremely important from the prognostic standpoint. Diffusely infiltrating tumors (linitis plastica) have a very poor outlook. The gastric submucosa is richly supplied with vessels, both blood vessels and lymphatics, leading to the early spread of epithelial neoplasm. It is well recognized that some gastric cancers can spread through the mucosa in a diffuse fashion without penetrating the wall, and these have a better outlook. Paneth cells and argentaffin cells have been described in addition to mucous cells in the intestinal and diffuse types; chief cells and parietal cells appear very infrequently (ROSAI 1981).

3.3.1.2 Squamous Cell Carcinoma

This is an uncommon tumor in the stomach and is more commonly found as an extension or metastasis from the esophagus (FEIN et al. 1985). Lesions of the cardia should probably be excluded from consideration for this reason (ANTONIADES 1982i). The tumors are usually exophytic and bulky at the time of resection. Despite this, primary surgical resection has been successful because the lesion is often confined to the stomach. Histologically, they tend to be well differentiated and to show keratin pearl formation and distinct intercellular bridges (ANTONIADES 1982i). Nonetheless, they are aggressive neoplasms, with all patients being dead in 7 months in one series (see ANTONIADES 1982i). The majority of squamous cell tumors have foci of glandular differentiation (adenosquamous).

3.3.1.3 Undifferentiated and Unclassified Carcinoma

Epithelial tumors without distinct glandular or squamous differentiation have been described under terms such as medullary and trabecular carcinoma (OOTA and SOBIN 1977). As mentioned previously, however, the use of markers has greatly increased accuracy of diagnosis, especially in differentiating undifferentiated carcinomas from lymphoma of various types. Mucin stains are also helpful in categorizing poorly differentiated malignancies as adenocarcinomas (BOLAND et al. 1982). Primary choriocarcinoma, teratocarcinoma, and malignant melanoma have been reported as isolated cases in the stomach.

3.3.2 Carcinoids

These tumors are rare as primary lesions of the stomach, with only 3%–4% of all GI carcinoids arising in this site (ANTONIADES 1982a). They will be discussed more extensively below (see Sect. 3.4.2). In the stomach, they arise from the basal gastric mucosa, and are more frequently found in the antrum. They may be either argentaffin (Fontana stain, reduced silver) or argyrophil (Bodian stain, reduced silver). The cells are small, have a uniform appearance with round or oval nuclei and occur in nests or cords, occasionally showing rosettes. A tendency toward glandular differentiation has been noted to have a somewhat worse prognosis. Electron microscopy shows typical granules of the neurosecretory type, and a variety of endocrine activities has been associated with these neoplasms (ANTONIADES 1982a) (see Sect. 3.4.2). They tend to be less aggressive than adenocarcinoma, and primary gastric carcinoids are somewhat more indolent than their counterpart in small intestine though they are more aggressive than those of the appendix. Behavior is somewhat unpredictable because of the rarity of this neoplasm. In one series, 93% of patients with localized tumors survived 5 years while only 23% with regional lymphnode involvement were alive and overall 5-year survival was approximately 52% (ANTONIADES 1982a).

3.3.3 Nonepithelial Malignancies

A wide variety of nonepithelial primary malignancies have been reported in the stomach as they have elsewhere. Whereas the spectrum of tumors includes oddities such as primary malignant fibrous histiocytoma (SHIBUYA et al. 1985), the only malignancies to warrant further discussion here are those arising from smooth muscle and the lymphomas. The latter are neoplasms whose frequency is increasing (SANDLER 1984) (see also reviews: AOZASA et al. 1985; DRAGOSICS et al. 1985; KAUFMAN et al. 1984; PAPADIMITRIOU et al. 1985; PHILBEN et al. 1983; SKUDDER and SCHWARTZ 1985; WEINGRAD et al. 1982).

3.3.3.1 Smooth Muscle Derived Malignancies Including Leiomyoblastoma

Smooth muscle derived malignancies are the commonest nonepithelial malignant tumors exclusive of lymphoma, and constitute about 0.5% of gastric malignancies (AKWARI et al. 1978). Leiomyosarcoma of the stomach is usually a solitary, well localized neoplasm composed of spindle cells or round cells, but may be extremely pleomorphic. Well differentiated lesions are difficult to distinguish from their benign counterparts, but criteria for malignancy have been described (RANCHOD and KEMPSON 1977) which rely on size, cellularity, cellular atypia, and particularly the mitotic activity in the neoplasm, as they do in other sites. The majority arise in the 5th and 6th decades of life. They may present with a variety of symptoms including bleeding (melena) and a mass, and are difficult to recognize radiographically. Symptoms tend to be prolonged prior to diagnosis, and overall survival is reasonably good.

A distinctive neoplasm of smooth muscle has been designated leiomyoblastoma in the belief that they arise from smooth muscle stem cells (ABRAMSON 1973; FAEGENBURG et al. 1975; KLIFTO et al. 1983). They are thought to arise from vascular smooth muscle or from the muscle of the gastric wall. About 80% arise in the stomach, and 75% of these are found in the antrum (ABRAMSON 1973; FAEGENBURG et al. 1975). While they generally have a favorable prognosis, with 80% surviving, they do metastasize, mostly to regional lymph nodes.

3.3.3.2 Lymphoma

Non-Hodgkin's lymphoma is an increasingly frequent gastric primary neoplasm, constituting 1%-4% of all gastric malignancies (SANDLER 1984). Hodgkin's disease is rare in the stomach as it is in other gastrointestinal sites. As noted previously, the gastrointestinal tract is now recognized to play an important role in cellular and humoral immunity, and to be a major site of altered immune response. While the incidence of gastric adenocarcinoma appears to be declining in this country, lymphomas are increasing. The stomach is the most frequent primary site of lymphoma arising in the gastrointestinal tract (see reviews: AOZASA et al. 1985; KAUFMAN et al. 1984; PAPADIMITRIOU et al. 1985; PPHILBEN et al. 1983; SKUDDER and SCHWARTZ 1985; WEINGRAD et al. 1982).

The clinical features of this neoplasm are similar to those of adenocarcinoma, making diagnosis on clinical grounds difficult prior to biopsy. Grossly, they may present as polypoid, fungating, or ulcerating lesions, and they often involve the stomach diffusely (linitis plastica type) (Fig. 2). Up to 25% are multicentric in origin.They arise from the lamina propria and involve the mucosa secondarily (Fig. 4); the muscle wall is usually spared.

Histologic typing of lymphomas, in general, is undergoing major reconsideration at present (NON-HODGKIN'S PATHOLOGIC CLASSIFICATION PROJECT 1982). While more will be said about this below, a detailed discussion is beyond the scope of this review. Immunohistochemistry demonstrating monoclonality of the cells is receiving increasing emphasis, and is extremely important in differentiating malignant from benign – so-called pseudolymphomatous – lesions (LERMAN-SAGIE et al. 1985; BROOKS and ENTERLINE 1983; SEO et al. 1982). Another difficulty is that disseminated lymphomas frequently involve the GI tract, with the stomach being involved in up to 50% of patients with non-Hodgkin's lymphoma at autopsy (SANDLER 1984). Large cells and cells of B cell origin predominate, though histologic classification is confusing because of the different schemes currently in use (NON-HODGKIN'S LYMPHOMA PATHOLOGIC CLASSIFICATION PROJECT 1982). Overall prognosis is better than for carcinoma. Involvement of regional lymph nodes is the usual route of spread.

3.3.3.3 Others

Although a spectrum of nonepithelial malignancies has been reported as arising in stomach, most of these are so rare as not to warrant further comment. The so-called malignant fibrous histiocytoma, which is now the commonest malignant soft tissue tumor, has been reported to arise from the stomach (SHIBUYA et al. 1985). Other malignancies such as those arising from nerve, blood vessels, and fat may be found extremely infrequently. A granular cell tumor of the stomach has been reported recently (ABDELWAHAB and KLEIN 1983).

3.3.4 Miscellaneous Malignancies

As noted above, choriocarcinomas, melanomas, and other tumors have been reported to arise in the stomach. An entity which we have mentioned before as arising in the esophagus (see Sect. 3.2.3), termed *carcinosarcoma*, has also been described in the stomach (HANADA et al. 1985). This is composed of spindle cells and epithelial cells, and is a rather aggressive lesion.

3.3.5 Secondary Malignancies Involving the Stomach

The stomach is richly supplied with blood and lymph vascular channels. It is, therefore, a more frequent site of secondary malignant involvement than the esophagus, and a wide variety of neoplasms has been described as spreading to the stomach secondarily. Among the commonest are those arising in the esophagus (SAITO et al. 1985), which may also extend directly to large bowel, or in the lung; moreover, as noted above, secondary involvement of the stomach by generalized non-Hodgkin's lymphoma is described in up to 50% of cases at autopsy.

3.4 Malignancies of the Small Intestine

The small intestine is the least frequent site of primary neoplasia in the gastrointestinal tract (RYAN and LIVSTONE 1985). Only 600–700 individuals die annually from primary malignancies arising in the small intestine, which makes deaths from neoplasms of this site 50- to 60-fold less frequent than deaths from neoplasms of the large intestine. Primary malignancies of the mucosa are more frequent proximally (duodenum, periampullary) though these are often considered separately from jejunal or ileal malignancies.

Primary lymphoma is more frequent distally, with the distal ileum being the commonest site of origin. The third most frequent tumor is the small intestinal carcinoid. Adenocarcinoma, lymphoma, and carcinoids make up over 90% of primary small intestinal malignancies. Which of these is more frequent depends on the series, but adenocarcinomas usually are considered most frequent, especially if the ampullary tumors are included (ADOTEY 1985). Among nonepithelial tumors, leiomyosarcoma is most frequent (PACELLA and CALEEL 1985). Malignancies of the small intestine are symptomatic in 90% of the cases (ADOTEY 1985). Symptoms frequently suggest obstruction, and include pain, weight loss, and vomiting or diarrhea. The peak incidence of these tumors varies between 40 and 70 years of age, and symptoms tend to be present for 6 months to a year prior to diagnosis (ADOTEY 1985).

3.4.1 Epithelial Malignancies

Adenocarcinoma constitutes about 50% of small intestinal malignant neoplasms in most series (ADOTEY 1985; HAQ et al. 1985). The earliest example was a duodenal carcinoma recorded in 1746 (ADOTEY 1985). The lack of carcinomas in this site is noteworthy in view of their prevalence elsewhere in the GI tract. Though many theories have been advanced, none is totally accepted. The rapid transit of semiliquid material, lack of significant bacterial colonization, and rapid turnover of mucosal cells have been invoked (ADOTEY 1985). Two predisposing factors appear to be long-standing Crohn's disease (RIDDELL 1985) in relationship to ileal lesions and celiac sprue (gluten enteropathy) in relationship to both epithelial neoplasia and lymphoma (ADOTEY 1985; COOPER and READ 1985).

3.4.1.1 Adenocarcinoma and Its Histologic Variants

As noted, adenocarcinoma is the most frequent small intestinal malignancy in most series, and may constitute up to 50% of primary neoplasms in this site (HAQ et al. 1985). The incidence varies, because carcinomas arising in the periampullary region (Fig. 5) may be reported along with those arising from bile ducts, and because the true incidence of lymphoma of the small intestine is difficult to assess because of the frequency with which this site is secondarily involved by disseminated non-Hodgkin's lymphoma. Well differentiated (Fig. 7) *papillary adenocarcinomas* and *mucinous adenocarcinomas* are among the commonest variants. A rare *signet ring* type of carcinoma has been reported as arising in the small intestine, particularly the duodenum, but they are more likely to be an extension of a lesion arising in the stomach.

Adenocarcinomas arise from the mucosa, but rapidly penetrate the bowel wall. They may be fungating or polypoid, and they tend to become annular and constricting (Fig. 6). It is common for them to present as a palpable mass, but this may also reflect intussusception rather than the bulk of the tumor. Jejunal tumors are said to produce left upper quadrant pain, while ileal tumors produce periumbilical or right lower quadrant discomfort. Jaundice may be the presenting sign with primary duodenal lesions. The lack of specificity of symptoms and signs has led to the fact that many are unexpected findings at laporotomy, and thus they tend to have metastasized to regional lymph nodes by the time they are discovered and to carry a generally poor prognosis. The liver is also a common site of spread (ADOTEY 1985; HAQ et al. 1985; ROSAI 1981).

3.4.1.2 *Undifferentiated and Unclassified Carcinoma*

As noted previously, unclassified epithelial neoplasms are found in the small intestine as they are elsewhere, but with diligent investigation with modern techniques their number is diminishing. Their numbers are too small to document distinctive features.

3.4.2 Carcinoids

The small intestine is a frequent site of primary carcinoids, although the carcinoid is the most frequent malignant neoplasm in the appendix. Small intestinal carcinoids are more aggressive than those of the appendix, though those arising in the large intestine are said to be the most malignant (ANTONIADES 1982a). These tumors have excited a great deal of interest (ANTONIADES 1982a; BLACK 1968; HOUGH et al. 1983; VAN SICKLE 1972; YANG et al. 1983), but only a few features can be presented here.

Carcinoids of the small intestine arise in the submucosa from the enterochromaffin (Kulchitsky cells) (BLACK 1968) (Fig. 11). The origin of these cells has been debated, but the suggestion by Pearse that they are amine producing and also take up and decarboxylate amines (APUD) has received wide acceptance. The argyrophilic reaction is the best general stain (YANG et al. 1983). Many of the tumors are *nonargentaffin*. The argentaffin reaction is related to amine content, while the argyrophil reaction appears to reflect sialoglycopeptide content. The latter indicates the presence of many different hormone substances, including gastrin, somatostatin, motilin, secreting glucagon, and the like. Gastrin secretion has been associated with the Zollinger-Ellison syndrome. Metastatic lesions may secrete the same or different hormones (YANG et al. 1983). The classical carcinoid syndrome is related to serotonin production and consists of flushing, diarrhea, palpitations, and intermittent hypertension. It is almost invariably related to hepatic metastasis. Rare complications of these tumors include obliterative endovascular elastosis and right-sided heart lesions as well as scleroderma-like skin lesions.

The tumors may be multiple in up to 25% of cases. They frequently excite a fibroblastic reaction in the adjacent bowel wall, and may produce local obstruction. Although penetration of the bowel wall is present in up to 50% of both small intestinal and appendiceal carcinoids, metastasis tends to occur predominantly in local lymph nodes and prognosis is relatively good. However, liver metastasis carries a bad prognosis, with less than 5% 5-year survival (ANTONIADES 1982a). Nonetheless, resection of metastatic lesions may be warranted to reduce symptoms related to the hormonal effects.

With the advent of immunohistochemical techniques, it seems better to try to characterize the tumors hormonally (BLACK 1968) rather than to cling to silver staining and descriptive classification such as argentaffin, nonargentaffin, and composite which are those used in the WHO classification (MORSON and SOBIN 1976). A relationship to other neoplasms such as in MEN syndrome and von Recklinghausen's disease has been reported (HOUGH et al. 1983).

3.4.3 Nonepithelial Malignancies

As noted above, primary malignancy of the small intestine represents less than 1% of all malignancies and 3%–6% of gastrointestinal cancers (ADOTEY 1985). Two nonepithelial neoplasms, leiomyosarcoma and primary intestinal lymphoma (PSIL), however, are among the most frequent neoplasms arising in this segment of the intestine. While this is an interesting pattern, its significance may merely be to underscore the lack of epithelial malignancies arising in small intestine. That is, these two malignancies are the most frequent nonepithelial neoplasms in the remainder of the gastrointestinal tract, too (ROSAI 1981; MORSON and SOBIN 1976).

3.4.3.1 *Leiomyosarcoma*

Malignancies arising in smooth muscle constitute approximately 10% of small intestinal malignancies. They almost always arise from muscularis propria, and are bulky tumors, with the majority of the tumor projecting outward. They may, however, compress the lumen sufficiently to obstruct. Bleeding is a common presenting sign because they are highly vascular and they may ulcerate through the mucosa. They tend to produce intra-abdominal metastases but extra-abdominal metastases are infrequent. Liver involvement is reported (PACELLA and CALEEL 1985).

3.4.3.2 *Lymphoma*

While the stomach is the most frequent primary site for gastrointestinal lymphoma, the small intestine is the next most common (ADOTEY 1985; COOPER and READ 1985). The physiologic role of the extensive lymphoid tissue of the small intestinal submucosa is currently being elucidated. Hodgkin's disease is uncommon in the small intestine as it is elsewhere in the GI tract, and is almost always secondary to diffuse involvement when found in this location. Of the non-Hodgkin lymphomas, PSIL follows two types: the "Western" and the "Mediterranean" (COOPER and READ 1985). The latter belongs to a group of immunoproliferative disorders (IPSID) consisting of B cell proliferation involving the small intestine diffusely. The fact that intestinal parasites including protozoans are associated and they response to antibiotic therapy, along with the geographic distribution of this disorder in third world countries, may call into question the neoplastic nature of the disease (COOPER and READ 1985).

In western Europe, PSIL constitutes up to 18% of small intestinal malignancy, and a similar proportion has been reported in the United States. There are two age peaks, under 15 and in the 5th and 6th decades, with a male to female preponderance of 2:1 (COOPER and READ 1985).

Grossly, PSIL can present in a variety of fashions from a polypoid lesion or diffuse involvement of a segment of the intestinal wall to so-called aneurysmal dilatation proximal to an area of constriction. About 20% of the patients present with multiple lesions, and spread occurs first to regional mesenteric nodes, and then to distant sites. It may be impossible in the later stages to determine whether the small intestinal involvement was primary or secondary.

The microscopic classification of PSIL is fraught with the same difficulties as are encountered in other sites. In the Rappaport classification about 60% would be large diffuse "histiocytic" cells, and 25% lymphocytic lymphoma (COOPER and READ 1985). Despite the "histiocytic" misnomer, the majority of these neoplasms are of B cell origin. Prognosis is much more dependent on stage than on histologic type.

Two other topics warrant comment: the association of lymphoma with celiac sprue and Mediterranean lymphoma. Lymphoma in celiac disease is reported in adults, and appears to have its onset in patients diagnosed over the age of 50 with a 1 in 10 chance that these older patients will develop it within 4 years after diagnosis. Patients with celiac disease and lymphoma may respond to gluten-free diet after treatment of the lymphoma. About 90% are "histiocytic" lymphoma and may be true histiocytoma; extraintestinal lymphoma is also reported (COOPER and READ 1985).

Mediterranean lymphoma is associated with IgA producing B cells which proliferate diffusely throughout the intestine. There is a spectrum of cells from benign plasma cells producing alpha-chains to malignant cells which may or may not produce alpha-chains. The geographic concentration of cases in the Middle East and North Africa accounts for the name. The malignancy is multifocal and may produce large masses with ulceration, narrowing, or infiltration of the wall of the intestine (COOPER and READ 1985) (see also reviews: AOZASA et al. 1985; DRAGOSICS et al. 1985; KAUFMAN et al. 1984; PAPADIMITRIOU et al. 1985; PHILBEN et al. 1983; SKUDDER and SCHWARTZ 1985; WEINGRAD et al. 1982).

3.4.3.3 *Others*

Sporadic cases of other nonepithelial malignancies such as rhadomyosarcoma, neurofibrosarcoma, and the like have been reported to arise in the small intestine. Their frequency is such that generalization is hazardous (ADOTEY 1985).

3.4.4 Secondary Malignancies Involving the Small Intestine

Because of its rich blood and lymphatic involvement, the small intestine is a frequent site of metastasis. Involvement of the intestinal wall usually comes from the mesenteric nodes, and colon cancer probably contributes to the majority of cases. Secondary involvement by diffuse lymphoma has been mentioned above.

3.5 Malignancies of the Appendix

The appendix is an infrequent site of primary malignancy with the exception of two lesions: carcinoids and mucus-producing adenocarcinomas. The former are relatively indolent and may invade the appendiceal wall but rarely spread beyond the regional lymph nodes (THIRLBY et al. 1984). The latter are associated with so-called mucocele of the appendix, which was formerly thought to be due principally to inflammatory obstruction of the appendiceal lumen (ROSAI 1981).

3.5.1 Epithelial Malignancies

Several variants of adenocarcinoma are found to arise in the appendix, but almost all are rare.

3.5.1.1 Adenocarcinoma and Its Histologic Variants

As noted, a variety of adenocarcinomas arise in the appendix, ranging from those indistinguishable histologically from those arising elsewhere in the bowel with well formed glands (see Sect. 3.6.1.1) to a very rare *signet ring* form. Mucocele of the appendix is now recognized to have several etiologies (ROSAI 1981) ranging from obstruction of the lumen with hyperplasia of the mucosa to benign and malignant neoplasia. The latter is designated *mucinous cystadenocarcinoma* and is similar to tumors arising in the ovary, with which it may coexist. Penetration of the wall of the appendix may lead to gelatinous intraperitoneal deposits in which malignant cells can be found. This condition, "pseudomyxoma peritonei", is confined to the abdominal cavity and may lead to intestinal obstruction and death of the patient (ROSAI 1981).

3.5.1.2 Others

Although isolated instances of undifferentiated carcinoma and adenocanthoma do occur, they are extremely infrequent.

3.5.2 Carcinoids

These tumors are the commonest appendiceal tumors. Estimates range from 1 per 500 surgically removed appendices to 1 per thousand autopsies (THIRLBY et al. 1984; ROSAI 1981). They show all of the varieties seen elsewhere in the intestine, but they rarely metastasize (THIRLBY et al. 1984) and many are an incidental finding. The majority are found at the tip of the appendix and the peak incidence is the third and fourth decades. The classic type consists of small nests of uniform cells with small round nuclei and scanty cytoplasm (Fig. 11), and there is a tubular type sometimes showing cells resembling Paneth cells, and a mucinous or goblet cell type. The latter two forms may be difficult to distinguish from adenocarcinomas without silver staining. Carcinoids of the appendix rarely if ever lead to the carcinoid syndrome, largely because of their lack of aggressive behavior. A tumor producing ACTH with an associated Cushing's syndrome has been reported (ROSAI 1981).

3.5.3 Nonepithelial Malignancies

Though a variety of nonepithelial neoplasms can arise from the appendix, they are vanishingly rare. A lesion consisting of granular cells similar to granular cell tumors, but probably arising from smooth muscle cells, has been reported. Its malignant potential is uncertain (ROSAI 1981). Malignant lymphomas have been reported to arise in the appendix (DRAGOSICS et al. 1985; PHILBEN et al. 1983).

3.5.4 Secondary Malignancies Involving the Appendix

It is not uncommon for bulky cecal adenocarcinomas to extend to the appendix. While the reverse is possible, it is uncommon.

3.6 Malignancies of the Large Intestine

Adenocarcinoma arising in the colon and rectum represents the second leading cause of death from cancer in the United States. Of the leading cancer killers, it is the one with an almost equal incidence in both sexes. The incidence is greater than the others, but because early detection is possible with a simple test (occult stool blood) and because they tend to be well differentiated neoplasms, the cure rate is high, but closely related to the stage at the time of resection.

3.6.1 Epithelial Malignancies

The vast majority of colon and rectum cancers are adenocarcinomas. A few squamous cell tumors and carcinoids also arise in this site. The antecedent presence of a benign neoplasm (polyp) in the majority of instances is well established, and multiple polyposis syndromes such as the familial polyposis syndrome give rise to at least one focus of malignant transformation in almost 100% of cases (ROSAI 1981; MORSON and SOBIN 1976).

3.6.1.1 Adenocarcinoma and Its Histologic Variants

This very common tumor of Western civilization, while prevalent in Europe and North America, is uncommon in Africa. Multiple etiologic factors are involved (HILL 1985), but diet appears to play a central role. The relationship to polyps has already been mentioned, though unquestionably tumors do arise in nonpolypoid epithelium (BOLAND et al. 1982; DATTA 1985; RIDDELL 1985). The majority are well differentiated tumors (Fig. 8) which are bulky masses separated sharply from the surrounding normal mucosa. Cecal carcinoma tends to present as a large mass with insidious bleeding often leading to anemia as the presenting symptom. This is due to the tendency for central necrosis and ulceration of these large lesions. In the sigmoid and rectum, the tumors tend to grow into the wall and be annular and constricting (Fig. 6) with intermittent constipation and diarrhea being a presenting sign. The dictum that the majority of adenocarcinomas of the colon are in reach of the examining finger is changing, as right-sided lesions become more common. Nonetheless, the rectum and the sigmoid colon are still the most common sites of origin, followed by the cecum. The prognosis is clearly linked to stage (CHAPUIS et al. 1985; DAVIS et al. 1985), with the modified Dukes classification (A, B_1, B_2, C_1, C_2) still correlating well with outcome. As noted, since most are well differentiated, grading of these carcinomas has not correlated well with prognosis. Recently, DNA and chromosome analysis has shown promise as a prognostic tool, with aneuploidy predicting a worse prognosis for tumors of the same stage (ALITALO et al. 1983; KERR et al. 1986). A variety of antigenic markers, principally the carcinoembryonic antigen (CEA) have become useful, particularly in following the absence and detecting recurrence of colorectal cancers (LADENSON and MCDONALD 1980; KOPROWSKI et al. 1981).

Histologic variants include mucinous and *signet ring* (poor prognosis) or *linitis plastica* variants. Some show villous structures, and squamous metaplasia is described (adenoacanthoma).

3.6.1.2 Undifferentiated and Unclassified Carcinoma

As noted above, the majority of colon carcinomas are well differentiated adenocarcinomas. Some of these show malignant squamous differentiation (adenosquamous carcinoma) and squamous carcinoma can be found arising in the cecum as well as elsewhere in the colon (ROSAI 1981). The linitis plastica type of signet ring tumor referred to above is also an extremely aggressive lesion showing altered differentiation. Totally undifferentiated or unclassifiable epithelial malignancies are uncommon in the colon and rectum, and careful workup sometimes reveals that they are metastatic from sites such as the stomach. Oat cell-like carcinomas have occasionally been reported (ROSAI 1981; MORSON and SOBIN 1976).

3.6.2 Carcinoids

Carcinoids are rare in the colon, tend to be solitary, small lesions, and are the most aggressive of the tumors arising from enterochromaffin cells of the intestine (VAN SICKLE 1972). They are somewhat more common in the rectum than elsewhere in the large bowel. Histologically, they are similar to those described previously (Sects. 3.4.2 and 3.5.2) but tend to invade the bowel wall and spread to adjacent lymph nodes even when the primary lesion is small.

3.6.3 Nonepithelial Malignancies

A variety of nonepithelial tumors has been reported to arise in the large bowel. These range from lymphoma and leiomyosarcoma (POSEN and BAR-MAOR 1983), the commonest, to rareties such as Kaposi's sarcoma, which may, on occasion, present with intestinal symptomatology (ROSAI 1981).

3.6.3.1 Leiomyosarcoma

Leiomyosarcomas are more common than their benign counterpart (leiomyoma). They do not differ

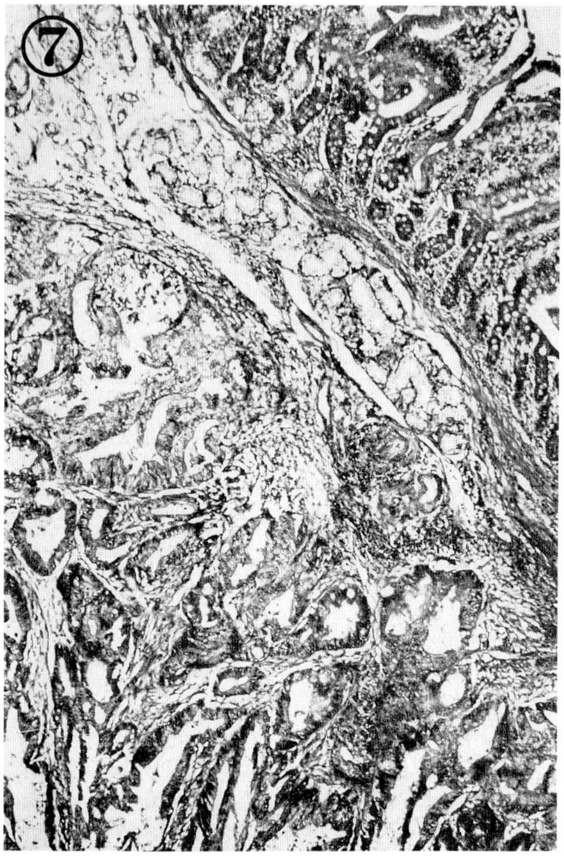

⑦

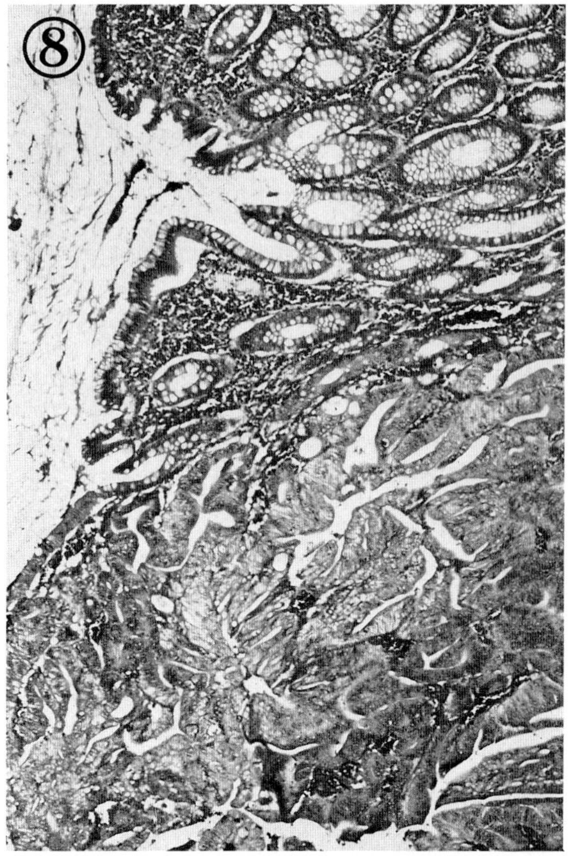

⑧

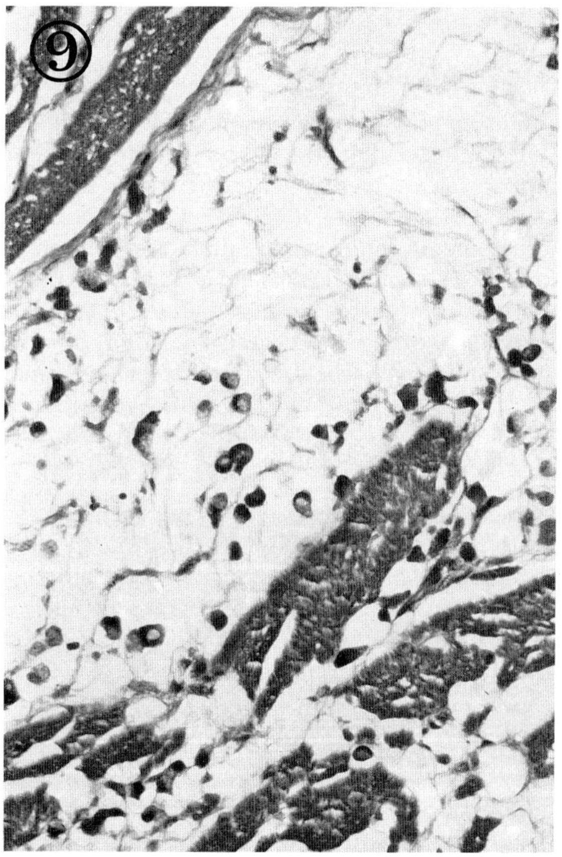

⑨

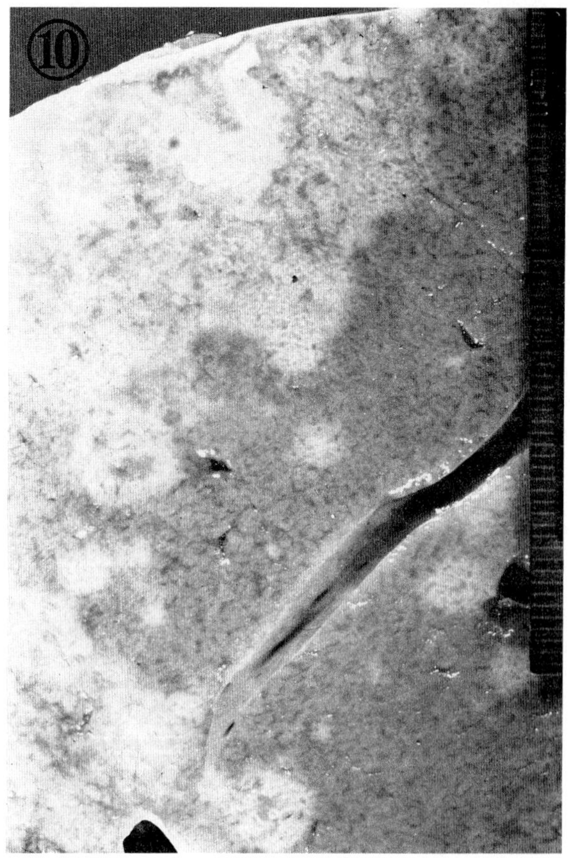

⑩

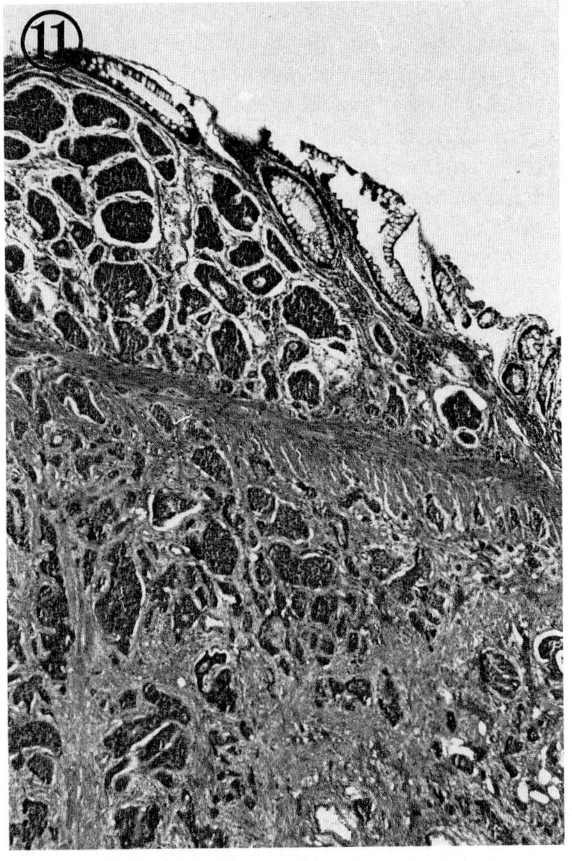

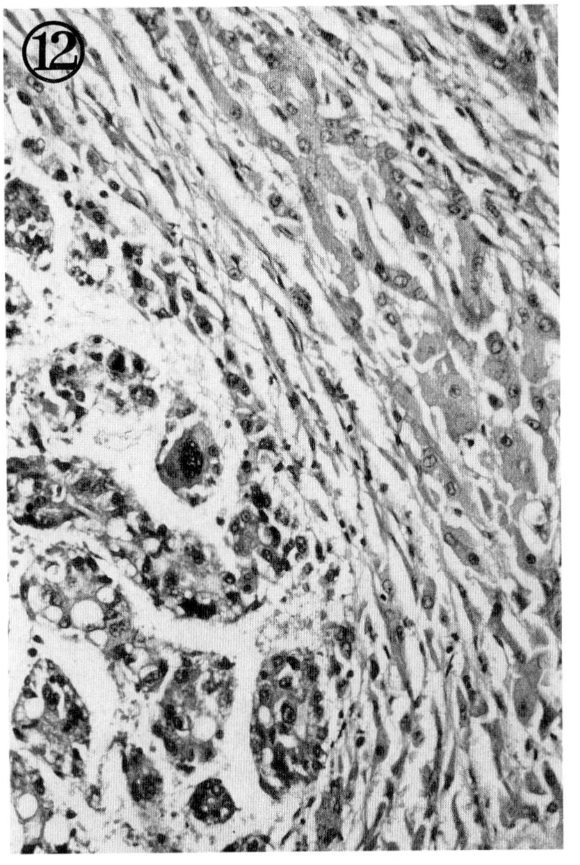

Fig. 7. Photomicrograph of adenocarcinoma of periampullary region of duodenum (see Fig. 5). Normal duodenal mucosa is seen *(top right)* separated by muscularis mucosa from submucosal Brunner's glands, which in turn are separated from the neoplastic glands of the well differentiated adenocarcinoma *(lower left)*. ×50

Fig. 8. Photomicrograph of adenocarcinoma of colon. Colonic mucosa is seen *(top)* sharply demarcated from the proliferating neoplastic glands *(bottom)* of the well differentiated adenocarcinoma. ×50

Fig. 9. Photomicrograph of signet ring cell carcinoma of stomach infiltrating fibroadipose tissue in the stomach wall. The round single neoplastic cells show central clear spaces which contain mucus. ×200

Fig. 10. Photograph of liver diffusely involved by a primary cholangiolar carcinoma. Necrosis is seen in the tumor *(upper left)*. The tumor invades the normal liver *(lower right)*. No cirrhosis is seen

Fig. 11. Photomicrograph of carcinoid of small intestine. Note mucosa at *upper right* and nests of small neoplastic cells in the submucosa and invading the muscularis. Although muscular invasion is frequent with carcinoids arising in the site, metastasis rarely occurs. ×125

Fig. 12. Photomicrograph of primary hepatocellular carcinoma. Compressed hepatic parenchyma is seen at the *upper right*. This tumor which was principally of the trabecular type was multifocal and arose in a cirrhotic liver. ×125

significantly in their pathologic characteristics or clinical presentation from those found elsewhere in the GI tract. They have also been reported in infants (Posen and Bar-Maor 1983).

3.6.3.2 Lymphoma and Others

Lymphoma is not as common a primary tumor in the large bowel as it is in the stomach or small intestine. It has a variable presentation from a large mass with or without ulceration to a diffuse enlargement of mucosal folds. As in other areas of the GI tract, the non-Hodgkin varieties, particularly diffuse "histiocytic" and poorly differentiated lymphocytic, are the most common types.

Sporadic reports of nonepithelial malignancies of various types exist, but are too rare to warrant further comment, with the exception that Kaposi's sarcoma has been reported as mentioned above. Whether primary Kaposi's sarcoma of the rectum will increase in incidence in the near future because of the AIDS problem, remains to be determined.

3.6.4 Secondary Malignancies Involving the Large Intestine

Metastatic spread can occur from a variety of sites including primary malignancies of the colon itself. Gastric carcinomas have a tendency to intra-abdominal spread with seeding of the pouch of Douglas and creation of a rectal shelf. Direct extension of a prostatic adenocarcinoma or of a squamous cell carcinoma of the cervix can at times simulate a rectal primary tumor.

3.7 Malignancies of the Anal Canal and Anal Margin

As a transition zone between the columnar epithelium of the rectum and the keratinizing squamous epithelium of the skin, the anal region is prone to some rather unique primary malignancies (Antoniades 1982j; Merlini and Eckert 1985). Squamo-columnar junctions in other sites such as the cervix are zones of active proliferation which tend to be susceptible to neoplastic transformation.

3.7.1 Epithelial Malignancies

As in other sites, various epithelial malignancies constitute the majority of malignant neoplasms arising within the anal canal and at the margin (Merlini and Eckert 1985). Nonetheless, anal carcinoma is a relatively infrequent lesion which presents with bleeding or pain as the principal clinical finding. There is a rich vascular supply to the anus, and a double lymphatic supply which permits spread to both perirectal and inguinal lymph nodes. This point should be kept in mind in either surgical or radiation therapy of anal epithelial malignancies.

3.7.1.1 Squamous Cell Carcinoma and Its Histologic Variants

There are two types of squamous tumor related to the anus (Greenall et al. 1985a, b). The first, *epidermoid carcinoma,* is similar in appearance and behavior to those arising in other sites such as the esophagus, and constitutes 70% or more of the anal carcinomas. The second is designated basaloid or cloacogenic because of its resemblance to basal or basosquamous cell carcinoma. Occasional glandular differentiation is seen in these tumors and mucin-producing cells can be demonstrated. They are thus thought to arise from the transition zone or anal gland ducts. Palisading around tumor cell nests gives the histologic resemblance to basal cell carcinomas of skin, and focal squamous or mucinous differentiation is also seen as mentioned previously. The resemblance to basal cell carcinoma is only histologic because they are aggressive and frequently metastasize (Rosai 1981; Wood 1967). Other variants such as *mucoepidermoid* carcinoma are presently considered to be encompassed by the designation "cloacogenic." Survival has been linked to differentiation, with 90% survival in well differentiated lesions and almost no 5-year survivors with undifferentiated tumors. The tumor has been reported to be twice as frequent in women (Rosai 1981). True basal cell carcinoma can occur at the anal margin. Bowen's disease (intraepithelial neoplasia, carcinoma in situ) also occurs in this region (Rosai 1981; Greenall et al. 1985a, b).

3.7.1.2 Adenocarcinoma and Its Histologic Variants

Various adenocarcinomas have been reported in the anal region, but care must be taken to separate rectal carcinomas which involve the anal canal secondarily from those arising in the anus itself. On occasion, mucinous adenocarcinomas have been reported as arising in the anus and have been thought to arise from anal gland ducts. Malignant glandular tumors related to sweat glands and apocrine glands are also found in this region. The anus is a site for extramammary Paget's disease, and these cells, which can be distinguished by their content of acid mucopolysaccharides, are believed to arise in apocrine gland ducts and may spread to the skin and mucous membranes. Paget's disease presents as an erythematous scaling lesion which spreads first to regional lymph nodes but frequently disseminates and kills the patient (Rosai 1981; Wood 1967).

3.7.1.3 Undifferentiated and Unclassified Carcinoma

The point made above that the degree of differentiation of anal epithelial malignancies correlates with the prognosis bears repeating. Thus highly undifferentiated carcinomas carry an almost uniformly fatal prognosis.

3.7.2 Nonepithelial Malignancies

The one mesenchymal neoplasm which is frequent enough to warrant discussion is embryonal rhabdomyosarcoma (sarcoma botryoides). The perianal region of infants and children is involved, and they may present as soft grape-like clusters. The outcome is usually fatal. The cells are small with dense nuclei, and electron microscopic or immunocytochemical techniques may be required to establish their relationship to skeletal muscle.

3.7.3 Melanoma

The anal canal must be considered as a potential primary site when disseminated melanoma is discovered without a known primary tumor. Indeed, primary melanoma of the anus is a fairly common melanoma of this region which may present as polypoid nodules beneath the mucosa. Rectal bleeding is frequently the first sign, and the outcome is almost uniformly fatal (ANTONIADES 1982j; ROSAI 1981).

3.7.4 Secondary Malignancies Involving the Anal Canal

The anus is subject to extension of tumors from the rectum, principally, but may also be involved secondarily by tumors of the prostate, bladder, cervix, and vagina. Metastasis to this area is uncommon except from rectal carcinomas. Primary epidermal carcinoma of adjacent skin may secondarily involve the anus.

3.8 Malignancies of the Liver and Intrahepatic Biliary System

The liver is the largest gland of the body and 60% of its mass is made up of functional parenchymal cells, or hepatocytes. The remainder contains numerous ducts which begin as canaliculi and converge until they exit to join the extrahepatic biliary system. Embryologically both bile ductal cells and hepatocytes probably derive from the same entodermal evagination which enters the mesenchyme of the septum transversum and arborizes (ARIAS et al. 1982). From this, it is not surprising that the principal hepatic malignancies are adenocarcinomas derived from hepatocytes, bile ducts, or combinations thereof (ROSAI 1981; GIBSON and SOBIN 1978). Worldwide, hepatoma (hepatocellular carcinoma) is a very common tumor, perhaps the most frequent neoplasm. Its association in the Middle East and Far East with parasites and particularly in Africa with the hepatitis B antigens is widely accepted. In the United States, primary liver cancer is increasing somewhat and well over 80% of cases are associated with cirrhosis. While alcoholic cirrhosis accounts for the highest *number* of these cases, so-called postnecrotic (macronodular) cirrhosis shows a greater tendency to give rise to malignant tumors.

3.8.1 Epithelial Malignancies

As mentioned above, the hepatocytes and bile ducts account for the vast majority of hepatic malignancies. A few unusual tumors of the supporting scaffolding of the liver such as angiosarcoma will be described below because of their known relationship to etiologic agents. In this regard, there is mounting evidence of an association between hepatitis B virus infection and hepatocellular carcinoma. While this association includes recovery of viral DNA and/or antigens from tumors (ARIAS et al. 1982), it rests largely on the circumstantial evidence of higher incidence of hepatocellular malignancy in antigen or antibody positive individuals in endemic areas of hepatitis B infection. Without meaning to denigrate the accumulating evidence, it is fair to say that the vast majority of cancers of the uterine cervix were linked to herpes type II infection until recently, when mounting evidence pointed to the papilloma virus as a more probable cause! Aflatoxin B, a potent hepatocellular carcinogen in animals, has also appeared to be a favored agent in high incidence areas, suggesting the possibility of cocarcinogenesis (ARIAS et al. 1982).

3.8.1.1 Hepatocellular Carcinoma

This is the most frequent of primary hepatic neoplasms, and shows a variety of patterns grossly and microscopically (ROSAI 1981; GIBSON and SOBIN 1978; EDMONDSON 1958). They may present as rapid decompensation of liver function in a patient with known cirrhosis who was seemingly doing well, or as an intra-abdominal mass sometimes leading to ascites. The oncofetal antigen, alphafetoprotein, is frequently and sometimes markedly elevated. The tumors are frequently multicentric, especially in individuals with cirrhosis.

The principal histologic variant of hepatocellular carcinoma is the *trabecular* form composed of cords of atypical cells (Fig. 12). A *pseudoglandular* pattern is seen as a pure form or admixed with other histologic forms. This pattern may be difficult to distinguish from that of a bile duct carcinoma (cholangiocarcinoma) arising in the liver. Other variants include: scirrhous, compact, pleomorphic clear cell, and others such as a small cell lesion which is difficult to establish as having a liver cell origin. Bile production is seen, but is not a common feature. Tumor cells may contain various inclusions such as Mallory's (alcoholic) hyaline and ground glass cytoplasm (hepatitis B surface antigen). A globular body has been described in a particular type of neoplasm, fibrolamellar carcinoma, which occurs in younger people and has a somewhat better prognosis (ROSAI 1981).

3.8.1.2 Cholangiocarcinoma (Intrahepatic Bile Duct Carcinoma)

The intrahepatic bile ducts give rise to tumors which are designated *cholangiolar* or *cholangiocarcinoma* (Fig. 10) depending on the size of the bile ducts from which the tumor is thought to arise. Rare adenosquamous, mucoepidermoid, or frankly squamous carcinomas have been described as arising in the liver (ROSAI 1981; EDMONDSON 1958). A variant of bile duct carcinoma designated bile duct cystadenocarcinoma has been described as a multilocular and multifocal cystic tumor which may be well defined and therefore surgically separable from the liver with an accompanying better prognosis (ROSAI 1981). Combined hepatocellular and cholangiocarcinomas in which both elements are clearly present are well recognized.

3.8.1.3 Hepatoblastoma

This is a rare malignant tumor which occurs with an age peak between 1 and 2 years of age. It is composed of small "embryonal" and larger "fetal" cells resembling their normal counterparts. Elements such as extramedullary hematopoiesis, loose fibrous connective tissue, and cartilage and osteoid are regularly found (ROSAI 1981).

3.8.2 Nonepithelial Malignancies

While in rare instances, leiomyosarcoma and fibrosarcoma have been recorded, there are two tumors arising in supporting stroma which warrant comment. These are angiosarcoma (hemangiosarcoma) and embryonal sarcoma. The former is a malignant vascular tumor which usually forms multiple, poorly defined masses and histologically shows cells infiltrating the sinusoids and even bizarre giant cells. A clear association with Thorotrast or vinyl chloride exposure has been documented (ARIAS et al. 1982).

Embryonal sarcoma presents between the ages of 5 and 10 years and is composed of a variety of malignant mesenchymal elements. The occasional differentiation toward rhabdomyosarcomas has been seen. The prognosis is poor (ROSAI 1981).

3.8.3 Miscellaneous Malignancies

A variety of rare tumors such as teratoma and carcinosarcoma have been reported sporadically in the liver (EDMONDSON 1958). The latter may show carcinomatous elements which arise from either hepatocytes or bile ducts.

3.8.4 Hematopoietic and Lymphoid Neoplasms

The liver is one of the principal hematopoietic organs during fetal life, and one of the first to regain this capacity when there is a need for extramedullary hematopoiesis. It is frequently involved secondarily in leukemia and lymphomas of a variety of types. Leukemia arising in the liver is difficult to document. Primary lymphomas of the liver have been reported (BIEMER 1984; OSBORNE et al. 1985).

3.8.5 Secondary Malignancies Involving the Liver

The liver is possibly the most frequent site of metastatic involvement in the body, closely followed by the lung. All of the common neoplasms frequently metastasize to the liver. The most frequent primary tumors arise in the colon, lung, and breast.

3.9 Malignancies of the Gallbladder and Extrahepatic Biliary System

Of the extrahepatic biliary tree, the gallbladder is the most common site of primary malignancy. There is a strong association with calculi, but whether this is cause or effect remains to be established (DUNBAR et al. 1983; WEEDON 1984).

3.9.1 Epithelial Malignancies

Tumors of the extrahepatic bile ducts show no association with cholelithiasis. The periampullary region has already been mentioned with the small intestine (see Sect. 3.4.1.2), and while it is the commonest tumor of this region, it is usually an adenocarcinoma, and it will not be discussed further here. Carcinomas of these ducts are uncommon, tend to arise from the larger ducts (common duct, junction of cystic with common duct, hepatic ducts in that order) and to spread rapidly, even though they are detected early because of their tendency to obstruct and to produce jaundice (SCHOENFIELD 1977). Increased evidence of bile duct malignancy is seen in ulcerative colitis patients and *Clonorchis sinensis* and *Giardia lamblia* infestation (SSCHOENFIELD 1977). Carcinoma of the extrahepatic biliary system occurs more frequently in women.

3.9.1.1 Adenocarcinoma and Its Histologic Variants

The majority of tumors of the gallbladder and of the bile ducts are adenocarcinomas (DUNBAR et al. 1983; SCHOENFIELD 1977). An association with polyps has been noted in the gallbladder, and interestingly many tumors are unsuspected findings in people operated upon for gallstones (DUNBAR et al. 1983). Nonetheless, survival is usually less than 1 year because of rapid extension to lymph nodes and to the liver. A variety of histologic variants may be found in either the gallbladder or the ducts. These include papillary tumors, which are somewhat more common, mucinous, signet ring cell, and adenosquamous carcinomas. They have a tendency to invade perineural lymphatics, which accounts for the frequency of abdominal pain as a presenting symptom. Well differentiated lesions of the bile duct may be extremely difficult to distinguish from normal epithelium. The latter has a tendency to burrow into the muscle wall, and unequivocal invasion must be demonstrated to avoid misdiagnosis. A well differentiated lesion which is slowly growing called sclerosing carcinoma may resemble sclerosing cholangitis on X-ray (ALTEMEIER et al. 1957; PECK et al. 1974).

3.9.1.2 Squamous Cell Carcinoma

True squamous cell carcinomas with keratin pearls and intercellular bridges do arise as primary tumors both in the gallbladder and, rarely, in the extrabiliary duct system. However, it is more frequent for adenocarcinomas to show squamous metaplasia (adenoacanthoma) or foci of squamous malignancy accompanying malignant glands (adenosquamous carcinoma) (ROSAI 1981). Anaplastic carcinomas are somewhat more common than squamous carcinomas in these sites (SCHOENFIELD 1977).

3.9.2 Nonepithelial Malignancies

Since epithelial malignancies of this system constitute about 0.01% of malignancies found at autopsy and constitute over 90% of malignant tumors of this system, nonepithelial malignancies are extremely rare in this site. Two unusual tumors that have been reported are granular cell tumor and embryonal rhabdomyosarcoma (sarcoma botryoides) (ROSAI 1981).

3.9.3 Miscellaneous Malignancies

Malignancies such as melanoma (ANTONIADES 1982h; WEEDON 1984), carcinoids (WEEDON 1984), and carcinosarcoma (WEEDON 1984) have been reported as arising in the gallbladder or extrahepatic biliary system.

3.9.4 Secondary Malignancies Involving the Gallbladder

It is common for primary tumors of the gallbladder to extend to the liver and metastasize to regional nodes early in their course. Pancreatic primary tumors, since they are most common in the head of the pancreas, frequently extend to or spread around the bile ducts and produce obstruction. Metastatic melanoma involving the gallbladder is said to be more common than primary melanoma of this organ as illustrated by Rosai (ROSAI 1981).

3.10 Malignancies of the Exocrine Pancreas

Pancreatic malignancies are one of the increasing health problems of the United States and parts of the Western world (GORDIS and GOLD 1986). The majority of these (80%–90%) are thought to arise from pancreatic ducts, and acinar tumors constitute less than 2% of primary pancreatic neoplasms (GORDIS and GOLD 1986). It is now the fourth leading cause of death from cancer, and generally the prognosis is poor with a rapid demise (less than 1-year survival in 90% of patients). Fortunately, there seems to be a plateauing of the incidence recently, but despite numerous suggestions including coffee and cigarettes, the etiology of these malignancies remains obscure. Almost two-thirds are found in the head and only one-third in the body and tail together. Although this may lead to early detection through the production of jaundice, cure is still infrequent. Perineural invasion is characteristic and can account for the frequency of pain as a presenting symptom (FAINTUCH and LEVIN 1986).

3.10.1 Epithelial Malignancies

As stated above, the vast majority of pancreatic primary malignant tumors arise from the ducts (KLOPPEL and FITZGERALD 1986). Occasional cystic tumors and more acinar lesions are found. Anaplastic tumors, some which contain giant cells, are also seen.

3.10.1.1 Adenocarcinoma and Its Histologic Variants

Ductal carcinomas are included in the group of adenocarcinomas (KLOPPEL and FITZGERALD 1986). They are usually fairly well differentiated tumors, and can be very difficult to recognize on frozen section, particularly when there is an associated chronic pancreatitis. An obstructing carcinoma may produce dilatation of the duct, accompanying fibrosis and atrophy of acinar elements with preservation of the islets. The gross tumor may be ill defined, adding to the difficulty at times. Therefore, close inspection of cytologic details is frequently required to make the proper diagnosis. Additional variants include a cystadenocarcinoma (ANTONIADES 1982f; KLOPPEL and FITZGERALD 1986) which may present as a large cyst filled with mucinous material which can be difficult to distinguish from benign cystic lesions unless unequivocal invasion of the cyst wall is demonstrated, although the malignant potential of benign appearing lesions is debated (COMPAGNO and OERTEL 1978). True acinar tumors can be recognized by their abundant cytoplasm which usually contains demonstrable granules. Zymogen granules can usually be demonstrated by electron microscopy (KERN et al. 1986) or histochemistry in these tumors. The tumors are usually composed of solid nests, and a trabecular arrangement suggests an islet cell origin rather than an acinar tumor. Secretion of lipase and amylase by these tumors has been reported (ROSAI 1981). A papillary malignancy of low metastatic potential has also been reported most often in young females. The multiplicity of variants has been emphasized in a recent discussion (KLOPPEL and FITZGERALD 1986), and the size of the primary tumor has been related to prognosis (TSUCHIYA et al. 1985).

3.10.1.2 Anaplastic and Undifferentiated Carcinomas

Most poorly differentiated tumors probably are variants of ductal carcinoma. Tumors composed of large multinucleated giant cells may be seen and are sometimes designated giant cell tumors. Another variant is composed of spindle cells and a third form is composed of small round cells (ROSAI 1981).

3.10.2 Nonepithelial Malignancies

These are all extremely rare and when they arise do not differ significantly in their pathologic features from those described in other sites.

3.10.3 Miscellaneous Malignancies

A tumor which presents in young children or infants has been described as an infantile pancreatic carcinoma (pancreaticoblastoma). It consists of solid nests of well differentiated cells next to well differentiated glands. It may progress slowly (ROSAI 1981). Many other varieties, particularly those with endocrine function, could be described (CH'NG et al. 1986).

3.10.4 Secondary Malignancies Involving the Exocrine Pancreas

Involvement of the pancreas by metastatic disease of any type which spreads intra-abdominally is relatively common. The pancreas has a rich blood and lymphatic supply and tumors such as adenocarcinomas of the colon frequently find their way to one of the lymph nodes in or around the pancreas.

3.11 Malignancies of the Endocrine Pancreas

These tumors, although fascinating, are relatively uncommon (CH'NG et al. 1986). The majority of islet cell tumors are benign, but it is difficult to predict behavior on the basis of histopathologic feature. The tumors are usually well circumscribed solitary nodules, but multiple tumors can be found in both the benign and malignant varieties. They may also be found in association with the multiple endocrine neoplasia (MEN) syndrome (type I). Histologically, they may be difficult to distinguish from carcinoids, which can arise in the pancreas. All of the hormones known to be released by the islets have been found associated with these tumors, with syndromes such as hypoglycemia (insulin) (CH'NG et al. 1986), hyperglycemia (glucogen) (ANTONIADES 1982e), watery diarrhea (secretin, VIP) (ANTONIADES 1982d), and the ulcerogenic syndrome of Zollinger and Ellison (gastrin) (ANTONIADES 1982c). Even when malignant, they tend to run a slowly progressive course, and the patients may suffer more from hormonal imbalance than tumor involvement (CH'NG et al. 1986).

Acknowledgments. The authors are grateful to Mr. William Gunning for his assistance with the illustrations and Ms. Marilyn Cline for her superb secretarial support.

References

Abdelwahab IF, Klein MJ (1983) Granular cell tumor of the stomach: A case report and review of the literature. Am J Gastroenter 78: 71-76

Abramson DJ (1973) Leiomyoblastomas of the stomach. Surg Gynecol Obstet 136: 118-125

Adotey JM (1985) Primary malignant tumours of the small intestine. Br J Clin Pract 39: 54-58

Akiyama H, Tsurumaru M, Kawamura T, Ono Y (1981) Principles of surgical treatment for carcinoma of the esophagus: Analysis of lymph node involvement. Ann Surg 194: 438-446

Akwari OE, Dozois RR, Weiland LH, Beahrs OH (1978) Leiomyosarcoma of small and large bowel. Cancer 42: 1375-1384

Alitalo K, Schwab M, Lin CC, Varmus HE, Bishop JM (1983) Homogeneously staining chromosomal regions contain amplified copies of an abundantly expressed cellular oncogene (c-myc) in malignant neuroendocrine cells from a human colon carcinoma. Proc Natl Acad Sci USA 80: 1707-1711

Altemeier WA, Gall EA, Zinninger NM, Hoxworth PI (1957) Sclerosing carcinoma of the major intrahepatic bile ducts. Arch Surg 75: 450-461

Anderson LL, Lad TE (1982) Autopsy findings in squamous-cell carcinoma of the esophagus. Cancer 50: 1587-1590

Antoniades J (1982a) Carcinoid tumors of the gastrointestinal tract. In: Antoniades J (ed) Uncommon malignant tumors. Masson, New York, pp 162-172

Antoniades J (1982b) Carcinosarcoma of the esophagus. In: Antoniades J (ed) Uncommon malignant tumors. Masson, New York, pp 201-202

Antoniades J (1982c) Pancreatic gastrinoma (Zollinger-Ellison syndrome). In: Antoniades J (ed) Uncommon malignant tumors. Masson, New York, pp 325-328

Antoniades J (1982d) WDHA syndrome (watery diarrhea, hypokalemia, achlorhydria syndrome) (pancreatic cholera syndrome). In: Antoniades J (ed) Uncommon malignant tumors. Masson, New York, pp 329-333

Antoniades J (1982e) Pancreatic glucagonoma. In: Antoniades J (ed) Uncommon malignant tumors. Masson, New York, pp 334-337

Antoniades J (1982f) Cystadenocarcinoma of the pancreas. In: Antoniades J (ed) Uncommon malignant tumors. Masson, New York, pp 338-340

Antoniades J (1982g) Melanoma of the esophagus. In: Antoniades J (ed) Uncommon malignant tumors. Masson, New York, pp 77-78

Antoniades J (1982h) Melanoma of the gallbladder. In: Antoniades J (ed) Uncommon malignant tumors. Masson, New York, pp 79

Antoniades J (1982i) Squamous cell carcinoma of the stomach. In: Antoniades J (ed) Uncommon malignant tumors. Masson, New York, pp 8-11

Antoniades J (1982j) Melanoma of the anal canal. In: Antoniades J (ed) Uncommon malignant tumors. Masson, New York, pp 89-91

Aozasa K, Tsujimoto M, Inoue A, Nakagawa K, Hanai J, Kurata A, Nosaka J (1985) Primary gastrointestinal lymphoma: a clinicopathologic study of 102 patients. Oncology 42: 97-103

Arias IM, Popper H, Schachter D, Shafritz DA (1982) The liver. Raven, New York

Athanasoulis CA, Aral IM (1968) Leiomyosarcoma of the esophagus. Gastroenterology 53: 271-274

Biemer JJ (1984) Hepatic manifestations of lymphomas. Ann Clin Lab Sci 14: 252-260

Black WC (1968) Enterochromaffin cell types and corresponding carcinoid tumors. Lab Invest 19: 473-485

Boland CR, Montgomery CK, Kim YS (1982) Alterations in human colonic mucin occurring with cellular differentiation and malignant transformation. Proc Natl Acad Sci USA 79: 2051-2055

Bralow SP (1982) Diagnosis and staging of esophageal and gastric cancer. Cancer 50: 2566-2570

Brooks JJ, Enterline HT (1983) Gastric pseudolymphoma: Its three subtypes and relation to lymphoma. Cancer 51: 476-486

Cameron AJ, Beverly JO, Payne WS (1985) The incidence of adenocarcinoma in columnar-lined (Barrett's) esophagus. N Engl J Med 313: 857-859

Ch'ng JLC, Polak JM, Bloom SR (1986) Miscellaneous tumors of the pancreas. In: Go VLW, Brooks FP, DiMagno EP, et al. (eds) The exocrine pancreas: Biology, pathobiology, and diseases. Raven, New York, pp 763-771

Chapuis PH, Fisher R, Dent OF, Newland RC, Pheils MT (1985) The relationship between different staging methods and survival in colorectal carcinoma. Dis Colon Rectum 28: 158-161

Compagno J, Oertel JE (1978) Mucinous cystic neoplasms of the pancreas with overt and latent malignancy (cystadenocarcinoma and cystadenoma). A clinicopathologic study of 41 cases. Am J Clin Pathol 69: 573-580

Cooper BT, Read AE (1985) Small intestinal lymphoma. World J Surg 9: 930-937

Cooper EH, Giles GR (1984) Biochemical markers in gastrointestinal malignancies. In: DeCosse JJ, Sherlock P (eds) Clinical management of gastrointestinal cancer. Nijhoff, Boston, pp 1–34

Cubilla AL, Fitzgerald PJ (1984) Atlas of tumor pathology: tumors of the exocrine pancreas. Armed Forces Institute of Pathology, Washington, D.C.

Datta PK (1985) Inflammatory bowel disease and carcinoma. Practitioner 229: 465–469

Davis NC, Evans EB, Cohen JR, Theile DE, Job D (1985) Clinicopathological staging of colorectal cancer: has the time arrived? Br J Surg 72: S47–S52

Dragosics B, Bauer P, Radaszkiewics T (1985) Primary gastrointestinal non-Hodgkin's lymphomas: a retrospective clinicopathologic study of 150 cases. Cancer 55: 1060–1073

Dunbar LL, Adkins RB, Farringer J, Waterhouse G, O'Leary JP (1983) Carcinoma of the gallbladder and bile ducts. Am Surg 49: 94–104

Earlam R, Cunha-Melo JR (1980) Oesophageal squamous cell carcinoma. I. A critical review of surgery. Br J Surg 67: 381–390

Edmondson HA (1958) Atlas of tumor pathology: Tumors of the liver and intrahepatic bile ducts. Armed Forces Institute of Pathology, Washington, D.C.

Faegenburg D, Farman J, Dallemand S, Schechter LS, Rosen Y, Chiat H (1975) Leiomyoblastoma of the stomach. Radiology 117: 297–300

Faintuch J, Levin B (1986) Clinical presentation and diagnosis of exocrine tumors of the pancreas. In: Go VLW, Brooks FP, DiMagno RP, et al. (eds) The exocrine pancreas: Biology, pathobiology, and diseases. Raven, New York, pp 675–687

Fein R, Kelsen DP, Geller N, Bains M, McCormack P, Brennan MF (1985) Adenocarcinoma of the esophagus and gastroesophageal junction. Cancer 56: 2512–2518

Frantz VK (1959) Atlas of tumor pathology: Tumors of the pancreas. Armed Forces Institute of Pathology, Washington, D.C.

Gibson JB, Sobin LH (1978) Histological typing of tumours of the liver, biliary tract and pancreas. WHO, Geneva

Goldenberg DM, Deland FH (1984) Antimarker antibodies for the external imaging of gastrointestinal cancer. In: DeCosse JJ, Sherlock P (eds) Clinical management of gastrointestinal cancer. Nijhoff, Boston, pp 351–364

Gordis L, Gold EB (1986) Epidemiology and etiology of pancreatic cancer. In: Go VLW, Brooks FP, DeMagno EP, et al. (eds) The exocrine pancreas: biology, pathology, and diseases. Raven, New York, pp 621–636

Greenall MJ, Quan SHQ, DeCosse JJ (1985a) Epidermoid cancer of the anus. Br J Surg 72: S97–S103

Greenall MJ, Quan SHQ, Stearns MW, Urmacher C, DeCosse JJ (1985b) Epidermoid cancer of the anal margin. Am J Surg 149: 95–101

Hanada M, Nakano K, Ii Y, Takami M (1985) Carcinosarcoma of the stomach: A case report with light microscopic, immunohistochemical, and electron microscopic study. Acta Pathol Jpn 35: 951–959

Haq MM, Blumenthal BJ, Culotta RJ, Cain DN, McCleery JM (1985) Small bowel adenocarcinoma: report of three cases and review of the literature. Tex Med 81: 51–54

Hill MJ (1985) Cancer of the large bowel: Human carcinogenesis. Br J Surg 72: S37–S39

Hough DR, Chan A, Davidson H (1983) Von Recklinghausen's disease associated with gastrointestinal carcinoid tumors. Cancer 51: 2206–2208

Kaufman Z, Eliashiv A, Shpitz B, Witz M, Griffel B, Dinbar A (1984) Primary gastrointestinal lymphoma: a review of 21 cases. J Surg Oncol 26: 17–21

Kern HF, Rausch U, Mollenhauer J (1986) Fine structure of human pancreatic adenocarcinoma. In: Go VLW, Brooks FP, DiMagno EP, et al. (eds) The exocrine pancreas: biology, pathobiology and diseases. Raven, New York, pp 637–647

Kerr IB, Spandidos DA, Finlay IG, Lee FD, McArdle CS (1986) The relation of ras family oncogene expression to conventional staging criteria and clinical outcome in colorectal carcinoma. Br J Cancer 53: 231–235

Klifto EJ, Chase PJ, Metzman M, Kupersmit M, Allen S (1983) Leiomyoblastoma of the stomach: report of a case and review of the literature. J Am Osteopath Assoc 83: 30–33

Kloppel G, Fitzgerald PJ (1986) Pathology of nonendocrine pancreatic tumors. In: Go VLW, Brooks FP, DiMagno EP, et al. (eds) The exocrine pancreas: biology, pathobiology, and diseases. Raven, New York, pp 649–674

Koprowski H, Herlyn M, Steplewski Z (1981) Specific antigen in serum of patients with colon carcinoma. Science 212: 53–55

Ladenson JH, McDonald JM (1980) Colorectal carcinoma and carcinoembryonic antigen (CEA). Clin Carcinoma 26: 1213–1220

Lauren P (1965) The two histological main types of gastric carcinoma. Diffuse and so-called intestinal type carcinoma. Acta Pathol Microbiol Scand 64: 31–49

Lerman-Sagie T, Ziv Y, Rubin M, More C, Dintsman M (1985) Gastric lymphoma versus pseudolymphoma: the importance of immunological differentiation. Am J Gastroenterol 80: 763–766

Merlini M, Eckert P (1985) Malignant tumors of the anus. Am J Surg 150: 370–372

Ming S (1973) Atlas of tumor pathology: tumors of the esophagus and stomach. Armed Forced Institute of Pathology, Washington, D.C.

Morson BC, Sobin LH (1976) Histological typing of intestinal tumours. WHO, Geneva

Noguchi Y, Ohta H, Takagi K, et al. (1985) Synchronous multiple early gastric carcinoma: a study of 178 cases. World J Surg 9: 786–793

Non-Hodgkin's Lymphoma Pathologic Classification Project (1982) National Cancer Institute sponsored study of classifications of non-Hodgkin's lymphomas: summary and description of a working formulation for clinical usage. Cancer 49: 2112–2135

Northover JMA (1985) Carcinoembryonic antigen and recurrent colorectal cancer. Br J Surg 72: S44–S46

Oota K, Sobin LH (1977) Histological typing of gastric and oesophageal tumours. WHO, Geneva

Osborne BM, Butler JJ, Guarda LA (1985) Primary lymphoma of the liver: ten cases and a review of the literature. Cancer 56: 2902–2910

Pacella DA, Caleel RT (1985) Leiomyosarcoma of the jejunum: report of a case and review of the literature. J Am Osteopath Assoc 85: 330–334

Papadimitriou CS, Papacharalampous NX, Kittas C (1985) Primary gastrointestinal malignant lymphomas: a morphologic and immunohistochemical study. Cancer 55: 870–879

Peck JJ, Kern WH, Mikkelsen WP (1974) Sclerosis of the extrahepatic bile ducts. Arch Surg 108: 798–800

Philben VJ, Edney JA, Thompson J, Armitage JO (1983) Gastrointestinal lymphoma. Nebr Med J 68: 357–359

Posen JA, Bar-Maor JA (1983) Leiomyosarcoma of the colon in an infant. Cancer 52: 1458–1461

Ranchod M, Kempson RL (1977) Smooth muscle tumors of the gastrointestinal tract and retroperitoneum. A pathologic analysis of 100 cases. Cancer 39: 255–262

Riddell RH (1985) Dysplasia and cancer in inflammatory bowel disease. Br J Surg 72: S83–S86

Robbins SL, Cotran RS, Kumar V (1984) Pathologic basis of disease. Saunders, Philadelphia

Rosai J (1981) Ackerman's surgical pathology. Mosby, St Louis

Ryan M, Livstone EM (1985) Gastrointestinal cancer. In: Gitnick G (ed) Current gastroenterology. Year Book Medical, Chicago, pp 249–297

Saito T, Iizuka T, Kato H, Watanabe H (1985) Esophageal carcinoma metastatic to the stomach: A clinicopathologic study of 35 cases. Cancer 56: 2235–2241

Sandler RS (1984) Primary gastric lymphoma: A review. Am J Gastroenterol 79: 21–25

Schoenfield LJ (1977) Tumors of the biliary system. In: Dietschy JM (ed) Diseases of the gallbladder and biliary system. Wiley, New York, pp 280–311

Schottenfeld D (1984) Epidemiology of cancer of the esophagus. Semin Oncol 11: 92–100

Seo IS, Binkley WB, Warner TFCS, Warfel KA (1982) A combined morphologic and immunologic approach to the diagnosis of gastrointestinal lymphomas. I. Malignant lymphoma of the stomach (a clinicopathologic study of 22 cases). Cancer 49: 493–501

Shibuya H, Azumi N, Onda Y, Abe F (1985) Multiple primary malignant fibrous histiocytoma of the stomach and small intestine. Acta Pathol Jpn 35: 157–164

Silverberg E (1985) Cancer statistics. CA 35: 12–13

Skudder PA, Schwartz SI (1985) Primary lymphoma of the gastrointestinal tract. Surg Gynecol Obstet 160: 5–8

Takahashi T, Iwama N (1985) Three-dimensional microstructure of gastrointestinal tumors: gland pattern and its diagnostic significance. Pathol Annu 20: 419–440

Thirlby RC, Kasper CS, Jones RC (1984) Metastatic carcinoid tumor of the appendix. Dis Colon Rectum 27: 42–46

Tsuchiya R, Takashi O, Takatoshi N (1985) Size of the tumor and other factors influencing prognosis of carcinoma of the head of the pancreas. Am J Gastroenterol 80: 459–462

Vanden Heule B, Taylor CR, Terry R, Lukes RJ (1982) Presentation of malignant lymphoma in the rectum. Cancer 49: 2602–2607

Van Sickle DG (1972) Carcinoid tumors. Cleve Clin Q 39: 79–86

Weedon D (1984) Pathology of the gallbladder. Masson, New York

Weingrad DN, Decosse JJ, Sherlock P, Straus D, Lieberman PH, Filippa DA (1982) Primary gastrointestinal lymphoma: a 30-year review. Cancer 49: 1258–1265

Wood DA (1967) Atlas of tumor pathology: Tumors of the esophagus and stomach. Armed Forces Institute of Pathology, Washington, D.C.

Yang K, Ulich T, Cheng L, Lewin KJ (1983) The neuroendocrine products of intestinal carcinoids. Cancer 51: 1918–1926

Zalcberg JR, McKenzie IFC (1985) Tumor-associated antigens – an overview. J Clin Invest 3: 876–882

4 Radiological Examination of the Patient with Gastrointestinal Cancer

RONELLE A. DUBROW

CONTENTS

4.1 Introduction

Radiology has many roles to play in evaluating patients with malignancies of the gastrointestinal (GI) tract. These include diagnosis, staging, assessment of treatment response, detection of recurrence and complications as well as, recently, intervention in the management of complications and the palliation of disease. Detection of tumors of the hollow viscera is the province of barium examinations and endoscopy. Both modalities have advantages and disadvantages and in many respects are complementary. Detection of cancers of the liver, bile ducts, and pancreas is generally by ultrasound and/or computed tomography (CT) scanning as well as by endoscopic cholangiopancreatography and transhepatic cholangiography. Diagnosis is made by histologic analysis. Biopsies are obtained endoscopically of the hollow viscera and percutaneously of solid organ tumors, using radiologic guidance (CT, ultrasound, fluoroscopy). Percutaneous biopsy has proved to be a safe reliable procedure in many large series (FERRUCCI et al. 1980; HARTER et al. 1983; BERNARDINO 1984). It can be used to diagnose recurrence and metastasis as well as the primary tumor.

Staging in GI malignancies was traditionally performed by clinical examination and in some cases liver/spleen scanning and angiography. CT has provided another dimension in preoperative staging. Initial expectations that CT would eliminate all need for surgical staging have been largely unfulfilled. As more and careful studies are being done, CT is being recognized as relatively insensitive in a number of areas; for example in detecting tumor in normal sized lymph nodes, in detecting even advanced peritoneal seeding, in separating peritumoral nodes from the tumor itself, and often in deter-

RONELLE A. DUBROW, M.D., Associate Professor, Division of Diagnostic Imaging and Department of Diagnostic Radiology, The University of Texas System Cancer Center, M.D. Anderson Hospital and Tumor Institute, 1515 Holcombe Blvd., Houston, TX 77030, USA mining how far through the wall the tumor has grown. Even in the area of hepatic metastases, where CT has been the undisputed master of imaging techniques, newer modalities such as CT portography, ethiodized oil emulsion (EOE)-13 CT scanning, and magnetic resonance imaging (MRI) of the liver as well as careful correlation with resected specimens have revealed that the sensitivity of standard contrast enhanced CT scanning may be in the range of only 60%–70% (Matsui et al. 1987; Freeny et al. 1986). This will be discussed in more detail in the section on liver metastases.

In terms of preoperative staging, however, CT still frequently provides useful information. On the horizon are MRI and endoscopic ultrasound as staging tools. The latter seems to be particularly effective in documenting the lateral spread of the tumor through the wall and beyond and in documenting adjacent lymphadenopathy. Reference will be made to all of these modalities as they apply to the tumor in question.

Staging systems are frequently based on physical findings or on surgical/histologic findings and serve as prognostic guides. Moss et al. (1981 d) introduced a CT based staging system for malignant neoplasms of the alimentary tube (Table 1) which has been referred to frequently in the subsequent radiologic literature. It provides a common reference point for comparing different imaging studies and allows easy correlation with surgical and pathologic data.

In addition to detection, diagnosis, and staging, radiology has been increasingly involved in treatment both of complications and of the primary and metastatic disease itself. Examples of interventional procedures which are not aimed at direct tumor treatment would include percutaneous postsurgical abscess drainage, percutaneous gastrostomy placement (for feeding in the case of esophageal obstruction, for example), embolization of bleeding ulcers or tumors, and biliary or biliary-enteric anastomotic stricture dilatation. This type of radiologic intervention will not be discussed further as it applies to many clinical situations unrelated to gastrointestinal malignancy as well as to patients with tumor.

Interventional techniques aimed at tumor treatment would include percutaneous catheterization for intraarterial chemotherapy, transcatheter embolization of primary or metastatic tumor (e.g., islet cell tumors of the pancreas and carcinoid metastatic to the liver), and CT and ultrasound guided laser and cryotherapy of liver lesions. These will be discussed in more detail.

Table 1. CT-based staging system for gastrointestinal malignancies (adapted from Moss et al. 1981 d)

Stage I	– Intraluminal mass Wall or normal thickness
Stage II	– Focal or diffuse wall thickening No extension beyond the bowel wall
Stage III	– Wall thickening with extension beyond the wall into surrounding tissue. – Local or regional adenopathy may be present – No distant metastases
Stage IV	– Distant metastases irrespective of local tumor extent

4.2 Esophagus

4.2.1 Esophagography

4.2.1.1 Diagnosis

Barium esophagography is a sensitive tool in the detection of esophageal carcinomas (Halpert et al. 1985). The double contrast esophagram has been shown to be sensitive in demonstrating even small and/or superficial esophageal cancers (Koehler et al. 1976; Itai et al. 1978; Cozzi et al. 1987). However, most patients who are examined have symptoms, generally dysphagia, and have advanced lesions at presentation.

The radiographic appearance of advanced esophageal cancer can be predominantly polypoid, ulcerative, or infiltrating (Fig. 1). A multinodular mass may mimic varices on a single view (Fig. 2); however, the irregularly enlarged folds will not change from image to image as true varices will. Spindle cell carcinomas are unusual bulky intraluminal masses which expand the esophageal lumen without complete obstruction (Fig. 3) (Olmstead et al. 1983). A verrucous squamous cell carcinoma is another rare variant which has a small cauliflower-like appearance on barium studies. Esophagogastric junction (EGJ) carcinomas, although typically adenocarcinomas, have in general the same radiographic appearance as squamous cell or adenocarcinomas throughout the esophagus. A unique presentation occurs when the tumor infiltrates submucosally at the EGJ, causing a tapered narrowing and a proximal motility disturbance that simulates achalasia (Fig. 4). Although radiographic features to aid in differentiating this pseudoachalasia from true primary achalasia have been described (Marshak and Eliasoph 1957; Lawson and Dodds 1976; Dodds et al. 1986), distinction can occasionally be quite difficult. A high index of suspicion must be

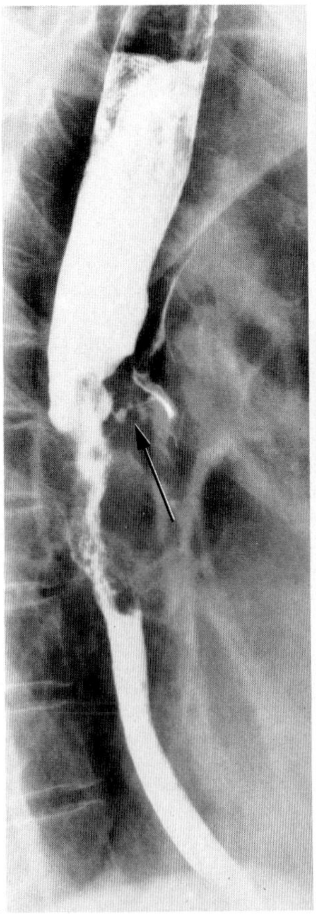

1

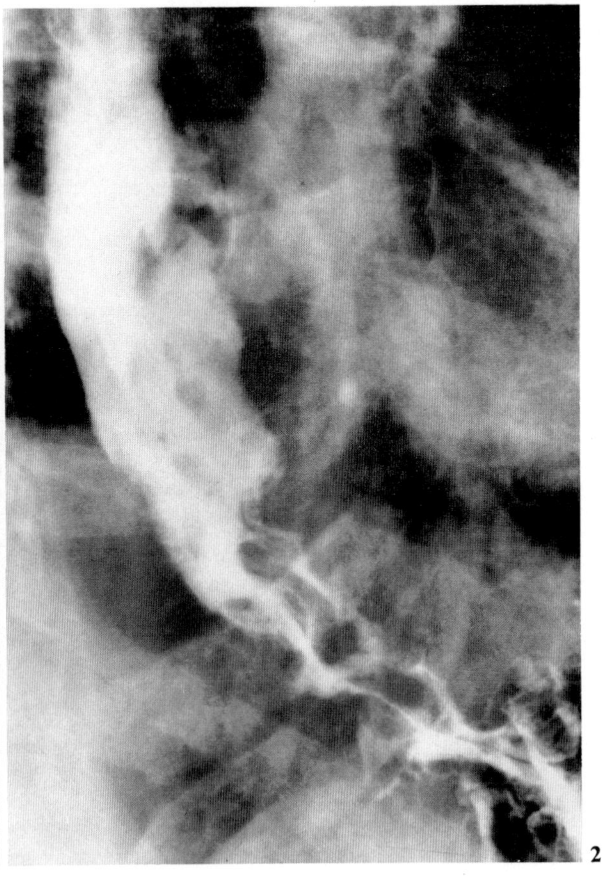

2

Fig. 1. Infiltrating carcinoma of the mid esophagus. Luminal obstruction is manifested by an air-fluid level proximal to the tumor. A fistula *(arrow)* to the left mainstem bronchus is also demonstrated

Fig. 2. Varicoid carcinoma of the distal esophagus. Multiple nodules of tumor mimic varices on this single view but did not change on other images as varices would

maintained whenever elderly patients with a recent onset of dysphagia demonstrate this pattern on barium examinations.

"Small" esophageal cancers are eccentric polypoid or plaque-like lesions less than 3.5 cm in length usually protruding no more than 1.5 cm into the lumen (KOEHLER et al. 1976; ZORNOZA and LINDELL 1980). Superficial spreading carcinoma is manifested by shallow depressions and tiny elevations of the mucosal surface which may be circumferential or longer than 3.5 cm (ITAI et al. 1978). Both of these patterns are detected more reliably by double contrast esophagography than by the standard examination. Although some authors have suggested prolonged survival in patients with these types of lesion (KOEHLER et al. 1976), others have shown no difference (ZORNOZA and LINDELL 1980). "Early" esophageal carcinoma is defined as carcinoma with invasion no deeper than the submucosa and no spread to lymph nodes (YAMADA 1979). "Early" carcinoma pathologically and "small" or "superficial" carcinoma radiologically are not synonymous. In one series, histologic early carcinoma was shown to be present in masses as large as 4.5 cm and in a case of pseudoachalasia (LEVINE et al. 1986).

In addition to detection, the esophagram is also helpful in defining the degree of obstruction resulting from a tumor and in detecting fistulae to the tracheobronchial tree (Fig. 1). Follow-up barium examinations can be helpful in assessing response to radiation or chemotherapy (LEVINE et al. 1987) and in detecting recurrence or complications after surgical or other therapy (Fig. 5) (HISHIKAWA et al. 1984; HAYNES et al. 1984; WOLF et al. 1986; JAFFE et al. 1987).

4.2.1.2 Staging

The length of an esophageal tumor does not correlate well with extent of invasion or metastasis. In one series, lesions less than 3 cm usually had not spread to regional nodes, but beyond 3 cm, there was no correlation (YAMADA 1979). Similarly no significant correlation has been found between the ra-

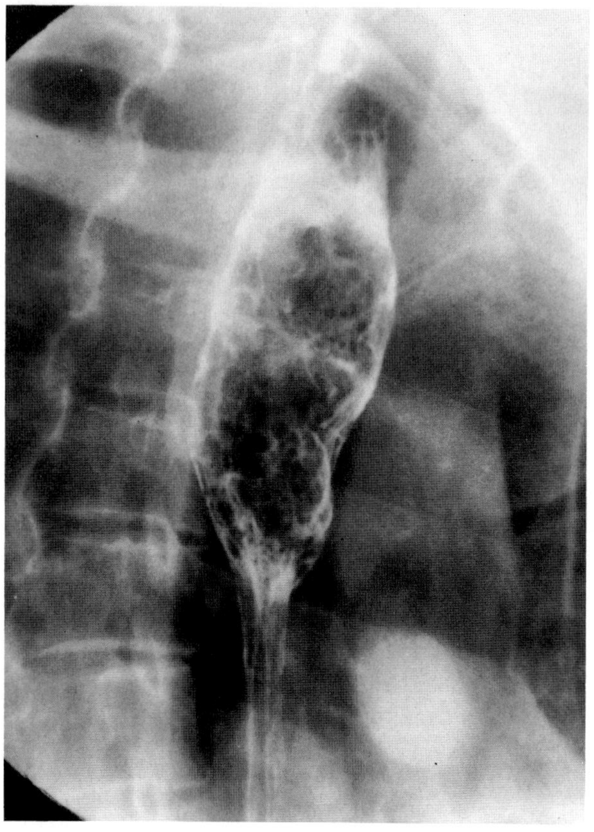

Fig. 3. Spindle cell carcinoma of the proximal esophagus. A bulky intraluminal mass expands the esphageal lumen, mimicking an impacted food bolus

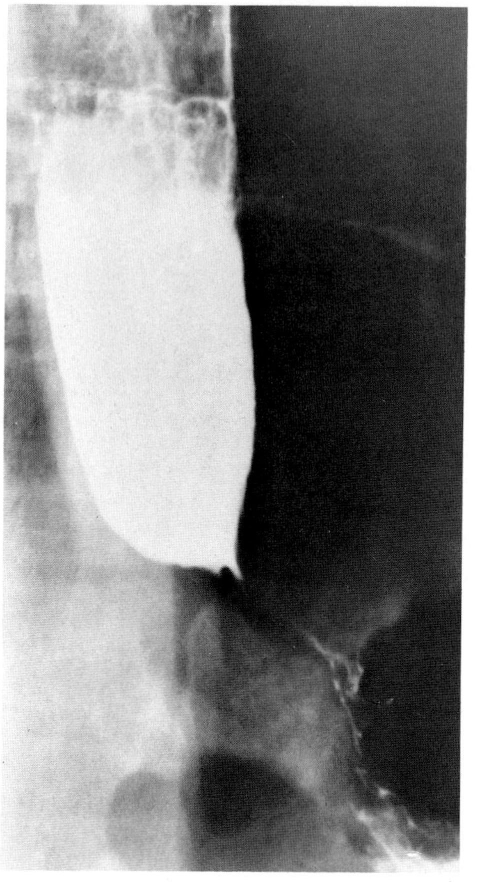

Fig. 4. Pseudoachalasia. It is difficult to differentiate this adenocarcinoma arising at the cardia and infiltrating submucosally along the distal esophagus from true primary achalasia. Dilatation of the esophagus with an air-fluid level and tapered narrowing at the EGJ can be seen in both

diographic pattern of a tumor and the stage or prognosis for the patient (Mori et al. 1979; Yamada 1979). Several investigators have evaluated deviations of the "esophageal axis" as a sign of mediastinal invasion or unresectability. While perhaps more sensitive than tumor length or pattern, there are 10%–20% false positive and 35% false negative results, making this unsuitable as a staging tool (Akiama et al. 1972; Mori et al. 1979).

4.2.2 Computed Tomography

Computed tomography is, by far, the most sensitive modality available for staging patients with esophageal carcinoma. Older techniques to assess local spread such as azygography and diagnostic pneumomediastinum have almost universally been replaced by CT. However, precise data on its accuracy in staging large numbers of patients are not yet available.

The CT scan can image the esophagus throughout its length. In 60% of scans the esophagus will be seen distended by air (Halber et al. 1979). The wall in this circumstance should be less than 3 mm thick (Reinig et al. 1983). When not distended, it is more difficult to assess the esophagus, as the overall diameter is variable. However, the external margins are normally clearly demarcated by mediastinal fat from surrounding structures, even in very thin patients. The blurring of those margins and the thickening of the wall beyond 5 mm are reliable indicators of esophageal disease. Most tumors are manifested by circumferential wall thickening with an overall diameter exceeding 2 cm (Reinig et al. 1983; Samuelsson et al. 1984) (Fig. 6). The appearance of mediastinal invasion of contiguous structures has been described. Displacement and compression of the tracheobronchial tree by esophageal mass when scans are obtained in full inspiration indicates invasion of the airway (Halvorsen and Thompson 1987) (Fig. 7). Obliteration of the fat

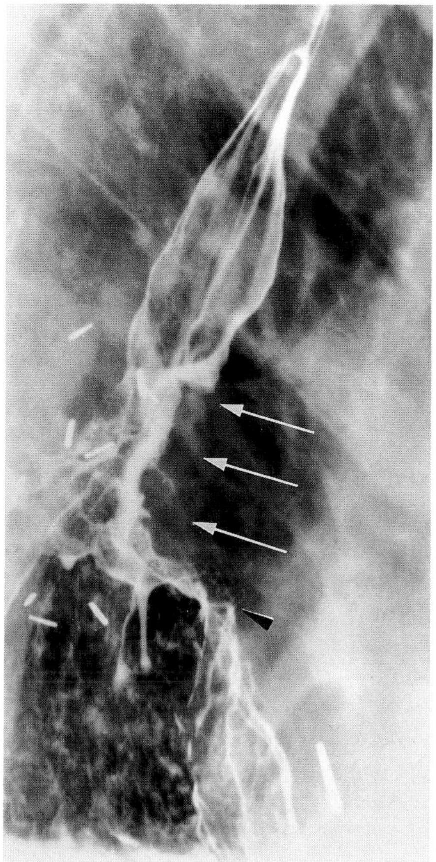

Fig. 5. Recurrence after esophagogastrectomy. Infiltrating mass with shouldered margins *(arrows)* is seen along the esophageal side of the intrathoracic esophagogastric anastomosis *(arrowhead)*

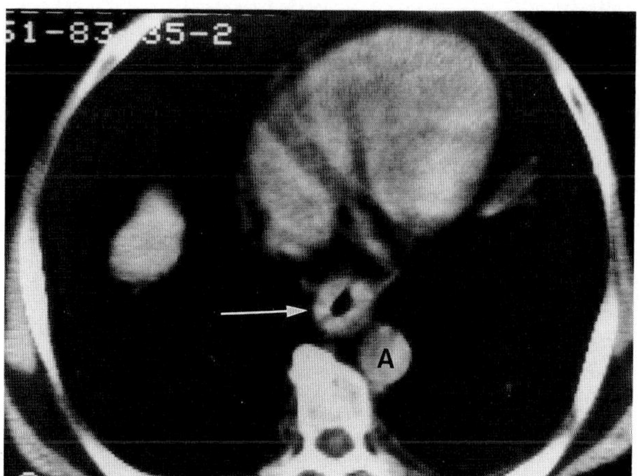

Fig. 6. Esophageal carcinoma on CT. Circumferential thickening of the wall of the esophagus can be seen *(arrow)*. There is preservation of the fat plane between the esophagus and aorta *(A)*

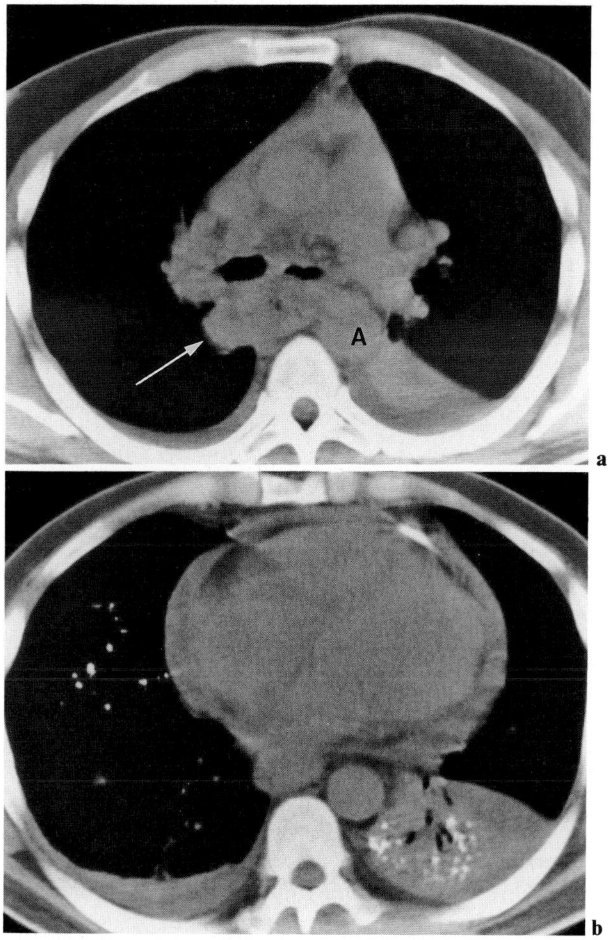

Fig. 7 a, b. Esophageal carcinoma with mediastinal invasion. **a** A large esophageal mass is identified at the level of the carina *(arrow)*. The left mainstem bronchus is narrowed and indented posteriorly. The fat plane between the mass and the aorta *(A)* is obliterated. **b** On a lower section barium can be seen in a collapsed left lower lobe due to a fistula between the esophagus and the left mainstem bronchus (same case as Fig. 1)

plane between esophagus and aorta for a distance equal to 90° of the circumference of the aorta indicates invasion of the aortic wall (Picus et al. 1983). Obliteration of the fat plane for less than 45° of the circumference indicates a lack of invasion and 45°–90° is indeterminate. Extension of mass to the prevertebral soft tissues with erosion of the vertebral bodies would indicate invasion and unresectability. Obliteration of the fat plane between the esophagus and the crura, azygos vein, or pericardium is more difficult to assess and less well accepted as a criterion of invasion and unresectability.

Various authors have evaluated the accuracy of CT in staging esophageal carcinoma. The results are summarized in Table 2. Most authors agree on the insensitivity of CT in detecting small periesophageal and upper abdominal lymph nodes as well as in detecting tumor in normal size (< 1 cm) mediastinal and upper abdominal nodes. Mediastinal invasion of contiguous structures was reliably detected in several reports, when tumors at the EGJ were excluded, and correlates well with long-term survival (HALVORSEN et al. 1986). Evaluation of the periesophageal soft tissue planes around the EGJ yields sufficiently high false positive and negative diagnoses of invasion that CT staging cannot be considered reliable in this region (THOMPSON et al. 1983; FREENY and MARKS 1982; TERRIER and SCHAPIRA 1984). Overall, however, the numbers of patients evaluated by CT with surgical correlation have been small and further studies are needed to assess its predictive accuracy in both staging and in the more limited task of simply determining resectability.

Computed tomography is also useful in following patients whose cancers have been resected (HEIKEN et al. 1984b; GROSS et al. 1985; BECKER et al. 1987). Local recurrence in the mediastinum and distant metastasis are not uncommon. While not sensitive at detecting mucosal recurrence at the anastomosis, which is better evaluated by esophagography and endoscopy, CT is better than these modalities in detecting mediastinal adenopathy or mass and in detecting distant metastases. Percutaneous biopsy can be performed under CT guidance if needed to document the metastatic spread.

4.2.3 Endoscopic Ultrasound

Using a specially adapted side-viewing endoscope with an attached ultrasound transducer, one can now visualize the wall of the esophagus from the inside out. The advantage of this compared with CT is greatly improved resolution. The ultrasound probe can be applied directly to the wall or more commonly wrapped in a balloon which is inflated with water and in contact with the esophageal wall (BOLONDI et al. 1986). Five layers of the esophageal wall have been described corresponding to the interface between the transducer and mucosa, the mucosa, the submucosa, the muscularis propria, and the adventitia (BOLONDI et al. 1986; TIO and TYTGAT 1986). A tumor is seen as a mass of homogeneous echogenicity with disruption of the normal layers of the esophageal wall. In one study, resectability was accurately predicted in 24 of 26 patients with esophageal cancer (TIO et al. 1986). Local and distant lymph nodes were assessed correctly in 17 of 19 patients with resectable tumors. Only one cancer could not be imaged because of stenosis. On the other hand, in another study at least 50% of all esophageal tumors could not be fully evaluated because of stenosis, and assessment of local invasion was therefore not as good by endoscopic ultrasound (EUS) as by CT (HEYDER and LUX 1986). Lymph node involvement by EUS was, however, superior to CT assessment. The technique shows great promise, particularly for the preoperative assessment of resectability, and further evaluation of its role in esophageal cancer staging will be important.

Table 2. Staging accuracy of preoperative CT in esophageal carcinoma

	No. of patients with surgical or autopsy correlation	Overall staging accuracy
MOSS et al. (1981c)	17	100%
PICUS et al. (1983)	30	80%
THOMPSON et al. (1983)	61	84%
BECKER et al. (1986)	50	70%
QUINT et al. (1985a)	33	39%
QUINT et al. (1985b)	10	70%

4.2.4 Magnetic Resonance Imaging

Magnetic resonance imaging (MRI) is in its infancy especially in relation to body imaging. Its application to the esophagus is unknown although an examination of one patient with esophageal cancer was first reported in 1981 (SMITH and HUTCHISON 1981). Preliminary work has shown that MRI can detecte tumors but is less accurate than CT in detecting periesophageal spread (QUINT et al. 1985b). Its usefulness in this area will need to be investigated further.

4.3 Stomach

4.3.1 Barium Examination

In Japan, where the prevalence of gastric cancer is eight times that of the United States, double contrast radiography and endoscopy are considered complementary tools in screening programs to detect early gastric cancer. In the western world, exa b

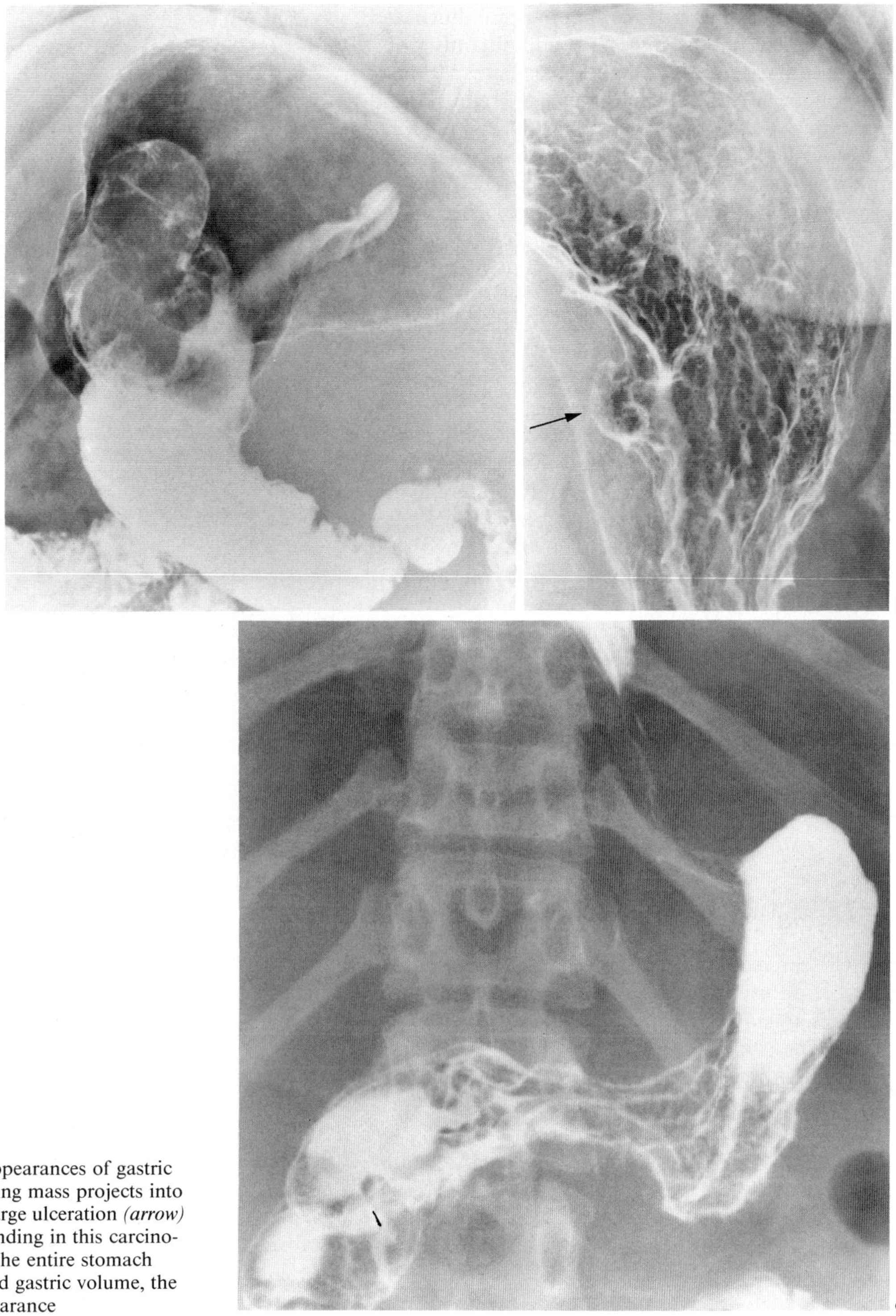

c

Fig. 8 a–c. Various appearances of gastric carcinoma. **a** Fungating mass projects into the gastric body. **b** Large ulceration *(arrow)* is the predominant finding in this carcinoma. **c** Infiltration of the entire stomach produces a diminished gastric volume, the "linitis plastica" appearance

aminations are performed on symptomatic patients and the overwhelming majority of lesions diagnosed are advanced gastric cancers (WHITE et al. 1985). In detecting advanced cancers, endoscopy and radiology, using state-of-the-art biphasic examination techniques, are equally sensitive (SHAW et al. 1987; WILJASALO et al. 1980). The biphasic examination uses the advantages of the double contrast technique combined with the advantages of the single contrast compression technique in one examination (GELFAND et al. 1987; OP DEN ORTH 1979).

Some advanced gastric carcinomas are illustrated in Fig. 8. The tumors may be predominantly polypoid, ulcerated, or infiltrating. In the latter instance, the volume of the stomach is usually reduced and maintains a fixed appearance on all radiographs, the so-called linitis plastica appearance (Figs. 8c and 9a). Occasionally, the rugal folds will be enlarged (infiltrated) without decreased distensibility or lack of peristalsis, and differentiation from lymphoma or benign gastropathies may be difficult (BALTHAZAR and DAVIDIAN 1981). Scirrhous carcinoma (linitis plastica) involving the distal antrum may occasionally be confused with adult pyloric hypertrophy or scarring due to peptic ulcer disease (BALTHAZAR et al. 1980). Carcinoma in which ulceration is the predominant feature will only rarely simulate a classically benign ulcer but may fall in an indeterminate category - not radiographically benign or malignant. In all of these instances, a high index of suspicion must be maintained that carcinoma is present and endoscopy, with repeated biopsies if necessary, be performed.

The appearances of early gastric cancer in the Japanese experience are well described and categorized in the literature and will not be repeated here. The reader is referred to SHIRAKABE and KHIKAWA'S (1966) *Atlas of X-ray Diagnosis of Early Gastric Cancer* and to the chapter by Maruyama on early gastric cancer in LAUFER'S (1979) *Atlas of Double Contrast Gastrointestinal Radiology*

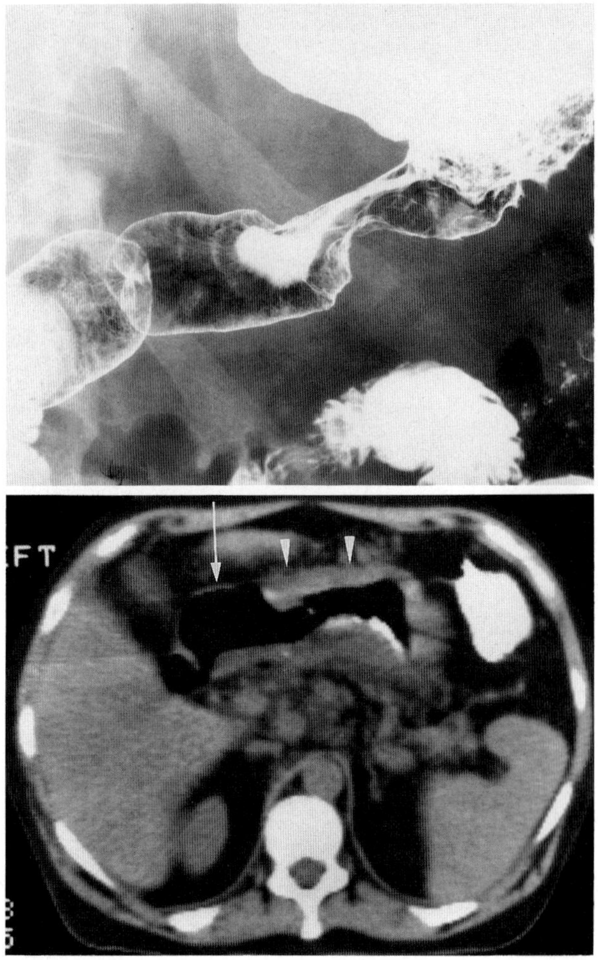

Fig. 9 a, b. Gastric carcinoma on CT; correlation with UGI (upper gastrointestinal examination). **a** The barium examination demonstrates marked constriction of the lumen beginning in the proximal antrum. **b** The CT scan demonstrates normal wall thickness of the antrum *(arrow)* abruptly becoming thickened towards the body *(arrowheads)*

4.3.2 Computed Tomography

Gastric carcinoma is manifested on CT scanning as focal or diffuse wall thickening or as a polypoid intraluminal mass (BALFE et al. 1981; MOSS et al. 1980). The stomach must be adequately distended with dilute oral contrast or air for the wall to be evaluable. Tumors generally show wall thicknesses greater than or equal to 1 cm (Fig. 9b).

Obliteration of the fat planes between the stomach and the liver, spleen, pancreas, and gastrohepatic and gastrocolic ligaments is suggestive of local invasion of these structures. CT may also detect hepatic and nodal metastases as well as metastases in the pelvis, particularly in the cul-de-sac or involving the ovary(ies) (Fig. 10). When metastatic lesions are suspected, percutaneous biopsy can be performed if necessary for documentation.

Only limited studies have evaluated the accuracy of CT staging of gastric malignancies (Table 3). Although specificity was high, in later studies, sensitivity and overall accuracy were not. The false negative rate was high for detecting (1) local invasion (both macroscopic and microscopic), (2) perigastric lymph nodes, (3) tumor in normal size (<1 cm) upper abdominal lymph nodes and (4) peritoneal seeding. In tumors arising at the EGJ both false negative and false positive diagnoses of local tumor extension are reported, making CT even less reliable in this region. It might be concluded from these results that CT scanning is not warranted in the routine preoperative evaluation of patients with gastric cancer. However, since there are very few false positive scans, except around the EGJ, patients would not be denied potentially curative surgery by preoperative CT staging; but patients with incurable lesions by virtue of metastasis might be spared needless surgery. Further studies need to be

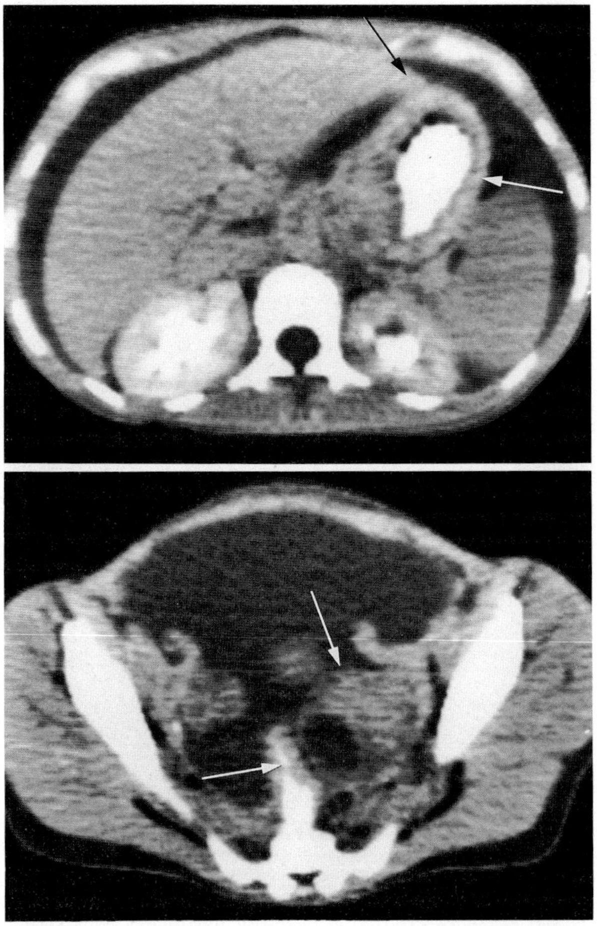

Fig. 10 a, b. Gastric carcinoma with ovarian metastases (Krukenberg tumor). This patient presented with ascites and a left pelvic mass and was presumed to have ovarian carcinoma. Biopsies subsequently proved signet ring carcinoma of gastric origin. **a** CT scan through the stomach demonstrates a thickened gastric wall *(arrows)* and ascites. **b** The pelvic CT confirms a left ovarian mass *(arrows)* and ascites

performed. In patients who have had their tumors resected, CT scanning has detected the development of local recurrence and distant metastasis, occasionally before patients became symptomatic (MULLIN and SHIRKHODA 1985).

4.3.3 Endoscopic Ultrasound

As in the esophagus, the technique of introducing an ultrasound transducer, attached to a special endoscope, into the gastric lumen to visualize the wall and surrounding structures has tremendous potential in the staging of gastric malignancies. The transducer can be wrapped in a water filled balloon as in the esophagus, or the stomach itself can be filled with water (BOLONDI et al. 1986). The same five layers of the gastric wall can be delineated as described in the section on EUS of the esophagus. Disruption of those layers by a mass is indicative of invasion. In one study, EUS correctly staged 30 out of 36 patients with gastric cancer with respect to resectability and local and distant lymph node metastasis (TIO et al. 1986). Although still in the early stages of development, this technique shows great promise as a preoperative staging tool.

Table 3. Staging accuracy of preoperative CT in gastric carcinoma

	No. of patients with surgery or autopsy correlation	Overall staging accuracy
LEE et al. (1979)	7	100%[a]
MOSS et al. (1981b)	13	100%
FREENY and MARKS (1982)	15	80% (EGJ cancers only)
THOMPSON et al. (1983)	12	42% (EGJ cancers only)
TERRIER and SCHAPIRA (1984)	25	68% (EGJ cancers only)

[a] Does not include detection of all sites of disease nor of two early cancers confined to the wall

4.4 Small Bowel

4.4.1 Barium Enema

Despite the fact that the small bowel makes up 75% of the length of the gastrointestinal tract, primary malignant tumors are very uncommon. Patients invariably have symptoms and advanced lesions at presentation. The small bowel examination, performed with a high volume of medium density barium and with frequent compression spot films or by the enteroclysis technique (SELLINK 1974) is the procedure of choice in detecting malignant lesions of the small bowel.

Adenocarcinomas tend to occur in the proximal small bowel. In the duodenum, they are often polypoid or ulcerated (Fig. 11). In the jejunum and ileum they are most often annular constricting lesions which obstruct the lumen at presentation (Fig. 12). Overhanging edges and destruction of the mucosa can usually be seen. Obstruction due to metastasis to the small bowel, which is more common,

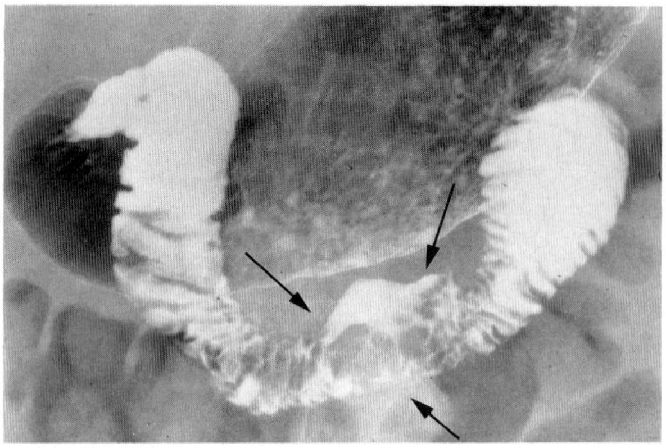

Fig. 11. Duodenal carcinoma. There is an ulcerated mass *(arrows)* involving the third portion of the duodenum which is not obstructing the flow of barium

can be seen as tapered narrowing with intact mucosa because of the intramural or serosal origin of the tumor. However, not infrequently, metastatic disease will invade the mucosa and be indistinguishable from a primary lesion. Occasionally inflammatory lesions, such as Crohn disease or tuberculosis, can mimic carcinoma.

Carcinoid tumors of the small bowel tend to occur in the distal ileum. They are the most common small bowel neoplasm found at autopsy (BANCKS et al. 1975; BALTHAZAR 1978). At this stage, in which they represent an incidental finding at surgery or autopsy, they are small intramural masses and may be seen as such in a small bowel examination. Occasionally, they may serve as the lead point of an intussusception and cause symptoms of small bowel obstruction. As they grow through the bowel wall, carcinoid tumors incite an intense hyperplastic response in the muscle layers of the bowel and marked desmoplasia in the invaded mesentery. This causes thickening of the small bowel folds in the involved loop, separation of loops, and ultimately fixation and tethering of multiple loops often toward a central point in the right lower quadrant. After resection of the primary tumor, they may recur in the right lower quadrant with similar radiographic signs of mesenteric fibrosis.

4.4.2 Computed Tomography

Very little has been written about the use of CT in patients with small bowel malignancies. One of the likely reasons is that they are so uncommon. Another is that they often require surgical intervention because of obstruction and can be staged surgically.

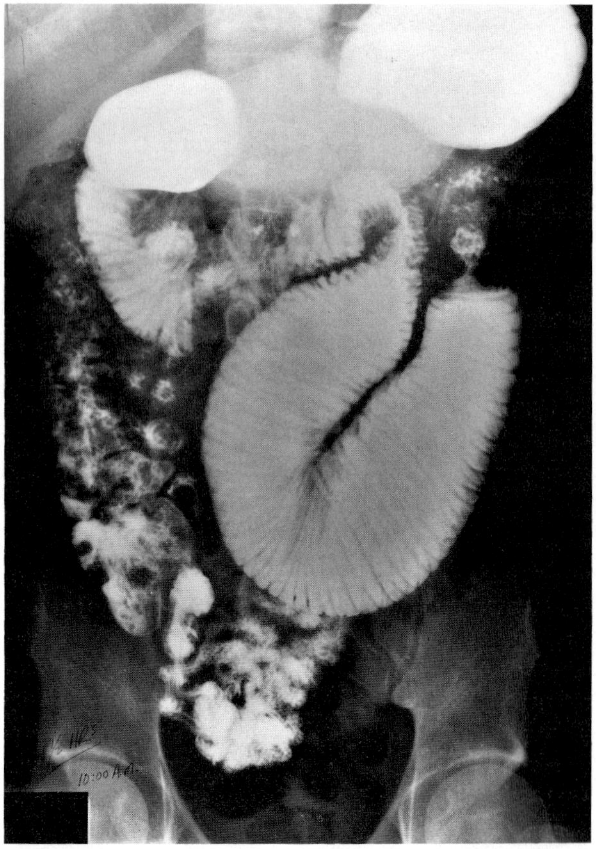

Fig. 12. Jejunal carcinoma. A proximal small bowel obstruction is evident. At the transition between dilated and normal caliber small bowel, there is abrupt narrowing with shouldered margins compatible with an "apple-core" primary carcinoma of the jejunum

One might generalize, however, from results in other tumors, that CT is the best imaging modality available for assessing the local extent of disease and the presence and extent of metastasis. In one study of the role of CT in duodenal neoplasms, there was some confusion in diagnosis between primary carcinoma of the duodenum and pancreatic and metastatic carcinoma invading the duodenum (FARAH et al. 1987). However, the report showed some duodenal lesions clearly separate from the pancreas, and it is therefore possible that CT may help, in some cases, in the differentiation of the origin of the tumor. CT is likely, also, to be of help in following patients, after surgical resection, for the development of metastasis.

4.4.3 Angiography and Hepatic Artery Embolization

Angiography as an adjunct in the preoperative diagnosis of small intestinal carcinoids is widely accepted. The angiographic findings include localized kinking and obstruction of small arteries in the mesenteric arcade and staining of the primary tumor mass. When hypervascular liver metastases are present as well the findings are considered quite specific for carcinoid tumors (KINKHABWALA and BALTHAZAR 1978).

Some patients with small bowel carcinoids and hepatic metastases will have the carcinoid syndrome due to the production of pharmacologically active compounds by the tumor in the liver. Palliation of this syndrome and a decrease in the extent of hepatic metastases can be achieved by hepatic artery embolization (ODURNY and BIRCH 1985; CARRASCO et al. 1986). In one series (CARRASCO et al. 1986), there was an 87% response rate, but also 9% mortality from the procedure. The latter may have been due to poor patient selection as these patients already had severely compromised liver function from metastasis and died of hepatic failure after embolization. The impact of embolization on patient survival is difficult to determine as the disease usually pursues a somewhat indolent course without any treatment.

4.5 Colon and Rectum

4.5.1 Barium Enema

A single contrast barium enema employs a low density (20% weight/volume) barium suspension administered in a controlled retrograde fashion with manual compression and often spot films of the entire colon and rectum. It also includes a postevacuation view to assess the mucosal pattern. The double contrast enema uses high density (85%-100% weight/volume) barium to coat the mucosal surfaces and a large volume of air to distend the barium coated colon and rectum. Both methods are quite sensitive in detecting advanced colonic carcinoma. The detection of early colon cancer is basically a matter of detecting small polyps (SKUCAS and SPATARO 1981). While still a matter of debate, it is generally accepted that the double contrast enema is more sensitive in the detection of small, less than 2 cm polyps (THOENI and MENUCK 1977; OTT et al. 1980; DE ROOS et al. 1985) and should be used in patients who are at risk for colon

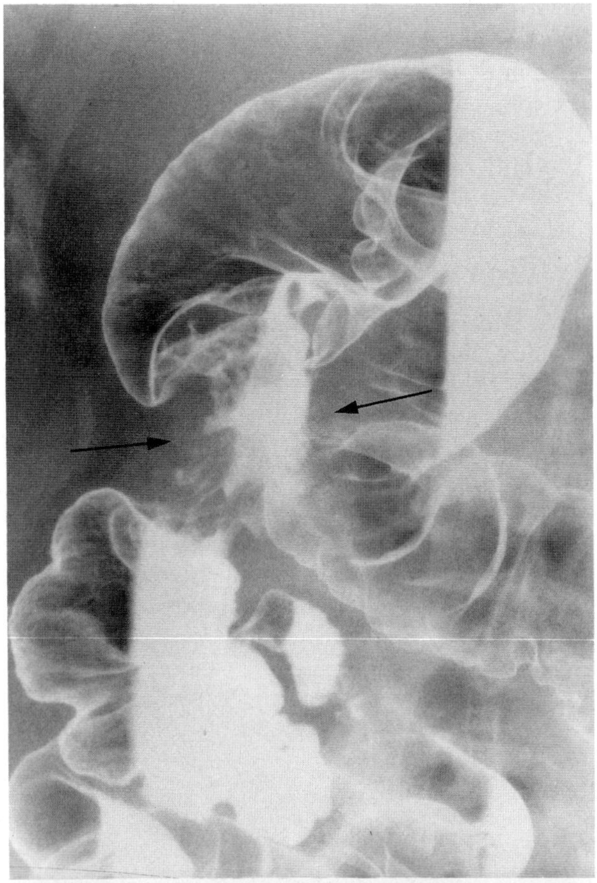

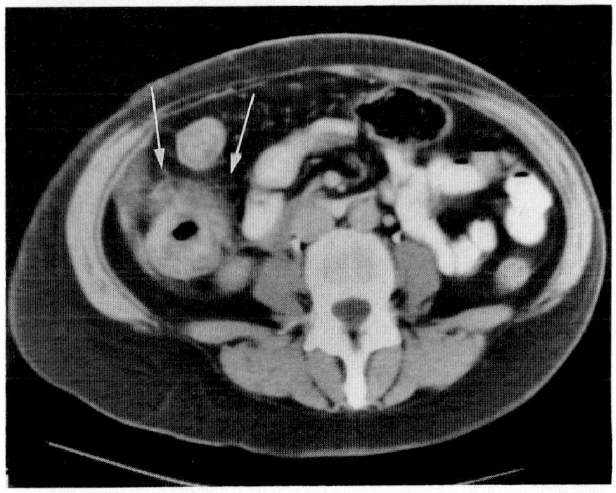

Fig. 13. a, b. Carcinoma of the ascending colon. **a** The double contrast enema examination demonstrates an annular constricting tumor with overhanging margins and mucosal destruction in the ascending colon *(arrows)*. **b** The CT scan shows circumferential thickening of the right colon with extension of tumor into the pericolic fat anteriorly *(arrows)*

polyps and cancer, predominantly patients over 50 years of age.

In most cases of small polypoid filling defects in the colon, it is not possible to tell whether the tumor is benign or malignant (SKUCAS and SPATARO 1982). The only helpful criterion is size: polyps less than 1 cm harbor carcinoma 1% of the time, polyps 1–2 cm have a 10% incidence of malignancy, and polyps over 2 cm are malignant 33%–46% of the time (MUTO et al. 1975). Criteria other than size have largely been unreliable. However, as tumors advance, they assume the more classic appearance of fungating intraluminal masses or annular constricting lesions with overhanging margins and mucosal destruction (Fig. 13a). Some cancers are predominantly ulcerated. Many are only partially circumferential (Fig. 14). The differential diagnosis is usually not difficult. Occasionally perforated tumors may simulate acute peridiverticulitis if the inflammatory component masks the tumor itself and diverticular disease may present as an annular constriction mimicking carcinoma. Amebomas, carcinoid, lymphoma, and other rare entities can occasionally be confused.

Because of the risk of synchronous primary cancers and polyps, the entire colon should be careful-

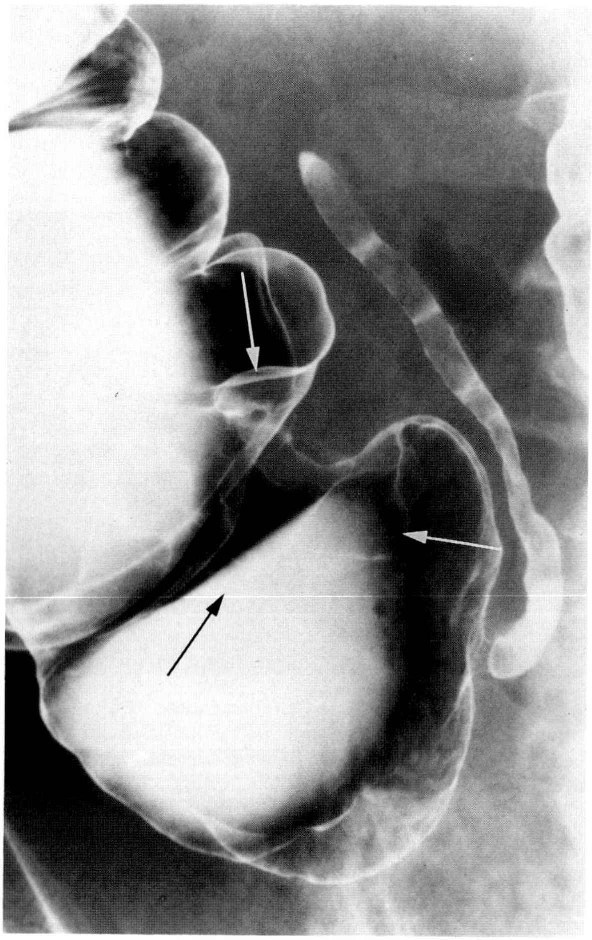

Fig. 14. Cecal carcinoma. There is a saddle-shaped mass arising in the region of the ileocecal valve *(arrows)*

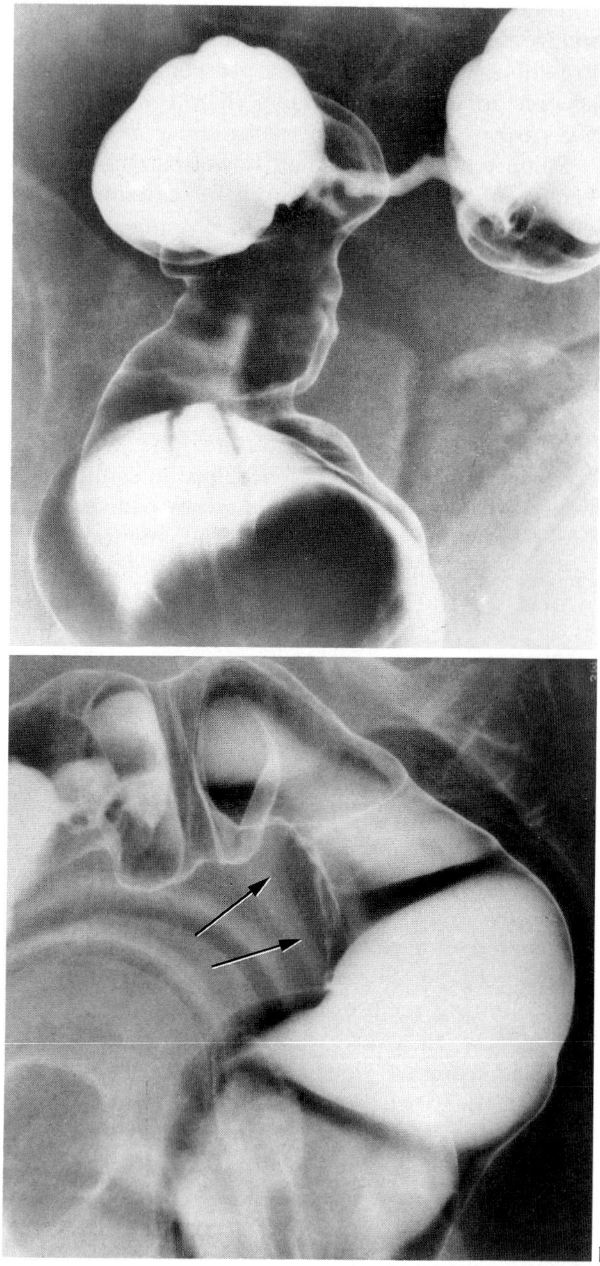

Fig. 15 a, b. Peritoneal metastasis from sigmoid carcinoma. **a** An annular tumor involves the sigmoid colon. **b** There is a mass effect with pleating of the overlying mucosa at the rectosigmoid junction *(arrows)* representing metastasis to the most inferior recess of the peritoneal cavity and extrinsic involvement of the bowel. Clinically, a "rectal shelf" would be palpable ly examined when a primary tumor is found. Often the barium examination will be possible even when the colonoscopist cannot pass through a narrowed lumen. In the event of more complete obstruction, the colon should be carefully examined after the detected primary is resected. The contrast enema also serves postoperatively to evaluate for complications such as leak and in the long term for evidence of anastomotic recurrence and metachronous polyps and cancers. On rare occasions, the barium enema is helpful in preoperative staging by demonstrating peritoneal metastases either in the pelvic cul-de-sac impressing upon the anterior rectum or as one or more serosal masses involving the colon (Fig. 15).

4.5.2 Computed Tomography

Computed tomography is not a primary modality for detecting colon cancer. However, detection of known cancers is possible in 82%-85% of cases (ADALSTEINSSON et al. 1985; FREENY et al. 1986; BALTHAZAR et al. 1988). The tumor generally appears as a focal mass of homogeneous soft tissue density or as circumferential wall thickening (> 5 mm) (Fig. 13b). Sensitivity rates vary with the technique employed. They are highest when the colon is clean and distended with air or contrast. The use of CT as a preoperative staging tool has been disappointing. Overall staging accuracy has been limited by the inability of CT scans to assess subtle extension of tumor through the wall of the colon and to detect nodal spread (Table 4). Nodes under 1.5 cm are generally considered "normal" by CT criteria. Perirectal and pericolonic lymph node involvement tends to occur in much smaller nodes. False negative diagnoses of tumor involvement predominate, ranging from 20%-60% of cases evaluated (GRABBE et al. 1983; FREENY et al. 1986; THOMPSON et al. 1986). Even setting 1 cm as the upper limits of normal in one series did not significantly affect the overall staging accuracy (BALTHAZAR et al. 1988), and the number of false positive diagnoses of nodal involvement increased - from 3% in the first three series to 13%.

A diagnosis of pericolonic or perirectal spread is made by CT when the outer margin of the tumor is spiculated or irregular and strands of soft tissue density extend into the pericolonic fat or when contiguous organs or structures are invaded (Fig. 13b). The inaccuracy of CT scanning with respect to local extension relates to a high rate of false negative diagnoses, 20%-40% in three series (GRABBE et al. 1983; FREENY et al. 1986; BALTHAZAR et al. 1988). The conclusion of most authors is that CT is not sufficiently reliable to warrant routine preoperative scanning (THOMPSON and HALVORSEN 1987).

In contrast, CT has proved to be of considerable value in postoperative follow-up, particularly in detecting locally recurrent rectosigmoid carcinoma. In patients who have had an AP resection and in whom clinical examination for local recurrence is limited, CT offers a way of looking into the pelvis to detect recurrence. Unfortunately on any one examination the appearance of postoperative fibrosis in the presacral space is not always distinguishable from locally recurrent tumor. Although in one series (LEE et al. 1981) patients without recurrences after AP resection had no presacral masses, most authors have found variable amounts of soft tissue in the presacral space. Much of the time, these postoperative masses stay the same or become smaller and better defined over time (KELVIN et al. 1983). When recurrence is present, the soft tissue abnormality is seen to enlarge, become ill-defined, or invade local structures (KELVIN et al. 1983; REZNEK et al. 1983). This is the argument for obtaining a baseline CT scan of the pelvis within 2-4 months of surgery. If radiation therapy is given after surgery, the tissue

Table 4. Staging accuracy of preoperative CT in colorectal carcinoma

	No. of patients with surgical/autopsy correlation	Overall staging accuracy (%)
GRABBE and WINKLER (1985[a])	155	56-79[b]
ADALSTEINSSON et al. (1985)[a]	127	61
FREENY et al. (1986)	80	48
THOMPSON et al. (1986)	25	60
BALTHAZAR et al. (1988)	76	64

[a] Local spread only; distant metastasis not assessed
[b] 56% = staging accuracy for nodal involvement; 79% = staging accuracy for perirectal extension. Combined statistics not available

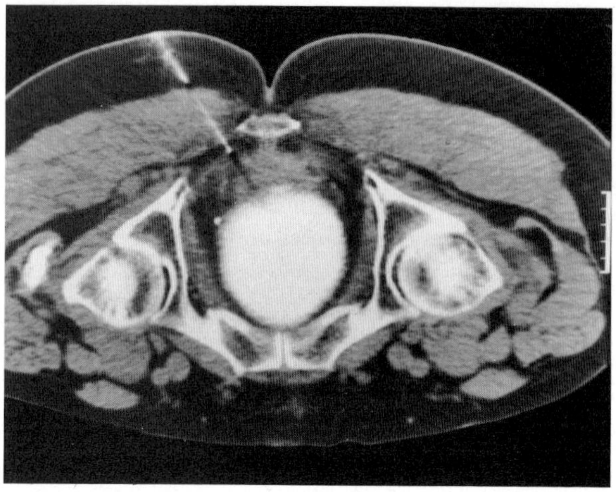

Fig. 16. CT guided biopsy of presacral mass after AP resection. A needle is in place at the outer margin of a small soft tissue mass between the bladder and sacrum. Cytology demonstrated recurrent rectal carcinoma

planes may become somewhat less well defined and the need for a baseline study after therapy is again of paramount importance. If a baseline is not available for comparison and a presacral mass is present, it is often advisable to perform a percutaneous biopsy to exclude recurrence (BUTCH et al. 1985) (Fig. 16).

In patients who have had anterior resections, CT is also helpful in detecting local recurrence (GRABBE and WINKLER 1985). Although barium examinations and endoscopy can detect mucosal recurrence, CT can document extraluminal extension and can detect perianastomotic recurrence which has not altered the mucosa (Fig. 17). In addition, CT is the most sensitive and available modality we have so far for detecting liver metastases in the colorectal cancer patient at risk for their development. In general, CT is considered close to 90% accurate in detecting recurrent or metastatic cancer (FREENY et al. 1986; THOMPSON et al. 1986; MCCARTHY et al. 1985; CHEN et al. 1987).

4.5.3 Endosonography

Transrectal ultrasonography has shown promise in evaluating lateral extension of rectal tumors within reach of the rectal probe. A colonoscopic ultrasound system for evaluating more proximal lesions is not yet commercially available. The transrectal approach can image lesions within approximately 12 cm of the anal verge. The examination is limited if a tight stenosis is present which does not admit

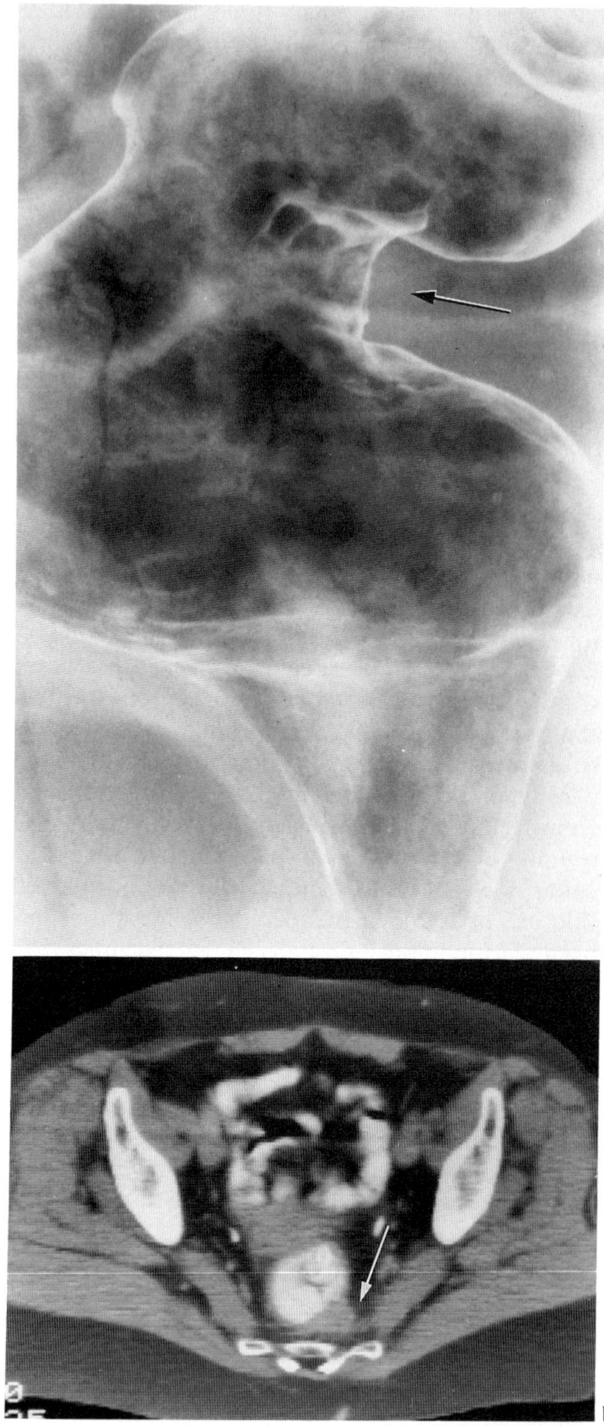

a

b

Fig. 17 a, b. Anastomotic recurrence after anterior resection. **a** The barium enema demonstrates subtle mucosal nodularity *(arrow)* along the left lateral aspect of the anastomosis which could represent postoperative deformity or recurrent tumor. **b** A CT scan at the same time demonstrates a definite extramural soft tissue mass on the left *(arrow)*, confirming the suspicion of recurrent tumor. *(Shirkhoda 1988)* the probe. Longitudinally oriented linear-array and transversely oriented "radial" scanning devices are available with transducers in the 5–7.5 mHz range. The normal rectal wall appears as a five-layer structure with alternating hyperechoic and hypoechoic layers. It is generally agreed that the fourth hypoechoic layer corresponds to the muscularis propria. The third hyperechoic layer is probably the submucosa and the fifth hyperechoic layer corresponds to the serosa and interface with the perirectal fat. Neoplasms are hypoechoic masses. Lateral spread is assessed by visualization of the integrity or disruption of the various layers deep to the neoplasm. Several preliminary reports have assessed the accuracy of endosonographic staging of rectal tumors within 12 cm of the anal verge compared to histopathology after resection. When only lateral spread is assessed, overall accuracy has been reported to be close to 90% (RIFKIN and WECHSLER 1986; BEYNON et al. 1986; DI CANDIO et al. 1987). A study correlating specimen ultrasound with histopathologic analysis found accuracies of 93% and 88% for invasion of the submucosa and serosa respectively, but only 77% for invasion of the muscularis propria (WANG et al. 1987). The authors believe this relates to artifacts from attenuation and shadowing by the bulky tumor mass. Evaluation of perirectal nodes has not proven to be as reliable again because nodes containing tumor cannot be differentiated from hyperplastic nodes and because nodes which are not enlarged are not seen. The potential of endosonography and its clinical usefulness will require further study.

4.5.4 Magnetic Resonance Imaging

Little is known about the impact of MRI on staging and follow-up of patients with colorectal carcinoma. A single study has addressed the comparative efficacy of staging rectal cancer by MRI versus CT in 16 patients (BUTCH et al. 1986). Eighty percent (80%) of these patients had locally advanced lesions and MRI proved to be as good or better in staging accuracy as CT. This was particularly true in relation to the assessment of perirectal invasion because of the higher contrast, compared to CT, on T1-weighted MRI sequences between tumor and perirectal fat, allowing easier discrimination of invasion. However, subtle degrees of invasion were not identified. Similarly, MRI did no better at telling cancerous from hyperplastic lymph nodes than CT. MRI requires bowel cleansing prior to scanning because stool and tumor have the same signal intensity. CT can usually be performed satisfactorily with no preparation. Both techniques require adequate rectal distention with air or appropriate contrast material.

Because of the different magnetic resonance characteristics of fibrotic tissue and tumor, MRI has potential in evaluating patients after AP resection (GOMBERG et al. 1986). This has yet to be evaluated.

4.5.5 Monoclonal Antibodies

The use of monoclonal antibodies for in vivo detection of tumor and metastasis and for therapy is in its infancy but is progressing rapidly. Currently monoclonal 1gG or its fragments are being used most often, labelled with ^{131}I or ^{111}In. Scanning is performed 1–7 days after injection. Preliminary reports suggest that colon tumors or metastases greater than 1.5 cm can be detected with fair sensitivity (ABDEL-NABI et al. 1987; LARSON 1987). Intraperitoneal injection, in one study, detected peritoneal implants secondary to gastrointestinal malignancies with greater sensitivity than either CT or MRI (CARRASQUILLO et al. 1988). Future directions include development of antibodies with less normal-tissue cross-reactivity, two-stage injection techniques allowing the antibody to equilibrate before a shorter acting (more easily imaged) radiolabel is given, and use of monoclonal antibodies for therapy in radiosensitive tumors (LARSON 1987).

4.6 Anus

4.6.1 Barium Enema

The anal canal is generally best examined by inspection, digital examination and anoscopy. The role of barium enema in anal cancers is primarily to assess the rectum and colon for other incidental lesions and when tumors are large, to evaluate the proximal extent of the lesion. Generally, the tumors are small irregular polypoid or plaque-like defects at the anorectal junction (THOENI and VENBRUX 1982). Larger submucosal masses may be seen, frequently with subtle areas of ulceration. When small, polypoid tumors may be confused with hemorrhoids. Perirectal absesses and distal adenocarcinomas of the rectum may occasionally be in the differential diagnosis.

4.6.2 Computed Tomography

Because anal cancers are uncommon, experience with CT staging and follow-up is limited. In one series (COHAN et al. 1985), CT detected unsuspected paraaortic lymphadenopathy in one of seven patients which altered the treatment plan for that patient. Rectal masses were seen on CT in all seven patients evaluated preoperatively (out of 25 patients). This may reflect the selection for scanning of only those patients with large tumors. Postoperative and/or postradiation scanning was found to be useful in detecting the high incidence of locally recurrent disease. As discussed in the section on rectal carcinomas, a baseline scan 2–4 months after surgery will undoubtedly be of help in differentiating recurrence from post-treatment changes. Percutaneous biopsy can be easily performed to provide tissue confirmation.

4.7 Liver Metastases

4.7.1 Computed Tomography

The detection of liver metastases is of great importance in patients with gastrointestinal malignancies as these tumors frequently metastasize to the liver. CT has replaced liver scintigraphy as the primary screening modality for metastasis detection because of its greater sensitivity (ALDERSON et al. 1983) and its ability to image extrahepatic sites also at risk for metastasis. The accuracy of liver CT is very dependent on the way the scans are performed. It has been shown that tumor to normal liver contrast is increased by the use of iodinated water soluble contrast, especially if the contrast is administered as a rapid intravenous bolus and rapid sequential scans are obtained of the entire liver within 3–4 min of injection (YOUNG et al. 1980; MOSS et al. 1982; BERLAND et al. 1982; BURGENER and HAMLIN 1983; ALPERN et al. 1986). This has become the "gold standard" of enhanced CT scanning.

At longer intervals, there is more chance that contrast will have diffused into the extravascular space of the tumor (which it does more slowly than in normal hepatic parenchyma) and will have washed out of the normal liver (which happens more rapidly than in most tumors) such that the tumor and liver become relatively isodense (BURGENER and HAMLIN 1981). Thus in early reports pre- and postcontrast scanning were often equally sensitive (MOSS et al. 1979) or insensitive (SUGARBAKER and VERMESS 1984). However, in patients with hypervascular tumors, there is danger that lesions will become isodense with dynamic scanning. In one series 37% of metastases became relatively undetectable (isodense) after contrast but were easily detected on precontrast scans (BRESSLER et al. 1987). Noncontrast CT scans are therefore recommended in patients with primary islet cell tumors, carcinoids, and other vascular tumors.

Another technique to enhance the lesion to liver contrast and therefore improve detectability involves delayed CT scanning 4–6 h after administration of 60 g intravenous Iodine. In one study comparing this technique to dynamic CT scanning, 58% of patients had lesions detected with greater confidence and 27% had more lesions detected (BERNARDINO et al. 1986). One other group has also shown high sensitivity for lesion detection with this method (MILLER et al. 1987). The technique is based on the fact that the liver secretes 1%–2% of an iodine load into the biliary system. Another technique based on reticuloendothelial system uptake of contrast is EOE-CT scanning. EOE is an iodinated ester of poppyseed oil in an emulsified form. Initial reports were very enthusiastic as sensitivity for the detection of hepatic metastasis seemed quite good (SUGARBAKER and VERMESS 1984; VERMESS et al. 1982; LEWIS et al. 1982). However, the emulsion has proved to have limited stability and the frequency of side-effects has not been acceptable. It is therefore unlikely that it will become commercially available.

The final method of contrast enhancement involves arterial injection of iodinated contrast either into the hepatic artery, CT angiography (CTA), or into the superior mesenteric artery to enhance the liver through the portal venous flow, CT arterial portography (CTAP). The former has demonstrated improved detection of liver lesions (FREENY and MARKS 1986a) but is limited by the multiple variations of hepatic arterial anatomy seen commonly necessitating catheter manipulations and repeated injections to visualize the entire liver. This is impractical. However, CTAP has gained rapid acceptance and has been shown to be highly sensitive (MILLER et al. 1987; MATSUI et al. 1987), but the false positive rate has been high: 18% in one study (MATSUI et al. 1987). In any case, it is an invasive procedure and has generally been reserved for special indications such as the preoperative evaluation before hepatic resection.

The sensitivity of the various modalities discussed above can only be assessed in comparison to resected specimens examined histologically. The results of four studies which performed such analyses

Table 5. Liver lesion detectability by various scanning techniques compared to resected specimens

SUGARBAKER and VERMESS (1984)	129 metastases 53 patients	NC CT	40.6%
		WSC CT	33.6%
		EOE CT	76.7%
GUNVEN et al. (1985)	113 metastases 31 patients	CT*	80%
		US	66%
MILLER et al. (1987)	56 metastases 14 patients	CTAP	77%
		DS CT	83%
		EOE CT	82%
MATSUI et al. (1987)	45 metastases 22 patients	Dynamic WSC CT	63%
		CTAP	84%
		US	58%

NC CT = non contrast enhanced CT
WSC CT = water soluble contrast enhanced CT
EOE CT = Ethiodized oil emulsion enhanced CT
CTAP = CT arterial portography
DS CT = 4–6 h delayed scanning WSC enhanced CT
US = ultrasound
CT-* = combination of dynamic bolus WSC CT, CTA, and CTAP

are summarized in Table 5. Although delayed scanning, EOE-CT, and CTAP have the highest sensitivity, all are impractical in most settings where large numbers of patients need to be screened for metastatic disease. The recommended technique for the majority of scanning will therefore remain dynamic incremental CT with rapid bolus infusion of contrast, with other techniques reserved to answer specific questions.

The discussion of lesion detectability has not so far discussed specificity. Cysts, hemangiomas, and focal fatty infiltration can occasionally masquerade as metastases on CT. Criteria to differentiate hemangiomas from metastases have been described (ITAI et al. 1983b; FREENY and MARKS 1986b). If a lesion has a typical appearance of hemangioma, there is an 86% chance that it is actually a hemangioma (FREENY and MARKS 1986b). Ultrasound, technetium-red blood cell radionuclide scanning, and MRI imaging may be of further help as all of these modalities have a characteristic appearance of hemangiomas. Cysts can be diagnosed conclusively by CT if they are as large or larger than the thickness of the CT slice and have CT numbers known to represent water density. Occasionally this may mean scanning at very thin slice thicknesses to be able to measure the lesion without partial volume effects (SHIRKHODA and FALLONE 1987). Ultrasound can also be of help in equivocal cases (BRICK et al. 1987). Focal fatty infiltration is only rarely rounded (HALVORSEN et al. 1982; YATES and STREIGHT 1986) and usually has a characteristic wedge shaped appearance. In troublesome cases, liver scintigraphy will resolve any questions by demonstrating a normal liver (HALVORSEN et al. 1982).

The CT scan is also of use in following metastases for response to therapy. Although attempts have been made to calculate liver and tumor volume quantitatively (HENDERSON et al. 1981; MOSS et al. 1981a; COSTELLO et al. 1983), the techniques are cumbersome and comparison is still generally made on a lesion by lesion analysis. It is important that the scans be performed in a similar fashion from one study to the next.

4.7.2 Ultrasound

Ultrasound is a safe, easy, and inexpensive way to screen the liver for the presence of metastasis. There is no characteristic pattern for metastases except that they are hypo- or hyperechoic with respect to normal liver parenchyma. They can easily be distinguished from simple cysts, and reasonably reliable criteria have been developed for their distinction from classic hemangiomas. The sensitivity of ultrasound in detecting metastatic disease has not been carefully evaluated. Early reports (SNOW and GOLDSTEIN 1979; TAYLOR et al. 1976) reported accuracies of 85%–90%. Sensitivity was 75% in one (SNOW and GOLDSTEIN 1979). However, there was no good histopathologic correlation in these cases nor was there modern dynamic CT or MR for comparison. In MATSUI et al.'s series (1987), ultrasound detected 58% of 45 metastatic lesions in 22 patients undergoing hepatic resection. Many of the lesions missed by ultrasound were less than 1.5 cm in diameter. In GUNVEN et al.'s series (1985) ultrasound detected 66% (see Table 5).

The resolution and hence the sensitivity of ultrasound are greatly increased by applying a specialized transducer directly to the liver in the operating room. Lesions undetectable by preoperative studies and by palpation by the surgeon were detected by intraoperative ultrasound in 15% of 33 patients in one series (MACHI et al. 1986). Intraoperative ultrasound has also proven to be of value in demonstrating vascular anatomy in relation to tumor during hepatic resections and in guiding biopsies of nonpalpable masses (RIFKIN et al. 1987). It is likely that it will have an even greater role to play in the future.

Ultrasound has also been used to guide and monitor various experimental therapies for hepatic metastases. Intraoperative monitoring of hepatic cryosurgery has allowed proper placement of a cryoprobe in a hepatic tumor and demonstration of the extent of freezing in relation to the tumor margin (ONIK et al. 1986). Percutaneous placement of 14 gauge needles in hepatic metastases, using ultrasound guidance, allowed placement of an iridium 192 source in the needle tip for high-intensity interstitial radiation therapy (DRITSCHILO et al. 1986). The development of these and other innovative local therapies for hepatic metastases will most likely continue to rely on ultrasound or CT for localization.

4.7.3 Magnetic Resonance Imaging

The field of MRI is developing very rapidly at this point in time, particularly with respect to the liver and abdomen. New faster scanning techniques and the use of MRI contrast agents will make the data we have so far obsolete. However, there are already studies that suggest that MRI is going to be a sensitive technique for the detection of liver metastases (HEIKEN et al. 1985; REINIG et al. 1987; STARK and WITTENBERG 1987; GLAZER 1988). MR scanning appears also to have some specificity for tumor compared to cysts and hemangiomas. In a recent study (GLAZER 1988), 98% of hemangiomas could be differentiated from colorectal metastases by their MRI characteristics. With hypervascular tumors, this was possible only 61% of the time.

4.8 Hepatocellular Carcinoma

4.8.1 Computed Tomography

The appearance of hepatocellular carcinoma (HCC) on CT scans depends on the method of contrast administration and scanning. On nonenhanced scans, tumors may be hypodense, isodense, or hyperdense with respect to normal liver (KUNSTLINGER et al. 1980; ITAI et al. 1981). They may be solitary, multifocal, or diffuse. Uncommonly calcifications are observed. This is more often seen in the fibrolamellar variety (FRIEDMAN et al. 1985). In one series, the presence of a solitary mass greater than 10 cm and of a mass or masses which were hyperdense relative to liver parenchyma was strongly suggestive of hepatoma (ITAI et al. 1981). Other characteristic features include an isodense mass with a thin rim of lower density (representing the pseudocapsule present in many hepatomas) and protrusion of the mass from the expected contour of the liver. The CT signs of cirrhosis in the presence of a mass are very suggestive as well.

After the injection of iodinated contrast, masses may become better defined but they may also become isodense with respect to normal liver and escape detection (KUNSTLINGER et al. 1980; ITAI et al. 1979). Even though this is less likely with a dynamic sequential bolus technique, it may still be true of some smaller nodules scanned late in the sequence, and therefore both pre- and postcontrast scans are recommended. On dynamic studies, the tumor most often becomes hyperdense during the arterial phase and then hypodense during the portal venous enhancement phase. At this point the capsule and any internal septa characteristically appear hyperdense (ITAI et al. 1986b). If abundant necrosis is present the tumor will not become hyperdense initially (HOSOKI et al. 1982). Dynamic bolus CT also allows determination of arteriovenous shunting and tumor thrombus in some cases (KUNSTLINGER et al. 1980; MATHIEU et al. 1984) (Fig. 18a). In one series, 82% of arterioportal shunts involving large veins and 32% of shunts involving small veins were detected (ITAI et al. 1986a).

In Japan, iodized oil has been infused, along with chemotherapeutic agents, into the hepatic artery and CT scans obtained in the weeks after infusion (OHISHI et al. 1985). The iodized oil is taken up by even small tumors by mechanisms unknown and has remained for more than 1 year. Although hemangiomas initially are opacified, that contrast has disappeared by 20 days after infusion (YUMOTO et al. 1985). The diagostic potential of this technique is only beginning to be evaluated (HAYASHI et al. 1987).

The accuracy of CT in detecting hepatocellular carcinomas has increased as contrast enhancement techniques have changed. This is particularly true for small tumors less than 3 cm in diameter. Dynamic CT detected 56%-58% of all such tumors in two studies (TAKASHIMA et al. 1982; MATSUI et al. 1985). Almost all tumors over 3 cm are easily detected. In another series, CT after iodized oil infusion detected 67% of all tumor nodules (HAYASHI et al. 1987). CT arterial portography demonstrated 95% of small tumors (MATSUI et al. 1985); however, the false positive rate was somewhat higher than for other techniques.

4.8.2 Ultrasound

Ultrasound is an inexpensive, safe, and sensitive tool for the detection of HCCs. In Japan, where the prevalence of hepatoma is high, ultrasound and alpha fetoprotein have been recommended as com-

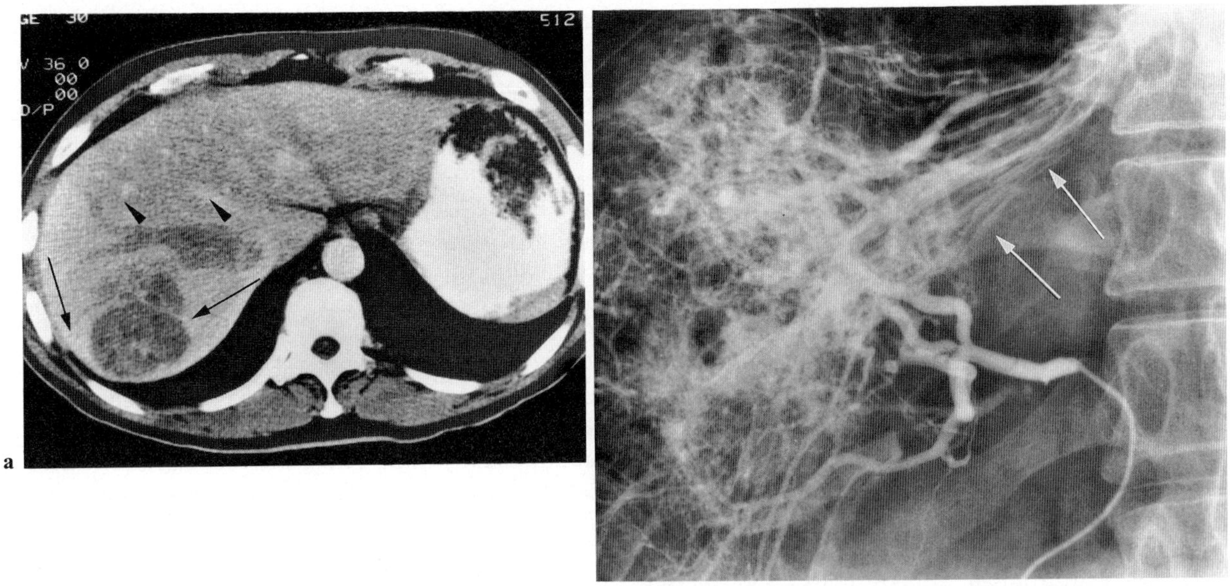

Fig. 18 a, b. Hepatoma with vascular invasion. **a** The CT scan shows a low density tumor mass in the right lobe of the liver *(arrows)* and a dilated, unopacified right hepatic vein *(arrowheads)*. Note the normal hyperdense appearance of other vascular structures for comparison. **b** This view from a selective right hepatic arteriogram demonstrates shunting into a enlarged right hepatic vein containing extensive tumor thrombus *(arrows)*. The lines and streaks of contrast are in blood spaces running longitudinally in and around the large tumor cast. (OKUDA et al. 1975)

plementary screening tools (TAKASHIMA et al. 1982). Hepatomas on ultrasound are seen as echogenic or echo-poor masses with respect to normal liver. However, in patients with severe cirrhosis, the liver itself may have many regenerating nodules which may be confused with tumors (HAYASHI et al. 1987). In addition, in one series ultrasound was particularly insensitive in the detection of small tumors (less than 3 cm) in the superolateral protein of the right lobe of the liver (TAKASHIMA et al. 1982). Ultrasound of the liver is definitely more operator dependent than is CT.

The use of pulsed Doppler ultrasound has recently been assessed to differentiate HCCs from other tumors (TAYLOR et al. 1987). Of 12 hepatomas, 10 had Doppler shifts of 5 kHz or above while no other tumor showed signals above 4 kHz. Hemangiomas showed signals below 0.7 kHz. The large Doppler shifts were correlated with large pressure gradients due to arteriovenous shunting as seen on angiography. This has future potential application.

The accuracy of ultrasound is close to 100% for lesions over 3 cm. For smaller tumors, sensitivity is in the range of 56%–63% (TAKASHIMA et al. 1982; MATSUI et al. 1985). However, intraoperative ultrasound was shown in one series to have a sensitivity of 94% for all nodules (HAYASHI et al. 1987).

4.8.3 Magnetic Resonance Imaging

The MRI scan shows great promise for the detection and evaluation of HCCs. Preliminary reports suggest that it may be as good or better than CT scanning (VERMESS et al. 1985; EBARA et al. 1986; ITOH et al. 1987). Tumors are generally hyperintense with respect to normal liver on T2-weighted sequences. About half are hypointense on T1-weighted sequences and the others are isointense or hyperintense. Frequently the lesions demonstrate better contrast compared to surrounding liver than on CT. Sensitivity at various sizes may be comparable to CT. In one study 97.5% of tumors greater than 2 cm were detected and 33.3% of tumors less than 2 cm (EBARA et al. 1986). In all reports, a thin hypointense rim, the pseudocapsule, was seen more frequently on MRI than on CT. Venous involvement can be evaluated without the use of iodinated contrast because of the signal void seen on MRI with flowing blood. The lack of contrast, however, precludes evaluation of arteriovenous shunting. The rapid advances in MRI technology, including fast scan techniques and paramagnetic contrast agents, may make even that possible in the future. Preliminary work with bolus injection of Gd-DTPA (gadolinium) has been reported (OHTOMO et al. 1987).

4.8.4 Angiography and Hepatic Artery Chemoperfusion and Embolization

Standard celiac or selective hepatic arteriography was the imaging modality of choice for detecting and evaluating primary tumors of the liver before the advent of real time ultrasound and CT. The characteristic features include a hypervascular mass or masses with arterioportal or venous shunting and often tumor thrombus in portal or hepatic veins (Fig. 18b). Occasionally, diffuse involvement of almost the whole liver can be seen. Signs of hepatic cirrhosis are often present.

Rarely large tumors are hypovascular. However, small (less than 2 cm) masses are often hypovascular and are not always detected by conventional angiography (TAKASHIMA et al. 1982). These may be primary (solitary masses) or daughter nodules in association with a larger primary mass. A technique was developed to permit hypovascular lesions to be seen as vascular blushes, which greatly improves detectability (TAKASHIMA and MATSUI 1980). This involves prolonged infusion of arterial contrast and prolonged filming such that tumor uptake relative to liver becomes enhanced.

Reports concerning the sensitivity of angiography in HCC, relative to ultrasound or CT, are variable in the literature, depending in part on the technique with which all three modalities are performed. In one series concerned with detection of small daughter nodules, arteriography had a sensitivity of 61% in patients who had surgery and histologic analysis of nodules. Specificity was 100% (HAYASHI et al. 1987). Often daughter nodules were obscured by the hypervascular main tumor mass. In terms of arteriovenous shunting and invasion of portal or hepatic veins, although CT and MRI have some success in demonstrating these features (ITAI et al. 1986a; ITOH et al. 1987), angiography remains the most sensitive modality. This is an important aspect in evaluation of resectability.

Most HCCs are not resectable. Angiography has a role to play in the treatment of tumors which are not resected. Both hepatic artery infusion of chemotherapy by percutaneous catheterization and hepatic artery embolization have shown promise in inducing tumor regression and prolonging survival (CHARNSANGAVEJ et al. 1983).

4.9 Carcinoma of the Gallbladder and Bile Ducts

4.9.1 Computed Tomography

Most patients with biliary carcinoma present with jaundice. CT will differentiate obstructive from nonobstructive jaundice with a high degree of accuracy (PEDROSA et al. 1981a; BARON et al. 1982). Evaluation of the site and cause of biliary obstruction has been reported in the literature to be 80%–97% and 63%–94% respectively (PEDROSA et al. 1981a, b; BARON et al. 1982; GIBSON et al. 1986; REIMAN et al. 1987). In only one series (GIBSON et al. 1986) was ultrasound more sensitive than CT, predominantly in detecting the cause of obstruction and tumor resectability. This has not been the general consensus otherwise.

The CT appearance of gallbladder carcinoma corresponds closely to that of ultrasound (YEH 1979; ITAI et al. 1980; WEINER et al. 1984). A large mass replacing the gallbladder, an intraluminal mass, and/or thickening of the wall greater than 5 mm are frequent findings. In one study, the thickened gallbladder wall enhanced markedly with contrast (THORSEN et al. 1984). In another series, gallbladder wall thickening was entirely nonspecific but a low density "halo" around the gallbladder was specific for cholecystitis when the wall was thick (SMATHERS and LEE 1984). Other findings specific for malignancy include invasion of the liver, liver metastases (uncommon), biliary obstruction (secondary to direct invasion or adenopathy), and enlarged pericholedochal, peripancreatic, or para-aortic lymph nodes. Prospective studies evaluating the sensitivity of CT in diagnosing gallbladder carcinoma have not been reported.

The appearance of cholangiocarcinoma on CT depends in part on the site of origin of the tumor. Lesions which arise in peripheral intrahepatic bile ducts usually present as low attenuation intrahepatic masses often with adjacent focal biliary dilatation (ITAI et al. 1983a; THORSEN et al. 1984). These findings are nonspecific and may be seen in primary liver tumor or metastasis. Well-defined cystic masses with internal papillary projections were described in one series (ITAI et al. 1983) and are somewhat more specific, representing biliary cystadenoma, cystadenocarcinoma, or papillary cholangiocarcinoma. In these cases, extrahepatic biliary dilatation was seen due to excessive formation of mucinous material by the tumors. Tumors at the confluence or in the extrahepatic system may or may not have a mass demonstrated on CT. The appearance of a mass can be nonspecific and can represent primary central liver tumor, metastasis to liver or nodes, or pancreatic tumor. Biliary dilatation without a mass needs to be differentiated from benign stricture or radiolucent stone disease. Several authors have reported finding ipsilateral lobar atrophy in patients with hilar cholangiocarcinoma and have suggested that it may be a relatively specific sign of bile duct tumor (VAZQUEZ et al. 1985; CARR et al. 1985; TAKAYASU and MURAMATSU 1986). Atrophy is known to occur in chronic biliary obstruction and is enhanced when the portal vein is obstructed (TAKAYASU et al. 1986). In another series, however (REIMAN et al. 1987), atrophy was seen most commonly in benign causes of chronic biliary obstruction and was not seen in any of their six patients with cholangiocarcinoma.

The CT scan is of help in guiding percutaneous biopsy and in assessing the extent of disease when the diagnosis is known. The number of isolated biliary segments can be determined by following the two major right and left ducts to their expected confluences. If these second order ducts are involved by tumor bilaterally, resection will not be possible (VOYLES et al. 1983). Evaluation of the biliary communications may aid in planning a percutaneous drainage approach as well (REIMAN et al. 1987). Invasion of the portal vein, when seen, and involvement of the caudate lobe can be helpful in documenting unresectability. As is the case in gallbladder cancer, prospective studies evaluating the accuracy of CT in detecting, diagnosing, and staging carcinomas of the bile ducts are not available.

4.9.2 Ultrasound

Ultrasound is often the first imaging modality requested in patients who ultimately are found to have carcinomas of the gallbladder or bile ducts. Patients frequently present with jaundice or symptoms mimicking gallstone disease. Ultrasound has been proven very effective in differentiating obstructive from nonobstructive jaundice (KOENIGSBERG et al. 1979; BARON et al. 1982) and is so sensitive in detecting gallstones that it is almost always the modality of choice to assess their presence.

Ultrasound findings in gallbladder carcinoma have been described (YEH 1979; RAGHAVENDRA 1980; DALLA PALMA et al. 1980; ALLIBONE et al. 1981; WEINER et al. 1984). A large mass replacing the gallbladder, a fungating mass or masses protruding into the gallbladder, and diffuse thickening of the gallbladder wall are the most common findings. The wall thickening is often difficult to distinguish from inflammatory disease unless there is marked asymmetry or irregularity and/or hepatic or nodal metastases are identified. Tumor mass has been identified extending into the extrahepatic bile ducts in a few cases and into the liver. Gallstones or calcifications are frequently present. When a large mass is present involving the liver, the findings may be nonspecific as to the primary origin of the tumor. An interesting observation in several cases was the appearance of a pancreatic mass with extrahepatic biliary obstruction masquerading as carcinoma of the pancreas (YEH 1979; WEINER et al. 1984). Metastasis to peripancreatic nodes is not uncommon in primary gallbladder cancer. The gallbladder should be carefully evaluated when pancreatic masses are seen to exclude this possibility.

In cholangiocarcinomas of the intra- and extrahepatic biliary tree ultrasound evaluation is helpful but not specific. The site of biliary obstruction can be detected by ultrasound in 33%–95% of patients and the cause of obstruction has been determined in 29%–88% (KOENIGSBERG et al. 1979; BARON et al. 1982; GIBBONS et al. 1983; HONICKMAN et al. 1983; GIBSON et al. 1986). In some cases, however, cause refers to benignity or malignancy rather than to a specific diagnosis. Cholangiocarcinomas may present with no evidence of a mass, precluding definitive diagnosis even of malignancy. Evidence of abrupt rather than tapered narrowing of the duct can suggest a diagnosis of malignancy (JONES et al. 1983). When a mass is present, differentiation from a primay liver tumor, metastasis to periportal nodes, or a primary pancreatic tumor may be impossible.

4.9.3 Magnetic Resonance Imaging

A single report on MRI of nine patients with cholangiocarcinoma compared to CT failed to find a significant advantage to MRI (DOOMS et al. 1986b). Some masses were more apparent because of the better contrast resolution of MRI. However, there was no more specificity to the diagnosis and perhaps less because bile duct dilatation was less obvious than on CT. Imaging the dilated biliary tract has been reported to be complicated and variable depending on the parameters employed (DOOMS et al. 1986a). It is possible that MRI will prove to be more sensitive than CT in demonstrating portal vein involvement and so be complementary in staging patients with bile duct tumors.

4.9.4 Angiography

Celiac and selective hepatic arteriography as well as indirect portography have important roles to play in determining resectability of biliary tumors. Angiography may also play a role in differentiating cholangiocarcinoma from hepatoma. Typical findings include encasement of central hepatic arteries and portal veins and small neoplastic vessels at the liver hilus (KAUDE and RIAN 1971; WALTER et al. 1976). Knowledge of the extent of vascular involvement is essential in preoperative planning.

4.9.5 Cholangiography

Percutaneous transhepatic cholangiography involves the percutaneous opacification of the biliary tree with iodinated contrast to delineate the site or sites of biliary obstruction. Standardly, this is performed from a right lateral approach. In central or left side tumors, the left ductal system may not be opacified by injection on the right. A separate approach to opacify the left ducts will be necessary (MUELLER et al. 1982b) (Fig. 19 b). In some cases, endoscopic retrograde cholangiography (ERCP) producing opacification of the distal bile duct will give complementary information about the distal extent of a tumor (Fig. 19 a).

The cholangiogram provides the best preoperative determination of the proximal extension of the cancer into the liver (VOYLES et al. 1983). If the diagnosis is unknown, the cholangiogram can reliably exclude stone disease and often document malignant obstruction. The various signs described in this disease are not entirely specific. They include abrupt termination of a dilated duct, irregular tapered narrowing, a branching infiltrating pattern at the hilus, and a polypoid intraluminal mass asso-

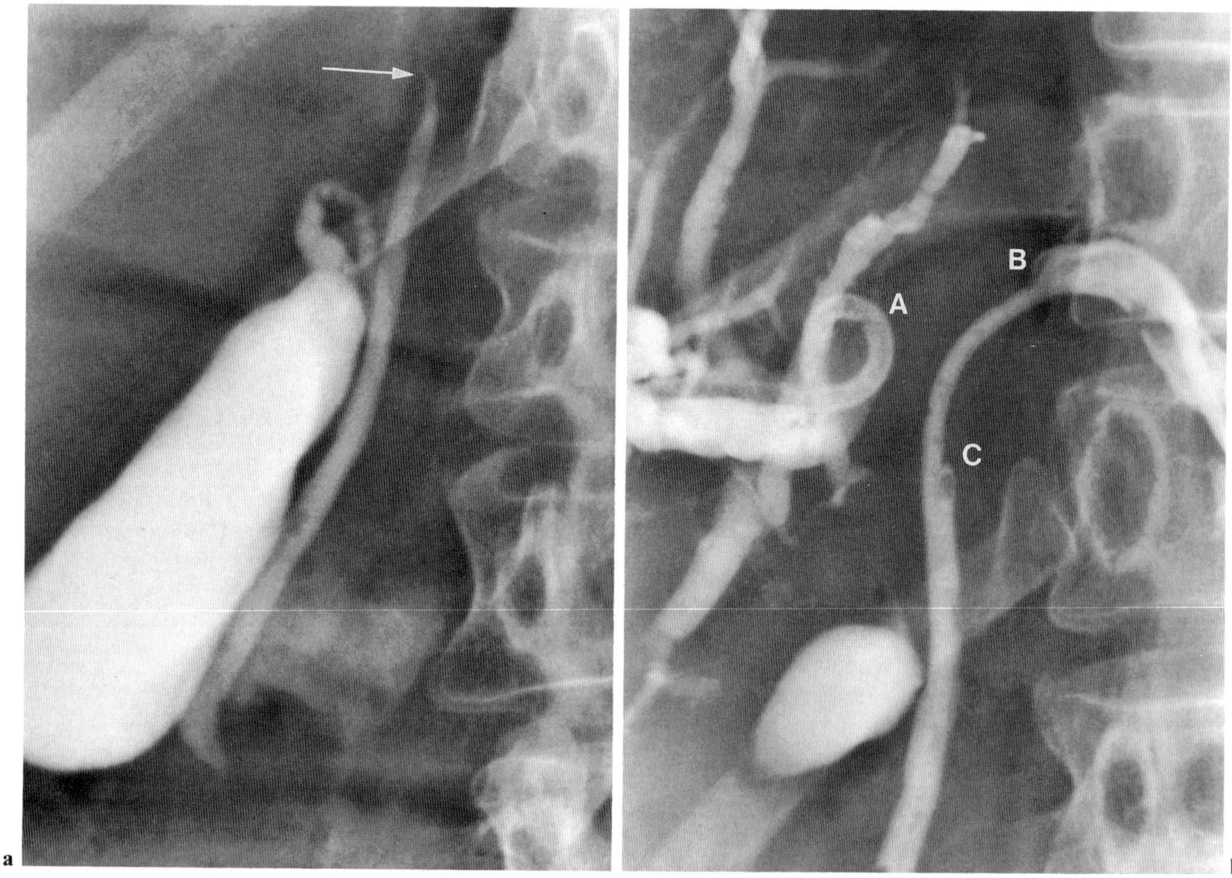

Fig. 19 a, b. Cholangiocarcinoma at the confluence of right and left hepatic ducts (Klatskin tumor). **a** ERCP demonstrates a normal caliber common bile duct with tapered narrowing of the high common hepatic duct *(arrow)* and filling of the gallbladder. **b** A pigtail catheter is in the right intrahepatic duct. A second catheter enters a left intrahepatic duct and passes through tumor to the common bile duct and duodenum (not shown). The tumor, by inference, fills the space between the obstructed right hepatic duct, left hepatic duct, and common hepatic duct. (*A, B,* and *C* respectively)

ciated with obstruction (LEGGE and CARLSON 1972; DILLON et al. 1981).

In addition to diagnosis, the percutaneous approach to the bile ducts is useful in palliative, nonoperative biliary drainage and in preoperative decompression. It has been suggested that the preoperative relief of obstruction improves hepatic function and decreases the morbidity and mortality from the ensuing surgery (VOYLES et al. 1983; GOBIEN et al. 1984). Palliative drainage may be performed percutaneously with external or internal drainage systems or endoscopically with internal stents. The choice depends in part on the location of the lesion and in part on the relative expertise of the endoscopists or radiologists. Serious complications of the percutaneous approach have included sepsis, intraabdominal bleeding, and death. These are reported in 8% and 16% of two series (MUELLER et al. 1982; BERQUIST et al. 1981). Minor complications are common and include long-term catheter malfunctions and dislodgements.

4.10 Pancreas

4.10.1 Computed Tomography

Computed tomography is a sensitive imaging modality for evaluating the pancreas and has had a major impact on the evaluation of patients with suspected pancreatic disease. In one study (FREENY et al. 1982), the utilization of ERCP and angiography was decreased by 68% and 54%, respectively, after introduction of a CT scanner. CT alone enabled an accurate diagnosis in 75% of 300 patients. In another series (HESSEL et al. 1982) CT had a sensitivity of 87% in detecting pancreatic lesions in 279 patients. The specificity was 90%. This was compared with ultrasound, which had a sensitivity of 69% and specificity of 82%. This is true in part because ultrasound has difficulty in routinely vizualizing the entire pancreas and is frequently hampered by overlying bowel gas. The general consensus has been that CT is the initial imaging modality of choice in detecting pancreatic disease and in assessing the extent of abnormality as well.

The appearance of pancreatic ductal adenocarinoma on CT is most commonly that of a focal mass or bulging of the contour of the normal pancreas. Pancreatic masses are usually isodense to the normal pancreas on nonenhanced and standard iodine enhanced CT scans unless small areas of necrosis are present. Dynamic CT scanning after bolus infusion of iodinated contrast has shown tumors to be hypodense with respect to the enhancing normal pancreatic tissue (MARCHAL et al. 1979; HOSOKI 1983; FREENY et al. 1988). This has enabled detection of small tumors which do not alter the contour of the pancreas and therefore should be the routine method of scanning the pancreas (FREENY et al. 1988). Pancreatic ductal dilatation (>5 mm in the head; >3 mm in the tail) (FREENY et al. 1988) is common in tumors of the head and body. Dilatation of the biliary tree can be seen in tumors of the head. Occasionally, ductal dilatation will be present with no evidence of a mass. At times diffuse enlargement of the pancreas can be seen mimicking pancreatitis (WITTENBERG et al. 1982). Inflammatory pancreatic masses can occasionally be focal as well, simulating tumor (NEFF et al. 1984).

A number of ancillary signs have been observed which not only help to make a diagnosis of cancer but also help to assess the degree of tumor extension or metastasis. These include the obvious signs of hepatic metastases, ascites, and retroperitoneal lymphadenopathy. More subtle signs include infiltration of the peripancreatic fat and vascular involvement of the major splanchnic vesses. Encasement of the major spanchnic vessels is a frequent occurrence in carcinoma of the pancreas and is an important determinant of nonresectability. Vascular involvement is most often seen as soft-tissue tumor surrounding the superior mesenteric and/or celiac artery or obliteration of the normal fat plane around those vessels or obstruction of the superior mesenteric, splenic, or portal veins with evidence of collateral flow (MEGIBOW et al. 1981; ITAI et al. 1982a, JAFRI et al. 1984).

Computed tomography has also been useful in evaluating patients with other more uncommon neoplastic lesions of the pancreas. Ten to fifteen percent of cystic masses are neoplasms (cystadenoma, cystadenocarcinoma) and can occasionally be confused with the more common pseudocysts. More often there will be solid components in the wall which help in the differentiation (WOLFMAN et al. 1982, ITAI et al. 1982b).

Functioning islet cell tumors are small (less than 2 cm) hypervascular tumors which can frequently be identified as hyperdense nodules on dynamic bolus CT (KRUDY et al. 1984; STARK et al. 1984b; ROSSI et al. 1985). Nonfunctioning islet cell tumors present late and are usually much larger. In one series (EELKEMA et al. 1984) 31% were over 10 cm. They often contain coarse focal deposits of calcium (Fig. 20) and characteristically enhance at least focally after contrast. The liver metastases, also being hypervascular, may become hyperdense or isodense

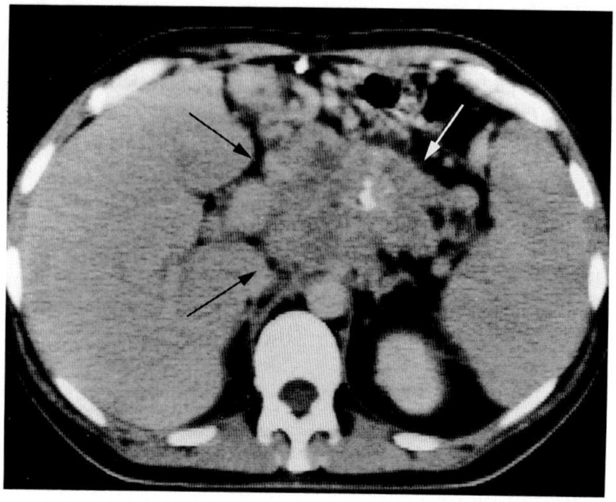

Fig. 20. Nonfunctioning islet cell tumor of the pancreas. A large mass *(arrows)* is present in the body of the pancreas with a coarse deposit of calcium centrally

to normal liver on enhanced scans. As with other hypervascular tumors, a nonenhanced sequence of the liver must be a part of the scanning technique (BRESSLER et al. 1987).

The accuracy of CT in detecting pancreating carcinoma is dependent in part on the method of scanning. In an early study (MOSS et al. 1980) CT had a sensitivity of 75% in detecting carcinoma. In a recent prospective study (FREENY et al. 1988) in which 161 patients were identified with pancreatic carcinoma, there were two (1%) false negative CT scans and 13 (8%) false positive. In this latter study, CT scanning was routinely performed with bolus dynamic contrast enhancement through the pancreas. The false positive examinations were all found to have pancreatic abnormalities, often pancreatitis or other tumor. This emphasizes the need to obtain tissue confirmation when a mass is identified in the pancreas.

Staging of pancreatic cancer, as well as detection, has been highly accurate by CT scanning (JAFRI et al. 1984; FREENY et al. 1988). Compared to angiography, involvement of the celiac or superior mesenteric arteries was detected more frequently by CT (MEGIBOW et al. 1981; JAFRI et al. 1984; FREENY et al. 1988) although more and different vessels were seen to be involved by angiography compared to CT. Overall, because of the ability of CT to detect other sites of disease precluding resection such as hepatic metastases or paraaortic nodes, the staging accuracy of CT and angiography are at least comparable. It is recommended that angiography, as the more invasive procedure, be used only when CT demonstrates resectability.

Despite the reliability of the CT signs of unresectability, the opposite is not true. In FREENY et al.'s study (1988) the negative predictive value of a diagnosis of resectable tumor was only 54%. In addition, of those patients resected for cure, median survival was only 19 months. In a study looking at the impact of all of our major diagnostic tools (angiography, ERCP, CT, and ultrasound) on the detection of pancreatic cancer at a curable stage, no difference in resectability was found before and after their introduction (SAVARINO et al. 1983). However, the number of exploratory laparotomies performed for diagnosis was substantially reduced. Now, with refinement of the equipment and techniques of CT scanning, use of the more invasive modalities (ERCP and angiography) has also been reduced significantly, shortening the time and expense for diagnosis (FREENY et al. 1988).

CT is also of value in guiding percutaneous biopsy and in following patients after resection for the development of postoperative complications or tumor recurrence (HEIKEN et al. 1984a).

4.10.2 Ultrasound

A pancreatic carcinoma is generally imaged on ultrasound as a solid focal mass with low level echoes and increased sound absorption (GOLDBERG 1984). A dilated pancreatic duct can often be identified. Comparisons with CT have generally demonstrated a greater sensitivity of CT compared to ultrasound for the detection of pancreatic disease (FOLEY et al. 1980b; HESSEL et al. 1982). Overlying bowel gas, especially in the left upper quadrant, is particularly troublesome in the routine evaluation of the body and tail of the pancreas. For this reason, CT is often preferred when there is a strong clinical suspicion of pancreatic tumor.

Ultrasound guided percutaneous opacification of the pancreatic duct is a relatively new imaging tool (MATTER et al. 1987; LEES and HERON 1987). It has proven to be a safe and helpful procedure when ERCP fails or is not conclusive because of obstruction and inability to vizualize the proximal duct. It has also been used successfully to guide fine needle aspiration biopsy of the pancreas, improving the success rate in their hands from 66.7% to 77.5% of all biopsies (HALL-CRAGGS and LEES 1986).

4.10.3 Magnetic Resonance Imaging

Magnetic resonance imaging of the pancreas, like all body imaging in MRI, is evolving rapidly with the development of new techniques and new equipment. Initial attempts at imaging were limited by poor spatial resolution and by motion artifacts as well as the inability of the system to differentiate bowel from pancreas (STARK et al. 1984a). Surface coil imaging improved the signal-to-noise ratio, allowing greater spatial resolution and superior imaging of the pancreas (SIMEONE et al. 1985). Other technical innovations as well have improved the quality of MRI body images. In a recent study on pancreas imaging (TSCHOLAKOFF et al. 1987), MRI provided equivalent information to CT in 69% of cases and added information in 14%. Good visualization of the pancreas was achieved in 93% of 58 patients in this report. However, the resolving power is still less than with CT, and ductal dilatation, small pseudocysts, and other small abnormalities were not reliably detected. Pancreatic calcifications are also not imaged well.

At low field strengths (0.35 T), the normal pancreas has a signal intensity similar to normal liver on both T1 and T2 weighted sequences (TSCHOLAKOFF et al. 1987). Good contrast of pancreatic tissue against surrounding retroperitoneal fat is seen on T1 weighted images but there is only poor delineation from bowel. On T2 weighted sequences, the pancreas shows good delineation from the fluid in the stomach and duodenum but less contrast with surrounding fat. The signal intensity of tumors is variable. Tissue characterization of abnormalities of the pancreas has not proven to be specific in differentiating types of tumor nor even in discriminating inflammatory from neoplastic disease. Staging has not yet been evaluated. However, possible advantages of MRI in staging pancreatic tumors include its potential ability to detect subtle invasion into peripancreatic fat and its potentially increased sensitivity over CT in detecting vascular invasion.

4.10.4 Angiography

The decrease in the use of angiography for the detection and staging of prancreatic carcinoma has been discussed previously in the section on CT. Generally, angiography is limited to patients in whom CT scanning is normal or inconclusive with respect to the presence of tumor and to patients with known tumors in whom no ancillary signs are present on CT to document nonresectability. The major findings consist of encasement of pancreatic and peripancreatic arteries and obstruction of the superior mesenteric, splenic, and/or portal veins. The tumors themselves are generally hypovascular.

Although catheter techniques offer little in the treatment of primary ductal carcinoma of the pancres or its metastases, some value has been demonstrated with respect to metastatic islet cell tumors. Sequential percutaneous embolization of hepatic metastases from pancreatic islet cell tumors has been shown to produce significant response rates and palliation of symptoms (AJANI et al. 1988).

4.10.5 Cholangiopancreatography

Endoscopic retrograde cholangiopancreatography was the major diagnostic tool in detecting pancreatic carcinoma before the development of ultrasound and CT. It is still considered extremely useful and is propably the most sensitive examination for the detection of small (<2 cm) tumors (TSUCHIYA et al. 1986).

Initial analysis of the observed morphological changes in the pancreatic and biliary ducts at ERCP suggested that there were highly specific signs for the diagnosis of malignancy (FREENY et al. 1976). These have been reassessed over the years and found to be much less specific (RALLS et al. 1980; PLUMLEY et al. 1982). The most specific sign is complete obstruction of the pancreatic or bile duct especially with an abrupt transition. In the presence of pancreatitis, pancreatic carcinoma can be a very difficult diagnosis to make by this modality or by ultrasound guided pancreatography (RALLS et al. 1980; LEES and HERON 1987).

References

Abdel-Nabi HH, Schwartz AN, Higano CS, Wechter DG, Unger MW (1987) Colorectal carcinoma: detection with indium-111 anticarcino-embryonic-antigen monoclonal antibody ZCE-025. Radiology 164; 617–621

Adalsteinsson B, Glimelius B, Graffman S, Hemmingsson, Pahlman L (1985) Computed tomography in staging of rectal carcinoma. Acta Radiol [Diagn] (Stockh) 26 (1): 45–55

Ajani JA, Carrasco CH, Charnsangavej C, Samaan A, Levin B, Wallace S (1988) Islet cell tumors metastatic to the liver: effective palliation by sequential hepatic artery embolization. Ann Intern Med 108: 340–344

Akiyama H, Kogure T, Itai Y (1972) The esophageal axis and its relationship to the resectability of carcinoma of the esophagus. Ann Surg 176: 30–36

Alderson PO, Adams DF, McNeil BJ, et al. (1983) Computed tomography, ultrasound, and scintigraphy of the liver in patients with colon or breast carcinoma: a prospective comparison. Radiology 149: 225–230

Allibone GW, Fagan CJ, Porter SC (1981) Sonographic features of carcinoma of the gallbladder. Gastrointest Radiol 6: 169–173

Alpern MB, Lawson TL, Foley WD, et al. (1986) Focal hepatic masses and fatty infiltration detected by cnhanced dynamic CT. Radiology 158: 45-49

Balfe DM, Koehler, RE, Karstaedt N, Stanley FJ, Sagel SS (1981) Computed tomography of gastric neoplasms. Radiology 140: 431-436

Balthazar EJ (1978) Carcinoid tumors of the alimentary tract. I. Radiographic diagnosis. Gastrointest Radiol 3: 47-56

Balthazar EJ, Davidian MM (1981) Hyperrugosity in gastric carcinoma: radiographic, endoscopic, and pathologic features. AJR 136: 531-535

Balthazar EJ, Rosenberg H, Davidian MM (1980) Scirrhous carcinoma of the pyloric channel and distal antrum. AJR 134: 669-673

Balthazar EJ, Megibow AJ, Hulnick D, Naidich DP (1988) Carcinoma of the colon: detection and preoperative staging by CT. AJR 150: 301-306

Bancks NH, Goldstein HM, Dodd GD (1975) The roentgenologic spectrum of small intestinal carcinoid tumors. AJR 123 (2): 274-280

Baron RL, Stanley RJ, Lee JKT, et al. (1982) A prospective comparison of the evaluation of biliary obstruction using computed tomography and ultrasonography. Radiology 145: 91-98

Becker CD, Barbier P, Porcellini B (1986) CT evaluation of patients undergoing transhiatal esophagectomy for cancer. J Comput Assist Tomogr 10 (4): 607-611

Becker CD, Barbier TA, Terrier F, Porcellini B (1987) Patterns of recurrence of esophageal carcinoma after transhiatal esophagectomy and gastric interposition. AJR 148: 273-277

Berland LL, Lawson TL, Foley WD, Melrose BL, Chintapalli KN, Taylor AJ (1982) Comparison of pre- and postcontrast CT in hepatic masses. AJR 138: 853-858

Bernardino ME (1984) Percutaneous biopsy. AJR 142: 41-45

Bernardino ME, Erwin BC, Steinberg HV, Baumgartner BR, Torres WE, Gedgaudas-McClees RK (1986) Delayed hepatic CT scanning: increased confidence and improved detection of hepatic metastases. Radiology 159: 71-74

Berquist TH, May GR, Johnson CM, Adson MA, Thistle JL (1981) Percutaneous biliary decompression: internal and external drainage in 50 patients. AJR 136: 901-906

Beynon J, Mortensen NJM, Foy DMA, Channer JL, Virjee J, Goddard P (1986) Colorectal disease: endorectal sonography: laboratory and clinical experience in Bristol. Int J Colorect Dis 1: 212-215

Bolondi cL, Zani L, Labo G (1986) Technique of endoscopic ultrasonography investigation: esophagus, stomach and duodenum. Scand J Gastroenterol [Suppl 123] 21: 1-5

Bressler EL, Alpern MB, Glazer GM, Francis IR, Ensminger WD (1987) Hypervascular hepatic metastases: CT evaluation. Radiology 162: 49-51

Brick SH, Hill MC, Lande IM (1987) The mistaken or indeterminate CT diagnosis of hepatic metastases: the value of sonography. AJR 148: 723-726

Burgener FA, Hamlin DJ (1981) Contrast enhancement in abdominal CT: bolus vs. infusion. AJR 137: 351-358

Burgener FA, Hamlin DJ (1983) Contrast enhancement of hepatic tumors in CT: comparison between bolus and infusion techniques. AJR 140: 291-295

Butch RJ, Wittenberg J, Mueller PR, Simeone JF, Meyer JE, Ferrucci JT (1985) Presacral masses after abdominoperineal resection for colorectal carcinoma: the need for needle biopsy. AJR 144: 309-312

Butch RJ, Stark DD, Wittenberg J, et al. (1986) Staging rectal cancer by MR and CT. AJR 146: 1155-1160

Carr DH, Hadjis NS, Banks LM, Hemingway AP, Blumgart LH (1985) Computed tomography of hilar cholangiocarcinoma: a new sign. AJR 145: 53-56

Carrasco CH, Charnsangavej C, Ajani J, Samaan NA, Richli W, Wallace S (1986) The carcinoid syndrome: palliation by hepatic artery embolization. AJR 147: 149-154

Carrasquillo JA, Sugarbaker P, Colcher D, et al. (1988) Peritoneal carcinomatosis: imaging with intraperitoneal injection of I-131-labeled B72.3 monoclonal antibody. Radiology 167: 35-40

Charnsangavej C, Chuang VP, Wallace S, Soo CS, Bowers T (1983) Work in progress: transcatheter management of primary carcinoma of the liver. Radiology 147: 51-55

Chen YM, Ott DJ, Wolfman NT, Gelfand DW, Karsteadt N, Bechtold RE (1987) Recurrent colorectal carcinoma: evaluation with barium enema examination and CT. Radiology 163: 307-310

Cohan RH, Silverman PM, Thompson WM, Halvorsen RA, Baker ME (1985) Computed tomography of epithelial neoplasms of the anal canal. AJR 145: 569-573

Costello P, Duszlak EJ, Lokich J, Matelski H, Clouse ME (1983) Assessment of tumor response by computed tomography liver volumetry. CT 7: 323-326

Cozzi G, Bellomi M, Gariboldi M, et al. (1987) Esophageal carcinoma: radiologic appearance of minimal lesions. Acta Radiol (Stockh) 28 (2): 177-180

Dalla Palma L, Rizzatto G, Pozzi-Mucelli, Bazzocchi M (1980) Grey-scale ultrasonography in the evaluation of carcinoma of the gall bladder. Br J Radiol 53: 662-667

De Roos A, Hermans J, Shaw PC, Kroon H (1985) Colon polyps and carcinomas: prospective comparison of the single- and double-contrast examination in the same patients. Radiology 154: 11-13

Di Candio G, Mosca F, Campatelli A, Cei A, Ferrari M, Basolo F (1987) Endosonographic staging of rectal carcinoma. Gastrointest Radiol 12: 289-295

Dillon E, Peel ALG, Parkin GJS (1981) The diagnosis of primary bile duct carcinoma (cholangiocarcinoma) in the jaundiced patient. Clin Radiol 32: 311-317

Dodds WJ, Stewart ET, Kishk SM, Kahrilas PJ, Hogan WJ (1986) Radiologic amyl nitrite test for distinguishing pseudoachalasia from idiopathic achalasia. AJR 146: 21-23

Dooms GC, Fisher MR, Higgins CB, Hricak H, Goldberg HI, Margulis AR (1986a) MR imaging of the dilated biliary tract. Radiology 158: 337-341

Dooms GC, Kerlan RK, Hricak H, Wall SD, Margulis AR (1986b) Cholangiocarcinoma: imaging by MR. Radiology 159: 89-94

Dritschilo A, Grant EG, Harter KW, Holt RW, Rustgi SN, Rodgers JE (1986) Interstitial radiation therapy for hepatic metastases: sonographic guidance for applicator placement. AJR 146: 275-278

Ebara M, Ohto M, Watanabe Y, Watanabe Y, et al. (1986) Diagnosis of small hepatocellular carcinoma: correlation of MR imaging and tumor histologic studies. Radiology 159: 371-377

Eelkema EA, Stephens DH, Ward EM, Sheedy PF (1984) CT features of nonfunctioning islet cell carcinoma. AJR 143: 943-948

Farah MC, Jafri SZH, Schwab RE, et al. (1987) Duodenal neoplasms: role of CT. Radiology 162: 839-843

Ferrucci JT, Wittenberg J, Mueller PR, et al. (1980) Diagnosis of abdominal malignancy by radiologic fine-needle aspiration biospy. AJR 134: 323-330

Foley WD, Berland LL, Lawson TL, Smith DF, Thorsen MK (1980a) Contrast enhancement technique for dynamic hepatic computed tomographic scanning. Radiology 147: 797–803

Foley WD, Stewart ET, Lawson TL, et al. (1980b) Computed tomography, ultrasonography, and endoscopic retrograde cholangiopancreatography in the diagnosis of pancreatic disease: a comparative study. Gastroint Radiol 5: 29–35, 51

Freeny PC, Marks WM (1982) Adenocarcinoma of the gastroesophageal junction: barium and CT examination. AJR 138: 1077–1084

Freeny PC, Marks WM (1986a) Hepatic perfusion abnormalities during CT angiography: detection and interpretation. Radiology 159: 685–691

Freeny PC, Marks WM (1986b) Patterns of contrast enhancement of benign and malignant hepatic neoplasms during bolus dynamic and delayed CT. Radiology 160: 613–618

Freeny PA, Bilbao MK, Katon RM (1976) "Blind" evaluation of endoscopic retrograde cholangiopancreatography (ERCP) in the diagnosis of pancreatic carcinoma: the "double duct" and other signs. Radiology 119: 271–274

Freeny PC, Marks WM, Ball TB (1982) Impact of high-resolution computed tomography of the pancreas on utilization of endoscopic retrograde cholangiopancreatography and angiography. Radiology 142: 35–39

Freeny PC, Marks WM, Ryan JA, Bolen JW (1986) Colorectal carcinoma evaluation with CT: preoperative staging and detection of postoperative recurrence. Radiology 158: 347–353

Freeny PC, Marks WM, Ryan JA, Traverso LW (1988) Pancreatic ductal adenocarcinoma: diagnosis and staging with dynamic CT. Radiology 166: 125–133

Friedman AC, Lichtenstein JE, Goodman Z, Fishman EK, Siegelman SS, Dachman AH (1985) Fibrolamellar hepatocellular carcinoma. Radiology 157: 583–587

Gelfand DW, Chen YM, Ott DJ (1987) Multiphasic examinations of the stomach: efficacy of individual techniques and combinations of techniques in detecting 153 lesions. Radiology 162: 829–834

Gibbons CP, Griffiths GJ, Cormack A (1983) The role of percutaneous transhepatic cholangiography and grey-scale ultrasound in the investigation and treatment of bilt duct obstruction. Radiology 164: 43–47

Gibson RN, Yeung E, Thompson JN (1986) Bile duct obstruction: radiologic evaluation of level, cause, and tumor resectability. Radiology 160: 43–47

Glazer GM (1988) MR imaging of the liver, kidneys, and adrenal glands. Radiology 166: 303–312

Gobien RP, Stanley JH, Soucek CD, Anderson MC, Vujic I, Gobien BS (1984) Routine preoperative biliary drainage: effective on management of obstructive jaundice. Radiology 152: 353–356

Goldberg BB (1984) Abdominal ultrasonography, 2nd edn. Wiley, New York

Gomberg JS, Friedman AC, Radecki PD, Grumbach K, Caroline DF (1986) MRI differentiation of recurrent colorectal carcinoma from postoperative fibrosis. Gastrointest Radiol 11: 361–363

Grabbe E, Winkler R (1985) Local recurrence after sphincter saving resection for rectal and rectosigmoid carcinoma: value of various diagnostic methods. Radiology 155: 305–310

Grabbe E, Lierse W, Winkler R (1983) The perirectal fascia: morphology and use in staging of rectal carcinoma. Radiology 149: 241–246

Gross BH, Agha FP, Glazer GM, Orringer MB (1985) Gastric interposition following transhiatal esophagectomy: CT evaluation. Radiology 155: 177–179

Gunven P, Makuuchi M, Takayasu K, Moriyama N, Yamasaki S, Hasegawa H (1985) Preoperative imaging of liver metastases: comparison of angiography, CT scan, and ultrasonography. Ann Surg 2 (5): 573–579

Halber MD, Daffner RH, Thompson WM (1979) CT of the esophagus. I. Normal appearance. AJR 133: 1047–1050

Hall-Craggs, Lees WR (1986) Fine-needle aspiration biopsy: pancreatic and biliary tumors. AJR 147: 399–403

Halpert RD, Feczko PJ, Feczko PJ, Spickler EM, Ackerman LV (1985) Radiological assessment of dysphagia with endoscopic correlation. Radiology 157: 599–602

Halvorsen RA, Thompson WM (1987) Computed tomographic staging of gastrointestinal tract malignancies. I. Esophagus and stomach. Invest Radiol 22: 2–16

Halvorsen RA, Korobkin M, Ram PC, Thompson WM (1982) CT appearance of focal fatty infiltration of the liver. AJR 139: 277–281

Halvorsen RA, Magruder-Habib K, Foster WL, Roberts L, Postlethwait RW, Thompson WM (1986) Esophageal cancer staging by CT: long-term follow-up study. Radiology 161: 147–151

Harter LP, Moss AA, Goldberg HI, Gross BH (1983) CT-guided fine-needle aspirations for diagnosis of benign and malignant disease. AJR 140: 363–367

Hayashi N, Yamamoto K, Tamaki N (1987) Metastatic nodules of hepatocellular carcinoma: detection with angiography, CT, and US. Radiology 165: 61–63

Haynes JW, Miller PR, Steiger Z, Leichman LP, Kling GA (1984) Celestin tube use: radiographic manifestations of associated complications. Radiology 150– 41–44

Heiken JP, Balfe DM, Picus D, Scharp DW (1984a) Radical pancreatectomy: postoperative evaluation by CT. Radiology 153: 211–215

Heiken JP, Balfe DM, Roper CL (1984b) CT evaluation after esophagogastrectomy. AJR 143: 555–560

Heiken JP, Lee JKT, Glazer HS, Ling D (1985) Hepatic metastases studied with MR and CT. Radiology 156: 423–427

Henderson JM, Heymsfield SB, Horowitz J, Kutner MH (1981) Measurement of liver and spleen volume by computed tomography. Radiology 141: 525–527

Hessel SJ, Siegelman SS, McNeil BJ, et al. (1982) A prospective evaluation of computed tomography and ultrasound of the pancreas. Radiology 143: 129–133

Heyder N, Lux G (1986) Malignant lesions of the upper gastrointestinal tract. Scand J Gastroenterol [Suppl 123] 21: 47–51

Hishikawa Y, Tanaka S, Miura T (1984) Esophageal ulceration induced by intracavitary irradiation for esophageal carcinoma. AJR 143: 269–273

Honickman SP, Mueller PR, Wittenberg J (1983) Ultrasound in obstructive jaundice: prospective evaluation of site and cause. Radiology 147: 511–515

Hosoki T (1983) Dynamic CT of pancreatic tumors. AJR 140: 959–965

Hosoki T, Chatani M, Mori S (1982) Dynamic computed tomography of hepatocellular carcinoma. AJR 139: 1099–1106

Itai Y, Kogure T, Okuyama Y, Akiyama H (1978) Superficial esophageal carcinoma: radiological findings in double-contrast studies. Radiology 126: 597–601

Itai Y, Nishikawa J, Tasaka A (1979) Computed tomography in the evaluation of hepatocellular carcinoma. Radiology 131: 165–170

Itai Y, Araki T, Yoshikawa K, Furui S, Yashiro N, Tasaka A (1980) Computed tomography of gallbladder carcinoma. Radiology 137: 713–718

Itai Y, Araki T, Furui S, Tasaka A (1981) Differential diagnosis of hepatic masses on computed tomography, with particular reference to hepatocellular carcinoma. J Comput Assist Tomogr 5 (6): 834-842

Itai Y, Araki T, Tasaka A, Maruyama M (1982a) Computed tomographic appearance of resectable pancreatic carcinoma. Radiology 143: 719-726

Itai Y, Moss AA, Ohtomo K (1982b) Computed tomography of cystadenoma and cystadenocarcinoma of the pancreas. Radiology 145: 419-425

Itai Y, Araki T, Furui S, Yashiro N, Ohtomo K, Iio M (1983a) Computed tomography of primar intrahepatic biliary malignancy. Radiology 147: 485-490

Itai Y, Ohtomo K, Araki T, Furui S, Iio M, Atomi Y (1983b) Computed tomography and sonography of cavernous hemangioma of the liver. AJR 141: 315-320

Itai Y, Furui S, Ohtomo K, et al. (1986a) Dynamic CT features of arterioportal shunts in hepatocellular carcinoma. AJR 146: 723-727

Itai Y, Ohtomo K, Kokubo T, et al. (1986b) CT of hepatic masses: significance of prolonged and delayed enhancement. AJR 146: 729-733

Itoh K, Nishimura K, Togashi K, et al. (1987) Hepatocellular carcinoma: MR imaging. Radiology 164: 21-25

Jaffe MH, Fleischer D, Zeman RK, Benjamin SB, Choyke PL, Clark LR (1987) Esophageal malignancy: imaging results and complications of combined endoscopic-radiologic palliation. Radiology 164: 623-630

Jafri SZH, Aisen AM, Glazer GM, Weiss CA (1984) Comparison of CT and angiography in assessing resectability of pancreatic carcinoma. AJR 142: 525-529

Jones TB, Dubuisson RL, Hughes JJ, Robinson AE (1983) Abrupt termination of the common bile duct: a sign of malignancy identified by high-resolution real-time sonography. J Ultrasound Med 2: 345-348

Kaude J, Rian R (1971) Cholangiocarcinoma. Radiology 100: 573-580

Kelvin FM, Korobkin M, Heaston DK, Grant JP, Akwari O (1983) The pelvis after surgery for rectal carcinoma: serial Ct observations with emphasis on nonneoplastic features. AJR 141: 959-964

Kinkhabwala M, Balthazar EJ (1978) Carcinoid tumors of the alimentary tract. II. Angiographic diagnosis of small intestinal and colonic lesions. Gastrointest Radiol 3: 57-61

Koehler RE, Moss AA, Margulis AR (1976) Early radiographic manifestations of carcinoma of the esophagus. Radiology 119: 1-5

Koenigsberg M, Wiener SN, Walxer A (1979) The accuracy of sonography in the differential diagnosis of obstructive jaundice: a comparison with cholangiography. Radiology 133: 157-165

Krudy AG, Doppman JL, Jensen RT, et al. (1984) Localization of islet cell tumors by dynamic CT: comparison with plain CT, arteriography, sonography, and venous sampling. AJR 143: 585-589

Kunstlinger F, Federle MP, Moss AA, Marks W (1980) Computed tomography of hepatocellular carcinoma. AJR 134: 431-437

Larson SM (1987) Lymphoma, melanoma, colon cancer: diagnosis and treatment with radiolabeled monoclonal antibodies. Radiology 165: 297-304

Laufer (1979) Atlas of double contrast gastrointestinal radiology. Double contract gastrointestinal radiology with endoscopic correlation. W.B. Saunders Company, Philadelphia, PA

Lawson TL, Dodds WJ (1976) Infiltrating carcinoma simulating achalasia. Gastrointest Radiol 1: 245-248

Lee JKT, Stanley RJ, Sagel SS, Levitt RG, McClennan BL (1981) CT appearance of the pelvis after abdomino-perineal resection for rectal carcinoma. Radiology 141: 737-741

Lee KR, Levine E, Moffat RE, Biogongiari LR, Hermreck AS (1979) Computed tomographic staging of malignant gastric neoplasms. Radiology 133: 151-155

Lees WR, Heron CW (1987) US-guided percutaneous pancreatography: experience in 75 patients. Radiology 165: 809-813

Legge DA, Carlson HC (1972) Cholangiographic appearance of primary carcinoma of the bile duct. Radiology 102: 259-266

Levine MS, Dillon EC, Saul SH, Laufer I (1986) Early esophageal cancer. AJR 146: 507-512

Levine MS, Langer J, Laufer I, Kligerman MM (1987) Radiation therapy of esophageal carcinoma: correlation of clinical and radiographic findings. Gastrointest Radiol 12: 99-105

Lewis E, auf der Heide JF, Bernardino ME, Barnes PA, Thomas JL (1982) CT detection of hepatic metastases with ethiodized oil emulsion 13. J Comput Assist Tomogr 6 (6): 1108-1114

Machi J, Isomoto H, Kurohiji T, et al. (1986) Detection of unrecognized liver metastases from colorectal cancers by routine use of operative ultrasonography. Dis Colon Rectum 29: 405-409

Marchal G, Baert AL, Wilms G (1979) Intravenous pancreaticography in computed tomography. J Comput Assist Tomogr 3 (6): 727-732

Marshak RH, Eliasoph J (1957) Cardiospasm or carcinoma? The roentgen findings. Am J Dig Dis 2 (1): 11-25

Mathieu D, Grenier P, Larde D, Vasile N (1984) Portal vein involvement in hepatocellular carcinoma: dynamic CT features. Radiology 125: 127-132

Matsui O, Takashima T, Kadoya M (1985) Dynamic computed tomography during arterial portography: the most sensitive examination for small hepatocellular carcinomas. J Comput Assist Tomogr 9 (1): 19-24

Matsui O, Takashima T, Kadoya M, et al. (1987) Liver metastases from colorectal cancers: detection with CT during arterial portography. Radiology 165: 65-69

Matter D, Bret PM, Bretagnolle, Valette PJ, Fond A (1987) Pancreatic duct: US-guided percutaneous opacification. Radiology 163: 635-636

McCarthy SM, Barnes D, Deveney K, Moss AA, Goldberg HI (1985) Detection of recurrent rectosigmoid carcinoma: prospective evaluation of CT and clinical factors. AJR 144: 577-579

Megibow AJ, Bosniak MA, Ambos MA, Berambaum ER (1981) Thickening of the celiac axis and/or superior mesenteric artery: a sign of pancreatic carcinoma on computed tomography. Radiology 141: 449-453

Miller DL, Simmons JT, Chang R, et al. (1987) Hepatic metastasis detection comparison of three CT contrast enhancement methods. Radiology 165: 785-790

Mori S, Kasai M, Watanabe T, et al. (1979) Preoperative assessment of resectability for carcinoma of the thoracic esophagus. I. Esophagogram and azygogram. Ann Surg 190: 100-105

Moss AA, Schrumpf J, Schnyder P, Korobkin M, Shimshak RR (1979) Computed tomography of focal hepatic lesions. A blind clinical evaluation of the effect of contrast enhancement. Radiology 131: 427-430

Moss AA, Federle M, Shapiro HA, et al. (1980) The combined use of computed tomography and endoscopic retrograde cholangiopancreatography in the assessment of suspected pancreatic neoplasm: a blind clinical evaluation. Radiology 134: 159-163

Moss AA, Cann CE, Friedman MA, Marcus FS, Resser KJ, Berninger W (1981a) Volumetric CT analysis of hepatic tumors. J Comput Assist Tomogr 5 (5): 714-718

Moss AA, Schnyder P, Marks W, Margulis R (1981b) Gastric adenocarcinoma: a comparison of the accuracy and economics of staging by computed tomography and surgery. Gastroenterology 80 (1): 45-50

Moss AA, Schnyder P, Thoeni RF, Margulis AR (1981c) Esophageal carcinoma: pretherapy staging by computed tomography. AJR 136: 1051-1056

Moss AA, Margulis AR, Schnyder P, Thoeni RF (1981d) A uniform, CT-based staging system for malignant neoplasms of the alimentary tube. AJR 136: 1251-1253

Moss AA, Dean PB, Axel L, Goldberg HI, Glazer GM, Friedman MA (1982) Dynamic CT of hepatic masses with intravenous and intraarterial contrast material. AJR 138: 847-852

Mueller PR, van Sonnenberg E, Ferrucci JT (1982a) Percutaneous biliary drainage: technical and catheter-related problems in 200 procedures. AJR 138: 17-23

Mueller PR, Ferrucci JT, van Sonnenberg E (1982b) Obstruction of the left hepatic duct: diagnosis and treatment by selected fine-needle cholangiography and percutaneous biliary drainage. Radiology 145: 297-302

Mullin D, Shirkhoda A (1985) Computed tomography after gastrectomy in primary gastric carcinoma. J Comput Assist Tomogr 9 (1): 30-33

Muto T, Bussey HJR, Morson BC (1975) The evolution of cancer of the colon and rectum. Cancer 36: 2251-2270

Neff CC, Simeone JF, Wittenberg J, Mueller PR, Ferrucci JT (1984) Inflammatory pancreatic masses. Radiology 150: 35-38

Nicholls RJ, Chir M (1983) The appearances on computed tomography after abdominoperineal resection for carcinoma of the rectum: a comparison between the normal appearances and those of recurrence. Br J Radiol 56: 237-240

Odurny A, Birch SJ (1985) Hepatic arterial embolisation in patients with metastatic carcinoid tumours. Clin Radiol 36: 597-602

Ohishi H, Uchida H, Yoshimura H, et al. (1985) Hepatocellular carcinoma detected by iodized oil: use of anticancer agents. Radiology 154: 25-29

Ohtomo K, Itai Y, Yoshikawa K, et al. (1987) Hepatic tumors: dynamic MR imaging. Radiology 163: 27-31

Okuda K, Musha H, Yoshida T, et al. (1975) Demonstration of growing casts of hepatocellular carcinoma in the portal vein by celiac angiography: the thread and steaks sign. Radiology 117: 303-309

Olmsted WW, Lichtenstein JE, Hyams VJ (1983) Polypoid epthelial malignancies of the esophagus. AJR 140: 921-925

Onik G, Kane R, Steele G, et al. (1986) Monitoring hepatic cryosurgery with sonography. AJR 147: 665-669

Op den Orth JO (1979) The standard biphasic-contrast examination of the stomach and duodenum: method, results, and radiological atlas. Nijhoff, The Hague, pp 7-31

Ott, DJ, Gelfand DW, Wu WC, Kerr RM (1980) Sensitivity of double-contrast barium enema: emphasis on polyp detection. AJR 135: 327-330

Pedrosa CS, Casanova R, Rodriguez R (1981a) Computed tomography in obstructive jaundice. I. The level of obstruction. Radiology 139: 627-634

Pedrosa CS, Casanova R, Lezana AH, Fernandez MC (1981b) Computed tomography in obstructive jaundice. II. The cause of obstruction. Radiology 139: 635-645

Picus D, Balfe DM, Koehler RE, Roper CL, Owen JW (1983) Computed tomography in the staging of esophageal carcinoma. Radiology 146: 433-438

Plumley TF, Rohrmann CA, Freeny PC, Silverstein FE, Ball TJ (1982) Double duct sign: reassessed significance in ERCP. AJR 138: 31-35

Quint LE, Glazer GM, Orringer MB, Gross BH (1985a) Esophageal carcinoma: CT findings. Radiology 155: 171-175

Quint LE, Glazer GM, Orringer MB (1985b) Esophageal imaging by MR and CT: study of normal anatomy and neoplasms. radiology 156: 727-731

Raghavendra BN (1980) Ultrasonographic features of primary carcinoma of the gallbladder: report of five cases. Gastrointest Radiol 5: 239-244

Ralls PW, Halls J, Renner I, Juttner H (1980) Endoscopic retrograde cholangiopancreatography (ERCP) in pancreatic disease. Radiology 134: 347-352

Reiman TH, Balfe DM, Weyman PJ (1987) Suprapancreatic biliary obstruction: CT evaluation. Radiology 163: 49-56

Reinig JW, Stanley JH, Schabel SI (1983) CT evaluation of thickened esophageal walls. AJR 140: 931-934

Reinig JW, Dwyer AJ, Miller DL, et al. (1987) Liver metastasis detection: comparative sensitivities of MR imaging and CT scanning. Radiology: 162: 43-47

Reznek RH, White FE, Young JWR, Frey IK (1983) The appearance on computed tomography after abdomino-perineal resection for carcinoma of the rectum: A comparison between the normal appearances and those of recurrence. Br J Radiol 56: 237-240

Rifkin MD, Wechsler RJ (1986) A comparison of computed tomography and endorectal ultrasound in staging rectal cancer. Int J Colorect Dis 1: 219-223

Rifkin MD, Rosato FW, Branch HM (1987) Intraoperative ultrasound of the liver: an important adjunctive tool for decision making in the operating room. Ann Surg 205: 466-471

Rossi P, Baert A, Passariello R, Simonetti G, Pavone P, Tempesta P (1985) CT of functioning tumors of the pancreas. AJR 144: 57-60

Samuelsson L, Hambraeus GM, Mercke CE, Tylen U (1984) CT staging of oesophageal carcinoma. Acta Radiol [Diagn] (Stockh) 25 (1): 7-11

Savarino V, Mansi C, Bistolfi L, Zentilin P, Celle G (1983) Failure of new diagnostic aids in improving detection of pancreatic cancer at a resectable stage. Dig Dis Sci 28 (12): 1078-1082

Sellink JL (1974) Radiologic examination of the small intestine by duodenal intubation. Acta Radiol (Stockh) 15: 318

Shaw PC, van Romunde LKJ, Griffiven G, Janssen AR, Kreunig J, Eilers GAM (1987) Peptic ulcers and gastric carcinoma: diagnosis biphasic radiography compared with fiberoptic endoscopy. Rad 163: 39-42

Shirakabe H, Ichikawa H (1966) Atlas of X-ray diagnosis of early gastric cancer. Lippincott, Philadelphia

Shirkhoda A (ed) (1988) Radiological oncology of the abdomen and pelvis: An atlas and text. Year Book Medical Publishers, Chicago

Shirkhoda A, Fallone BG (1987) CT diagnosis of hepatic metastases. AJR 149

Simeone JF, Edelman RR, Stark DD, et al. (1985) Surface coil MR imaging of abdominal viscera: Part III. The pancreas. Radiology 157: 437-441

Skucas J, Spataro R (1981) The radiologic appearance of small colon carcinomas. Radiographics 1 (3): 66-72

Skucas J, Spataro RF (1982) The radiographic features of small colon cancers. Radiology 143: 335-340

Smathers RL, Lee JKT (1984) Differentiation of complicated cholecystitis from gallbladder carcinoma by computed tomography. AJR 143: 225-259

Smith FW, Hutchison JMS (1981) Oesophageal carcinoma demonstrated by whole-body nuclear magnetic resonance imaging. Br Med J 282: 510-512

Snow JH, Goldstein HM (1979) Comparison of scintigraphy, sonography, and computed tomography in the evaluation of hepatic neoplasms. AJR 132: 915-918

Stark DD, Wittenberg J (1987) Hepatic metastases: randomized controlled comparison of detection with MR imaging and CT. Radiology 165: 399-406

Stark DD, Moss AA, Goldberg HI, Davis PL, Federle MP (1984a) Magnetic resonance and CT of the normal and diseased pancreas: a comparative study. Radiology 150: 153-162

Stark DD, Moss AA, Goldberg HI, Deveney CW (1984b) CT of pancreatic islet cell tumors. Radiology 150: 491-494

Sugarbaker PH, Vermess M (1984) Improved detection of focal lesions with computerized tomographic examination of the liver using ethiodized oil emulsion (EOE-13) liver contrast. Cancer 54: 1489-1495

Takashima T, Matsui O (1980) Infusion hepatic angiography in the detection of small hepatocellular carcinomas. Radiology 136: 321-325

Takashima T, Matsui O, Suzuki M, Ida M (1982) Diagnosis and screening of small hepatocellular carcinomas. Radiology 145: 635-638

Takayasu K, Muramatsu T (1986) Hepatic lobar atrophy following obstruction of the ipsilateral portal vein from hilar cholangiocarcinoma. Radiology 160: 389-393

Taylor KJW, Carpenter DA, Hill CR, McCready VR (1976) Gray scale ultrasound imaging: the anatomy and pathology of the liver. Radiology 119: 415-423

Taylor KJW, Ramos I, Morse SS, Fortune KL, Hammers L, Taylor CR (1987) Focal liver masses: differential diagnosis with pulsed Dopper US. Radiology 164: 643-647

Terrier F, Schapira CL (1984) CT assessment of operability in carcinoma of the oesophagogastric junction. Eur J Radiol 4: 114-117

Thoeni RF, Menuck L (1977) Comparison of barium enema and colonoscopy in the detection of small colonic polyps. Radiology 124: 631-635

Thoeni RF, Venbrux AC (1982) Work in progress. The anal canal: distinction of internal hemorrhoids from small cancers by double-contrast barium enema examination. Radiology 145: 17-19

Thompson WM, Halvorsen FA (1987) Computed tomographic staging of gastrointestinal malignancies. II. The small bowel, colon and rectum. Invest Radiol 22: 96-105

Thompson WM, Halvorsen RA, Foster WL, Williford ME, Postlewait RW, Korobkin M (1983) Computed tomography for staging esophageal and gastroesophageal cancer: reevaluation. AJR 141: 951-958

Thompson WM, Halvorsen RA, Foster WL, Roberts L, Gibbons R (1986) Preoperative and postoperative CT staging of rectosigmoid carcinoma. AJR 146: 703-710

Thorsen MK, Quiroz F, Lawson TL, Smith DF, Foley WD, Stewart ET (1984) Primary biliary carcinoma: CT evaluation. Radiology 152: 479-483

Tio TL, Tytgat GNJ (1986) Endoscopic ultrasonography of normal and pathologic upper gastrointestinal wall structure: Comparison of studies in vivo and in vitro with histology. Scand J Gastroenterol [Suppl 123] 21: 27-33

Tio TL, den Hartog Jager FCA, Tytgat GNJ (1986) The role of endoscopic ultrasonography in assessing local resectability of oesophagogastric malignancies: accuracy, pitfalls, and predictability. Scand J Gastroenterol [Suppl 123] 21: 78-86

Tscholakoff, Hricak H, Thoeni R, Winkler ML, Margulis AR (1987) MR imaging in the diagnosis of pancreatic disease. AJR 148: 703-709

Tsuchiya R, Noda T, Harada N, et al. (1986) Collective review of small carcinomas of the pancreas. Ann Surg 203 (1): 77-81

Vasquez JL, Thorsen MK, Dodd WJ, Foley WD, Lawson TL (1985) Atrophy of the left hepatic lobe caused by a cholangiocarcinoma. AJR 144: 547-548

Vermess M, Doppman JL, Sugarbaker PH, et al. (1982) Computed tomography of the liver and spleen with intravenous lipoid contrast material: review of 60 examinations. AJR 138: 1063-1071

Vermess A, Leung AWL, Bydder GM, Steiner RE, Blumgart LH, Young IR (1985) MR imaging of the liver in primary hepatocellular carcinoma. J Comput Assist Thomogr 9 (4): 749-754

Voyles CR, Bowley NJ, Allison DJ, Benjamin IS, Blumgart LH (1983) Carcinoma of the proximal extrahepatic biliary tree: radiologic assessment and therapeutic alternatives. Ann Surg 197 (2): 188-194

Walter JF, Bookstein JJ, Bouffard EV (1976) Newer angiographic observations in cholangiocarcinoma. Radiology 118: 19-23

Wang KY, Kimmey MB, Nyberg DA, et al. (1987) Colorectal neoplasms: accuracy of US in demonstrating the depth of invasion. Radiology 165: 827-829

Weiner SN, Koenigsberg M, Morehouse H, Hoffman J (1984) Sonography and computed tomography in the diagnosis of carcinoma of the gallbladder. AJR 142: 735-739

White RM, Levine MS, Eterline HT, Laufer I (1985) Early gastric cancer: recent experience. Radiology 155: 25-27

Wiljasalo M, Tallroth K, Korhola O, Ihamaki T (1980) A comparison of double contrast barium meal and endoscopy. Diagn Imaging 49: 1-5

Wittenberg J, Simeone JF, Ferrucci JT, Mueller PR, van Sonnenberg E, Neff CC (1982) Non-focal enlargement in pancratic carcinoma. Radiology 144: 131-135

Wolf EL, Frager J, Brandt LJ, Frager DH, Bernstein LH, Beneventano TC (1986) Radiographic appearance of the esophagus and stomach after laser treatment of obstructing carcinoma. AJR 146: 519-522

Wolfman NT, Ramquist NA, Karstaedt N, Hopkins MB (1982) Cystic neoplasms of the pancreas: CT and sonography. AJR 138: 37-41

Yamada A (1979) Radiologic assessment of resectability and prognosis in esophageal carcinoma. Gastrointest Radiol 4: 213-218

Yates CK, Streight RA (1986) Focal fatty infiltration of the liver simulating metastatic disease. Radiology 159: 83-84

Yeh HC (1979) Ultrasonography and computed tomography of carcinoma of the gallbladder. Radiology 133: 167-173

Young SW, Turner RJ, Castellino RA (1980) A strategy for the contrast enhancement of malignant tumors using dynamic computed tomography and intravascular pharmacokinetics. Radiology 137: 137-147

Yumoto Y, Jinno K, Tokuyama K, Araki Y, Ishimitsu T, Maeda H, Konno T, Iwamoto S, Ohnishi K, Okuda K (1985) Hepatocellular carcinoma detected by iodized oil. Rad 154: 19-24

Zornoza J, Lindell MM (1980) Radiologic evaluation of small esophageal carcinoma. Gastrointest Radiol 5: 107-111

5 Esophageal Cancer

WILLIAM M. MENDENHALL and RODNEY R. MILLION

CONTENTS

WILLIAM M. MENDENHALL, M.D., Associate Professor of Radiation Therapy[1], RODNEY R. MILLION, M.D., Professor of Radiation Therapy[2], Department of Radiation Therapy, University of Florida College of Medicine, Box J-385, J. Hillis Miller Health Center, Gainsville, FL 32610-0385, USA

Source of Support
[1] American Cancer Society Junior Faculty Clinical Fellowship; [2] American Cancer Society Ashbel C. Williams, M.D., Memorial Professor of Clinical Oncology

5.1 Introduction

Cancer of the esophagus was noted more than 2000 years ago in the Honan province of northern China. In the Linhsien county of Honan province, where the "hard-of-swallowing disease" had been quite prevalent for many years, there existed a "Throat God Temple" until 1927, when it was destroyed during the war (YANG 1980). Cancer of the esophagus was also described by Jurgani, a Persian medical writer, as early as 1100 A.D. It is of interest to note that the highest incidence rates of esophageal cancer in the world are currently found in these two regions.

5.2 Anatomy

The esophagus begins at the cricopharyngeus at the level of C6 (approximately 16 cm from the incisors) and extends inferiorly to the esophagogastric junction (approximately 43 cm from the incisors). It is usually arbitrarily divided into three or four segments, although there is no anatomical change in the esophagus between these various segments.

For practical purposes, the esophagus can be divided into (1) the cervical esophagus, which extends from the cricopharyngeus (C6 vertebra) to the sternal notch (level of T4 vertebra, 23 cm from the incisors); (2) the upper thoracic esophagus, which lies posterior to the trachea and extends from the sternal notch to the carina; and (3) the lower thoracic esophagus, which extends from the carina to the esophagogastric junction.

The esophagus is composed of squamous epithelium, an inner circular layer of muscle, and an outer longitudinal layer of muscle. There is no serosa.

The cervical esophagus, which is 6–8 cm in length, lies posterior to the trachea and recurrent laryngeal nerves, medial to the lateral lobe of the thyroid gland and carotid sheaths, and anterior to the prevertebral fascia. The thoracic esophagus lies in the posterior mediastinum, behind the trachea, mainstem bronchi, and the heart and great vessels, and anterior to the prevertebral fascia.

The blood supply to the cervical esophagus is from the inferior thyroid artery and vein. The thoracic esophagus receives its blood supply from the branchial arteries, direct branches of the aorta, phrenic arteries, and the left gastric artery.

The lymphatics of the esophagus are abundant and initially drain to the periesophageal lymph nodes. From the cervical esophagus and the upper one-third of the thoracic esophagus, lymph may subsequently drain to the peritracheal, mediastinal, lower internal jugular, supraclavicular, and upper internal jugular lymph nodes. Lymph from the lower two-thirds of the thoracic esophagus may drain to the mediastinal lymph nodes and then superiorly to the lower internal jugular and supraclavicular nodes, or it may drain inferiorly to the celiac axis, parapancreatic, and paragastric nodes.

5.3 Natural History

Carcinomas of the esophagus, which constitute the vast majority of esophageal cancers, arise from the surface epithelium and probably represent the end stage of the spectrum of epithelial disorders that includes dysplasia and carcinoma in situ. The time required for a lesion to progress from severe dysplasia to invasive carcinoma and the percentage of dysplastic lesions that do so are not well defined and may vary from one geographic area to another. For instance, YANJIN et al. (1981) reported a series of 23 patients diagnosed as having early esophageal cancer during a mass screening program in Linhsien county in northern China who subsequently refused treatment. All 23 patients had either carcinoma in situ or invasive carcinoma confined to the mucosa or submucosa. Eleven patients (48%) developed advanced esophageal cancer in 18–116 months (mean 55.5±29 months) from diagnosis. In 12 patients (52%) the disease had not progressed with follow-up of 42–148 months (mean 74.4± 27.3 months), suggesting that the latency period can be fairly long (although it may be short, also). The 5-year survival rate in this group of patients was 18/23 (78%). It has been estimated that it may take 7–8 years to progress from esophagitis through dysplasia to carcinoma in situ and another 3–4 years for carcinoma in situ to develop into an advanced cancer (LIGHTDALE and WINAWER 1984). There is some suggestion that the time course may be shorter in some areas, such as Iran (LIGHTDALE and WINAWER 1984).

Once an invasive cancer has formed, it may grow longitudinally and circumferentially. As the cancer enlarges, it will eventually produce an annular lesion and progressively occlude the lumen of the esophagus. Submucosal spread of tumor is common with esophageal malignancies. KAKEGAWA et al. (1985) reported that 18.6% of 64 patients with carcinoma of the cervical esophagus had intramural "skip metastasis" upon pathological examination of the esophagectomy specimens. Of note is that most of these patients received preoperative irradiation, so that the reported incidence of such spread in this series probably represents a conservative estimate of the true risk of submucosal extension. Submucosal spread of tumor has been reported to occur up to 8 cm from the edge of the primary tumor (HEIMLICH 1970). As the tumor progresses, it will extend into and through the muscular layers of the wall of the esophagus and eventually extend to neighboring structures, such as the trachea, mainstem bronchi, recurrent laryngeal nerve, and aorta. The incidence of contiguous extension of the primary esophageal tumor to various adjacent structures as a function of location of the lesion in the esophagus as reported in an autopsy series of 171 patients is shown in Table 1. Because these data are obtained from patients who have died of their disease, one would expect the likelihood of contiguous extension of the primary tumor to be less at the time of diagnosis and to depend on size of the lesion.

The likelihood of lymph node metastasies will depend on the extent of the primary tumor. Lesions

Table 1. Incidence of contiguous infiltration of adjacent structures by the primary tumor[a] (modified from SONS and BORCHARD 1984)

Location of the primary tumor	No. of patients	Percent involvement					
		Mediastinum	Trachea	Bronchus	Pleura and Lung	Pericardium	Aorta
Proximal third	16	12.5%	62.5%	0%	0%	0%	6.3%
Middle third	87	19.5%	21.8%	16.1%	12.6%	1.2%	5.8%
Distal third	68	20.6%	10.3%	8.8%	19.1%	2.9%	5.9%
Total	171	19.3%	21.0%	11.7%	14.0%	1.8%	5.9%

[a] Autopsy series.

Table 2. Incidence of location of positive nodes at operation in 205 patients with squamous cell carcinoma of the esophagus[a] (modified from AKIYAMA et al. 1981)

Node group	Location of primary lesion		
	Upper thoracic (24 pts.)	Middle thoracic (116 pts.)	Lower thoracic (65 pts.)
Thoracic:			
Superior mediastinal	29.4%	11.4%	9.8%
Middle mediastinal	27.3%	20.7%	14.3%
Lower mediastinal	28.6%	18.0%	27.4%
Abdominal:			
Superior gastric	31.8%	32.8%	61.5%
Celiac axis	0%	4.4%	21.2%
Common hepatic artery	0%	2.0%	9.8%
Splenic artery	0%	6.3%	15.0%

[a] Overall, 121/205 patients (59%) had positive nodes.

Table 3. Risk factors for esophageal carcinoma (modified from LIGHTDALE and WINAWER 1984)

Factor	Risk
Alcohol and tobacco consumption	Unknown
Plummer-Vinson syndrome	Unknown
Nutritional factors	Unknown
Tylosis	95%
Achalasia	1.7%– 8.2%
Lye strictures	3.5%– 5.5%
Celiac disease	3%
Head and neck cancer	2% – 4%
Barrett's esophagus	2% –10%

less than 5 cm in length have approximately a 50% incidence of positive nodes as compared with those longer than 5 cm, in which the incidence of positive nodes is 90% (EARLAM and CUNHA-MELO 1980a). RIMIN et al. (1981) reported the incidence of positive regional nodes in a surgical series of 504 patients as a function of extension into the wall of esophagus to be as follows: submucosa, 0/1; muscularis, 52/175 (30%); outer esophageal surface, 118/273 (43%); and adjacent structures, 38/55 (69%). The authors state that "a few patients" received preoperative irradiation. The incidence and location of positive nodes in a series of 205 patients treated surgically for squamous cell carcinoma of the esophagus by AKIYAMA et al. (1981) are presented in Table 2.

The incidence of distant metastasis is relatively high. ANDERSON and LAD (1982) reported that 67/79 patients (85%) in an autopsy series had distant metastasis. Of note is that 65/67 of those patients (97%) who had distant metastasis also had evidence of locally recurrent or persistent tumor. The locations of the metastatic lesions are as follows: lung (52%), liver (47%), adrenals (20%), bone (14%), and brain (1%).

5.4 Epidemiology and Risk Factors

Esophageal cancer is usually found in the midthoracic portion of the esophagus. YIN et al. (1983) reported the location of the primary lesion, in a series of 1212 patients treated primarily with irradiation, to be as follows: cervical, 44 (3.6%); upper third thoracic, 282 (23.3%); middle third thoracic, 742 (61.2%); and lower third thoracic, 144 (11.9%). The incidence of esophageal cancer increases with age, and the disease is usually more common in men than in women, although there is an increased incidence of lesions in the cervical and upper third of the thoracic esophagus in women from Scandinavia, Great Britain, and Ireland. The male to female ratio varies from approximately 1.2:1 to 9.1:1 (WYNDER and BROSS 1961).

The average age-adjusted incidence rates per 100000 in the United States from 1973 to 1977 were as follows: white men, 4.8; white women, 1.6; black men, 16.9; and black women, 4.5 (SHOTTENFELD 1984). Esophageal cancer is particularly common in black men in the coastal regions of the Carolinas, Georgia, and northeastern Florida. There is a great deal of international variation in the incidence of esophageal cancer. Areas with a particularly high incidence of esophageal cancer include the Honan province in northern China, the Gonbad region of northern Iran, India, and the Transkei region of South Africa. The age-specific incidence rates in the Gonbad region of Iran for people 35–64 years of age are 262.9 per 100000 for women and 206.4 per 100000 for men (SHOTTENFELD 1984). The adjusted average incidence in Linhsien county of northern China, where esophageal cancer accounts for 20% of all deaths, is 108.56 per 100000 (YANG 1980). Esophageal cancer is relatively rare in Norway, Denmark, Sweden, and in the white population of the United States.

Some of the risk factors associated with esophageal cancer are listed in Table 3. Ethanol and tobacco abuse are known to be associated with esophageal cancer and probably account for many of the cases seen in the United States. The risk of esophageal cancer is thought to increase with increased ethanol consumption and to be higher for whiskey drinkers as opposed to heavy beer or wine drinkers (SHOTTENFELD 1984).

The Plummer-Vinson syndrome (which consists of esophageal webs, anemia, atrophy of the mucous membranes, brittle nails, and achlorhydria) is associated with an increased incidence of cancer of the esophagus, hypopharynx, and oral cavity. It is though to be secondary to a nutritional deficiency and is particularly common in British and Scandinavian women.

Esophageal cancer is thought to be related to some nutritional deficiences as well as the presence of nitrosamines and their precursors in food or drinking water. Molybdenum, a cofactor of nitrate reductose, affects the nitrite and nitrate levels in plants. It is found in relatively low levels in the soil of regions in northern China where the incidence of esophageal cancer is quite high. In addition, the content of molybdenum was found to be low in samples of serum, urine, and hair from men in Linhsien county in northern China, where the incidence of esophageal cancer is high, as compared with nearby areas where the incidence of esophageal cancer is low (YANG 1980). The levels of nitrites in well water and nitrites and nitrosamines in food were also found to be increased in Linhsien county, as compared with areas where the incidence of esophageal cancer is low. The high levels of nitrosamines in food, particularly pickled vegetables, from high-incidence areas of northern China are thought to be secondary to fungal contamination, which is probably related to the method used to preserve and store the food (YANG 1980).

Tylosis is an autosomal-dominant disease characterized by keratosis of the hands and feet and associated with a 95% risk of esophageal cancer by the age of 65 (HARPER et al. 1970) Other genetic conditions associated with esophageal cancer include Torre's syndrome and Fanconi's anemia (SHOTTENFELD 1984).

Achalasia, which is a motility disorder of the esophagus characterized by abnormal peristalsis and incomplete relaxation of the esophagogastric junction, is thought to be associated with an increased risk of esophageal cancer, with a latency period of 18-28 years (LIGHTDALE and WINAWER 1984).

Celiac disease (which consists of intermittent glossitis, hematopoietic disorders, radiographic abnormalities of the small bowel, disorders of calcium and carbohydrate metabolism, steatorrhea, and amelioration of symptoms with a gluten-free diet) is also associated with an increased risk of esophageal cancer. HARRIS et al. (1967) studied 202 patients with adult celiac disease or idiopathic steatorrhea who had been observed for an average of 8.2 years and found that 6/202 patients (3%) developed esophageal cancer after a mean duration of symptoms of 50 years (range 38-68 years). They concluded that celiac disease was associated with an increased risk of esophageal cancer in men but not in women, and that the risk may possibly be decreased if the patient is placed on a gluten-free diet.

Patients with head and neck cancer are at an increased risk for the development of esophageal cancer. GOLDSTEIN and ZORNOZA (1978) evaluated approximately 10000 patients treated at the M. D. Anderson Hospital for squamous cell carcinoma of the head and neck who were over the age of 40 years and found that 89 patients developed esophageal cancers, 16 synchronous and 73 metachronous (latency period of 2 months to 16 years, average 46 months). Of note is that 71/89 patients (80%) had developed two or more head and neck cancers. They concluded that the risk of esophageal cancer in this population of patients was substantially higher than the risk for the normal population.

5.5 Diagnostic Workup and Staging

5.5.1 Screening

Because of the dismal prognosis of patients who present with symptomatic esophageal cancers, there is considerable interest in the development of screening programs to detect early lesions in asymptomatic patients who are in high-risk groups. The screening program generally involves obtaining a brushing from the mucosa of the esophagus for cytologic examination (SHU 1983; NABEYA 1983; DOWLATSHAHI et al. 1985). The Chinese employ an inflatable balloon that is covered with a fine mesh and is swallowed, inflated, and then retrieved back up the esophagus. Contraindications to this procedure include esophageal varices, active gastric ulcer disease, severe cardiac disease with hypertension, acute laryngitis, and an esophagogram within 2 days of the procedure. The accuracy of this procedure is said to be 80% in mass screening programs and 90% in outpatient clinics (SHU 1983). In a mass screening of 81187 patients from 1970 to 1975, 880 patients (1.1%) were found to have esophageal cancer, of whom 649 (73.8%) had either carcinoma in situ or microinvasive carcinoma. The 5-year survival rate in this group of patients with early lesions was 90.3% (SHU 1983).

5.5.2 History and Physical Findings

The majority of patients with esophageal cancer present with weight loss and dysphagia. Less frequently, patients may present with dyspnea (secondary to tracheal involvement), hoarseness (secondary to recurrent laryngeal nerve involvement), cough or aspiration pneumonia (secondary to a tracheoesophageal fistula), chest pain, hematemesis, or a neck mass. Patients with carcinoma in the cervical esophagus tend to present with earlier lesions than those with lesions in the thoracic esophagus.

Physical findings, in addition to weight loss, may include vocal cord paralysis, pooling of secretions in the pyriform sinuses, direct extension of tumor to the hypopharynx or larynx, direct extension of tumor into the soft tissues of the neck, anterior bowing of the thyroid cartilage and trachea with loss of the thyroid click, and clinically positive neck nodes (MENDENHALL 1984). The most common location for a metastatic neck node is in the low internal jugular chain, under the insertion of the sternocleidomastoid muscle. Clinically positive nodes may also be found in the upper internal jugular chain, particularly for lesions of the cervical esophagus that directly involve the hypopharynx.

The differential diagnosis of carcinoma of the esophagus includes primary lung cancer extending to the esophagus, thymoma, thyroid cancer, mediastinal lymphoma, sarcoma, germ cell malignancies, metastatic disease to the mediastinal lymph nodes, and stomach cancer extending superiorly to involve the distal esophagus.

5.5.3 Biopsy Procedures

The diagnosis of esophageal cancer is usually established by esophagoscopy and biopsy. Infrequently, it may be difficult to obtain diagnostic tissue by endoscopy owing to constriction of the esophageal lumen above the tumor or because of extensive tumor necrosis. If it is not possible to perform a biopsy by

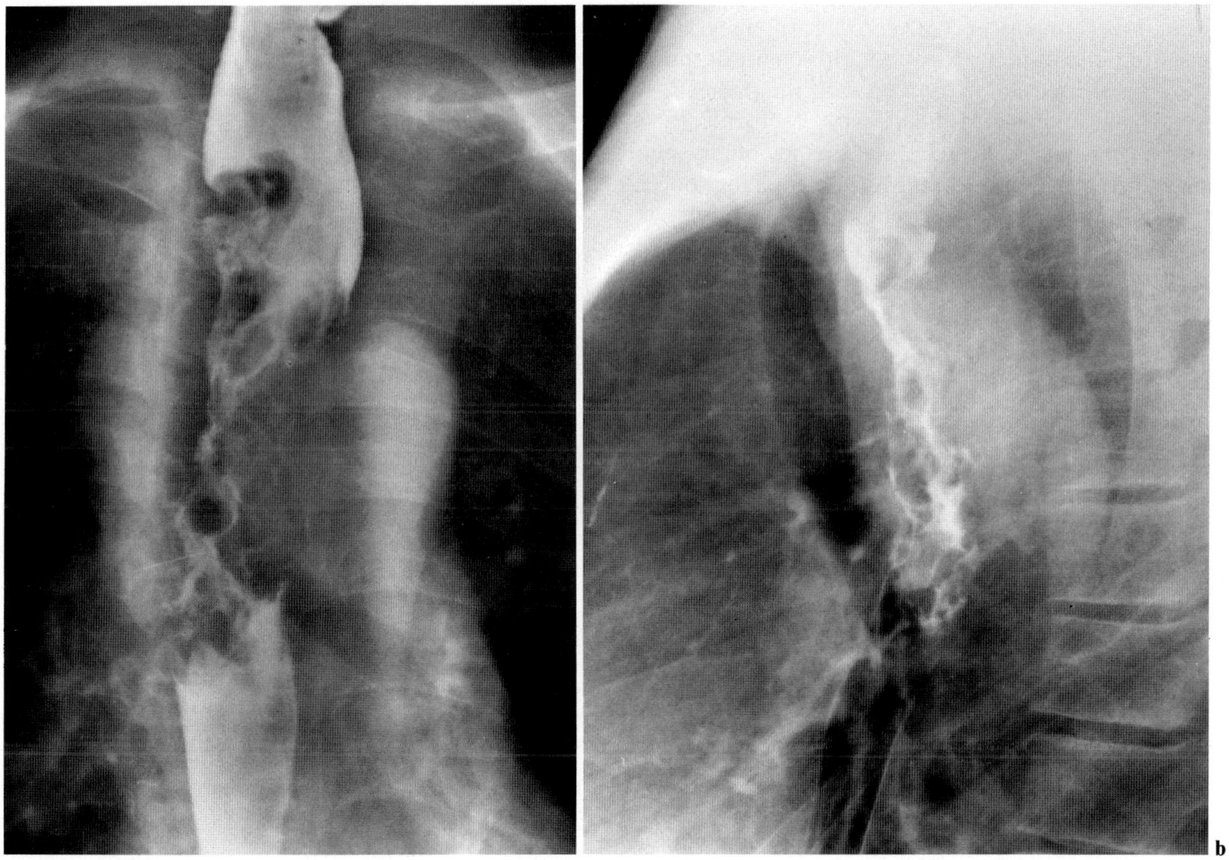

a b

Fig. 1 a, b. Esophagogram of a carcinoma of the upper third of the thoracic esophagus. **a** Anterior view. **b** Lateral view. Note the proximity of the esophageal cancer to the trachea endoscopy, cytologic examination of washings of the esophagus may reveal the diagnosis. Occasionally it may be necessary to excise a clinically positive neck node or perform a CT-directed needle biopsy of an intrathoracic lesion. As a last resort, it may be necessary to explore the patient surgically in an effort to arrive at a diagnosis.

5.5.4 Laboratory and Radiographic Evaluation

Once the diagnosis of esophageal cancer has been established, it is necessary to evaluate the extent of local-regional disease as well as the presence or absence of distant metastasis. Evaluation includes a complete blood cell count (to detect the presence of anemia) and liver function tests (which may be abnormal if liver metastases are present). Chest roentgenograms are obtained routinely and are used to detect distant metastasis to the lung and to determine the local extent of disease. Findings on the chest X-ray films may include tracheal deviation, a widened mediastinum, a widened tracheoesophageal line (>4 mm), and an abnormal azygoesophageal line (THOMPSON 1983). An esophagogram is used to determine the location and length of the lesion, the degree of obstruction, and the presence of a tracheoesophageal fistula (Figs. 1, 2). Tortuosity of the esophagus indicates fixation to mediastinal structures. A double-contrast esophagogram is probably preferable to a single-contrast study.

Computed tomography (CT) of the neck, chest, and upper abdomen may be used to define the extent of the primary lesion, to detect regional lymph node metastasis, and to detect distant metastasis (e.g., liver metastasis). CT defines the extramucosal extent of the lesion better than an esophagogram, whereas an esophagogram is more accurate in defining the mucosal extent of the cancer (Figs. 3, 4). The CT results from four series of 186 patients are summarized in Table 4. Of these, 132 patients (71%) had surgical, autopsy, or other confirmation of CT findings in the chest, and 96 patients (52%) had confirmation of findings in the abdomen. The overall accuracy of CT, based on these four series, was 90% for mediastinal invasion, 78% for enlarged abdominal nodes, and 98% for liver metastasis (THOMPSON 1983). Other studies, such as bone scans, are indicated for detecting bone metastasis in symptomatic patients, but are not routine staging studies. The value of magnetic resonance imaging is unknown at this time. The various imaging studies and their usefulness are outlined in Table 5.

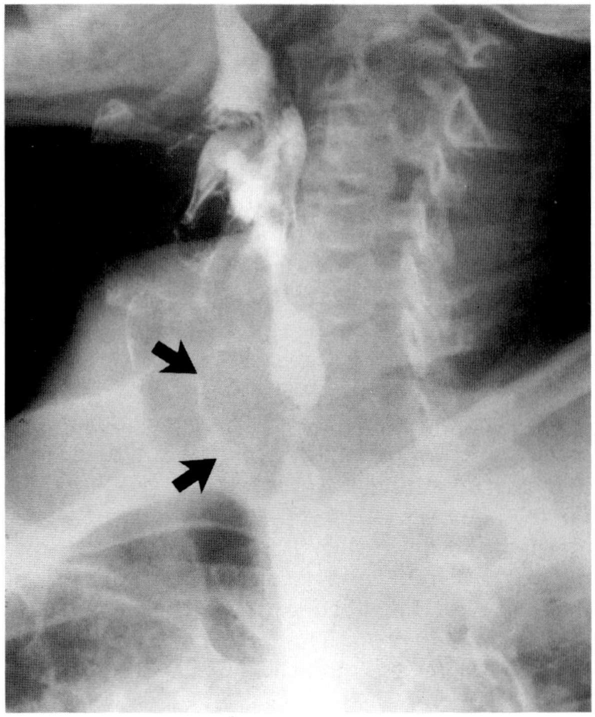

Fig. 2. Esophagogram of a carcinoma of the cervical esophagus. Note the widened space between the trachea and the esophagus *(arrows)*

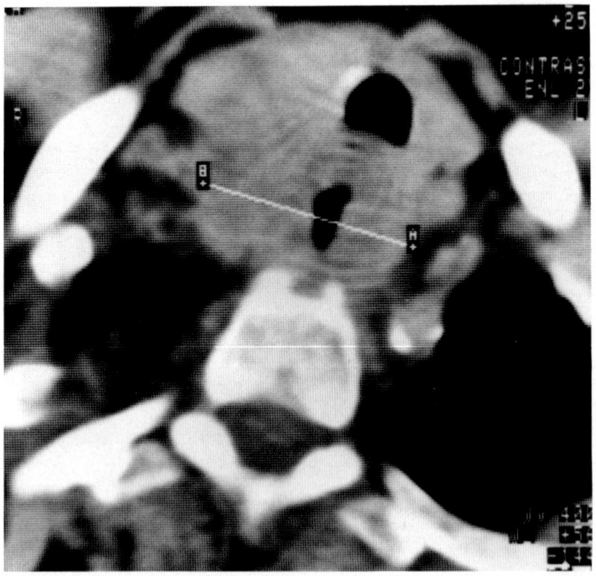

Fig. 3. Extensive carcinoma of the cervical esophagus extending to and surrounding the trachea

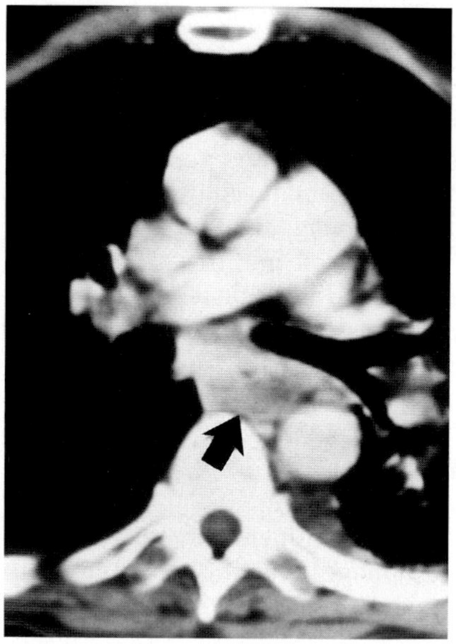

Fig. 4. Advanced carcinoma of the midthoracic esophagus, posterior to the mainstem bronchi

Table 4. Accuracy of computed tomography in 186 patients with esophageal cancer (modified from THOMPSON 1983)

Result	Mediastinal invasion	Abdominal node metastasis	Liver metastasis
False positive	7	10	0
True positive	85	32	9
False negative	7	9	2
True negative	33	45	85

5.5.5 Special Diagnostic Procedures

Bronchoscopy should be performed in patients with lesions at or above the level of the carina in order to detect possible direct extension of esophageal cancer to the tracheobronchial tree. CHOI et al. (1984) reported a series of patients with esophageal cancer who underwent bronchoscopy between 1978 and 1981. Abnormalities were noted in 33.9% of patients; 17.3% had impingement and 16.6% had invasion. The trachea was most frequently involved (usually by lesions in the cervical esophagus and upper third of the thoracic esophagus), followed by the left mainstem bronchus (usually by lesions of the middle third of the thoracic esophagus). The findings of this study are summarized in Table 6.

5.5.6 Staging

The staging system proposed by the American Joint Committee on Cancer (AJCC) (BEAHRS and MYERS 1983) is presented in Table 7. The AJCC staging system for the primary lesions has the disadvantage of being complicated and impractical. Very few patients present with T1 lesions, and it is quite difficult clinically to distinguish a T2 lesion from a T3 lesion. A staging system based solely on the length of the lesion would be easier to apply and would probably be more reliable. Lesions less than 2 cm long are often curable, those 2–5 cm long are occasionally curable, and those longer than 5 cm are almost never curable. Unfortunately, in practice, most

Table 5. Imaging studies for evaluating esophageal cancer (modified from THOMPSON 1983)

Modality	Staging capability	Benefit/cost ratio	Recommended routine staging procedure
A. Noninvasive			
Chest roentgenogram	(1) Detect metastasis (2) Detect second primary lesion	High	Yes
Bone roentgenogram	(1) Confirm bone metastasis	Low	No
Double-contrast esophagogram	(1) Detect and define primary tumor (2) Detect second primary lesion	High	Yes
Bone scan	(1) Detect bone metastasis	Low	No
Ultrasound	(1) Detect suspected abdominal metastasis	Fair	No
Magnetic resonance imaging	(1) Evaluate primary lesion (2) Detect metastasis	Unknown	Unknown
B. Minimally invasive			
CT with IV contrast material	(1) Define extent of primary lesion (2) Detect distant metastasis	High	Yes
C. Invasive			
Azygous venography	(1) Determine resectability	Fair	CT better
Transtracheal lymphography	(1) Detect mediastinal node metastasis	Low	CT better

Table 6. Incidence of abnormal bronchoscopic findings (modified from CHOI et al. 1984)

Length of lesion (cm)	No. of patients	Percent impingement on trachea or bronchus	Percent invasion of trachea or bronchus
1–3	48	16.7%	4.2%
4–6	174	16.1%	10.9%
7–9	119	17.6%	25.2%
10–12	55	18.2%	20.0%
>12	21	23.8%	47.6%

Table 7. American Joint Committee on Cancer (AJCC) staging system (modified from BEAHRS and MYERS 1983)

Stage	Definition
Primary tumor	
Tis	Carcinoma in situ
T1	Tumor ≤5 cm long, produces no obstruction, no circumferential involvement, and no extraesophageal spread
T2	Tumor >5 cm long, tumor of any size that is annular and/or produces obstruction; no evidence of extraesophageal spread
T3	Any tumor with evidence of extraesophageal spread
Nodes: cervical esophagus	
N0	No clinically palpable nodes
N1	Movable, unilateral palpable nodes
N2	Movable, bilateral palpable nodes
N3	Fixed nodes
Thoracic esophagus	
N0	Negative nodes
N1	Positive nodes
Distant metastasis	
M0	No distant metastasis
M1	Distant metastasis

local-regional control and survival data are reported without using *any* staging system, making it next to impossible to compare treatment results between institutions or even within one institution.

5.6 Pathology

Approximately 95% of esophageal cancers are squamous cell carcinomas, 4%–5% are adenocarcinomas, and other histologies represent 1% or less of esophageal neoplasms (SHOTTENFELD 1984; GUNNLAUGSSON et al. 1970; BOSCH et al. 1980). Most adenocarcinomas arise in the distal esophagus and may be associated with columnar metaplasia of the epithelium ("Barrett's esophagus"). Some of the infrequently observed esophageal malignancies include minor salivary gland tumors (such as mucoepidermoid carcinoma or adenoid cystic carcinoma), sarcomas, melanoma, small cell undifferentiated (oat cell) carcinoma, and pseudosarcoma. The clinical course and prognosis for small cell undifferentiated carcinoma of the esophagus appear to parallel those observed in small cell undifferentiated carcinoma of the lung, in that both disseminate rapidly and widely and are associated with a very poor prognosis (SARMA 1982).

5.7 Prognostic Factors

Various factors that may influence the prognosis of the patient with esophageal carcinoma are listed in Table 8. Each of these factors will be discussed briefly.

As in many other tumor systems, the prognosis for women with esophageal cancer is better than for men (KINOSHITA et al. 1982; GATZINSKY et al. 1985). NEWAISHY et al. (1982) reported a series of 444 patients treated between 1956 and 1974 with radical irradiation and found the 5-year survival to be 5.7% for men, versus 11.6% for women ($P<0.05$).

Age has been reported by some to have no influence on survival following treatment for esophageal cancer (GIULI and GIGNOUX 1980), whereas others have reported age to have a significant impact on the 5-year survival rate (GU 1984). It is probable that if one corrected for stage and death from intercurrent disease, age would have little impact on prognosis, except that the very elderly might not tolerate a course of radical treatment.

The influence of tumor site on prognosis is probably dependent on the treatment administered. RIMIN et al. (1981) reported a series of 575 patients

Table 8. Prognostic factors

Category	Factor
Host-related	Sex
	Age
Tumor-related	Site
	Macroscopic appearance
	Degree of obstruction
	Length of lesion
	Depth of penetration
	Lymph node metastasis
	Degree of histologic differentiation
	Blood vessel invasion
	Lymphatic invasion

who had undergone attempted curative resection for squamous cell carcinoma of the esophagus in Beijing, China, and found that the 5-year survival rates were 3/28 (10.7%) for upper-third lesions, 87/327 (26.6%) for middle-third lesions, and 72/220 (32.7%) for lower-third lesions. In contrast, GU (1984) reported the following 5-year survival rates for a series of patients treated with radical irradiation in China: cervical esophagus lesions, 10/31 (32.3%); upper thoracic lesions, 24/153 (15.7%); midthoracic lesions, 46/469 (9.8%); and lower thoracic lesions, 3/58 (5.2%) ($P < 0.01$). In general, surgery seems to yield the best results for distal esophageal cancers, whereas radiotherapy yields better results for proximal lesions.

The macroscopic appearance of the tumor has been reported to have an impact on survival, with the best prognosis for fungating tumors, an intermediate prognosis for ulcerative tumors, and the poorest prognosis for infiltrative tumors (GIULI and GIGNOUX 1980). The degree of obstruction, which is a reflection of the size of the lesion, has also been reported to have a significant influence on the 5-year survival rate (GU 1984).

The length of the lesion has been found by most investigators to have an impact on prognosis (GU 1984; GUOJUN et al. 1981a). GU (1984) reported a series of patients treated with radical irradiation in China and found the 5-year survival rates to decrease with increasing length of lesion (Table 9).

Depth of penetration and regional lymph node involvement have both been reported to influence prognosis significantly (GU 1984; BARKLEY et al. 1981; GUOJUN et al. 1981a; KINOSHITA et al. 1982; NABEYA 1983; SKINNER 1983). The 5-year survival rates as a function of these two factors in a series of 575 patients treated surgically between 1953 and 1973 in Beijing, China, are shown in Table 10 (RIMIN et al. 1981).

The degree of histologic differentiation has been reported to influence survival, particularly when comparing the survival rates of well-differentiated tumors and those that are moderately or poorly differentiated (KINOSHITA et al. 1982). In addition, pathologic evidence of lymphatic invasion has been found to correlate with the incidence of lymph node involvement, and pathologic evidence of blood vessel invasion correlates with the incidence of both regional node involvement and distant metastasis (SUGIMACHI et al. 1983). One would therefore expect these two pathologic findings to correlate with prognosis, as well.

Table 9. Five-year survival as a function of length of lesion in patients treated with radical irradiation (Number alive/number treated) (GU 1984)

Length (cm)	5-year survival[a]
<5	63/ 411 (15.3%)
5.1-7.0	108/1167 (9.3%)
7.1-9.0	80/1177 (6.8%)
>9.0	30/ 584 (5.2%)

[a] $P < 0.005$

Table 10. Impact of regional node involvement and depth of penetration on 5-year survival (modified from RIMIN et al. 1981)

Regional nodes	Depth of penetration	No. of patients	5-year survival
Negative	Submucosa	1	1
	Muscular layer	123	60 (48.8%)
	Outer surface of esophagus	155	75 (48.4%)
	Surrounding tissues	17	2 (11.8%)
	Undetermined	9	8
Positive	Muscular layer	52	4 (7.7%)
	Outer surface of esophagus	118	7 (5.9%)
	Surrounding tissues	38	2 (5.3%)
Undetermined	All	62	3 (4.8%)

5.8 General Management

There is really no general consensus as to how carcinoma of the esophagus should be treated. Each of the treatment modalities in turn will be discussed in order to offer a reasonable treatment approach to the patient with esophageal cancer. Treatment results and complications will be presented in a later section.

5.8.1 Surgery

Czerny performed the first resection of carcinoma of the cervical esophagus in 1877, and Torek, the first successful thoracic esophagectomy for cancer in 1913 (MACCOMB et al. 1967; TOREK 1913). Mikulicz is credited with having resected a carcinoma of the cervical esophagus and reconstructed the defect with skin flaps in 1884 (MACCOMB et al. 1967). Roux, in 1907, described the use of a jejunal graft to reconstruct the gullet, and in 1911, Kelling and Vulliet independently described the use of colon for esophageal reconstruction (BURDETTE and JESSE 1972).

Curative resection for carcinoma of the esophagus entails an en bloc resection of the primary lesion with an adequate margin of normal tissue. For cancers of the thoracic esophagus, this usually entails a total esophagectomy with resection of adjacent tissues as well as the vascular and lymphatic drainage of the tumor. For limited lesions of the cervical esophagus, a partial esophagectomy may be performed, at times in conjunction with a total laryngectomy, resection of one or both lobes of the thyroid gland, and/or neck dissection(s), depending on the extent of the lesion. There is some controversy over the use of partial versus total esophagectomy, because of the risk of submucosal skip metastases that may be found up to 8 cm from the primary lesion. Other areas of controversy for cervical esophageal cancer include the use of bilateral elective neck dissections and elective dissection of the anterior mediastinal lymph nodes. Selection criteria for radical surgery, as set forth by SKINNER (1983), are as follows: localized disease that is apparently resectable, no hepatic cirrhosis, no severe pulmonary disease, no symptomatic angina or cardiac valvular disease, and no generalized debility for patients over 70 years of age.

The methods used to reconstruct the esophagus following partial or total esophagectomy depend on the extent of the resection, the philosophy of the surgeon, and the fact that there is no optimal method to reconstruct the esophagus safely and simply (MENDENHALL 1984). The various methods of reconstruction, as well as their application, advantages, and disadvantages, are listed in Table 11. The method of choice to reconstruct lesions of the cervical esophagus is probably gastric transposition in healthy patients and myocutaneous flap reconstruction in patients of borderline operability (SURKIN et al. 1984). For lesions of the thoracic esophagus, gastric transposition is probably preferable to the other procedures listed.

Palliative surgical procedures, such as a bypass graft, gastrostomy, and/or pyriformostomy, are probably not indicated until a course of palliative irradiation has failed. Palliative radiation therapy is associated with a lower treatment-related morbidity and mortality than a bypass graft and with better palliation, when it works, than gastrostomy and/or pyriformostomy.

5.8.2 Radiation Therapy

Radiation therapy may be used with curative intent for esophageal cancer at any site within the esophagus. From a practical standpoint, however, it is easier and safer to deliver high-dose irradiation to the cervical esophagus and upper third of the thoracic esophagus than the middle and lower third of the thoracic esophagus, where it is difficult to avoid irradiation to a substantial volume of the lung, pericardium, and heart. There is no evidence that the use of neutrons, particles (e.g., helium), or the combination of irradiation and misonidazole offers any advantage over conventional radiation therapy (SCHWADE et al. 1984; CASTRO et al. 1983; LARAMORE et al. 1983; BARKLEY et al. 1981).

Radiation therapy has been used as an adjunct to surgery, either preoperatively or postoperatively (LAUNOIS et al. 1981; GUOJUN et al. 1981a, c; MARKS et al. 1976; FRASER et al. 1978). Most of the data are from nonrandomized studies; selected patients with lesions that tended to be more advanced were treated with the combination of irradiation and surgery, and the results were compared with those of patients with earlier-stage lesions treated with surgery alone. In some instances, the number of patients receiving adjuvant irradiation and the dose used were not specified. In conclusion, however, there is no convincing evidence to support the routine use of adjuvant irradiation with surgery for esophageal cancer. It may be indicated in the occasional patient with an advanced lesion of borderline resectability for whom the treatment of choice is thought to be surgery. In this instance, preoperative irradiation may render a lesion that would have been incompletely resectable into one that can be completely removed. We also recommend preoperative irradiation to patients with cancers of the cervical esophagus who are treated surgically, because it is our philosophy to treat advanced cancers in other head and neck sites with combined irradiation and surgery. In this situation, we prefer preoperative to postoperative irradiation in order to avoid irradiating the reconstructed gullet. Irradiation may be given postoperatively in patients treated initially by surgery for questionable or microscopically positive margins.

Radiation therapy may be used with palliative intent for patients with advanced local disease and/or distant metastasis. A relatively short course of treatment (e.g., 30 Gy in ten fractions over 2 weeks) will produce a good palliative response in about 50% of patients and is associated with relatively little morbidity, although there is a small risk of tracheoesophageal fistula or mediastinitis.

Table 11. Reconstructive procedures following resection for cancer of the esophagus

Procedure	Application	Advantages	Disadvantages
Pedicled skin grafts (WOOKEY 1942; WOOKEY 1948)	Limited lesions of the cervical esophagus	(1) Low operative mortality	(1) Fistulae and stricture are relatively common (2) Nonperistaltic (3) Total esophagectomy not performed
Bakamjian deltopectoral flap (BAKAMJIAN 1965)	Lesions of the cervical esophagus	(1) Low operative mortality	(1) Flap is nonperistaltic (2) Total esophagectomy not performed (3) Multistage reconstruction (4) Stenosis and fistulae are common
Pectoralis major flap (YAMAMOTO et al. 1985; NOZAKI et al. 1985; SURKIN et al. 1984; FABIAN 1984)	Lesions of the cervical esophagus	(1) Low operative mortality (2) One-stage procedure (3) Improved flap viability	(1) Flap is bulky and nonperistaltic (2) Stenosis and fistulae are common (3) Total esophagectomy not performed
Latissimus dorsi myocutaneous island flap (YAMAMOTO et al. 1985)	Lesions of the cervical esophagus	(1) Low operative mortality (2) One-stage procedure	(1) Total esophagectomy not performed (2) Flap is nonperistaltic
Laryngotracheal autograft (SOM 1956)	Limited lesions of the cervical esophagus	(1) Low operative mortality (2) One-stage procedure	(1) Rarely applicable (2) Total esophagectomy not performed (3) Fistulae and stricture are relatively common
Free jejunal graft (MCKEE and PETERS 1978; SOM 1956; JURKIEWICZ 1965; SEIDENBERG et al. 1959; HESTER et al. 1984; NOZAKI et al. 1985)	Lesions of the cervical esophagus	(1) One-stage procedure (2) Peristaltic graft (3) Same caliber as esophagus	(1) Total esophagectomy not performed (2) Relatively high incidence of graft necrosis and/or anastomotic leak (3) May not tolerate postoperative irradiation (4) Multiple anastomoses
Reversed gastric tube (HEIMLICH 1970)	Thoracic esophageal lesions	(1) Good blood supply (2) Long segment of esophagus may be resected (3) One-stage procedure (4) Stomach remains in abdomen (5) One anastomosis	(1) Graft necrosis may result in mediastinitis (2) May not be able to completely bridge large defect without moving the stomach
Gastric transposition (ONG and LEE 1960; LEQUESNE and RANGER 1966; HARRISON 1969)	All esophageal lesions	(1) Total esophagectomy (2) One anastomosis in neck (3) Good blood supply (4) One-stage procedure	(1) Gastric reflux (2) Graft may not tolerate postoperative irradiation (3) Stomach is no longer in normal location
Colon interposition (BRAIN and READING 1966; GOLIGHER and ROBIN 1954)	All esophageal lesions	(1) Good function (2) Stomach in normal position (3) Total esophagectomy may be performed	(1) Multiple anastomoses (2) Blood supply less dependable than for stomach (3) Relatively high operative mortality (4) Graft may not tolerate postoperative irradiation

5.8.3 Chemotherapy

Chemotherapy may be used as an adjunct to surgery and/or irradiation, or in the treatment of disseminated disease (SHIELDS et al. 1984; COOPERATIVE CLINICAL STUDY GROUP FOR ESOPHAGEAL CANCER 1983; SOGA et al. 1983; MARCIAL et al. 1980; KOLARIC et al. 1980; COIA et al. 1984; KEANE et al. 1985; EARLE et al. 1980; ANDERSEN et al. 1984; LEICHMAN et al. 1984; KELSEN et al. 1984; KELSEN 1984). Although it probably increases the acute toxicity of treatment, there is no convincing evidence thus far that adjuvant chemotherapy improves the rate of local-regional disease control or survival in esophageal cancer (KELSEN 1987;SEYDEL et al. 1988). Chemotherapy has also been combined with irradiation, hyperthermia, and surgery, although no long-term results are available at this time (SUGIMACHI et al. 1985).

5.8.4 Miscellaneous

Esophageal dilatation and placement of an esophageal prosthesis may be used for palliation in patients when a course of palliative irradiation is ineffective. Prostheses are not used for cervical esophageal lesions because they may cause tracheal compression, dyspnea, cough, or pressure necrosis of the esophageal wall (BOYCE 1982). A prosthesis may also be used to successfully occlude a tracheoesophageal fistula caused by an esophageal cancer (BOYCE 1982).

The use of the Nd:YAG laser has been reported to palliate the symptoms of locally advanced esophageal carcinoma (FLEISCHER and KESSLER 1983). The treatment is performed endoscopically with a sedative and topical anesthetic. The risk of esophageal perforation and/or tracheoesophageal fistula may be somewhat higher with this treatment than with palliative irradiation. It may be considered, however, in those patients whose disease progresses after a course of radiation therapy and in whom an esophageal prosthesis is not an option.

HAYATA et al. (1985) have reported the use of photodynamic therapy with a hematoporphyrin derivative, alone or combined with irradiation or surgery. The number of patients treated is small and the results are preliminary, but the authors suggest that it may be indicated in the medically inoperable patient with an early-stage lesion. They note that it should be combined with irradiation in all cases.

5.8.5 Summary

Radiation therapy and surgery yield equivalent results for patients with lesions of the cervical esophagus and those in the upper third of the thoracic esophagus. The advantage of radiation therapy is that it is associated with lower rates of acute morbidity and mortality than surgery. The advantage of surgery is that the long-term function of the transposed stomach is probably better than that of an irradiated esophagus, which tends to stenose over time. Therefore, patients with cancers of the cervical esophagus or upper third of the thoracic esophagus may be treated with either modality, depending on the stage of disease, medical condition of the patient, and the preference of the patient and the attending physicians. Radiation therapy and surgery also yield similar results for lesions of the middle third of the thoracic esophagus. Because it is necessary to treat a larger volume of the lung and heart in order to irradiate lesions in this area, and because it is very difficult to treat lesions to high doses and be assured that the spinal cord dose is limited to less than 50 Gy, our philosophy is to treat these patients with surgery alone. For lesions in the lower third of the thoracic esophagus, surgery offers results that are superior to radiation therapy. In addition, in order to radically irradiate lesions in this area, it is necessary to treat a substantial volume of the heart, lung bases, and pericardium. For these reasons, it is our philosophy to use surgery alone for distal-third cancers.

Adjuvant radiation therapy is not used routinely. Planned preoperative irradiation may be used in selected patients who are thought to be good surgical candidates but in whom the lesion is thought to be borderline resectable. We also recommend preoperative irradiation to patients with cervical esophageal cancer who are surgical candidates, as it is our philosophy to use a combination of irradiation and surgery for locally advanced carcinomas arising in other head and neck sites. Planned postoperative irradiation is used selectively in patients with questionable margins or microscopic residual disease following surgery. Adjuvant chemotherapy is regarded as an unproven treatment modality and is not used routinely at this point, outside of a study setting.

Patients who are not candidates for radical treatment because of locally advanced disease or distant metastasis are treated with low-dose, short-course radiation therapy (e.g., 30 Gy in ten fractions over 2 weeks). Those with progressive disease or no response following palliative irradiation are considered for placement of an esophageal prosthesis or gastrostomy tube.

5.9 Radiation Therapy Techniques

5.9.1 Radical Radiation Therapy

5.9.1.1 Cervical Esophagus

Small lesions high in the cervical esophagus are treated with parallel-opposed lateral portals. Larger lesions or those located more distally must be treated with some other technique because of the shoulders. At the University of Florida, a four-field box technique is used in which an anterior, a posterior, and parallel-opposed lateral fields are treated (Fig. 5) (MENDENHALL 1984; MENDENHALL et al. 1982). A beeswax bolus is used on each lateral field as a compensator to fill the gap created by the slope of the shoulders (Fig. 6). Customized Lipowitz's metal blocks are used on all four fields to minimize the volume of normal tissue included in the treatment volume. The fields are designed so as to irradiate the primary lesion with a 5-cm margin distally and at least a 3-cm margin proximally (Fig. 7). The anterior mediastinal, low internal jugular, and medial supraclavicular nodes are included in the primary treatment volume. In addition, a split anterior upper neck field is used to electively irradiate the

Fig. 5 a, b. Isodose curves for four-field box technique using 17 meVp X-ray beam. **a** Section at level of C6 vertebra. **b** Section at level of T2 vertebra

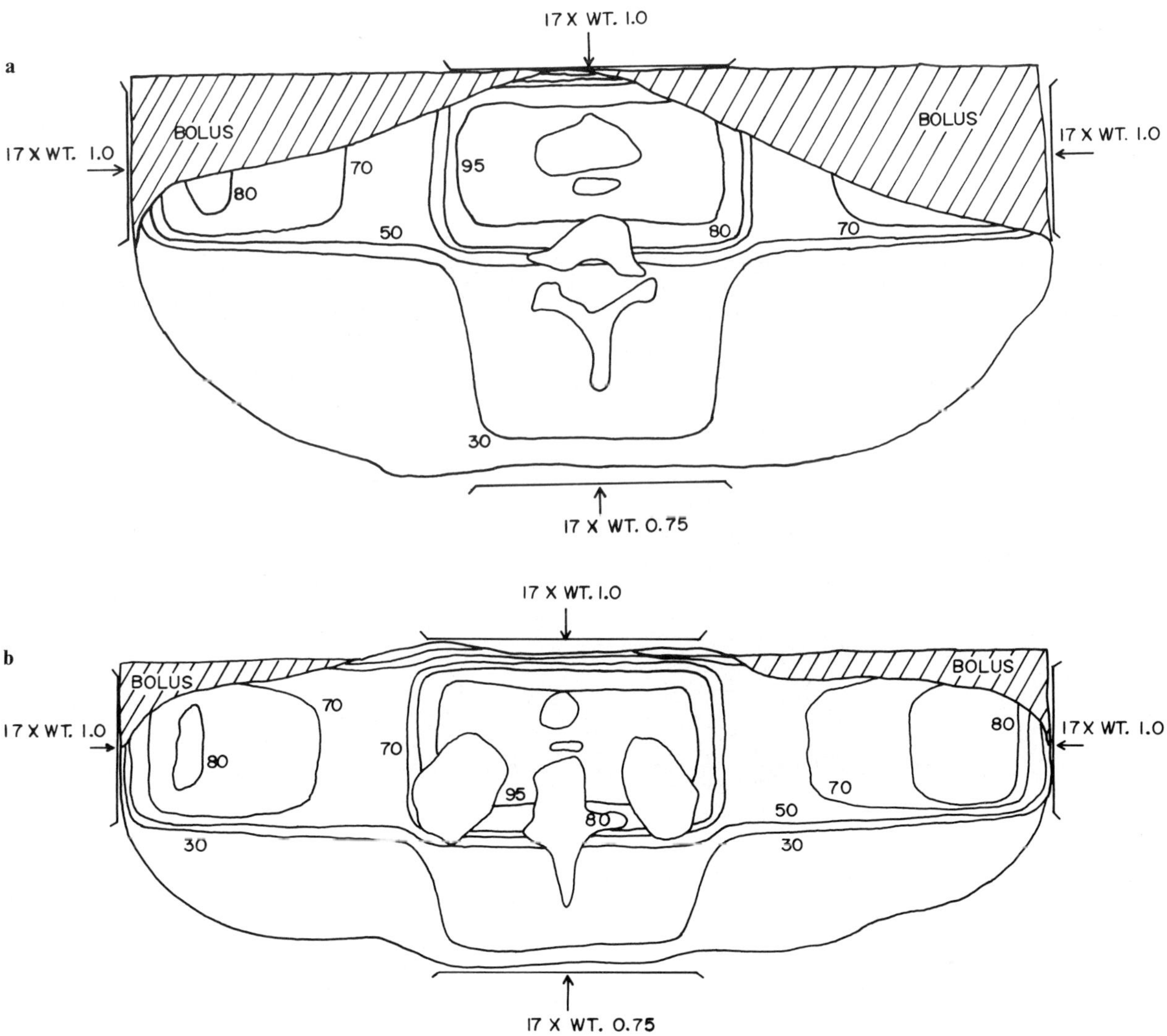

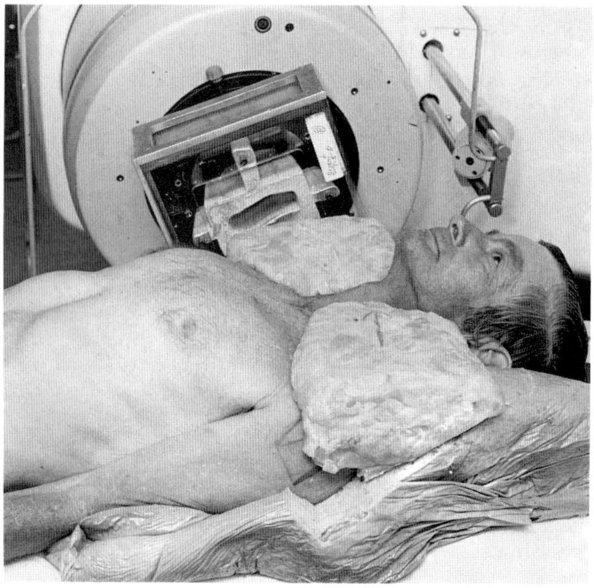

Fig. 6. Patient in position for treatment of right lateral portal. Note beeswax bolus, Lipowitz's metal blocks, and foam body mold. The portal is set up by an isocentric method. The beeswax bolus is placed in position after the portal is aligned. The lateral portal should be checked frequently by imaging films

midjugular and upper internal jugular nodes (Fig. 8). The advantages of the four-field box technique are that the dose to the spinal cord can be accurately calculated, port films are more easily interpreted because the fields are not angled, and field reductions (particularly off the spinal cord) are relatively simple. The disadvantage of the technique is that it is necessary to irradiate the shoulders and lung apices.

Treatment simulation is performed with the patient supine with the head and upper thorax supported in a customized foam mold to ensure accurate reproducibility of the treatment setup. The anterior border of the skin is outlined with lead wire, as are anatomical landmarks, such as the sternal notch, bottom of the cricoid cartilage, cricothyroid membrane, and notch of the thyroid cartilage. Any clinically positive nodes are also outlined by lead wire. The patient is asked to swallow contrast medium to define the extent of the primary lesion while simulation films are obtained, which are used to construct customized Lipowitz's metal blocks. Tattoos are used to mark the field corners and central axis so that, should the lines be inadvertently removed by the patient, the fields can be repro-

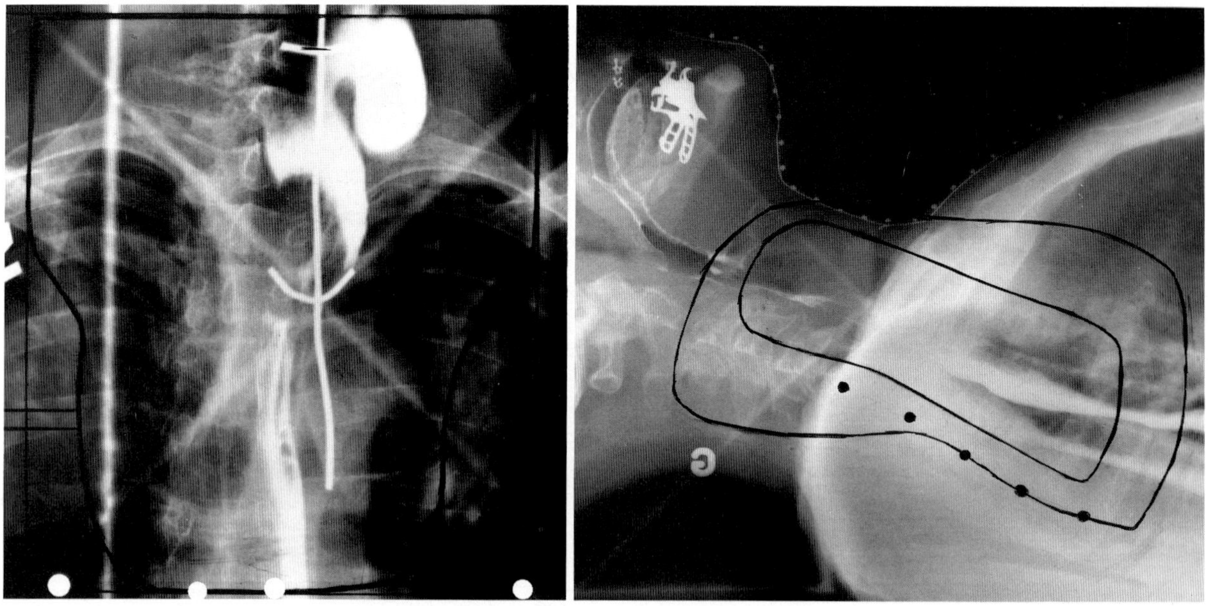

a b

Fig. 7. a Anterior and posterior fields. Note the esophageal diverticulum above the cancer. **b** Lateral and reduced lateral portals for a cancer of the cervical esophagus located just above at the thoracic inlet. A second and final reduction would just cover the gross lesion. The initial lower border is generous because of the inability to evaluate the mediastinal nodes by palpation. The initial lateral portal bows posteriorly on its superior aspect to cover the posterior cervical nodes, and bows anteriorly on its inferior aspect to cover the anterior mediastinal nodes. Note the proximity of the spinal cord to the esophagus. For this patient, it is 2–2.5 cm from the posterior wall of the esophagus to the anterior portion of the cervical spinal cord and about 3 cm to the thoracic spinal cord. *Black dots* mark the posterior aspect of the vertebral bodies. Solder wire with lead shot spaced at 1 cm outlines the midline anterior skin surface

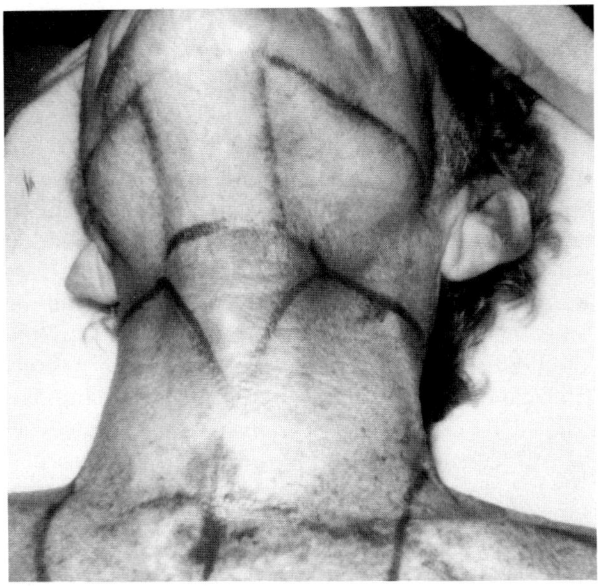

Fig. 8. Split anterior upper-neck field

duced. The fields are arranged isocentrically, and all four fields are treated during each treatment session. It is necessary to obtain frequent imaging films to ensure accurate treatment setup. This is particularly important after reducing the fields to avoid the spinal cord, when it is necessary to obtain daily imaging films of the lateral portals.

The large primary treatment volume is treated at 1.8 Gy per fraction, five fractions per week, to a dose of 45 Gy. At this point, the anterior and posterior fields are discontinued, and the treatment is continued at 2 Gy per fraction with reduced lateral portals that exclude the spinal cord. The final tumor doses are 65–70 Gy for T1 lesions and 75 Gy for T2–T3 lesions. High energy X-rays are preferable for this treatment. We currently use 17 meVp X-rays on the posterior and lateral fields and 8 meVp X-rays on the anterior field. Continuous-course irradiation is recommended over split-course techniques.

If the upper neck is clinically negative, it is electively irradiated using an anterior portal with a midline block to deliver 50 Gy given dose/25 fractions or 40.5 Gy given dose/15 fractions (Mendenhall and Million 1986). If clinically positive nodes are present, they are usually small and may be treated with radiation therapy alone, the dose to depend on the volume of the lymph node (Mendenhall et al. 1984). In the occasional patient with a clinically positive neck node(s) ≥ 3 cm in size, a neck dissection may be added following radiation therapy (Mendenhall et al. 1986).

5.9.1.2 Upper Third of the Thoracic Esophagus

Lesions in the upper third of the thoracic esophagus, near the sternal notch, may be treated with a modified four-field box technique. Treatment is initiated with anterior and posterior fields that include the primary tumor with at least a 5-cm proximal and distal margin, the anterior mediastinal lymph nodes, the low internal jugular nodes, and the medial supraclavicular nodes. This field arrangement is treated at 1.8 Gy per fraction to 45 Gy in 25 fractions over 5 weeks. It is necessary to calculate the spinal cord dose at the thinnest point in the treatment volume, usually the sternal notch, to make sure that the spinal cord dose not exceed 50 Gy. At this point, reduced lateral fields that exclude the spinal cord are used to boost the dose to the lesion to the desired final tumor dose. This may be accomplished with the patient in a supine position with the arms raised over the head, as described by Stevens et al. (1978).

5.9.1.3 Middle and Lower Thirds of the Thoracic Esophagus

Treatment is initiated with anterior and posterior fields that treat the primary lesion with a 5-cm proximal and distal margin, as well as the mediastinal nodes. A dose of 45 Gy in 25 fractions over 5 weeks is delivered, after which the fields are reduced to exclude the spinal cord, and either a rotational field or a three-field technique (anterior field and two posterior oblique fields) is used. It may be possible to complete the treatment with lateral fields if the lesion is small. If this technique is used, it is necessary that the volume of lung base irradiated be kept to a minimum. If a rotational boost technique is used, care should be exercised to correct for the difference between the axis of the esophagus and the axis of the body (Sewchand et al. 1978). The reduced fields are treated at 1.8 Gy per fraction to a final tumor dose of 65 Gy for T1 lesions and 70 Gy for T2–T3 lesions.

5.9.2 Preoperative Irradiation

In the event that preoperative irradiation is indicated, 45 Gy in 25 fractions over 5 weeks is delivered using anterior and posterior portals, followed by surgery in 4–6 weeks. The volume irradiated would depend on the site of the lesion, as previously described.

5.9.3 Postoperative Irradiation

The volume irradiated and treatment technique employed depend on the location of the lesion and site of known residual disease. Treatment is delivered at 1.8 Gy per fraction, five fractions per week, to a total dose of 60-65 Gy for questionable margins and 70 Gy for known microscopic residual disease. Patients with gross residual disease are rarely candidates for high-dose postoperative radiation therapy and are probably best treated with palliative irradiation.

5.9.4 Intracavitary Radiation Therapy

In the event that the lesion is very superficial (e.g., carcinoma in situ), an intracavitary boost may be used after 45-50 Gy of external beam irradiation has been delivered. Dummy sources are placed in a nasogastric tube, and the treatment is simulated to determine how far from the incisors the tube must extend in order to position the radium sources in the tumor volume, as previously defined by endoscopy. Radium tubes are then placed in a nasogastric tube, and the nasogastric tube is positioned as in the simulation, so that the sources are adjacent to the tumor (Fig. 9). SYED et al. (1987) have described an esophageal applicator that may be afterloaded with iridium 192, iodine 125, or micro-cesium 137. Because of the high surface dose and rapid fall-off inherent with intracavitary therapy, this technique should only be used for very superficial lesions (i.e., carcinoma in situ or microinvasive carcinoma).

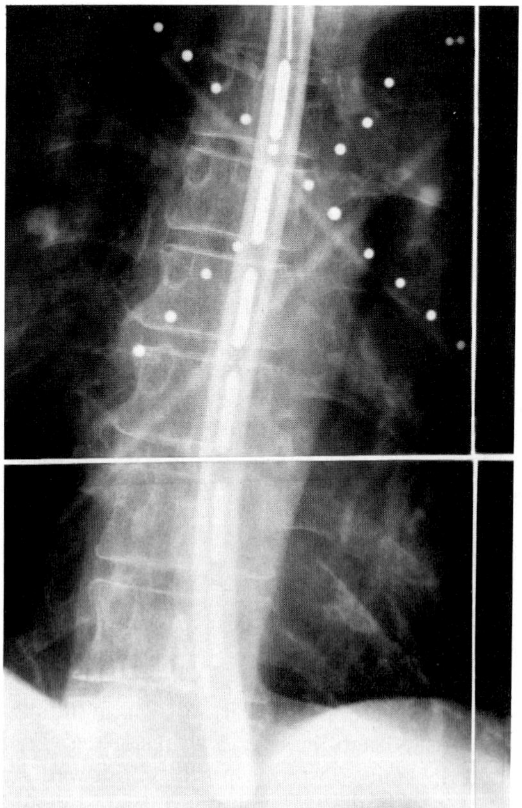

Fig. 9. Intracavitary boost with radium sources for a microinvasive carcinoma of the esophagus

5.9.5 Palliative Irradiation

Anterior and posterior fields are used to treat the lesion to a dose of 30 Gy in ten fractions over 2 weeks. Portals are arranged so as to treat the lesion with a 5-cm proximal and distal margin, as well as the mediastinal nodes. The low internal jugular and medial supraclavicular nodes may also be included for proximal lesions, provided that it does not require a significant increase in the treatment volume to do so.

5.10 Results of Standard Treatment

It is practically impossible to compare the results of radical radiation therapy and surgery, and even the results of a particular treatment modality from one institution to another. The vast majority of authors do not use a staging system or report local control rates. There is selection bias in favor of surgery, in that patients with locally advanced, unresectable lesions or who are medically unfit for surgery are referred for radiation therapy. Because there has been no prospective, randomized trial in which surgery and radiation therapy were compared, selected retrospective, nonrandomized trials will be presented.

The results of surgery are presented in Table 12. In general, approximately 70%-80% of patients taken to surgery are found to have resectable lesions (PARKER et al. 1982; GATZINSKY et al. 1985; WANG and CHIEN 1983; SKINNER 1983; GUOJUN et al. 1981a). The operative mortality varies from one institution to the next, as does the method of reconstruction, but the operative mortality rates have decreased in recent years. EARLAM and CUNHA-MELO (1980a) reviewed the results of surgery in 83783 pa-

Table 12. Results of surgery

Site, series	Minimum follow-up (years)	No. of patients treated	Operative mortality	3-year survival	5-year survival
Cervical					
Mayo Clinic (GUNNLAUGSSON et al. 1970)[a]	5	17	4/17 (24%)	No data	2/17 (12%)[b]
Tokyo (KAKEGAWA et al. 1985)[a]	<1	64	7/64 (11%)	No data	27%[c]
Padua (PERRACHIA et al. 1982)	<1	74	14/74 (19%)	15%[d]	No data
Edinburgh (PEARSON 1981)	5	25	No data	No data	4/25 (16%)
Upper third thoracic					
France (GIULI and GIGNOUX 1980)[a, e]	<1	106	31%	No data	14%[b, d]
Beijing (RIMIN et al. 1981)[e]	5	28	No data	No data	3/28 (10.7%)[b]
Midthoracic					
France (GIULI and GIGNOUX 1980)[a]	<1	926	31%	No data	10%[b, d]
Edinburgh (PEARSON 1981)[f]	5	126	No data	No data	12/126 (10%)
Mayo (GUNNLAUGSSON et al. 1970)[a, f]	5	103	15/103 (15%)	No data	7/103 (7%)[b]
Beijing (RIMIN et al. 1981)	5	327	No data	No data	87/327 (26.6%)[b]
Lower thoracic					
France (GIULI and GIGNOUX 1980)[a]	<1	789	30%	No data	17%[b, d]
Edinburgh (PEARSON 1981)	5	281	No data	No data	32/281 (11%)
Mayo (GUNNLAUGSSON et al. 1970)[a]	5	92	14/92 (15%)	No data	21/92 (23%)[b]
Beijing (RIMIN et al. 1981)	5	220	No data	No data	72/220 (33%)[b]
Overall					
Taiwan (WANG and CHIEN 1983)	5	387	5.1%	No data	46/387 (12%)
Chicago (SKINNER 1983)[a]	<1	80	9/80 (11%)	24%[b, d]	18%[b, d]
Beijing (RIMIN et al. 1981)	5	575	5.6%	No data	162/575 (28%)[b]

[a] Selected patients received adjuvant radiotherapy.
[b] Postoperative deaths included.
[c] No. of patients at risk not stated.
[d] Actuarial survival.
[e] Some cervical lesions included.
[f] Upper half of thoracic esophagus.

tients reported in 122 series between 1960 and 1979. They found average 5-year survival rates of 4% for all patients, 9% for those having an operation, and 12% for those whose lesion was resected. The mean postoperative mortality was 29% for those undergoing resection.

Patients with early-stage lesions have been found to have an excellent prognosis. GUOJUN et al. (1981b) reported a series of 237 patients with carcinoma in situ or invasive cancer confined to the submucosa with negative nodes. The resectability rate in this group of patients was 100%, the incidence of postoperative deaths was 2.5%, and the 5-year survival rate was 86%.

The results of radical irradiation in selected series are listed in Table 13. The large series reported from Edinburgh is of interest in that radiation therapy was the treatment of choice for all squamous cell carcinomas of the esophagus for a significant part of the study period (PEARSON 1981). Those not thought to be candidates for radical treatment were treated with palliative irradiation and were not included in the report.

GU (1984) reported the results of radical irradiation for a series of 79 patients with "early lesions" (not visible on X-ray studies) treated in China; 61/79 (77%) survived 5 years.

The results of planned preoperative irradiation and surgery are shown in Table 14. The first series listed is a randomized trial conducted in Rennes, France (LAUNOIS et al. 1981). Initially, 67 patients were randomized to the combined treatment and 57 patients to surgery alone. Fifteen patients were subsequently excluded from the analysis, five from the irradiated group and ten from the surgery-alone group, because they were thought to have lesions that were not completely resectable (14) or refused surgery (1). The EORTC study is a randomized trial comparing 33 Gy in ten fractions over 12 days followed by surgery with surgery alone (GIGNOUX et al. 1987). There was no difference in postoperative mortality or in the rate of long-term survival with a mean follow-up time of 3.6 years. The third series listed in Table 14, reported by GUOJUN et al. (1981c) from Beijing, was a nonrandomized trial in which the patients selected for preoperative irradia-

Table 13. Results of radical irradiation alone

Site, series	No. of patients	Minimum follow-up (years)	Local control	5-year survival
Cervical				
Univ. of Florida (MENDENHALL et al. 1988)	28	5	8/28 (29%)	4/28 (14%)
China (GU 1984)	31	5	No data	10/31 (32%)
Edinburgh (PEARSON 1981)	76	5	No data	19/76 (25%)
Upper-third thoracic				
Univ. of Wisconsin (BOSCH et al. 1980)[a]	21	5	No data	4/21 (19%)
China (GU 1984)	153	5	No data	24/153 (15.7%)
Midthoracic				
Edinburgh (PEARSON 1981)[b]	109	5	No data	17/109 (16%)
China (GU 1984)	469	5	No data	46/469 (9.8%)
Lower thoracic				
China (GU 1984)	58	5	No data	3/58 (5.2%)
Edinburgh (PEARSON 1981)[c]	103	5	No data	12/103 (12%)
Overall				
Edinburgh (PEARSON 1981)	288	5	50%	48/288 (17%)
Beijing (YIN et al. 1983)[d]	1212	5	No data	82/1212 (6.8%)

[a] May include some cervical lesions.
[b] Upper half of thoracic esophagus.
[c] Lower half of thoracic esophagus.
[d] Locally advanced, inoperable.

Table 14. Planned preoperative irradiation and surgery

Series	Minimum follow-up (years)	No. of patients	Radiation therapy	No. resected	Operative mortality	5-year survival
Rennes (LAUNOIS et al. 1981) (randomized)	Not stated	62	40 Gy	47/62 (76%)	13/47 (28%)	9.5%[a]
		47	None	33/47 (70%)	7/33 (21%)	11.5%[a]
EORTC (GIGNOUX et al. 1987) (randomized)	Not stated	102	33 Gy in 10 fx	75/102 (74%)	25%[e]	10%[a]
		106	None	87/106 (82%)	18%[e]	9%[a]
Beijing (GUOJUN et al. 1981c) (nonrandomized)	Not stated	408	Various	334/408 (82%)	13/334 (3.9%)	67/212 (32%)[b]
			None	736	22/736 (3.0%)	28%[b, c]
Charleston (MARKS et al. 1976)	Not stated	137	45 Gy in 18 fx (3½ weeks)	101/137 (74%)	18/101 (18%)	13.9%[b]
UCSF (FRASER et al. 1978)	5	11	50–60 Gy (6–8 weeks)[d]	8/11 (73%)	1/11 (9%)	4/11 (36%)[b]

fx, fraction
[a] Actuarial survival.
[b] Postoperative deaths included.
[c] No. of patients at risk not stated.
[d] One patient received 29 Gy.
[e] Includes patients explored but not resected.

tion were those with more proximal lesions and/or those in whom the resectability of the cancer was questionable. A wide range of doses was used, with 19% of patients receiving 21–30 Gy and 48% receiving 31–40 Gy. The incidence of anastomotic leaks following surgery was 15/334 (4.5%) for the preoperatively irradiated patients, compared with 27/736 (3.7%) for surgery alone. In addition, the incidence of positive nodes was 34% in the combined group, compared with 44% in the surgery-alone group. There also appeared to be an improved 5-year survival rate in the small group of patients receiving doses > 40 Gy, compared with those receiving lower doses of preoperative irradiation. The authors currently recommend preoperative doses of 40 Gy in 20 fractions/4 weeks and have initiated a randomized trial comparing preoperative radiation therapy and surgery to surgery alone (GUOJUN et al., 1981a, c). In the study reported by MARKS et al. (1976), 332 patients were treated with preoperative irradiation, of whom only 137 (41%) were explored.

5.11 Complications

5.11.1 Surgery

The complications of surgery will depend on the extent of the resection, the method of reconstruction, the general medical condition of the patient, and the experience of the surgeon. In general, the operative mortality has gradually declined over recent years. KINOSHITA et al. (1982) reported a series of 1329 patients who underwent resection of carcinomas of the upper and mid esophagus between 1946 and 1976. The incidence of postoperative deaths declined from 43/581 (7.4%) for 1946–1964 to 33/748 (4.4%) for 1965–1976. Similarly, MAILLET et al. (1982) reviewed 271 patients who were operated on for squamous cell carcinoma of the thoracic esophagus between 1961 and 1978. Although the overall postoperative mortality was 45/271 (16.6%), the subset with operations in 1977–78 had a postoperative mortality of 3/64 (4.7%). The rate of postoperative deaths was found to be highest for patients with lesions in the middle third of the esophagus and for those with more extensive local disease. The incidence of serious postoperative complications and deaths in this series is listed in Table 15. The incidence of postoperative complications and deaths as a function of the method of reconstruction for a series of patients who underwent resection of squamous cell carcinoma of the esophagus between 1952 and 1980 in Taiwan is shown in Table 16 (WANG and CHIEN 1983). The incidence of postoperative complications for a series of 82 patients undergoing reconstructive procedures for pharyngoesophageal lesions between 1972 and 1982 is shown in Table 17 (SURKIN et al. 1984). Reconstructive complications were defined as fistulae, stenosis, and flap necrosis, and operative complications would include, for example, pulmonary or cardiovascular problems.

HESTER et al. (1984), from Emory University, reported a series of 55 patients undergoing pharyngoesophageal reconstruction with a free graft of jejunum (50 patients) or colon (five patients) for carcinoma (41) or benign strictures (14). Of those patients treated for cancer, the rate of graft failure was 5/41 (12%) and the postoperative mortality was 3/41 (7%). Of those patients in whom the graft survived, function was excellent in 90%. GLUCKMAN et al. (1985) reported a similar experience with a series of 52 patients who had reconstruction with jejunal-free grafts. FISHER et al. (1985), from Duke, reported a series of 19 patients who underwent resection for previously untreated head and neck cancers and reconstruction with jejunal-free grafts. Three patients (16%) died postoperatively, seven patients (37%) experienced necrosis of the graft, and in three patients (16%) a stricture developed.

Table 15. Incidence of postoperative complications in 271 patients (MAILLET et al. 1982)

Complication	No. of severe complications	No. of postoperative deaths
Anastomotic leak or fistula	17 (6.2%)	10 (3.7%)
Respiratory	19 (7.0%)	8 (3.0%)
Cardiac	14 (5.2%)	10 (3.7%)
Hepatic	10 (3.7%)	7 (2.6%)
Chylothorax	5 (1.8%)	3 (1.1%)
Thromboembolic	6 (2.2%)	2 (0.7%)
Miscellaneous	–	5 (1.8%)

Table 16. Incidence of postoperative complications and death vs method of reconstruction (636 patients) (WANG and CHIEN 1983)

Reconstruction	Location of graft	No. of patients treated	No. of complications	No. of postoperative deaths
Gastric transposition	Retrosternal	154	49 (32%)	7 (4.5%)
Right colon interposition	Retrosternal	114	41 (36%)	8 (7.0%)
Gastric transposition	Transthoracic	39	3 (7.7%)	1 (2.6%)
Gastrostomy	No graft	329	38 (11.6%)	15 (4.6%)

Table 17. Postoperative complications[a] (modified from SURKIN et al. 1984)

Reconstruction	No. of patients	Postoperative mortality	Complications		Success rate of reconstruction
			Reconstructive	Operative	
Wookey repair	31	3 (10%)	18 (58%)	15 (48%)	21 (68%)
Chest flaps:					
Tubed deltopectoral	12	2 (17%)	8 (67%)	7 (58%)	6 (50%)
Tubed pectoralis major	17	1 (6%)	11 (65%)	12 (71%)	7 (41%)
Viscera:					
Stomach	12	1 (8%)	4 (33%)	6 (50%)	11 (92%)
Colon	10	2 (20%)	7 (70%)	7 (70%)	5 (50%)
Jejunum	2	0	2 (100%)	1 (50%)	1 (50%)

[a] Eighty-two patients with 84 reconstructive procedures.

5.11.2 Radiation Therapy

5.11.2.1 Acute

The main acute side-effect of radiation therapy is esophagitis, which starts after 2 weeks of treatment, reaches a plateau, and resolves 3-4 weeks following completion of radiation therapy. This may be managed symptomatically with acetaminophen elixir, Aspergum, and/or an antacid. Dry desquamation of the skin within the treatment portals is frequently noted at the end of the course of treatment; moist desquamation is unusual. If it is necessary to irradiate the upper neck, the patient may develop xerostomia and loss of taste. The xerostomia, if it develops, is usually minimal to moderate because most of the parotid tissue is not irradiated. Loss of taste generally resolves several months following the completion of treatment.

Radiation pneumonitis, developing 2-3 months following treatment, may occur in a small percentage of patients, particularly if a substantial portion of the lung bases is treated. This may be treated with steroids if it is severe. If it is necessary to place a patient on steroids, one should take care to taper the dose very slowly after the desired response has been obtained, in order to avoid a recurrence of the radiation pneumonitis. We have not seen any cases of symptomatic radiation pneumonitis in patients in whom only the lung apices were treated.

The most catastrophic complication of radiation therapy is tracheoesophageal fistula or hemorrhage, which may occur if the tumor has directly invaded the tracheobronchial tree or aorta. Bronchoscopy may help avoid the former problem by excluding those patients with apparent tumor invasion of the trachea or bronchi. If involvement of the tracheobronchial tree or aorta is superficial and the physician elects to proceed with irradiation, initiating treatment with a relatively low dose per fraction and gradually escalating to 1.8 Gy per fraction may decrease the risk of a serious complication. WARE et al. (1976) reported a series of 129 patients treated between 1950 and 1973 with doses of 50-60 Gy at 1.6-1.8 Gy per fraction. Six patients (4.7%) developed a tracheoesophageal fistula, and two patients (1.6%) died of hemorrhage during the course of radiation therapy. GU (1984) reported that the incidence of esophageal perforation was 3.2% and of fatal hemorrhage, 0.8% in a series of 1334 patients treated with radical radiation therapy in China.

5.11.2.2 Late

Fibrosis of the lung included in the treatment portals is one of the late complications of irradiation. The severity of the complication depends on the volume of lung treated, the dose delivered, and the part of the lung irradiated. Fibrosis of the lung apices, which contribute little to pulmonary function, rarely results in symptoms. However, if a large volume of the lung bases is treated to a high dose, serious late complications may result.

If a large volume of the heart is irradiated, a pericardial effusion may develop, and may be followed by constrictive pericarditis if the dose of irradiation is sufficiently high. Should the latter complication occur, it could necessitate pericardial stripping.

Almost all patients who are successfully treated for esophageal cancer with radical irradiation will have a persistent narrowing of the esophagus seen on esophagograms. If a symptomatic stricture develops, careful bougienage may be performed and may be required periodically. It is imperative that this procedure be performed with the utmost caution, as esophageal perforation may lead to fatal mediastinitis. PEARSON (1971) reported that, of 26

5-year survivors after high-dose irradiation, 23 (88%) were able to subsist on a normal diet, while three patients were on a liquid diet. Of the 23 patients who were on a normal diet, ten patients had required bougienage, five patients on one occasion and five patients on two or more occasions. PEARSON (1971) recommended waiting at least 3 months following irradiation to perform bougienage.

Hypothyroidism may occur if the thyroid gland is included in the irradiated volume.

Another complication of radiation therapy is Lhermitte's syndrome, which may appear several months following treatment; it usually lasts 2-3 months and spontaneously resolves. Lhermitte's syndrome is almost never the harbinger of radiation myelitis. Fortunately, radiation myelitis is quite rare, particularly if the spinal cord dose is limited to 50 Gy or less at 1.8-2 Gy per fraction. If it occurs, however, it may be fatal. GU (1984) reported an incidence of 5/1334 (0.4%) in patients treated with radical irradiation.

5.11.3 Preoperative Irradiation and Surgery

Although one would expect an increase in the incidence of operative complications in patients treated with planned preoperative irradiation, this has not been observed. LAUNOIS et al. (1981) reported a randomized series of 124 patients treated with preoperative irradiation (40 Gy in 8-12 days) followed by surgery, or with surgery alone. They found no significant difference in the postoperative mortality or in the incidence of anastomotic leaks or other serious postoperative complications.

GUOJUN et al. (1981c) reported a nonrandomized series of 334 patients treated with preoperative irradiation and surgery and 736 patients treated with surgery alone in Beijing, China. The dose of preoperative irradiation varied widely, with 48% receiving 31-40 Gy and 19% receiving 21-30 Gy. The 30-day operative mortality was 13/334 (3.9%) for the combined-treatment group, compared with 22/736 (3.0%) for the surgery-alone group. The incidence of anastomotic leaks was 15/334 (4.5%) for those preoperatively irradiated and 27/736 (3.7%) for those treated with surgery alone.

5.12 Clinical Trials

Despite improvements in perioperative care and radiation therapy equipment, there has been little change in the dismal survival rates for esophageal cancer over the last 20 years for either treatment modality (EARLAM and CUNHA-MELO 1980a, b). Those series reporting improved cure rates with various treatment protocols probably reflects selection of patient population rather than a real improvement resulting from the treatment technique utilized. The only promising accomplishment that has occurred has been the development of effective screening programs for high-risk groups so that patients may be diagnosed while the disease is in a very early stage and, therefore, much more amenable to cure. However, for the vast majority of patients diagnosed with esophageal cancer in the United States and Europe, the disease is locally advanced at presentation, and the prognosis is bleak. It is, therefore, not surprising that there is considerable interest in the possibility that chemotherapy, combined with irradiation and/or surgery, may offer an improvement in the results of radical treatment.

The overall response rates (complete plus partial) to single-agent chemotherapy as compiled in a review article by KELSEN (1984) are listed in Table 18. The author commented that the dose schedules used for mitomycin-C in the three series cited were probably prohibitively toxic. In the same article, 11 series using various combination chemotherapy regimens for 256 patients were reviewed. The mean response rate for the 11 series was 43% (with a range of 15% for cisplatin and bleomycin to 80% for cisplatin and 5-fluorouracil) with a median response duration that ranged from 4 to 8 months. The two largest series of patients yielded response rates of 15% in 61 patients treated with cisplatin and bleomycin (with a median response duration of 6 months) and 53% in 68 patients treated with cisplatin, vindesine, and bleomycin (with a median response duration of 7 months). In the author's opin-

Table 18. Response to single-agent chemotherapy (modified from KELSEN 1984)

Chemotherapeutic agent	No. of series	No. of patients	Response rate (CR + PR)
Bleomycin	7	80	12/80 (15%)
Mitomycin-C	3	58	15/58 (26%)
Adriamycin	2	33	6/33 (18%)
5-Fluorouracil	1	26	4/26 (15%)
Methotrexate	1	26	3/26 (12%)
CCNU	1	19	3/19 (16%)
Cisplatin	4	73	16/73 (22%)
Vindesine	4	83	28/83 (34%)
VP-16	2	30	2/30 (7%)

ion, there is no combination chemotherapy regimen that is clearly superior to other regimens tested.

EARLE et al. (1980) reported an Eastern Cooperative Oncology Group randomized study in which 77 evaluable patients with squamous cell carcinoma of the esophagus were given radiation therapy alone (37 patients) versus irradiation plus bleomycin (40 patients) between 1974 and 1979. The radiation dose was 50–60 Gy at 2 Gy per fraction over 5–6 weeks. Bleomycin, 3 mg, was given IV on days 1 and 2; if this was tolerated, the patient received 15 mg IV per day to a total dose of 210 mg. Treatment was fairly well tolerated, although three patients in the combined group developed symptomatic pulmonary fibrosis. With a minimum 1-year follow-up, there was no difference in the observed survival rates or the projected 5-year survival rates, which were ≤8% for both groups.

ANDERSEN et al. (1984) reported the results of a randomized Cooperative Scandinavian Trial conducted from 1977 to 1981. One hundred twenty-four evaluable patients with T1-2 N0 M0 lesions that were potentially resectable and located below the T5 vertebra were assigned to receive either preoperative irradiation (35 Gy in 4 weeks) plus surgery, or preoperative irradiation (30 Gy in 4 weeks) plus bleomycin (20 5-mg doses in 4 weeks) plus surgery. Patients with lesions found to be unresectable were given additional irradiation (28 Gy in 3 weeks) in the control arm or irradiation (25 Gy in 3 weeks) and maintenance bleomycin for 6 months. The authors found no difference in the median or 2-year survival rates between the two treatment arms. A second group of 82 evaluable patients with T1-T2 N0 M0 lesions that were either above the T5 vertebra or located below T5 in a medically inoperable patient were asigned in random fashion to receive split-course irradiation alone (35 Gy in 4 weeks, a 3-week rest, then 28 Gy in 3 weeks) or irradiation and bleomycin (30 Gy in 4 weeks plus bleomycin, 15 5-mg doses, then maintenance bleomycin for 6 months). There was also no difference in the median or 2-year survival rates between these two treatment groups.

KEANE et al. (1985), from the Princess Margaret Hospital in Toronto, reported a series of 35 patients with locally advanced squamous cell carcinomas of the esophagus treated between 1980 and 1983 with irradiation and combination chemotherapy, compared with a series of matched historical controls (70 patients) treated with radiation therapy alone. Radiation therapy consisted of either 45–50 Gy in 20 fractions over 4 weeks, or two courses of 22.5–25 Gy in ten fractions over 2 weeks, separated by a 4-week rest period. The chemotherapy consisted of mitomycin-C, 10 mg/m^2/24 h on day 1, and 5-fluorouracil continuous infusion, 1 g/m^2/24 h on days 1–4. Of the 35 patients entered on the protocol, three did not complete the planned treatment course, and one died at 5 months of radiation pneumonitis. There was a significant improvement in the 2-year actuarial local control and survival rates for the patients given combined treatment, compared with the matched historical controls treated with radiation therapy alone.

LEICHMAN et al. (1984), from Wayne State University, reported a series of 30 patients with potentially curable squamous cell carcinoma of the esophagus who were treated with a planned combination of chemotherapy, preoperative irradiation, and surgery between 1977 and 1979. Treatment consisted of concomitant preoperative irradiation (30 Gy in 15 fractions in 3 weeks) and chemotherapy (mitomycin-C, 15 mg/m^2 on day 1, and 5-fluorouracil, 1 g/m^2/24 h, continuous infusion, on days 1–4 and 29–32), followed by surgery. If no tumor was found in the pathologic specimen, no further treatment was given. If residual tumor was present, postoperative irradiation (20 Gy in ten fractions) was administered. Twenty-three of 30 patients (77%) underwent resection; no tumor was found in 6/23 specimens (26%). The in-hospital mortality was 7/23 (30%). Of the 16 patients who underwent resection and survived, 0/11 with residual tumor in the specimen survived 5 years disease-free, compared with 4/5 with no residual tumor. The 5-year survival rate was 4/30 (13%) for the entire group and 4/23 (17%) for those taken to surgery.

KELSEN et al. (1984), from Memorial Hospital in New York, reported a series of 34 patients with esophageal cancer treated with preoperative chemotherapy (cisplatin and bleomycin) followed by resection between 1976 and 1979. If the primary lesion proved to be stage T3 or if there were positive nodes, the patient received a second course of chemotherapy and postoperative irradiation (32 Gy at 4 Gy per fraction, two fractions per week). The observed operative mortality was 11%, and 9% of the patients survived for 3 or more years. In a second study, 34 patients received two courses of chemotherapy (cisplatin, vindesine, and bleomycin), followed by surgery. Postoperative irradiation (55 Gy in 5–6 weeks) was delivered if the lesion exhibited extraesophageal spread or if positive nodes were found. The mortality secondary to surgery was 5.6%, and the overall treatment-related mortality was 9%. With a minimum 2-year follow-up, 8/34 patients (24%) were alive and free of disease.

RESBEUT et al. (1985) reported a series of 28 patients with localized, unresectable esophageal cancer treated with combination chemotherapy, consisting of two cycles of vincristine, methotrexate with folic acid rescue, and cisplatin, followed by high-dose, split-course radiation therapy. They found that 24/28 patients (86%) were able to complete the planned course of treatment, and 3/28 (11%) were alive at 2 years following treatment.

In summary, although some of the results obtained with the combination of adjuvant chemotherapy with irradiation and/or surgery are promising, there is no solid evidence to support the routine use of chemotherapy in the radical management of esophageal cancer.

References

Akiyama H, Tsurumaru M, Kawamura T, Ono Y (1981) Principles of surgical treatment for carcinoma of the esophagus: analysis of lymph node involvement. Ann Surg 194: 438-445

Andersen AP, Berdal P, Edsmyr F, Hagen S, Hatlevoll R, Nygaard K, Ottosen P, et al. (1984) Irradiation, chemotherapy and surgery in esophageal cancer: a randomized clinical study. Radiother Oncol 2: 179-188

Anderson LL, Lad TE (1982) Autopsy findings in squamous cell carcinoma of the esophagus. Cancer 50: 1587-1590

Bakamjian VY (1965) A two-stage method for pharyngoesophageal reconstruction with a primary pectoral skin flap. Plast Reconstr Surg 36: 173-184

Barkley HT, Hussey DH, Saxton JP, Spanos WJ (1981) Radiotherapy in the treatment of carcinoma of the esophagus. In: Strohlein JR, Romsdahl MM (eds) Gastrointestinal cancer. Raven, New York, pp 171-187

Beahrs OH, Myers MH (eds) (1983) Manual for staging of cancer, 2nd edn. Lippincott, Philadelphia

Bosch A, Frias Z, Pellett JR (1980) Carcinoma of the esophagus: twenty-five years' experience at the University of Wisconsin hospitals. Wis Med J 79: 23-26

Boyce HW Jr (1982) Medical management of esophageal obstruction and esophageal-pulmonary fistula. Cancer 50: 2597-2600

Brain RHF, Reading PV (1966) Colon transplantation into the pharynx and cervical esophagus. Br J Surg 53: 933-942

Burdette WJ, Jesse R (1972) Carcinoma of the cervical esophagus. J Thorac Cardiovasc Surg 63: 41-53

Castro JR, Chen GT, Pitluck S, Cartigny A, Phillips TL, Saunders WM, Collier JM, et al. (1983) Helium charged-particle radiotherapy of locally advanced carcinoma of the esophagus, stomach and biliary tract. Am J Clin Oncol 6: 629-637

Choi TK, Siu KF, Lam KH, Wong J (1984) Bronchoscopy and carcinoma of the esophagus. I. Findings of bronchoscopy in carcinoma of the esophagus. Am J Surg 147: 757-759

Coia LR, Engstrom PF, Paul A, Gallagher MJ, Stoll D, Catalano R, Richter MP (1984) A pilot study of combined radiotherapy and chemotherapy for esophageal carcinoma. Am J Clin Oncol 7: 653-659

Cooperative Clinical Study Group for Esophageal Cancer (1983) Multidisciplinary treatment for esophageal carcinoma. Jpn J Clin Oncol 13: 417-424

Dowlatshahi K, Skinner DB, DeMeester TR, Zachary L, Bibbo M, Wied G (1985) Evaluation of brush cytology as an independent technique for detection of esophageal carcinoma. J Thorac Cardiovasc Surg 89: 848-851

Earlam R, Cunha-Melo JR (1980a) Oesophageal squamous cell carcinoma. I. A critical review of surgery. Br J Surg 67: 381-390

Earlam R, Cunha-Melo JR (1980b) Oesophageal squamous cell carcinoma. II. A critical review of radiotherapy. Br J Surg 67: 457-461

Earle JD, Gelber RD, Moertel CG, Hahn RG (1980) A controlled evaluation of combined radiation and bleomycin therapy for squamous cell carcinoma of the esophagus. Int J Radiat Oncol Biol Phys 6: 821-826

Fabian RL (1984) Reconstruction of the laryngopharynx and cervical esophagus. Laryngoscope 94: 1334-1350

Fisher SR, Cole TB, Meyers WC, Seigler HF (1985) Pharyngoesophageal reconstruction using free jejunal interposition grafts. Arch Otolaryngol 111(11): 747-752

Fleischer D, Kessler F (1983) Endoscopic Nd:YAG laser therapy for carcinoma of the esophagus: a new form of palliative treatment. Gastroenterology 85: 600-606

Fraser RW, Wara WM, Thomas AN, Mauch PM, Fishman NH, Galante M, Phillips TL, Buschke F (1978) Combined treatment methods for carcinoma of the esophagus. Radiology 128: 461-465

Gatzinsky P, Berglin E, Dernevik L, Larsson I, William-Olsson G (1985) Resectional operations and long-term results in carcinoma of the esophagus. J Thorac Cardiovasc Surg 89: 71-76

Gignoux M, Roussel A, Paillot B, Gillet M, Schlag P, Favre J-P, Dalesio O, et al. (1987) The value of preoperative radiotherapy in esophageal cancer: the results of a study of the E.O.R.T.C. World J Surg 11(4): 426-432

Giuli R, Gignoux M (1980) Treatment of carcinoma of the esophagus: Retrospective study of 2,400 patients. Ann Surg 192: 44-52

Gluckman JL, McDonough JJ, McCafferty GJ, Black RJ, Coman WB, Cooney TC, Bird RJ, Robinson DW (1985) Complications associated with free jejunal graft reconstruction of the pharyngoesophagus: a multiinstitutinal experience with 52 cases. Head Neck Surg 7: 200-205

Goldstein HM, Zornoza J (1978) Association of squamous cell carcinoma of the head and neck with cancer of the esophagus. AJR 131(5): 791-794

Goligher JC, Robin IG (1954) Use of left colon for reconstruction of pharynx and oesophagus after pharyngectomy. Br J Surg 42: 283-290

Gu XZ (1984) Radiotherapy for carcinoma of the esophagus. In: Huang GJ, K'ai WY (eds) Carcinoma of the esophagus and gastric cardia. Springer, Berlin Heidelberg New York, pp 258-274

Gunnlaugsson GH, Wychulis AR, Roland C, Ellis FH (1970) Analysis of the records of 1,657 patients with carcinoma of the esophagus and cardia of the stomach. Surg Gynecol Obstet 130: 997-1005

Guojun H, Dawei Z, Guoqing W, Hua L, Liangjun W, Jiasui L, Guiyu G, Xingjiang W (1981a) Surgical treatment of carcinoma of the esophagus: Report of 1647 cases. Chin Med J 94: 305-307

Guojun H, Lingfang S, Dawei Z, Zhangcai L, Guoqing W, Shuxian L, Fubao C (1981b) Diagnosis and surgical treatment of early esophageal cancer. Chin Med J 94: 229-232

Guojun H, Xianzhi G, Rugang Z, Lijun Z, Dawei Z, Yanjun M, Liangjun W, et al. (1981c) Combined preoperative irradiation and surgery in esophageal carcinoma: report of 408 cases. Chin Med J 94: 73-76

Harper PS, Harper RM, Howel-Evans AW (1970) Carcinoma of the oesophagus with tylosis. Q J Med 39: 317-333

Harris OD, Cooke WT, Thompson H, Waterhouse JAH (1967) Malignancy in adult coeliac disease and idiopathic steatorrhoea. Am J Med 42: 899-912

Harrison DFN (1969) Surgical management of cancer of the hypopharynx and cervical esophagus. Br J Surg 56: 95-103

Hayata Y, Kato H, Okitsu H, Kawaguchi M, Konaka C (1985) Photodynamic therapy with hematoporphyrin derivative in cancer of the upper gastrointestinal tract. Semin Surg Oncol 1: 1-11

Heimlich HJ (1970) Carcinoma of the cervical esophagus. J Thorac Cardiovasc Surg 59: 309-318

Hester TR, McConnel F, Nahai F, Cunningham SJ, Jurkiewicz MJ (1984) Pharyngoesophageal stricture and fistula: treatment by free jejunal graft. Trans South Surg Assoc 95: 272-278

Jurkiewicz MJ (1965) Vascularized intestinal graft for reconstruction of the cervical esophagus and pharynx. Plast Reconstr Surg 36: 509-517

Kakegawa T, Yamana H, Ando N (1985) Analysis of surgical treatment for carcinoma situated in the cervical esophagus. Surgery 97(2): 150-157

Keane TJ, Harwood AR, Elhakim T, Rider WD, Cummings BJ, Ginsberg RJ, Cooper JC (1985) Radical radiation therapy with 5-fluorouracil infusion and mitomycin-C for oesophageal squamous carcinoma. Radiother Oncol 4: 205-210

Kelsen D (1984) Chemotherapy of esophageal cancer. Semin Oncol 11(2): 159-168

Kelsen D (1987) Multimodality therapy of esophageal carcinoma: still an experimental approach (Editorial). J Clin Oncol 5(4): 530-531

Kelsen D, Bains M, Hilaris B, Martini N (1984) Combined modality therapy of esophageal cancer. Semin Oncol 11(2): 169-177

Kinoshita Y, Endo M, Nakayama K, Sato H, Sato H (1982) Clinical evaluation of ten-year survival cases after operation for upper and mid-thoracic esophageal cancer. Int Surg 67: 153-161

Kolaric K, Marivic Z, Roth A, Dujmovic I (1980) Combination of bleomycin and adriamycin with and without radiation on the treatment of inoperable esophageal cancer. A randomized study. Cancer 45: 2265-2273

Laramore GE, Davis RB, Olson MH, Cohen L, Raghaven V, Griffin TW, Rogers CC, et al. (1983) RTOG Phase I study on fast neutron teletherapy for squamous cell carcinoma of the esophagus. Int J Radiat Oncol Biol Phys 9: 465-473

Launois B, Delarue D, Campion JP, Kerbaol M (1981) Preoperative radiotherapy for carcinoma of the esophagus. Surg Gynecol Obstet 153: 690-692

Leichman L, Steiger Z, Seydel HG, Vaitkevicius VK (1984) Combined preoperative chemotherapy and radiation therapy for cancer of the esophagus: the Wayne State University, Southwest Oncology Group and Radiation Therapy Oncology Group experience. Semin Oncol 11(2): 178-185

LeQuesne LP, Ranger D (1966) Pharyngolaryngectomy, with immediate pharyngogastric anastomosis. Br J Surg 53: 105-109

Lightdale CJ, Winawer SJ (1984) Screening, diagnosis and staging of esophageal cancer. Semin Oncol 11(2): 101-112

MacComb WS, Healey JE Jr, McGraw JP, Fletcher GH, Gallager HS, Paulus DD (1967) Hypopharynx and cervical esophagus. In: MacComb WS, Fletcher GH (eds) Cancer of the head and neck. Williams and Wilkins, Baltimore, pp 213-240

Maillet P, Baulieux J, Boulez J, Benhaim R (1982) Carcinoma of the thoracic esophagus. Results of one-stage surgery (271 cases). Am J Surg 143: 629-634

Marcial VA, Velez-Garcia E, Cintron J, Ydrach AA (1980) Radiotherapy preceded by multidrug chemotherapy in carcinoma of the esophagus: a pilot study of the Radiation Therapy Oncology Group. Cancer Clin Trials 3: 127-130

Marks RD, Scruggs HJ, Wallace KM (1976) Preoperative radiation therapy for carcinoma of the esophagus. Cancer 38: 84-89

McKee DM, Peters CR (1978) Reconstruction of the hypopharynx and cervical esophagus with microvascular jejunal transplant. Clin Plast Surg 5(2): 305-312

Mendenhall WM (1984) Carcinoma of the cervical esophagus. In: Million RR, Cassisi NJ (eds) Management of head and neck cancer. Lippincott, Philadelphia, pp 393-406

Mendenhall WM, Million RR (1986) Elective neck irradiation for squamous cell carcinoma of the head and neck: analysis of time-dose factors and causes of failure. Int J Radiat Oncol Biol Phys 12(5): 741-746

Mendenhall WM, Million RR, Bova FJ (1982) Carcinoma of the cervical esophagus treated with radiation therapy using a four-field box technique. Int J Radiat Oncol Biol Phys 8(8): 1435-1439

Mendenhall WM, Million RR, Bova FJ (1984) Analysis of time-dose factors in clinically positive neck nodes treated with irradiation alone in squamous cell carcinoma of the head and neck. Int J Radiat Oncol Biol Phys 10(5): 639-643

Mendenhall WM, Million RR, Cassisi NJ (1986) Squamous cell carcinoma of the head and neck treated with radiation therapy: the role of neck dissection for clinically positive neck nodes. Int J Radiat Oncol Biol Phys 12(5): 733-740

Mendenhall WM, Parsons JT, Vogel SB, Cassisi NJ, Million RR (1988) Carcinoma of the cervical esophagus treated with radiation therapy. Laryngoscope 98: 769-771

Nabeya K (1983) Markers of cancer risk in the esophagus and surveillance of high-risk group. In: Sherlock P, Morson BC, Barbara L, Veronesi U (eds) Precancerous lesions of the gastrointestinal tract. Raven, New York, pp 71-86

Newaishy GA, Read GA, Duncan W, Kerr GR (1982) Results of radical radiotherapy of squamous cell carcinoma of the oesophagus. Clin Radiol 33: 347-352

Nozaki M, Huang TT, Hayashi M, Endo M, Hirayama T (1985) Reconstruction of the pharyngoesophagus following pharyngoesophagectomy and irradiation therapy. Plast Reconstr Surg 76(3): 386-392

Ong GB, Lee TC (1960) Pharyngogastric anastomosis after oesophagopharyngectomy for carcinoma of the hypopharynx and cervical esophagus. Br J Surg 48: 193-200

Parker EF, Gregorie HB, Prioleau WH, Marks RD, Bartles DM (1982) Carcinoma of the esophagus. Observations of 40 years. Ann Surg 195: 618-623

Pearson JG (1971) The value of radiotherapy in the management of squamous oesophageal cancer. Br J Surg 58(10): 794-798

Pearson JG (1981) Radiotherapy for esophageal carcinoma. World J Surg 5: 489-497

Peracchia A, Ancona E, Buin F (1982) The surgical treatment of cancer of the cervical esophagus: complications and preliminary results. Int Surg 67: 135-137

Resbeut M, Le Prise-Fleury E, Ben-Hassel M, Goudier MJ, Morice-Rouxel MF, Douillard JY, Chenal C (1985) Squamous cell carcinoma of the esophagus. Treatment by combined vincristine-methotrexate plus folinic acid rescue and cisplatin before radiotherapy. Cancer 56(6): 1246-1250

Rimin L, Yunkan L, Hongyi C, Yuzhi G, Jun Y, Jiaqi H (1981) Late results of surgical treatment in esophageal carcinoma and factors influencing prognosis. Chin Med J 94(11): 729-732

Sarma DP (1982) Oat cell carcinoma of the esophagus. J Surg Oncol 19: 145-150

Schwade JG, Kinsella TJ, Kelly B, Rowland J, Johnston M, Glatstein E (1984) Clinical experience with intravenous misonidazole for carcinoma of the esophagus: results in attempting radiosensitization of each fraction of exposure. Cancer Invest 2: 91-95

Seidenberg B, Rosenak SS, Hurwitt ES, Som ML (1959) Immediate reconstruction of the cervical esophagus by a revascularized isolated jejunal segment. Ann Surg 149: 162-171

Sewchand W, Jones TK, Khan FM, Levitt SH (1978) Spinal cord protection during cross-fire irradiation of the intrathoracic esophagus: tube-tilt vs. shielding. Radiology 126: 239-242

Seydel HG, Leichman L, Byhardt R, Cooper J, Herskovic A, Libnock J, Pazdur R, et al. (1988) Preoperative radiation and chemotherapy for localized squamous cell carcinoma of the esophagus: a RTOG Study. Int J Radiat Oncol Biol Phys 14(1): 33-35

Shields TW, Rosen ST, Hellerstein SM, Tsang T, Ujiki GT, Kies MS (1984) Multimodality approach to treatment of carcinoma of the esophagus. Arch Surg 119(5): 558-562

Shottenfeld D (1984) Epidemiology of cancer of the esophagus. Semin Oncol 11(2): 92-100

Shu YJ (1983) Cytopathology of the esophagus: an overview of esophageal cytopathology in China. Acta Cytol (Baltimore) 27(1): 7-16

Skinner DB (1983) En bloc resection for neoplasms of the esophagus and cardia. J Thorac Cardiovasc Surg 85: 59-71

Soga J, Fujimaki M, Tanaka O, Sasaki K, Kawaguchi M, Muto T (1983) Analysis of preoperative combined bleomycin and radiation therapy for esophageal carcinoma. World J Surg 7(2): 230-235

Som ML (1956) Laryngoesophagectomy: Primary closure with laryngotracheal autograft. Arch Otolaryngol 63: 474-480

Sons HU, Borchard F (1984) Esophageal cancer: autopsy findings in 171 cases. Arch Pathol Lab Med 108: 983-988

Stevens KR Jr, Fry R, Stone C (1978) A new technique for irradiating thoracic inlet tumors. Int J Radiat Oncol Biol Phys 4(7-8): 731-734

Sugimachi K, Inokuchi K, Kuwano H, Kai H, Okamura T, Okudaira Y (1983) Patterns of recurrence after curative resection for carcinoma of the thoracic part of the esophagus. Surg Gynecol Obstet 157(6): 537-540

Sugimachi K, Kai H, Inokuchi K (1985) Symposium on treatment of cancer. 5. Preoperative hyperthermo-chemo-radiotherapy of esophageal carcinoma. Analysis of 20 cases. Jpn J Med 24: 80-83

Surkin MI, Lawson W, Biller HF (1984) Analysis of the methods of pharyngoesophageal reconstruction. Head Neck Surg 6(5): 953-970

Syed AMN, Puthawala AA, Severance SR, Zamost BJ (1987) Intraluminal irradiation in the treatment of esophageal cancer. Endocuriether Hyperthermia Oncol 3: 105-113

Thompson WM (1983) Esophageal cancer. Int J Radiat Oncol Biol Phys 9(10): 1533-1565

Torek F (1913) The first successful resection of the thoracic portion of the esophagus for carcinoma; preliminary report. JAMA 60: 1533-1534

Wang P, Chien K (1983) Surgical treatment of carcinoma of the esophagus and cardia among the Chinese. Ann Thorac Surg 35(2): 143-151

Wara WM, Mauch PM, Thomas AN, Phillips TL (1976) Palliation for carcinoma of the esophagus. Radiology 121(3): 717-720

Wookey H (1942) Surgical treatment of carcinoma of the pharynx and upper esophagus. Surg Gynecol Obstet 75: 499-506

Wookey H (1948) Surgical treatment of carcinoma of hypopharynx and oesophagus. Br J Surg 35: 249-266

Wynder EL, Bross IJ (1961) A study of etiological factors in cancer of the esophagus. Cancer 14: 389-413

Yamamoto K, Yokota K, Higaki K (1985) Entire pharyngoesophageal reconstruction with latissimus dorsi myocutaneous island flap. Head Neck Surg 7: 461-464

Yang CS (1980) Research on esophageal cancer in China: A review. Cancer Res 40: 2633-2644

Yanjin M, Guangyi L, Xianzhi G, Wenheng C (1981) Detection and natural progression of early oesophageal carcinoma: preliminary communication. J R Soc Med 74: 884-886

Yin WB, Zhang L, Miao Y, Yu Z, Zhang Z, Zhueng C, Wang M, et al. (1983) The results of high-energy electron therapy in carcinoma of the oesophagus compared with telecobalt therapy. Clin Radiol 34(1): 113-116

6 Gastric Cancer

MICHAEL J. KATIN, VINCENT J. BELCASTRO, and DANIEL E. DOSORETZ

CONTENTS

6.1 Anatomy

Initially appearing as a dilatation of the digestive tube by the 4th week of embryonic life, the stomach begins to take on its more familiar position by the 10th week. The ventral mesogastrium at that point contributes to forming the lesser omentum, and the dorsal mesogastrium, the greater omentum. Deriving its name from the Greek, *"stomachos,"* meaning "opening," the stomach is the first intra-abdominal portion of the digestive tract and lies in the left upper quadrant, extending medially to the midline but with a caudal margin sometimes below the level of the umbilicus. The proximal and distal ends of the stomach are fixed by the cardia, 1 to 2 cm below the diaphragmatic hiatus, and by the duodenum, which becomes retroperitoneal just distal to the pylorus.

The stomach is usually described as being divided into four portions: the cardia, the fundus, the body, and the pyloric antrum.

MICHAEL J. KATIN, M.D., DANIEL E. DOSORETZ, M.D., Radiation Therapy Regional Center, Radiation Therapy Associates, 3680 Broadway, Fort Myers, FL 33901, USA

VINCENT J. BELCASTRO, M.D., Belcastro and Carrasquillo, Surgical Associates, Cape Coral Hospital, 708 Del Prado Blvd., Cape Coral, FL 33990, USA

The medial (right) border of the stomach is termed the lesser curvature, and the lateral (left) border, the greater curvature. The anterior and posterior surfaces of the stomach have vascular supply and lymphatic drainage related to the nearer of the curvatures, although there is considerable crossover. The stomach is covered by peritoneum with the exception of the gastrophrenic peritoneal reflection, a small space posterior to the cardia.

The arterial blood supply is from the celiac trunk, which gives rise to the left gastric artery, supplying the cardia and then turning caudally to supply the lesser curvature. The splenic artery arises from the celiac trunk and is partially embedded in the pancreas before reaching the splenic hilum. The short gastric artery and the left gastroepiploic artery arise from the splenic artery to supply the greater curvature. Finally, the common hepatic artery arises from the celiac trunk, then branches into the proper hepatic artery and the gastroduodenal artery. The former gives off the right gastric artery, supplying the pyloric area and anastomosing with the left gastric artery, the latter divides into the right gastroepiploic and anterior superior pancreatiduodenal arteries. The right gastroepiploic artery supplies the inferior portion of the greater curvature and anastomoses with the left gastroepiploic branches.

The venous drainage of the stomach corresponds to the arterial system. The left gastric vein arises from the lesser curvature, following a course with the left gastric artery to reach the portal vein. The right gastric vein also reaches the portal vein, having been formed from venous drainage from the pyloric area. Right and left gastroepiploic veins arise from the areas supplied by the corresponding arteries; the right vein ends in the superior mesenteric vein after having joined the gastrocolic vein, while the left vein ends in the splenic vein. Short gastric veins drain the fundus and the upper portion of the greater curvature and drain into the splenic vein. The lymphatic drainage follows a similar course: drainage from the lesser curvature forms a confluence in left gastric nodes, and in conjunction with paracardial nodes drainage the cardia, drains into the celiac nodes. Right gastric nodes drain the pyloric area, and hepatic nodes receive drainage from these along the course of the common hepatic artery. Pancreaticolienal nodes drain the distribution of the short gastric and gastroepiploic nodes, which then drain to the pyloric nodes and thereby to the hepatic nodes.

A cross-section of the stomach wall reveals four layers. From internal to external, these are the mucosal surface, the submucosa, the muscular layer (circular and longitudinal), and the serosa.

6.2 Background and Natural History

The rich venous and lymphatic drainage of the stomach, which has extensive interconnections within each system, explains the relatively high likelihood of local/regional and distant metastases with even relatively small malignancies. In addition, the fact that there can be spread along the submucosal and subserosal lymphatics without gross evidence of involvement can lead to inappropriate optimism at the time of surgery when negative margins have been achieved.

There is a dramatic worsening in prognosis in patients who are found to have any evidence of locally advanced carcinoma, i.e., involvement of lymph nodes other than those immediately adjacent, and invasion beyond the submucosa. For example, in the Memorial Sloan-Kettering Cancer Center (MSKCC) series (Shiu et al. 1980) only 6 of 157 patients with N2, N3, or M1 disease survived 5 years postoperatively compared with 23 of 56 with N0 or N1 disease, and in the University of Virginia series (Carter et al. 1984) there was 100% survival among patients with disease limited to the mucosa and submucosa, 42% survival among those with disease extending to the muscularis, and 0% and 12% survival, respectively, among those in whom disease went to the serosa and perigastric tissue. At the University of Minnesota, 107 patients with what had been thought to be curative initial surgery underwent a "second look" laparotomy and, of the 105 who were evaluable, 82 were found to have recurrent carcinoma and another four had been diagnosed prior to actually having had the second look procedure (Gunderson and Sosin 1982). Twenty-four had local/regional failure alone, and 72 had this as a component of failure. Only five had distant failure alone. These findings (Table 1) provide further impetus to pursue treatment in addition to surgery in patients with locally advanced gastric adenocarcinoma.

6.3 Epidemiology and Risk Factors

Malignant disease involving the stomach represents a major problem in many parts of the world. The highest national death rate for this disease is in Costa Rica, where the age-adjusted death rate is

Table 1. Extent of initial disease vs overall patterns of failure (Gunderson and Sosin 1982)

Extent of initial disease (LN and stomach lesion)	Total no. pts.	No. with failure	Patterns of failure					
			LF-RF		PS		DM	
			Alone	Component	Alone	Component	Alone	Component
Lymph nodes –								
a) Within wall (B_1)	6	4	1	4	0	3	0	0
b) Through wall (B_1 and B_3)	4	4	1	3	1	3	0	0
c) Unknown (B_1 vs B_2)	2	2	2	2	0	0	0	0
Lymph nodes +								
a) within wall (C_1)	20	12	1 (5%)	10 (50%)	1 (5%)	7 (35%)	1 (5%)	5 (25%)
b) Through wall (C_2 and C_3)	49	40 (43)[b]	14 (28.6%)	35 (71.4%)	1 (2.0%)	20 (40.8%)	2 (4.1%)	12 (24.5%)
c) Unknown (C_1 vs C_2)	20	16 (17)[b]	2 (10%)	14 (70%)	0	11 (55%)	2 (10%)	6 (30%)
Lymph nodes?								
a) Within wall (B_1 vs C_1)	1	1	1	1	0	0	0	0
b) Unknown	3	3	2	3	0	0	0	1
Totals	105[a]	82 (86)[b]	24	72	3	44	5	24

LN, lymph node; LF, local failure; RF, regional failure; PS, peritoneal seeding; DM, distant metastases
Percentages are of total group at risk.

[a] Of 107 patients in total group, two had unknown status, leaving 105 evaluable on failure basis.

[b] Numbers in parentheses include four patients in whom diagnosis of failure was made with nonoperative information, leaving 82 totally evaluable.

63.2 per 100000 for males and 27.8 for females (1982–1983), with Japan second at 58.8/27.5 (SILVERBERG and LUBERA 1989). The rate varies widely from country to counry; for example, it is 6.1 and 3.3, respectively, for males and females in the Dominican Republic, 26.6 and 14.2 for the Federal Republic of Germany, and 21.5 and 9.4 for England and Wales. In the United States, as in several of the Western industrialized nations, there has been a marked decrease in the death rate from this disease. In the United States, cancer of the stomach was the second leading cause of death from cancer in the 1930s; even in 1952–1954, the death rate per 100000 was 16.4 for males and 8.6 for women, whereas stomach cancer is now the seventh leading cause of death from cancer (death rate 7.8/3.7 for 1982–1983) and fourth among gastrointestinal malignancies. The estimated number of new cases in the United States for 1989 is 20000 (11900 males, 8100 females) (SILVERBERG and LUBERA 1989). Within individual nations, the incidence can vary from one area to the next, emphasizing the implication of environmental causation. The disease generally is more common in males than females (2:1 in the United States) and in lower socioeconomic groups (HAENSZEL and CORREA 1975).

Unfortunately, the lowering of the death rate from this disease in many countries does not represent a triumph of therapy but rather a decrease in incidence. The responsible changes in environment and diet have not been clearly defined. These factors may be more critical than genetic disposition: the mortality from stomach cancer among males of Japanese ancestry born in California is one-fifth of that among native Japanese males; for females the figure is one-third (KAWAI et al. 1980). The factors involved in the development of cancer of the stomach are suspected to be mainly dietary. Nitrates and nitrites are among the substances under investigation, with an intriguing relationship found in Colombia, in which inhabitants of certain communities have a much higher risk correlating to the nitrate content of the drinking water (CREASEY 1985). Other substances that have been associated with a higher risk of gastric carcinoma include asbestos, lead, and polycyclic hydrocarbons (CREASEY 1985). Vitamin C and vitamin A deficiencies have also been linked to higher incidences of carcinoma of the stomach (CREASEY 1985). It is possible that prevention of conversion of certain substances to nitrosamines in the stomach, or breakdown of nitrosamines (e.g., by vitamin C), reduces the risk of stomach cancer. This might also explain the increased risk in persons with atrophic gastritis, pernicious anemia, or partial gastrectomy, since acid inhibits nitrosation. Ironically, it may be that the addition of certain preservatives in foods, acting as antioxidants, as well as the decrease in preservation of foods by salting or smoking, has reduced the overall production of nitrosamines in vivo and been at least partially responsible for the drop in incidence in gastric carcinoma in the United States.

Persons with blood group A have been noted to be at increased risk for gastric cancer but these people are also at greater risk for pernicious anemia, which may be a more specific contributing factor (MACGREGOR 1974). A review of patients with gastric resections for benign disease demonstrated a 6.5% risk of developing cancer, with the relative risk being 2.46 in patients undergoing surgery before age 45 and only 0.65 in those undergoing surgery after age 45, which would suggest a long latent period (GIARELLI et al. 1983). Gastric polyps may also predispose to cancer. In one study malignant changes were noted in 17.0% of adenomatous polyps but in 0% of hyperplastic polyps; in polyps 2 cm or more in diameter, the incidence of malignant changes was 43%–59% (SUEHIRO et al. 1986).

Gastric cancer has also been found to be increased in atomic bomb survivors, the risk being four times higher among those exposed to more than 1 Gy than among those who received 0 Gy (SUEHIRO et al. 1986). Interestingly, this group had a higher incidence of females than males and an older age than the control population, and also a higher risk of second malignancies.

6.4 Diagnostic Workup and Staging

6.4.1 Diagnostic Workup

Symptoms of gastric carcinoma commonly include weight loss, abdominal pain, and vomiting. Early satiety is a worrisome but infrequently reported symptom. Physical examination rarely produces findings in early stage disease, although palpation of distant metastases can determine the presence of advanced cancer; relevant findings would include a Blumer's shelf, a Virchow's (left supraclavicular) node, an Irish's (left axillary) node, or a Sister Joseph's nodule (metastasis to the umbilicus). Metastases to the ovaries (Krukenberg's tumor) may yield positive findings on pelvic examination. Stool may be positive for occult blood. Laboratory studies are usually not helpful in diagnosing early disease, although the carcinoembryonic antigen (CEA) level may be elevated, a low hemoglobin on CBC may indicate blood loss, and elevated liver enzymes may indicate metastases to the liver.

X-ray evaluation is the most common nonoperative procedure for evaluation of the stomach. Unfortunately, some of the rules established in the past for differentiating between benign and malignant ulcerations have proved inadequate. Although benign ulcerations tend to have mucosal folds radiating toward the crater, tend to be located on the lesser curvature, and tend to have a Hampton's line (a smooth line of barium corresponding to the mucosal edge), these features are not absolute proof that the ulcerations in question are indeed benign.

The appearance of the stomach in terms of distensibility as seen on fluoroscopy can also give information. Patients with a linitis plastica (scirrhous) presentation of gastric malignancy will have a stomach that is rigid on examination, giving the appearance of a leather water bottle. This is caused by extensive submucosal involvement by carcinoma.

With the widespread availability of fiberoptic endoscopy, the effectiveness of radiographic evaluation can be estimated. The false-positive rate for gastric carcinoma is approximately 5%, but the false-negative rate can be as high as 25% (SEGAL et al. 1975).

Diagnostic accuracy with endoscopy is as high as nearly 100% for exophytic tumors, but only 80% for infiltrative lesions (TATSUTA et al. 1982). Attempts have been made to stain differentially the areas of the stomach at the time of endoscopy, e.g., by the use of Congo red dye, which turns blue-black in acid-secreting areas but not in atrophic, nonsecreting areas, and the use of methylene blue to stain areas of metaplasia. The use of hematoporphyrin derivatives has also been suggested to accentuate areas of abnormality, and potentially they have a role in therapy as well (AIZAWA et al. 1987).

The addition of brushing for cytological material can increase the chance of diagnosing carcinoma to 96% from approximately 85% with biopsy alone (WINAWER et al. 1976).

There can be difficulty in diagnosing gastric lymphoma by radiographic or endoscopic techniques. For example, widening of gastric folds can sometimes give the appearance of gastritis or hypertrophy rather than submucosal involvement by lymphoma. Stiffening of the gastric wall can resemble linitis plastica or scirrhous carcinoma.

There seems to be no evidence that benign ulcers can transform into carcinoma although in biopsy of an ulcerating carcinoma care must be taken that a representative tissue sample is obtained so that a false-negative reading does not prevent the diagnosis.

Table 2. Summary of diagnosis studies

Diagnostic studies	Category evaluable		
	T	N	M
Upper gastrointestinal series	×		
Chest radiographs			×
Computerized axial tomography	×	×	×
Endoscopy			
Biopsy	×		
Cytology (brushing)	×		
Laparoscopy	×	×	×
Bone scan			×
CBC	×		
CEA	×		
Chemistry profile			×

Computerized body section radiography has been investigated to try to help identify cases of locally advanced gastric carcinoma and thus avoid surgery in cases that will be found to be too advanced for resection. However, there has still been inadequate correlation for total accuracy. In one study, of 37 patients evaluated preoperatively, 19 were found to have more extensive disease than predicted, and three of six thought to have widespread disease instead had disease restricted to the local/regional area (COOK et al. 1986). It is possible that magnetic resonance imaging, particularly with the improvement in surface coils, will be able to provide more detailed information preoperatively to allow patients with occult advanced disease to be spared a laparotomy.

Laparoscopy has proven of value in assessing serosal infiltration, fixation of the primary tumor, intraperitoneal spread, and metastases to the liver, with the latter two parameters being assessed with 89.47% and 96.5% efficiency in a large study from Brazil (POSSIK et al. 1986).

A summary of diagnostic studies is shown in Table 2.

6.4.2 Staging

Staging systems previously in use have essentially been superseded by the TNM classification as approved by the International Union Against Cancer and the American Joint Committee on Cancer (AMERICAN JOINT COMMITTEE ON CANCER 1983) (Tables 3, 4).

Table 3. TNM staging

Tis	Mucosa
T1	Mucosa and submucosa
T2	To serosa
T3	Through serosa
T4A	Through serosa and extending to adjacent structures
T4B	Involvement of liver, diaphragm, pancreas, abdominal wall, retroperitoneum, small bowel, or via serosa to duodenum
N1	Involvement of perigastric nodes within 3 cm of tumor
N2	Involvement of perigastric nodes more than 3 cm from tumor
N3	Involvement of other intra-abdominal nodes and retroperitoneal or mesenteric nodes
M1	Distant metastasis

Table 4. Stage groups

0	Tis	N0	M0
I	T1	N0	M0
II	T2	N0	M0
	T3	N0	M0
III	T4	N0	M0
	T1-4	N1	M0
	T1-4	N1	M0
	T1-4	N2	M0
IV	T4	N0-3	M0
	T1-3	N3	M0
	T1-4	N0-3	M1

6.5 Pathological Classification and Significance in Management and Prognosis

Adenocarcinoma is by far the most common malignancy of the stomach. In most studies, the histological type represents 95% or more of all cases. Grossly, four presentations may occur: polypoid, ulcerative, infiltrative (superficial spreading), and infiltrating (scirrhous, linitis plastica). Multicentric carcinomas have been found in up to 22% of stomachs examined microscopically, but grossly in only approximately 2.2% (Collins and Gall 1952). The percentage of lesions arising in the proximal stomach, i.e., the cardia and fundus, is about 34% versus 46% in the distal stomach (antrum and body) (Cady and Choe 1980).

The second most common histological type is lymphoma. Gastric lymphomas can be subtyped but with changing criteria for classification, older reviews are difficult to interpret. A study from Indiana University indicated that 18 of 22 cases collected over 20 years showed immunoperoxidase staining indicative of B cell origin (Seo et al. 1982). Of 1394 patients with Non-Hodgkin's lymphoma seen at the Princess Margaret Hospital between 1967 and 1978, 150 had gastrointestinal involvement and 113 of these had involvement limited to the gastrointestinal tract (Gospodarowicz et al. 1983). Although the histology by site was not specified, the majority of these cases were classified as diffuse histiocytic lymphoma. Of 26 patients with primary gastric lymphoma seen at the Massachusetts General Hospital (MGH) between 1963 and 1980, 19 were classified as having histiocytic lymphoma; of the remaining patients, one had well differentiated lymphocytic lymphoma, three poorly differentiated lymphocytic lymphoma, one undifferentiated non-Burkitt, and two mixed lymphocytic/histiocytic lymphoma (Shimm et al. 1983). Mention should be made of the entity of pseudolymphoma of the stomach. This can be a difficult differential diagnosis pathologically, but although resection is usually necessary for diagnosis, it is also curative. A total of 116 cases in the English literature were reported as of 1981, and another series of 15 was recorded in 1987 (Mattingly et al. 1987). It is noteworthy that 16 cases have been reported of adenocarcinoma of the stomach developing following treatment for gastric lymphoma (Baron et al. 1987).

Benign leiomyomas may occur in the stomach, and it is estimated that, in the gastrointestinal tract, the ratio of benign to malignant lesions is 2:1 (Lee 1983). However, many leiomyomas are asymptomatic, and of those diagnosed because of symptoms, the ratio was reversed. It is thought that malignant degeneration of leiomyomas is rare. Sarcomas make up only 1%-4% of malignant gastric tumors (Lee 1983). Most are leiomyosarcomas, although mixed histologies occur (22 epithelioid leiomysarcomas and 20 spindle cell and pleomorphic sarcomas were reported by the Armed Forces Institute of Pathology). Sarcomas can spread by metastasis to lymph nodes, by the hematogenous route, and by intraperitoneal spread. Prior irradiation of the stomach has been linked to the subsequent development of sarcomas, although the small number of cases brings the causality into question. At the University of Chicago, of 2049 patients who had received radiation therapy to reduce acid secretion for peptic ulcer disease, three were later found to have sarcomas arising within the treatment fields: two had gastric leiomyosarcomas and one, malignant fibrous histiocytoma of the chest wall (Lieber et al. 1985). A gastric sarcoma was reported in a 14-year-old who had previously received radiation therapy for neuroblastoma (Schneider et al. 1986).

Neurofibromatosis has been reported to involve the stomach, although rarely. Such involvement can give rise to symptoms of intestinal pain and dyspepsia, and also cause upper gastrointestinal bleeding. In addition, a case has been reported of degeneration to a malignant schwannoma (PETERSEN and FERGUSON 1984).

Squamous cell carcinoma of the stomach is a rare entity, with fewer than 100 cases having been reported in the world's literature. It is not, of course, to be confused with squamous cell carcinoma of the esophagus extending distally through the cardia. Although the small number of cases makes analysis difficult, it appears that the lesion tends to be slow growing and exophytic and has a higher potential for cure than adenocarcinoma. The etiology is not established. A case has been described following luetic involvement of the stomach, and two cases were reported following treatment of myeloma and lupus with cyclophosphamide (VAUGHAN et al. 1977; MACLOUGHLIN et al. 1980). Gastric carcinoid tumors have been well described and have been identified to be associated with atrophic gastritis (MENDELSOHN et al. 1987).

Involvement of the stomach by granular cell myoblastoma has been described; eight cases were found in the records of the Armed Forces Institute of Pathology, and 16 other cases had been previously reported (JOHNSTON and HELWIG 1981). No record of recurrence or metastasis was noted, indicating a benign behavior.

The stomach can also be a site of involvement by Kaposi's sarcoma; approximately one-third of patients with acquired immunodeficiency syndrome have Kaposi's sarcoma, and gastrointestinal involvement may be present prior to cutaneous lesions. In one series, nine men were found to have this combination; of these, one had lesions in the stomach as the only location, and two others had stomach involvement in addition to other gastrointestinal sites (WALL et al. 1984).

Metastases of other malignancies to the stomach have been recognized, and the possibility always should be kept in mind that a lesion in the stomach may have arisen elsewhere. It is estimated that metastases to the stomach are found in 2% of autopsies in patients with multiple metastases (ADAMS et al. 1983). The most common sources are carcinoma of the breast, lung, pancreas, thyroid, prostate, liver, and melanoma. Even metastases from sarcomas have been described.

6.6 Prognostic Factors Influencing Choice of Treatment

As discussed previously, a significant number of patients will be found to have locally advanced disease or metastatic disease at the time of surgery. Carcinoma may be extensively present throughout the stomach, through the serosa to surrounding tissues, involving the adjacent nodes, or contaminating the peritoneal cavity, in addition to being able to spread hematogenously by way of the portal circulation. In a review of cases approached primarily by surgery at MSKCC between 1955 and 1975, SHIU et al. (1980) investigated the relationship between tumor size and extent and the positivity of lymph nodes and metastases. It was noteworthy that in 23% of the cases there was invasion of an adjacent organ by the primary tumor. A study at the University of Virginia retrospectively compared preoperative findings in patients with what proved to be early gastric carcinoma or locally advanced gastric carcinoma and found that there was no significant difference between the two groups in terms of symptoms, endoscopic evaluation, or radiographic findings (CARTER et al. 1984). A positive correlation has been found between the depth of penetration of the gastric wall and the chance of regional lymph node involvement. Survival has been inversely related to depth of gastric wall involvement (Fig. 1a, b).

6.7 General Management, Results, and Clinical Trials

In view of the potential for cure with surgery and the limitations of other modalities, surgical exploration remains the procedure of choice in any patient with gastric adenocarcinoma without obvious evidence of distant metastatic disease. Surgery is also the standard treatment for patients with localized gastric lymphoma. No studies have been performed on any sizeable number of patients to compare nonoperative techniques to surgery.

The best treatment of gastric adenocarcinomas involving the cardia seems to be an aggressive approach by esophagogastrectomy. A Lewis approach (combined abdominal-right thoracic approach) has been shown to produce a much higher cure rate than palliative resection it involves resection of the distal esophagus, proximal stomach, spleen, and celiac lymph nodes. A report from the Veterans Administration Wadsworth Hospital Center noted significantly better survival among patients treated in

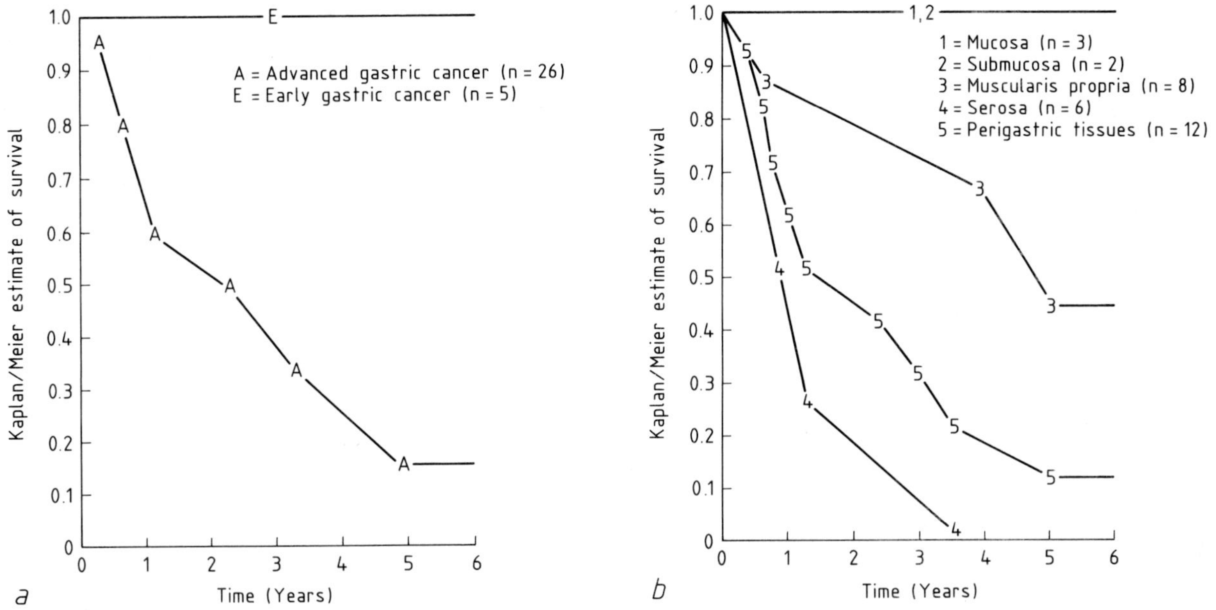

Fig. 1. **a** Survival of early gastric cancer vs advanced gastric cancer patients. **b** Survival by depth of gastric invasion. (CARTER et al. 1984)

this manner (mean survival 2.8 years) than among those treated with a palliative surgical procedure alone (10.7 months), although this series also included patients with squamous cell carcinoma of the esophagus invading the cardia (STONE et al. 1977). Another series, from the M.D. Anderson Hospital and Tumor Institute, dealing with 219 patients with adenocarcinoma, showed that 151 had been treated with curative total gastrectomies or esophagogastrectomies, 45 had undergone palliative resections, and 21 had been subjected to only exploration or bypass procedures (BODDIE et al. 1983). Survival in the first group was significantly better than in the group treated with palliative surgery, although the choice of surgical procedure was probably determined by the findings at surgery. It is to be recalled that operative mortality dropped significantly in the period 1970–1981 compared to 1941–1969 and thus allowed more aggressive procedures to be done with reasonable survival.

Several procedures are available for resection of carcinomas of the stomach itself. High distal subtotal gastrectomy involves transecting the stomach near the cardia, and removing the stomach with lymphatics of the hepatoduodenal ligament, both omenta, and the first portion of the duodenum. Total gastrectomy is the removal of the stomach, the abdominal portion of the esophagus, the first portion of the duodenum, the spleen, the lesser and greater omenta, and dissection of lymph nodes along the left gastric and hepatic arteries. For an extended radical total gastrectomy, the celiac lymph nodes are more completely dissected, the body and tail of the pancreas are removed, and the splenic and left gastric arteries are ligated (Fig. 2).

Multiple studies have addressed the question of the necessity of extended surgical procedures in carcinoma of the body of the stomach. One approach is to evaluate the results in patients proven to have carcinoma limited to the mucosa and submucosa. Analysis of patients from the University of Virginia showed 5 of 31 patients to fall into this category. Interestingly, subtotal gastrectomy was performed in four of these five patients and total gastrectomy in the fifth patient; all were alive and

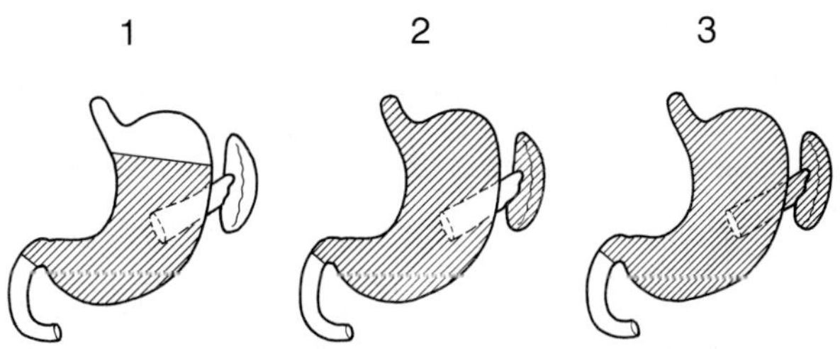

Fig. 2. The three types of gastric resection commonly performed for carcinoma of the midstomach: *1*, radical high distal subtotal gastrectomy; *2*, radical total gastrectomy, with splenectomy; *3*, extended total gastrectomy, with en bloc distal pancreatectomy and splenectomy (SHIU et al. 1980)

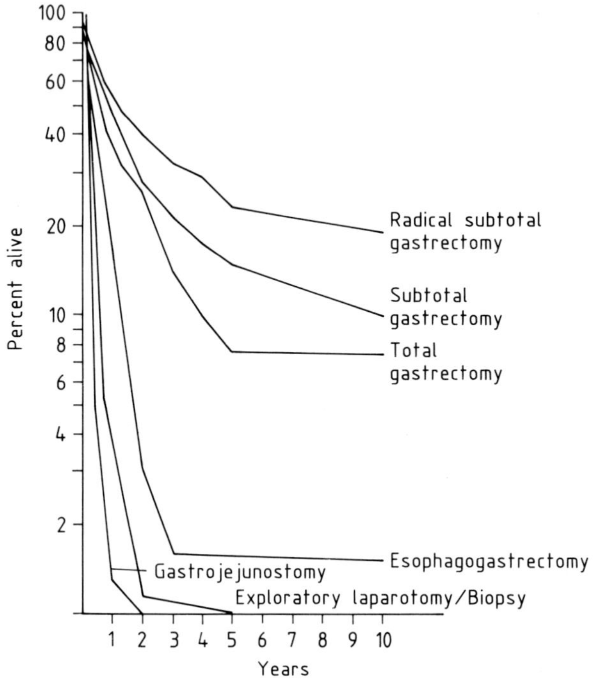

Fig. 3. Survival, computed by the life-table method, after various procedures for adenocarcinoma of the stomach (DUPONT et al. 1978)

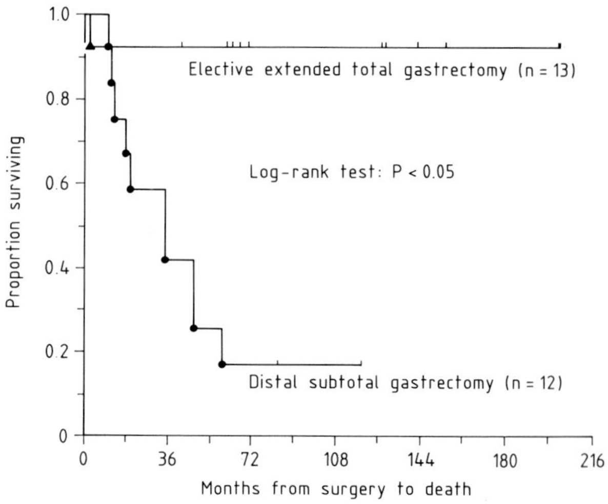

Fig. 4. Survival curves of patients treated by elective extended total gastrectomy and high distal subtotal gastrectomy: early TNM stage tumors only (SHIU et al. 1980)

free of disease at 5 years, although one patient had a suture line recurrence and was cured by near-total gastrectomy (CARTER et al. 1984). Two of the five had involvement of regional lymph nodes. In a large series reported from Charity Hospital the highest survival rate resulted from radical subtotal gastrectomy (22.1%) as opposed to subtotal gastrectomy (14.4%) and total gastrectomy (7.6%) (DUPONT et al. 1978). However, it needs to be noted that operative mortality for these three groups was 16.4%, 23.7%, and 37.2% respectively (Fig. 3). The conclusion of an analysis from MSKCC was that the highest cure rate was found in patients undergoing extended total gastrectomy (19%) as compared with total gastrectomy (16%) and high distal subtotal gastrectomy (10%) (SHIU et al. 1980). The behavior of early stage tumors was analyzed separately, and it is very interesting to note that in patients with T1–4 N0 and N1 tumors survival was markedly better when extended total gastrectomy was performed, as against distal subtotal gastrectomy (93% vs 17%) (Fig. 4).

Palliation as a result of surgery has also been evaluated. At the Medical College of Wisconsin, patients were assessed for relief of symptoms, and the conclusion was that a curative resection produced a longer interval of palliation of symptoms than did palliative resection or bypass (47.6 months vs 14.6 months vs 5.9 months) as well as increasing the survival at 3 years (38% vs 7% vs 0%) (EKBOM and GLEYSTEEN 1980). Similarly, a recent review of cases from the Birmingham Cancer Registry confirmed that palliative resection results in better length of survival than bypass, intubation, or observation (HALLISSEY et al. 1988).

The use of radiation therapy postoperatively has been done with a variety of techniques and equipment and therefore is somewhat difficult to assess. Several reports indicate that a benefit accrues from postoperative irradiation, usually in the palliative setting. For example, a paper from TAKAHASHI reported 9- to 10-month longer survival in patients who received irradiation with cobalt 60 following palliative gastric resection than in those who underwent surgery alone (TAKAHASHI 1964).

Moertel et al. reported in 1969 that administration of 45 Gy over 4–5 weeks in conjunction with 5-fluorouracil (45 mg/kg in three divided doses at the start of radiation therapy) produced an increase in average survival in patients with unresectable gastric carcinoma compared with treatment with radiation therapy alone, which provided no advantage in survival compared with supportive care (MOERTEL et al. 1969). Perhaps as a result of this study, as well as the fact that patients with gastric carcinoma have both distant and local failure, most adjuvant programs have relied on radiation therapy in conjunction with chemotherapy, or chemotherapy alone.

Most studies of postoperative therapy have combined patients who have undergone curative resection, palliative resection (because of either microscopic residual or gross residual disease), or nonresection, thus making it somewhat difficult to assess results. Nonetheless, a number of reports have analyzed adjuvant treatment given because the natural history of the disease indicates a high chance following an apparently curative prodecure. In 1979, a series from Groote Shuur Hospital described prospective randomization of patients with gastric carcinoma who had T1–3, N1–2 disease into control and treatment groups, with the treatment group receiving 20 Gy with cobalt 60 plus 5-fluorouracil, 500 mg for 4 days at the start of radiation therapy and then five further courses at 28-day intervals; patients with T4 and M1 disease were randomized to control, 20 Gy plus 5-fluorouracil as described above, or thiotepa alone (DENT et al. 1979). Analysis at 140 weeks did not reveal any significant differences between control and treatment groups, although having undergone curative rather than noncurative surgery was a determining factor regarding prognosis regardless of postoperative treatment.

In 1982, the Gastrointestinal Tumor Study Group (GITSG) reported a study that involved randomization of patients who had undergone curative resection between observation and chemotherapy with methyl-CCNU and 5-fluorouracil (GASTROINTESTINAL TUMOR STUDY GROUP 1982a). This study showed a significant survival advantage for the treated group (median survival of >48 months in the treated group vs 33 months in the control group). However, a Veterans Administration Surgical Oncology Group report from 1983 did not find an improvement in survival in patients treated with these drugs over controls, regardless of whether surgery had been thought to be curative or incomplete, or whether the patient was found to have nonresectable disease (HIGGINS et al. 1983). At 3.5 years, survival was 38.9% for the control group vs 37.8% for the treated group among those patients who had undergone curative surgery; at 1.5 years, it was 31.0% vs 29.0% among those who had undergone incomplete surgical resection; and at 1.0 years, it was 14.7% versus 13.9% among those who had not been able to have resection.

In 1984 MOERTEL reported the results of a randomized comparison of postoperative treatment with 5-fluorouracil and radiation therapy (37.5 Gy in 24 fractions) versus observation in patients who had undergone curative resection but who were considered to have a poor prognosis, i.e., patients with involvement of the cardia, regional node metastases, or invasion of adjacent structures (MOERTEL et al. 1984). Patients randomized to treatment had a statistically significantly better 5-year survival than those in the control group (23% vs 4%). However, it was noted that a sizeable number of patients in the treatment group had declined to receive treatment after randomization was made, and, in fact, when they were separated from the group actually receiving treatment, the 5-year survival rates were 20% for those receiving treatment and 30% for those randomized to the group but declining treatment. No specific reasons for the excellent results in this latter group were encountered.

Another attempt to administer aggressive postoperative treatment to patients who had a high expectation of recurrence was described by GUNDERSON et al. from MGH in 1983 (see also LINGOS et al. 1985). In a nonrandomized study, patients considered to have had curative procedures received postoperative chemotherapy 2–3 weeks following surgery; radiation therapy was then administered to a dose of 45–52 Gy, and 3–4 weeks afterward, maintenance chemotherapy was started. Initially the chemotherapy utilized was 5-fluorouracil and BCNU but this was later changed to FAM (5-fluorouracil, Adriamycin, and mitomycin C). Results were analyzed based on the patients' operative status. Those patients with disease resected but considered at high risk had a median survival time to progression of 22.2 months and a median survival time of 24 months. This study also included patients who had not undergone curative resection. Those with microscopic residual disease had a median time to progression of 18.6 months and a median survival time of 24 months; for those with gross residual disease the Figures were 8.4 and 15 months respectively, and for those with unresectable disease, 9.1 and 14 months respectively. It is noteworthy that local failure occurred in only 14% of those with curative resections, 13% of those with microscopic residual disease, 14% of those with gross residual disease, and 66% of those with unresectable tumors (Table 5). Those patients who received

Table 5. Survival and disease progression[a] (GUNDERSON et al. 1983)

Group	Median time to progression (mo)	Median survival (mo)
Resected, high risk	22.2	24
Microscopic residual disease	18.6	24
Gross residual disease	8.4	15
Unresectable	9.1	14

[a] Patients in the medically inoperable and recurrent groups were not included in this table.

FAM and radiation therapy continued to be followed: 20 of 24 expired with a median survival of 18.5 months; there was a 41% 2-year survival rate, and a 19% 3-year survival rate. Local control was maintained in 16 of the 20 patients who expired.

Many other reports have addressed the question of postoperative treatment of patients with known residual carcinoma (either microscopic or gross) or nonresectable gastric carcinoma. The GITSG compared Adriamycin alone, 5-fluorouracil, Adriamycin, and methyl-CCNU (FAMe), and 5-fluorouracil, mitomycin C, and cystosine arabinoside (FMC) given postoperatively in patients with residual or nonresected gastric carcinoma (GASTROINTESTINAL TUMOR STUDY GROUP 1979). An advantage in terms of response rate, interval to disease progression, and survival was found for FAMe as compared with Adriamycin alone and FMC. In 1982, the results of another GITSG protocol were reported, showing an advantage for treatment with FAMe or FAMi (5-fluorouracil, Adriamycin, and mitomycin C) over treatment with FIMe (5-fluorouracil, ICRF-159, and methyl-CCNU) or FMe (5-fluorouracil and methyl-CCNU) (GASTROINTESTINAL TUMOR STUDY GROUP 1982b). Median survivals were 34.4, 29.6, 22.9, and 17.4 weeks, respectively. In comparison, in 1982 another GITSG protocol was published, comparing the use of 5-fluorouracil and methyl-CCNU vs irradiation plus 5-fluorouracil (GASTROINTESTINAL TUMOR STUDY GROUP 1982c). A dose of 25 Gy was administered over 3 weeks, followed by a 2-week break and an additional 25 Gy; 5-fluorouracil was given as 500 mg/m^2 intravenously for the first 3 days of each course of irradiation. After 4 years of follow-up, the group which received combined modality treatment had a higher 5-year survival rate than that which received chemotherapy, with 8 of 45 patients alive as compared to 3 of 45 (Fig. 5). Separation of the group into those who had incomplete resection and those who had nonresection again showed a survival advantage for the former group. It was noted that the combined modality group had a higher initial death rate, possibly secondary to enhanced toxicity, and the survival advantage for combined modality treatment became evident only after the 2-year mark (Fig. 6).

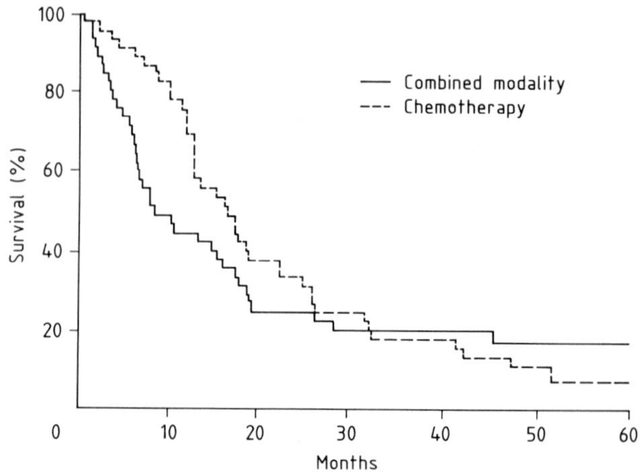

Fig. 5. Survival of patients in the GITSG study comparing chemotherapy and combined modality therapy administered postoperatively in patients with gross or microscopic residual gastric adenocarcinoma

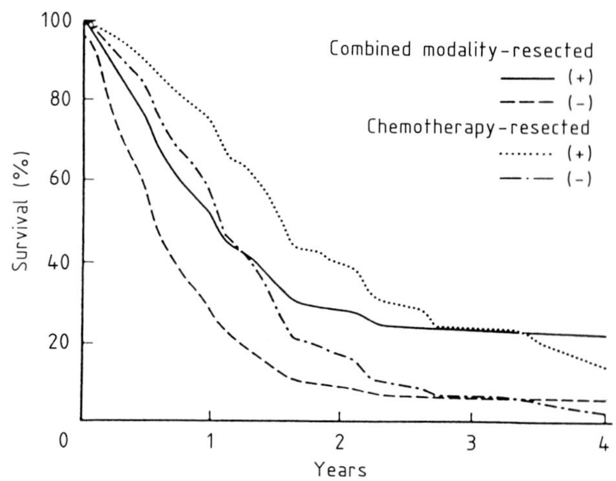

Fig. 6. Effect of resection of the primary tumor on survival in the GITSG study (GASTROINTESTINAL TUMOR STUDY GROUP 1982c)

To add further to the confusion regarding the optimum treatment for residual or nonresected gastric carcinoma, it is to be noted that the North Central Cancer Treatment Group published a comparison of 5-fluorouracil, 5-fluorouracil plus Adriamycin, and 5-fluorouracil, Adriamycin, and mitomycin C in patients who had incomplete or nonresection of gastric carcinoma and found no survival advantage of the latter two combinations over 5-fluorouracil alone (CULLINAN et al. 1985). Also, the Eastern Coopeative Oncology Group reported a study, EST 3275, which randomized 180 patients to receive 2 years of 5-fluorouracil plus methyl-CCNU versus observation alone, following curative resection; no survival advantage was found for chemotherapy vs observation, and, in addition, two treated patients died of leukemia (ENGSTROM et al. 1985).

In summary, the optimal treatment of patients with a high risk of recurrence and of those with known residual carcinoma remains undetermined.

Earlier indications of the benefit of chemotherapy in each of these groups have not been unequivocally confirmed, and the use of radiation therapy in conjunction with chemotherapy seems to provide a definite benefit. The techniques of radiation therapy, however, need to be defined since inadequate dosage and either excessive or inadequate portals can render the use of this modality ineffective. Definition of treatment dose and volume must also depend on knowledge of potential toxicity.

Treatment of gastric lymphoma should provide a different set of considerations compared with treatment of adenocarcinoma because of the relative responsiveness of these cell types to radiation therapy and chemotherapy. Nonetheless, surgical resection has been performed in the majority of cases available for evaluation in the literature, and the concentration therefore has been on the choice of postoperative treatment. In the series from the Princess Margaret Hospital, 48 patients were identified with lymphoma localized to the stomach out of a total of 1394 patients with Non-Hodgkin's lymphoma seen between 1967 and 1978 (GOSPODAROWICZ et al. 1983). All 48 patients received radiation therapy postoperatively, 12 to the left upper quadrant and para-aortic area and 36 to the whole abdomen. The treatment policy was to administer 20–25 Gy minimum tumor dose to the large field with a 15-Gy boost to areas of postoperative gross residual disease. These two groups had no difference in number of sites of failure, probably because nine of those treated with smaller fields were scored as having small bulk disease and six of these nine were listed as stage II. Patients with unresectable or bulky lymphoma following surgery were treated with chemotherapy.

A report of a series of patients from MSKCC described 51 patients treated for primary gastric lymphoma (SHIU et al. 1982). Another 61 patients had involvement of the stomach but also had disease in other areas. Seventeen of these lymphomas were located in the distal third of the stomach, eight in the middle third, and seven in the proximal third; eight were in the distal two-thirds and four in the proximal two-thirds; and the entirety of the stomach was occupied in three cases. In the remaining four cases the location was not specified. Between 1949 and 1959 surgery was the most common means of treatment, with 18 patients having subtotal gastrectomy and 12 having extended total gastrectomy; 18 patients received radiation therapy postoperatively with fields usually limited to the left upper portion of the abdomen. Between 1970 and 1976 less extensive surgical resections were done, mainly because

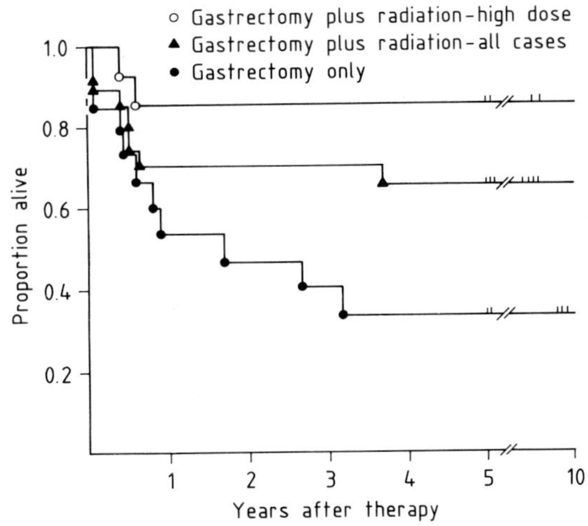

Fig. 7. Survival curves of patients after curative treatment of primary gastric lymphoma according to the method of treatment. At 5 years the survival rates were: 85% after gastrectomy and postoperative radiation of at least 20 Gy given in 2½–3 weeks, 67% after gastrectomy and postoperative radiation for all patients so treated, and 33% after gastrectomy alone without postoperative radiation. (SHIU et al. 1982)

of preoperative diagnosis, and radiation therapy was given postoperatively with larger fields and higher dosages. Overall, 42 of the 51 patients had complete surgical resection. Of 13 patients who received more than 20 Gy postoperatively, 11 were alive more than 5 years after treatment. However, in patients who received less than 20 Gy, there was 67% survival, and in patients who underwent surgery alone, only 33% survived 5 or more years. Chemotherapy was not used postoperatively at that time (Fig. 7).

A report from the MGH analyzed 26 patients with gastric lymphoma and assessed factors contributing to survival (SHIMM et al. 1983). Sixteen patients had undergone partial gastrectomy and nine, total gastrectomy. Nineteen patients received postoperative radiation therapy, one with whole abdominal irradiation, and the rest were treated to the gastric bed and regional lymph nodes. Fifteen patients received 40 Gy or less. Negative factors in terms of survival were involvement of regional lymph nodes, penetration of serosa, and the histological subtype of histiocytic lymphoma. Having positive margins did not affect survival, presumably because postoperative irradiation was administered. Judging from these factors, it was concluded that postoperative irradiation was not necessary for cure in those patients with favorable prognostic characteristics, namely, negative surgical margins, negative regional lymph nodes, and disease confined within the sero-

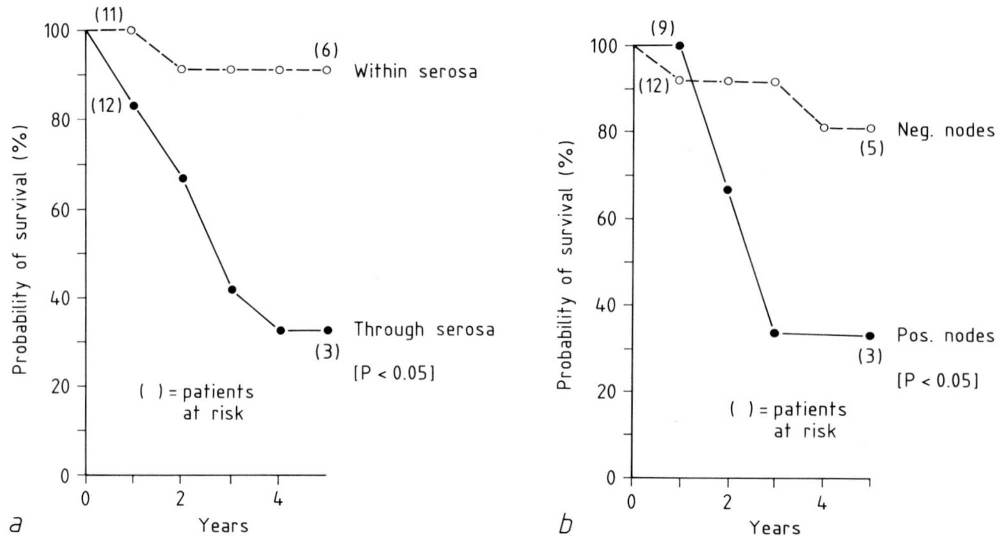

Fig. 8. Influence of (**a**) serosal penetration and (**b**) regional lymph node involvement on survival in patients with gastric lymphoma (Shimm et al. 1983)

sa. It was recommended that patients with positive surgical margins receive postoperative irradiation to the gastric bed and upper para-aortic nodes with a dose exceeding 40 Gy. Whole abdominal irradiation was not recommended because patients with a risk of dissemination would be receiving chemotherapy regardless (Fig. 8a, b).

6.8 Radiation Therapy Techniques

Sophisticated treatment planning is mandatory for treatment of the upper abdomen, since dosage to several critical structures needs to be kept in mind. Considering that the dosage to the extended fields will need to be kept in the range of 40–45 Gy, this portion of treatment can usually be carried out with parallel opposed anteroposterior and posteroanterior fields. With the use of the planning simulator, the stomach or gastric pouch can be localized by giving the patient barium to swallow, and the kidneys can be visualized following injection of contrast medium. Inclusion of the porta hepatis and pancreaticoduodenal nodes is indicated, requiring coverage of a portion of the right kidney and the liver. The location of the spinal cord can be obtained from a lateral simulation film and positions of the kidneys and spinal cord can also be correlated to findings on computerized body section radiography. No more than 50% of both kidneys should be included in the fields because of the risk of radiation-induced nephritis or nephrosclerosis, and custom blocking with Cerrobend can be used to ensure blockage of adequate amounts of both kidneys. Regardless, it is still likely that a majority of the left kidney will need to be irradiated. Care must also be taken to avoid excessive dosage to the liver and heart.

Illustrative Case. A 60-year-old male had been found to have moderately differentiated ulcerating adenocarcinoma of the stomach with invasion through the gastric wall into the spleen. He underwent gastrectomy, splenectomy, distal pancreatectomy, and pyloroplasty. He received adjuvant chemotherapy with 5-fluorouracil, Adriamycin, and mitomycin C, but refused external beam radiation therapy. He then developed an anastomotic recurrence, diagnosed by endoscopy, and was started on radiation therapy. At the time of treatment planning, barium sulfate was given orally to demonstrate the location of the gastric remnant. The position and function of the kidneys were checked by giving the patient 50% diatrizoate sodium solution intravenously and localization films were obtained to demonstrate the relationship of the kidneys to the treatment portals. Attention was directed to the right kidney being blocked out of the field since the left kidney was almost entirely included. Following administration of 44.4 Gy to the target volume, with a calculated spinal cord dose of 40.86 Gy (with a postero-anterior spinal cord block utilized to keep down the spinal cord dose), a field reduction was carried out for an additional five fractions to bring the dose to 53.4 Gy. There was no further gastrointestinal bleeding and on repeat endoscopy 6 months later, no evidence of cancer was encountered.

The use of intraoperative irradiation as a boost dose has been explored in several centers. In 1981, Abe and Takahashi published results from Kyoto University, including the treatment of 84 patients with gastric carcinoma. The actuarial survival rates for patients with stage II–IV disease were improved by the use of intraoperative irradiation of 28–40 Gy in addition to surgery. The results from this series were updated in 1988, with the conclusion that irradiation alone could not be considered adequate to cure a primary tumor but was effective postoperatively to treat residual disease (ABE et al. 1988). The conclusions for the use of intraoperative therapy were that the modality can be used most effectively in tumors of the middle portion of the stomach or antrum, that a single dose pf 28 Gy should be used for clinically undetected residual tumor, and that a single dose of 30–35 Gy should be used for macroscopic residual tumor. The electron energy was selected to encompass the entire lesion with the 90% isodose line. The most common setting for intraoperative radiotherapy was delivery of a boost dose to (a) the area of resection, (b) the pancreas, to which a posteriorly located tumor may have been adherent, or (c) the celiac axis in cases in which gross total resection was accomplished but in which there was significant risk of metastases to these nodes.

6.9 Complications

The short-term toxicity of radiation therapy to the upper abdomen is well-known; it includes nausea and vomiting requiring antiemetics in approximately 50% of patients, decreased appetite, and possible dehydration secondary to decreased oral intake. In the MGH series, 3 of 24 patients had a loss of more than 15% of their body weight (LINGOS et al. 1985). Thrombocytopenia or neutropenia is rare from radiation therapy alone but more common when combined modality treatment is given. Mucosal irritation from irradiation of the gastrointestinal tract may predispose patients to superinfection with herpes simplex or Candida.

The dose to areas of known residual disease should be boosted, if possible, to 50–55 Gy. However, it must be kept in mind that dosages of 55 Gy or more to the stomach may have a 50% risk of producing mucosal injury. Radiation-induced ulcers are rare and are usually located in the antrum, and perforations have been reported. Epithelial necrosis may lead to upper gastrointestinal hemorrhage.

Another problem associated with gastric irradiation is decreased gastric acidity; this can persist for years although it may return to normal with recovery of chief or parietal cells. In patients with resection of the bulk of the stomach, regardless of whether irradiation is administered, the possibility of vitamin B_{12} deficiency due to lack of intrinsic factor must be kept in mind as a problem that can arise several years later.

The use of chemotherapy in conjunction with radiation therapy provides the possibility of enhanced reactions to irradiation, and if Adriamycin has been utilized the dose to the heart should be minimized, if possible. Another potential problem is the induction of second malignancies, considering that methyl-CCNU has been implicated in producing leukemia and preleukemia changes.

Because of the need for limitation of the dose of radiation to structures in the vicinity of the target area, the use of intraoperative irradiation has been explored, as discussed earlier. The use of intraoperative irradiation can allow a high dose to be administered to an area of concern while other structures are displaced out of the field. Nonetheless, among the patients reported from the National Cancer Institute in the United States were four patients with carcinoma of the stomach, two of whom developed fistulas, one pancreatic and one biliary, within the intraoperative field. This indicates that in this area of the body the dosage that can be used for a boost may still need to be kept limited even with this technique (SINDELAR et al. 1983).

6.10 Summary

In summary, surgery remains the mainstay of treatment of malignancies of the stomach. Future directions of investigation will include further work on radiation sensitizers, the use of hematoporphyrin derivatives, the enhancement of radiation therapy and chemotherapy by hyperthermia, and use of new combinations of chemotherapy in conjunction with or sequentially with irradiation. In addition, better techniques for early detection and recognition of specific etiological factors may become available. In the meantime, therapy of this set of malignancies, and particularly that of locally advanced gastric adenocarcinoma, remains a topic for aggressive research.

References

Abe N, Takahashi M (1981) Intraoperative radiotherapy: the Japanese experience. Int J Radiat Oncol Biol Phys 7: 863–868

Abe N, Takahashi M, Ono K, Tobe T, Inamoto T (1988) Japan gastric trial in intraoperative radiation therapy. Int J Radiat Oncol Biol Phys 15: 1431–1433

Adams HW, Adkins JR, Rehak EM (1983) Malignant fibrous histiocytoma presenting as a bleeding gastric ulcer. Am J Gastroenterol 78: 212–213

Aizawa K, Okunaka T, Ohtani T et al. (1987) Localization of mono-L-aspartyl chlorin e6 (NPe6) in mouse tissues. Photochem Photobiol 46: 789–793

American Joint Committee on Cancer (1983) Manual for staging of cancer. Lippincott, Philadelphia

Baron BW, Bitter MA, Baron JM, Bostwick DG (1987) Gastric adenocarcinoma after gastric lymphoma. Cancer 60: 1876–1882

Berthrong M, Fajardo LF (1982) Alimentary tract. In: Fajardo FL (ed) Pathology of radiation injury. Masson, New York, pp 47–74

Boddie AW, McMurtrey MJ Giacco GG, McBride CM (1983) Palliative total gastrectomy and esophagogastrectomy: reevaluation. Cancer 51: 1195–1200

Boice JD, Greene MH, Killen JY et al. (1983) Leukemia and preleukemia after adjuvant treatment of gastrointestinal cancer with semustine (methyl-CCNU). N Engl J Med 309: 1079–1084

Cady B, Choe DS (1980) Changing patterns of gastric carcinoma. In: Nieburgs HE (ed) Proceedings of the third international symposium on the detection and prevention of cancer. Part II. Cancer detection in specific sites, vol II. Dekker, New York, pp 2041–2049

Carter KJ, Schaeffer HA, Ritchie WP (1984) Early gastric cancer. Ann Surg 199: 604–609

Collins WT, Gall EA (1952) Gastric carcinoma, multicentric lesion. Cancer 5: 62

Cook AO, Levine BA, Sirinek KR, Gaskill HV (1986) Evaluation of gastric adenocarcinoma. Arch Surg 121: 603–605

Creasey W (1985) Diet and cancer. Lea & Febiger, Philadelphia, p 125

Cullinan SA, Moertel CG, Fleming TR et al. (1985) A comparison of three chemotherapeutic regimens in the treatment of advanced pancreatic and gastric carcinoma. JAMA 253: 2061–2067

Dent DM, Werner ID, Novis B, Cheverton P, Brice P (1979) Prospective randomized trial of combined oncological therapy for gastric carcinoma. Cancer 44: 385–391

Dupont JB, Lee JR, Burton GR, Cohn I (1978) Adenocarcinoma of the stomach: Review of 1,497 cases. Cancer 41: 941–947

Ekbom GA, Gleysteen JJ (1980) gastric malignancy: resection for palliation. Surgery 88: 476–481

Engstrom PF, Lavin PT, Douglas HO, Brunner KW (1985) Postoperative adjuvant 5-fluorouracil plus methyl-CCNU therapy for gastric cancer patients; Eastern Cooperative Oncology Group Study (EST 3275). Cancer 55: 1868–1873

Gastrointestinal Tumor Study Group (1979) Phase II–III chemotherapy studies in advanced gastric cancer. Cancer Treat Rev 63: 1871–1876

Gastrointestinal Tumor Study Group (1982a) Controlled trial of adjuvant chemotherapy following curative resection for gastric carcinoma. Cancer 49: 1116–1122

Gastrointestinal Tumor Study Group (1982b) A comparative clinical assessment of combination chemotherapy in the management of advanced gastric carcinoma. Cancer 49: 1362–1366

Gastrointestinal Tumor Study Group (1982c) A comparison of combination chemotherapy and combined modality therapy for locally advanced gastric carcinoma. Cancer 49: 1771–1777

Giarelli L, Melato M, Stanta G, Bucconi S, Manconi R (1983) Gastric resection: a cause of high frequency of gastric cancer. Cancer 52: 1113–1116

Gospodarowicz MK, Bush RS, Brown TC, Chua T (1983) Curability of gastrointestinal lymphoma with combined surgery and radiation. Int J Radiat Oncol Biol Phys 9: 3–9

Gunderson LL, Sosin H (1982) Adenocarcinoma of the stomach: areas of failure in a reoperation series (second or symptomatic look). Clinicopathologic correlation and implications for adjuvant therapy. Int J Radiat Oncol Biol Phys 8: 1–11

Gunderson LL, Hoskins RB, Cohen AC, Kaufman S, Wood WC, Carey RW (1983) Combined modality treatment of gastric cancer. Int J Radiat Oncol Biol Phys 9: 965–975

Haenszel W, Correa P (1975) Developments in the epidemiology of stomach cancer over the past decade. Cancer Res 35: 3452–3459

Hallissey MT, Allum WH, Roginski C, Fielding J (1988) Palliative surgery for gastric cancer. Cancer 62: 440–444

Higgins GA, Amadeo JH, Smith D, Humphrey EW, Keehn RJ (1983) Efficacy of prolonged intermittent therapy with combined 5-FU and methyl-CCNU following resection for gastric carcinoma. Cancer 52: 1105–1112

Johnston MJ, Helwig EB (1981) Granular cell tumors of the gastrointestinal tract and perianal region: a study of 74 cases. Dig Dis Sci 26: 807–816

Kawai K, Kizu M, Miyaoka T (1980) Epidemiology and pathogenesis of gastric carcinoma. Front Gastrointest Res 6: 71

King RM, van Heerden JA, Weiland LH (1982) The management of gastric polyps. Surg Gynecol Obstet 155: 846

Lee Y-TNM (1983) Leiomyosarcoma of the gastro-intestinal tract: general pattern of metastasis and recurrence. Cancer Treat Rev 10: 91–101

Lieber MR, Winans CS, Griem ML, Moossa AR, Elner VM, Franklin WA (1985) Sarcomas arising after radiotherapy for peptic ulcer disease. Dig. Dis Sci 30: 593–599

Lingos T, Tepper JE, Gunderson LL, Orlow E, Kaufman SD, Younger J (1985) Adjuvant FAM-RAD-FAM after resection of high risk gastric carcinoma. Int J Radiat Oncol Biol Phys 11 [Suppl]: 116

MacGregor IL (1974) Carcinoma of the colon and stomach: A review with comment on epidemiologic associations. JAMA 227: 911–915

MacLoughlin GA, Cave-Bigley DJ, Tagore V, Kirkham N (1980) Cyclophosphamide and pure squamous-cell carcinoma of the stomach. Br Med J I: 524–525

Mattingly SS, Cibull ML, Ram MD, Hagihara PH, Griffen WO (1987) Pseudolymphoma of the stomach. Arch Surg 116: 25–29

Mendelsohn G, de la Monte S, Dunn JL, Yardley JH (1987) Gastric carcinoid tumors, endocrine cell hyperplasia and associated intestinal metaplasia. Cancer 60: 1022–1031

Moertel CG, Childs DS, Reitemeier RJ, Colby MY, Holbrook MA (1969) Combined 5-fluorouracil and supervoltage radiation therapy of locally unresectable gastrointestinal cancer. Lancet II: 865–867

Moertel CG, Childs DS, O'Fallon JR, Holbrook MA, Schutt AJ, Reitemeier RJ (1984) Combined 5-fluorouracil and radiation therapy as a surgical adjuvant for poor prognosis gastric carcinoma. J Clin Oncol 2: 1249–1254

Petersen JM, Ferguson DR (1984) Gastrointestinal neurofibromatosis. J Clin Gastroenterol 6: 529–534

Possik RA, Franco EL, Pires DR, Wohnrath DR, Ferreira EB (1986) Sensitivity, specificity, and predictive value of laparoscopy for the staging of gastric cancer and for the detection of liver metastasis. Cancer 58: 1–6

Schneider K, Dickerhoff R, Bertele RM (1986) Malignant gastric sarcoma: diagnosis by ultrasound and endoscopy. Pediatr Radiol 16: 69–70

Segal AW, Healy MJR, Cox AG (1975) Diagnosis of gastric cancer. Br Med J 3: 669

Seo IS, Brinkley WB, Warner TFCS, Warfel KA (1982) A combined morphologic and immunologic approach to the diagnosis of gastrointestinal lymphomas: malignant lymphoma of the stomach (a clinicopathologic study of 22 cases). Cancer 49: 493-501

Shimm DS, Dosoretz DE, Anderson T, Linggood RM, Harris NL, Wang CC (1983) Primary gastric lymphoma. Cancer 52: 2044-2048

Shiu MH, Papacristou DN, Kosloff C, Eliopolous G (1980) Selection of operative procedure for adenocarcinoma of the midstomach. Ann Surg 192: 730-737

Shiu MH, Karas M, Nisce L, Lee BJ, Filippa DA, Lieberman PH (1982) Management of primary gastric lymphoma. Ann Surg 195: 196-202

Silverberg E, Lubera J (1989) Cancer statistics, 1989. CA 36: 3-20

Sindelar WF, Kinsella T, Tepper J, Travis EL, Rosenberg SA, Glatstein E (1983) Experimental and clinical studies with intraoperative radiotherapy. Surg Gynecol Obstet 157: 205-219

Stone R, Rangel DM, Gordon HE, Wilson SE (1977) Carcinoma of the gastroesophageal junction; a ten year experience with esophagogastrectomy. Am J Surg 134: 70-76

Suehiro S, Nagasue N, Abe S, Ogawa Y, Sasaki Y (1986) Carcinomas of the stomach in atomic bomb survivors. Cancer 57: 1894-1898

Takahashi M (1964) Studies on preoperative and postoperative telecobalt therapy in gastric cancer. Nippon Acta Radiol 24: 129-132

Tatsuta M, Okuda S, Tamura H, Taniguchi H (1982) Endoscopic diagnosis of early gastric cancer by the endoscopic Congo red-methylene blue test. Cancer 50: 2956-2960

Tokunaga O, Watanabe T, Morimatsu M (1987) Pseudolymphoma of the stomach. Cancer 59: 1320-1327

Vaughan WP, Straus FH, Paloyan D (1977) Squamous carcinoma of the stomach after luetic linitis plastica. Gastroenterology 72: 945-948

Wall SD, Friedman SL, Margulis AR (1984) Gastrointestinal Kaposi's sarcoma in AIDS: radiographic manifestations. J Clin Gastroenterol 6: 165-181

Willett C, Tepper JE, Orlow EL, Shipley WU (1985) Renal complications secondary to radiation treatment of upper abdominal malignancies. Proceedings of the 27th Annual Astro Meeting. Int J Radiat Oncol Biol Phys 11 [Suppl]: 117

Winawer SJ, Sherlock P. Hajdu SI (1976) The role of upper gastrointestinal endoscopy in patients with cancer. Cancer 37 [Suppl]: 440-447

7 Pancreatic Cancer

RALPH R. DOBELBOWER, JR., STEVEN M. WAGNER, RONALD J. FADELL, JOHN M. HOWARD, and LIBERATO J.A. DIDIO

CONTENTS

RALPH R. DOBELBOWER[1], JR., M.D., Ph.D., F.A.C.R., Professor and Chairman, Professor of Neurological Surgery (Radiation Therapy), STEVEN M. WAGNER[3], M.D., Clinical Associate Professor of Medicine, RONALD J. FADELL[2], M.D., JOHN M. HOWARD[4], Professor of Surgery, M.D., LIBERATO J.A. DIDIO[5], M.D., D.Sc., Ph.D., Professor of Anatomy, Departments of [1] Radiation Therapy, [2] Radiology, [3] Medicine, [4] Surgery and [5] Anatomy, Medical College of Ohio, 3000 Arlington Avenue, C.S. 10008, Toledo, OH 43699-0008, USA

7.1 Anatomy

The pancreas is a hammer-like, extraparietal gland of the duodenum, closely related to this portion of the small intestine from the anatomic, physiologic, pathologic, clinical, and surgical standpoints. The pancreas is a lobulated, amphicrine organ (DIDIO 1970), that is, possessing both endocrine and exocrine functions. In the adult male the gland weighs approximately 100 g (85 g average in the female). Its dimensions are generally the following: 14-18 cm long (from right to left), 2-9 cm high (cephalocaudad direction), and 2-3 cm thick (ventrodorsal direction). It has four indistinctly separated portions: head (with the uncinate process), neck, body (with the omental tubercle), and tail (Fig. 1), although from the surgical standpoint it may be divided into a right (cephalocervical) and a left (corporocaudate) segment, separated by a paucivascular area or intersegmental plane, just to the left of the superior mesenteric artery (BUSNARDO et al. 1988).

The pancreas is retroperitoneal, its deep location and relative inaccessibility being responsible for the expression "hermit organ." Modern imaging techniques [computed body section radiography (CT), ultrasound, endoscopic retrograde cholangiopancreatography (ERCP), magnetic resonance imaging (MRI)] have revolutionized the in vivo study and appreciation of pancreatic anatomy (Fig. 2, 3).

The head, neck, and body of the pancreas are relatively fixed against the posterior abdominal wall, whereas the tail is more movable. The uncinate process extends cephalad and toward the left forming the incisura pancreatis, a notch for the superior mesenteric vein which joins the splenic vein to constitute the portal vein. The head is variable in location and blends into the neck, a slight narrowing followed by the body; the latter tapers into the tail, which generally nestles in the splenic hilus.

The peritoneum relates to the anterior aspect of the pancreas similarly to what occurs with the duodenum. There is a *pars tecta pancreatis,* that is, a portion of the anterior aspect of the pancreas hid-

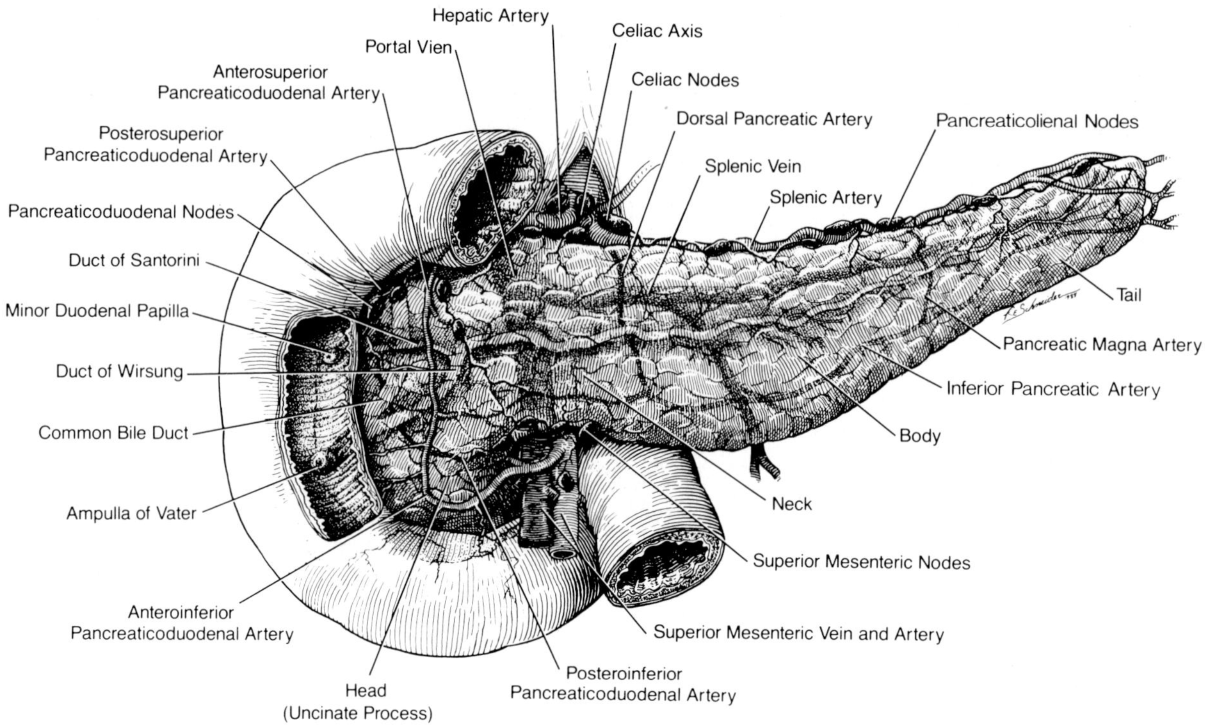

Fig. 1. Anatomy of the pancreas and certain peripancreatic structures

den (a) behind the root of the transverse mesocolon, (b) behind the root of the mesentery, and (c) behind the interposed coalescent ascending mesocolon (DIDIO 1948, 1956).

The main pancreatic duct (of Wirsung) runs along the long axis of the gland closer to the dorsal surface of the gland. Secondary ducts join at almost right angles in herringbone pattern. The duct of Wirsung often joins the common bile duct in the duodenal wall to form the hepatopancreatic ampulla (of Vater), contained in the major duodenal papilla, 8–10 cm beyond the gastroduodenal pylorus. The accessory pancreatic duct (of Santorini) anastomoses with the duct of Wirsung and drains into the duodenum through the minor duodenal papilla, generally approximately 2 cm before (cephalad to) the major duodenal papilla. In approximately 18% of patients, the duct of Wirsung and the common bile duct drain into the duodenum separately. Both papillae contain muscular mechanisms to open (dilator muscle) and close (sphincter muscle) the orifices of the ducts, called biliopancreatic pylorus, in the major duodenal papilla, and pancreatic pylorus, in the minor duodenal papilla (DIDIO and ANDERSON 1968).

The head of the pancreas lies in the arms of the duodenum. The body and tail are directed dorsocephalad as they course to the left toward the hilus of the spleen. The pancreas and duodenum are both located in the retroperitoneal space. The pancreatic peritoneum and the peritoneum of the dorsal wall of the stomach are in apposition across the lesser peritoneal sac (omental bursa). The dorsal surface of the pancreas is anatomically related to the kidneys, suprarenal glands, renal arteries, renal veins, aorta, and inferior vena cava. To the right of the head of the pancreas lies the second (vertical) portion of the duodenum and the liver. Loops of small intestine and the transverse colon are related to the ventral surface of the pancreas. The lumbar spine and spinal cord (or cauda equina) are dorsal to the pancreas. The exact location of the pancreas within the retroperitoneal space, the size of the gland, and its configuration are all rather variable. The gland may move several centimeters with respiration and may move as much as the height of two vertebral bodies when the body positon changes from supine to erect.

The pancreas is supplied by numerous arteries (Fig. 1). The circulation, especially in the head and body, is characterized by numerous anastomoses. In general, the arteries originate from the celiac trunk and the superior mesenteric artery. The splenic artery and hepatic artery both give branches to the

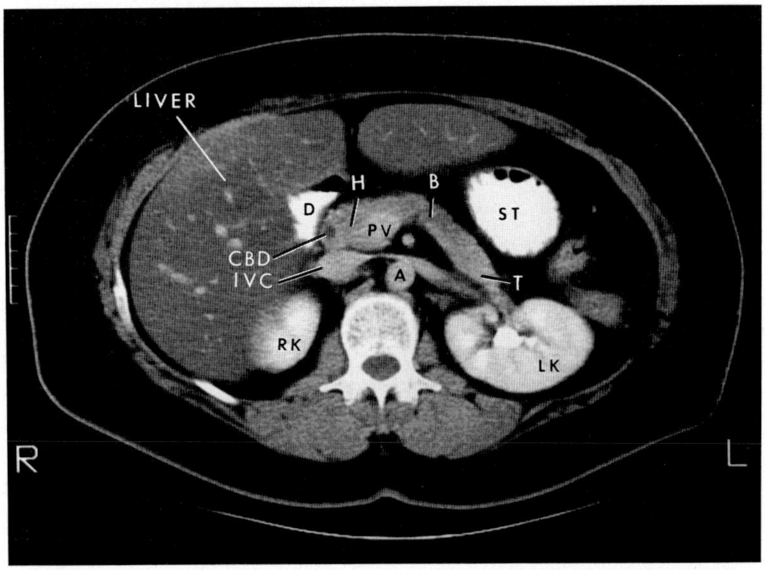

Fig. 2. Transverse computerized body section tomogram at the level of the pancreas. *RK*, right kidney; *IVC*, inferior vena cava; *CBD*, common bile duct; *D*, duodenum; *H*, head of pancreas; *PV*, portal vein; *A*, aorta; *B*, body of pancreas; *ST*, stomach; *T*, tail of pancreas; *LK*, left kidney

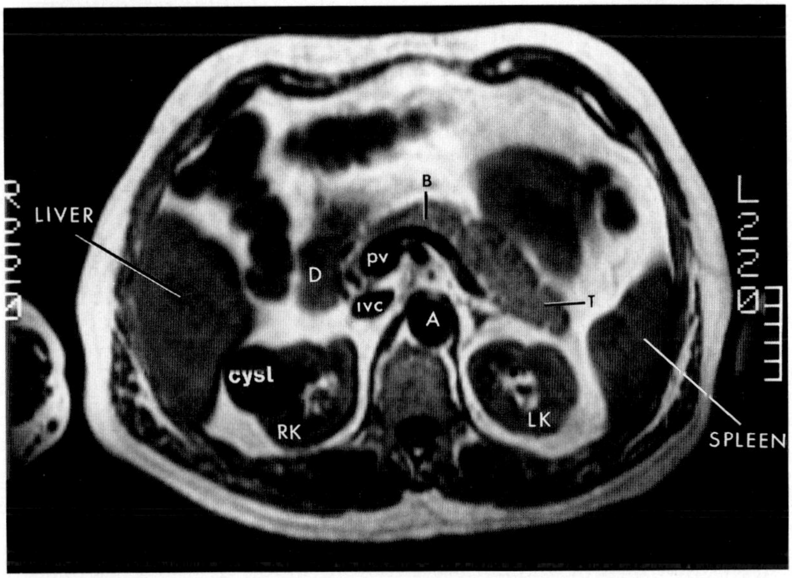

Fig. 3. Transverse magnetic resonance image at the level of the pancreas. *RK*, right kidney; *D*, duodenum; *IVC*, inferior vena cava; *PV*, portal vein; *B*, body of pancreas; *A*, aorta; *T*, tail of pancreas; *LK*, left kidney

pancreas. The former follows a serpiginous course along the posterocephalic border of the pancreas. The anterosuperior pancreaticoduodenal artery and the posterosuperior pancreaticoduodenal artery originate from the gastroduodenal artery and anastomose with the anteroinferior pancreaticoduodenal artery and the posteroinferior pancreaticoduodenal artery, both of which originate from the superior mesenteric artery. The superior mesenteric artery has an important relationship with the pancreas in that it often passes directly through the gland immediately after its origin from the aorta.

The pancreas is permeated by an extensive lymphatic network. Lymph nodes may be embedded in the gland itself, but usually lie adjacent to it along the paths of major arteries. The pancreaticolienal lymph nodes lie along the splenic artery and drain the body and tail of the gland and then drain into the celiac lymph nodes. Pancreaticoduodenal lymph nodes drain into the hepatic lymph nodes or the superior mesenteric nodes and thence to the celiac nodes. Lymphatic drainage from part of the head may be through subpyloric nodes and hepatic nodes to the celiac nodes. Both the celiac and the superior mesenteric nodes drain to the cisterna chyli, ventral to the second lumbar vertebra, and eventually to the left superior jugular venous confluence via the thoracic duct.

7.2 Background and Natural History

Cancer of the pancreas is an insidious disease that is almost always fatal (Jordan 1987). Except for lesions located fortuitously near the common bile duct, there are no early symptoms of the disease. Diagnosis is usually difficult and time-consuming. Definitive surgical treatment of the disease often lies at the limits of the surgeon's skill and the patient's endurance (Dobelbower 1979b). In the overwhelming majority of cases, the ultimate prognosis is death from cancer.

Approximately 60% of pancreatic cancers occur in the head of the gland, 20% in the body and/or tail, and another 20% involve the entire gland (Wynder et al. 1973). Microscopic foci of cancer are scattered throughout the gland in as many as 38% of cases (Tryka and Brooks 1979).

At the time of diagnosis, the disease extends beyond the capsule of the pancreas in at least 85% of cases (Cubilla et al. 1978a). The structures most commonly involved by contiguous extension of disease are, in order of decreasing frequency: peripancreatic fat, common bile duct, blood vessels (superior mesenteric artery and vein, vena cava, portal vein, aorta), mesentery, duodenum, and neighboring organs (stomach, small bowel, large bowel, liver, spleen, kidney, etc.).

Because of gross and histologic similarities, it is often difficult if not impossible to differentiate tumors of the pancreas from tumors of the distal biliary tract or the ampulla of Vater. Neoplasms of the biliary tract have a similar presentation, mode of spread, and natural history to those of the pancreas, whereas tumors of the ampulla of Vater are often diagnosed at a much earlier stage and, hence, have a much better prognosis. At the time of diagnosis, lymph nodes are involved with metastatic disease in at least 50% of cases of cancer of the pancreas (Cubillia and Fitzgerald 1978, 1984). The lymph node groups most commonly involved are, in decreasing order, the posterior pancreaticoduodenal group, the superior head groups, the inferior head group, the superior body groups, the anterior pancreaticoduodenal group and the inferior body group (Cubilla et al. 1978b). Nodes are also frequently found behind the pylorus and the common bile duct.

The liver and peritoneum are the most common sites of distant metastases in patients with cancer of the pancreas. Other sites of metastases, in decreasing order of frequency, include the lungs, adrenal glands, kidneys, skin, bone, and brain (Cubilla and Fitzgerald 1984; Klöppel 1984; Mikal and Campbell 1950). At autopsy, as many as 95% of patients will show evidence of hematogenous metastasis; however, at the time of diagnosis as many as 50% of patients will have disease apparently localized to the upper abdomen without evidence of hepatic or other hematogenous metastases.

The extent of disease at presentation has a major impact on the clinical approach to treatment. Patients with disease confined to the pancreas may be considered for surgical resection with or without adjuvant treatment. Patients with disease beyond the confines of the gland, but without distant hematogenous spread, may be considered for extended resection with or without adjuvant therapy or for radiotherapy with or without chemotherapy. This includes some patients with involved regional lymph nodes and some with local extensions of disease that involve adjacent structures such as the portal vein, inferior vena cava, liver, stomach, etc. Patients with widespread peritoneal disease, hepatic metastasis, or other areas of hematogenous spread are not candidates for definitive treatment and, in some cases, may be best managed with palliative and/or supportive care.

7.3 Epidemiology and Risk Factors

Pancreatic cancer is the fifth most common cause of cancer death in the United States today (Silverberg and Lubera 1989). It ranks second as a cause of death due to gastrointestinal tumors (Silverberg and Lubera 1989). The incidence and mortality rates for pancreatic cancer have increased steadily in the United States for the past five decades with virtually no difference between the two because of the remarkably high mortality of the disease (Aoki and Ogawa 1978). The highest incidence rates occur in western industrialized countries while the lowest incidence rates are found in Hungary, Nigeria, and India (Waterhouse et al. 1976). Especially high incidence of disease has been seen in black males in the United States, Maoris of New Zealand, and native Hawaiians.

Increasing mortality rates have been observed in both sexes in all countries; however, the rates have begun to level off in North America over the last decade. In this country, the increase in death rate has primarily occurred in age groups over 45 years (FONTHAM et al. 1987).

Eighty percent of cases of pancreatic cancer occur in the seventh and eighth decades of life (GORDIS and GOLD 1986). The disease is more common in men than in women, with a sex ratio of 1.1:1 to 2:1 (BERG and CONNELLY 1979; GRIEVE 1973; GO and DIMAGNO 1977). The disease tends to be diagnosed at an older age for women than for men (67 versus 63 years (MOLDOW and CONNELY 1968). A few cases of familial grouping of pancreatic cancer have been reported, suggesting a hereditary predisposition and pancreatic cancer in later life (DAT and SONTAG 1982; FRIEDMAN and FIALKOW 1976; MACDERMOTT and KRAMER 1973). Pancreatic cancer is unusual in individuals less than 40 years of age (TSUKIMOTO et al. 1973; FONTHAM et al. 1987; MOYNAN et al. 1964; SMITH et al. 1967; TAXY 1976).

The American Cancer Society estimates that in the year 1989 there will be 27000 new cases of, and 24500 deaths from pancreatic cancer (SILVERBERG and LUBERA 1989). It accounts for approximately 3% of all new cancer cases and 5% of all cancer deaths in the United States. In this country, the incidence and mortality rates for blacks of both sexes appear to exceed those for whites residing in the same geographic area (WATERHOUSE et al. 1976). The reason for this difference is not known.

The age-adjusted incidence rate for black males in the United States is one of the highest in the world, while one of the lowest incidence rates is in Nigeria (WATERHOUSE et al. 1976), suggesting the importance of environmental factors. Some studies have suggested that pancreatic cancer is more frequent in Jews than Catholics or Protestants. In Israel, Jews have incidence rates three times those of non-Jews (WATERHOUSE et al 1976) and in the United States the Jewish population shows a higher incidence rate than the non-Jewish population (WYNDER et al. 1973). Israeli Jews born in Europe or America show incidence rates four times those of Israeli Jews born in Asia or Africa (FONTHAM et al. 1987). Migrants to Australia from seven different European countries demonstrated higher mortality rates than those of the respective country of origin and higher than that of native Australians, suggesting the importance of major dietary change (MCMICHAEL et al. 1980).

The relationship between diabetes mellitus and pancreatic cancer is not clearly understood. The cancer incidence in diabetics is approximately half that of the nondiabetic population (FONTHAM et al. 1987); however, pancreatic cancer is proportionally three times more common in diabetic patients than in nondiabetic patients (BELL 1957; KESSLER 1970). Between 13% and 14% of patients with pancreatic cancer also have diabetes mellitus. Although recurrent pancreatitis often mimics pancreatic cancer, the only known association between the two is cause and effect, i.e., the pancreatitis associated with pancreatic cancer is thought to be the result of ductal obstruction by the latter rather than a predisposition toward the former caused by the latter. Islet cell tumors of the pancreas may be a feature of multiple endocrine neoplasia type I (MEN I), an autosomal dominant disorder consisting of benign or malignant tumors of pituitary, parathyroid, pancreas, and adrenal cortex. Members of families afflicted with hereditary pancreatitis also have an increased risk of pancreatic cancer (CASTLEMAN et al. 1972).

Cigarette smoking has been clearly established as an important etiologic factor with a relative risk of 1.6 associated with a 20 pack-year history (DOLL and PETO 1976). Mortality ratios from pancreatic cancer for cigarette smokers range from 1.6 to 3.1, as compared to nonsmokers (BEST 1966; CEDERLOF et al. 1975; DOLL and PETO 1976; HAMMOND 1966; HIRAYAMA 1972; KAHN 1966). Smoking pipe tabacco and chewing snuff have not been established as risk factors. It has been suggested that cancer of the pancreas may be associated with chemical and metal industries (WYNDER et al. 1973), naphthylamine (MANCUSO and EL-ATTAR 1967), the coke and gas industry, exposure to gasoline (LIN and KESSLER 1981), manufacture of radioactive substances (MANCUSO et al. 1977; HUTCHINSON et al. 1979), and asbestos exposure (SELIKOFF and SEIDMAN 1981). Coffee and alcohol consumption have also been linked to pancreatic cancer by some investigators, but the association between consumption of these beverages and cigarette smoking confounds interpretation of data. The role of hormones in the etiology of pancreatic cancer has been inadequately studied.

7.4 Diagnostic Evaluation

The clinical manifestations of cancer of the pancreas are often vague and ill-defined early in the course of the disease. During the common 4- to 9-month interval between beginning of symptoms and diagnosis, the disease inexorably progresses locally and often spreads. Except for lesions fortuitously located near the ampulla of Vater, the diagnosis is rarely made at an early stage.

The last two decades have seen an explosion of technology applicable to the diagnosis of pancreatic cancer, but this has yet to favorably impact the mortality from this dread disease. Furthermore, WEISS et al. (1985) have suggested that multiple pancreatic operative biopsies may be associated with rapid intra-abdominal spread of tumor. This then tempers the aggressiveness of diagnostic pursuit for those physicians who consider the disease universally fatal with no benefit from treatment. On the other hand, some prominent pancreatic surgeons believe that resection should be performed liberally for this disease even without histologic documentation of a clinical diagnosis of pancreatic cancer. Their philosophy is that the smallest tumors will be the most difficult to diagnose, hence a negative biopsy should not be considered a contraindication to resection. Until methods are developed to diagnose pancreatic cancer at an early stage, such disparate views are bound to exist.

7.4.1 History and Physical Findings

Weight loss, pain, and jaundice are the most common symptoms in patients presenting with cancer of the pancreas (Table 1). The onset of symptoms is usually insidious with mild weakness, abdominal pain, or anorexia being the initial symptom (in retrospect). Weight loss averages approximately 10 kg. This tends to be more common and more severe in patients with lesions located in the body or tail of the pancreas as compared to those with lesions in the head (Table 1). By the time of diagnosis, the pain is usually severe, requiring narcotic analgesia. It is typically a visceral pain, not associated with local tenderness and neither aggrevated by, nor relieved by, physical activity or ingestion of food. The pain radiates to the back in 25%–60% of patients (more common in lesions of the body). Back pain is present in the absence of abdominal pain in less than 5% of patients (WAY 1987). The pain is partially relieved by positions that flex the spine (curled up in a chair, fetal position, knee to chest position). Almost all patients will develop some abdominal pain during the course of the illness. It characteristically localizes in the epigastrium and the hypochondrium and seems more severe at night.

Jaundice is the only distinctive symptom (sign) associated with cancer of the pancreas. This is due to the anatomic relationship of the pancreas to the common bile duct. Approximately 80% of patients with a lesion in the head of the gland will present with jaundice (Table 1). Indeed, the development of jaundice is usually the symptom which prompts the patient to seek medical attention. Almost invariably the icterus is progressive and unremitting. The associated symptoms of acolic stools and dark urine accompany the jaundice in most cases (Table 1). Jaundice is uncommonly associated with lesions of the body or tail, except when the disease is very advanced locally or has produced extensive hepatic metastasis.

Psychiatric abnormalities, predominately emotional depression, often occur in patients with pancreatic cancer. This appears to be more common with lesions located in the body or tail of the gland, as compared to those with lesions in the head (FAINTUCH and LEVIN 1986).

The most common physical signs seen at presentation are jaundice and hepatomegaly, the former most common in patients with lesions located in the head of the gland and the latter most common in cases involving the body and tail (Table 2). The gallbladder is palpable (Courvoisier's sign) in approximately one-third of patients with lesions in the head of the gland at the time of presentation. An abdominal mass is palpable in 15%–25% of pa-

Table 1. Presenting symptoms of adenocarcinoma of the pancreas by site of tumor

Head		Body and tail	
Symptom	% patients	Symptoms	% patients
Weight loss	90	Weight loss	100
Jaundice	80	Pain	90
Pain	75	Weakness	45
Anorexia	65	Nausea	45
Dark urine	65	Vomiting	40
Light stool	60	Anorexia	35
Nausea	45	Constipation	25
Vomiting	40	Emotional depression	20
Weakness	35	Jaundice	10
Itching	25		
Diarrhea	15		
Emotional depression	10		

Table 2. Presenting signs of adenocarcinoma of the pancreas by site of tumor

Head		Body and tail	
Sign	% patients	Sign	% patients
Jaundice	80	Hepatomegaly	35
Hepatomegaly	75	Abdominal mass	25
Courvoisier's signs	30	Ascites	20
Abdominal mass	15	Jaundice	10
Ascites	5		

tients. It is more common in patients with lesions involving the body and tail of the gland and it often pulsates as the mass lies just ventral to (or actually invades, in which case a bruit may be heard) the aorta. An abdominal fluid wave, shifting dullness, and/or coronal resonance (signs of ascites) are present in 15%-20% of patients. A few patients will present with clear physical signs of distant spread of disease such as supraclavicular lymphadenopathy (Virchow's node) or subcutaneous tumor nodules.

Signs and symptoms of diabetes mellitus may be seen in as many as 15% of patients. Superficial thrombophlebitis (Trousseau's sign) is only rarely found at the time of presentation of pancreatic cancer. Gastrointestinal hemorrhage is occasionally the event precipitating medical consultation.

7.4.2 Laboratory Data

Routine laboratory tests are nonspecific and usually not sufficient to distinguish pancreatic cancer from pancreatitis or other pancreatitic diseases. In patients with lesions located in the pancreatic head, the serum bilirubin (predominately conjugated) level will usually be over 10 mg/dl and the serum alkaline phosphatase level will be markedly elevated. Elevations of other common liver function tests are more variable. A normochromic normocytic anemia is present in about 60% of patients and occult blood is found in the stool in as many as 90% of cases. The erythrocyte sedimentation rate is often elevated.

A variety of serologic tumor markers have been found to be associated with pancreatic cancer, principally the carcinoembryonic antigen (CEA) alphafetoprotein (AFP), and the pancreatic oncofetal antigen (POA). The CEA level will be elevated in the sera of as many as 85% of patients with pancreatic cancer (HOLYOKE et al. 1979; KLAVINS 1981; MOOSA and LEVIN 1981; ZAMCHECK and MARTIN 1981). Unfortunately, it is also elevated in 65% of patients with other cancers and nearly half of patients with benign pancreatic disease. The serum level of the tumor antigen CA 19-9 is often elevated in patients with cancer of the panreas or other gastrointestinal tumors. The sensitivity and specificity of this test with regard to distinguishing pancreatic cancer from chronic pancreatitis and from patients without pancreatic cancer have been reported to be 85% and 95%, respectively (DEL FAVERO et al. 1983; DEL VILLANO et al. 1983; SCHMIEGEL et al. 1984; STEINBERG et al. 1984). The serum AFP level is not a useful marker for pancreatic cancer because an elevated level is predictive of pancreatic cancer in only about a third of cases, and only 25% of patients with pancreatic cancer have elevated AFP levels (KLAVINS 1981; MCINTIRE et al. 1975; MOOSA and LEVIN 1981). The POA level is elevated in pancreatic cancer, lung cancer, colon cancer, and biliary cancer, so it is not a useful screening test for pancreatic cancer (BANWO et al. 1974; GELDER et al. 1978a, b; OGUCHI et al. 1984). Other antigens thought to be associated with pancreatic adenocarcinoma include CA-50, DU-PAN-2, 47DIO, SPAN-1, RA96, PCAA and PaA (METZGAR and ASCH 1988). In general, the lack of tumor specificity and pancreatic specificity make serum markers unreliable when used alone. The use of multiple antigen markers may enhance diagnostic sensitivity but will decrease specificity. Antigen markers are often useful as clinical management aids as indicators of progression of disease.

7.4.3 Radiographic Studies

The routine upper gastrointestinal series is rarely helpful in diagnosis of pancreatic cancer but a variety of abnormalities may be seen. Hypotonic duodenography will show some abnormality in about three-quarters of patients with pancreatic cancer. The most characteristic findings are duodenal narrowing, widening of the C-loop of the duodenum and the inverted or reversed 3-sign (due to duodenal distortion by fixation of the ampulla with indentation of the concave border of the duodenum), and indentation of the duodenum by a dilated common bile duct. Other findings include double contours on the duodenum or antrum ("antral pad" sign), distortion and spiculation of the duodenal mucosal folds on their medial aspect, and narrowing or obstruction of the duodenum. Lesions in the head of pancreas affect the second portion of the duodenum whereas those in the body affect the third portion and those in the tail, the fourth portion. On the upper gastrointestinal examination, as with clinical symptoms, lesions in the head of the pancreas produce radiographic signs much earlier than those in the body or tail, and a finding of diffuse widening of the C-loop generally is indicative of advanced disease in those cases caused by pancreatic malignancy. Routine chest radiographs will show evidence of pulmonary metastasis in approximately 5% of patients.

Selective celiac and superior mesenteric angiography can be helpful in diagnosis and staging. Tu-

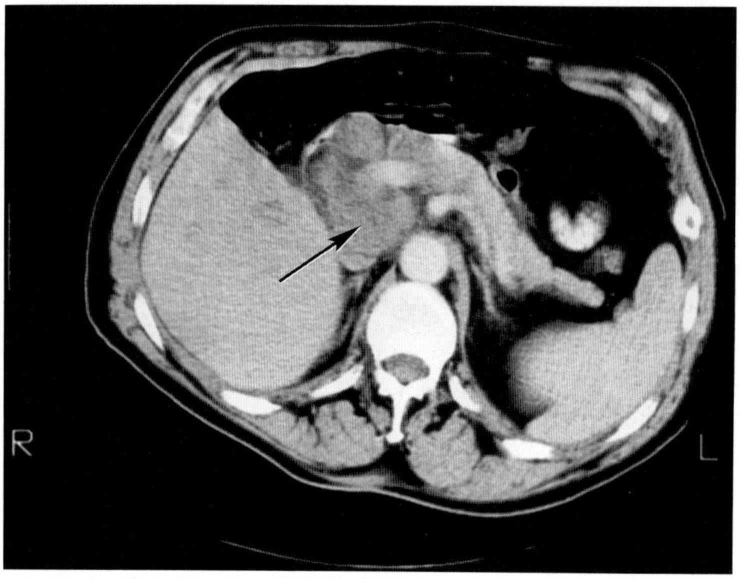

Fig. 4. Transverse computerized body section tomogram showing tumor of head of pancreas *(arrow)*

mor size and location may be assessable to some extent with such studies by virtue of vascular displacement, neovascularity, and tumor blush. Tumor neovascularity (small, irregular, tortuous vessels) is more common with islet cell carcinomas and cystadenocarcinomas and arteriography continues to be useful in the diagnosis of islet cell tumors, which can frequently be missed on CT scan. Arteriography is also useful in the determination of resectability and providing a "roadmap" for the surgeon in such cases. Obstruction of the portal vein and encasement of major arteries or veins (superior mesenteric, hepatic, portal, middle colic) reliably indicate unresectable lesions, whereas abnormal vasculature confined to the pancreas itself is often an indication of resectable disease. With superselective arteriographic techniques, the diagnostic accuracy of the procedure can be increased to 90% or more (PAUL et al. 1965; ROSCH and KELLER 1981).

The CT scan has become one of the most useful studies for diagnosis and staging of cancer of the pancreas (Fig. 4). A tumor mass or focal area of enlargement of the pancreas is characteristic of cancer of the pancreas as opposed to diffuse swelling of the gland in pancreatitis. Lesions in the pancreatic head are most difficult to detect on CT scan. With (and sometimes without) obstruction of the major pancreatic duct, this structure can be clearly visualized with CT scanning using proper scan techniques (5 mm slices through pancreas). A dilated pancreatic duct without pancreatic calcification is a finding highly suspicious for pancreatic adenocarcinoma even if no mass can be demonstrated. Other CT criteria used in the diagnosis of pancreatic cancer include change in shape and attenuation values, obliteration of peripancreatic fat planes, and loss of tissue boundaries. Sometimes CT scans in conjunction with endoscopic retrograde cholangiopancreatography (ERCP) will provide even more information and ERCP should be performed despite negative CT and ultrasound studies if strong clinical suspicion of pancreatic cancer is present. The sensitivity and specificity rates for CT scanning in the diagnosis of pancreatic cancer range from 60% to 80%, and 65% to 85%, respectively (see Table 3 for approximate values).

7.4.4 Special Diagnostic Procedures

Pancreatic ultrasonography is also a valuable study in the evaluation of the patient with suspected pancreatic cancer. Whether this study is more or less accurate than CT in the diagnosis of the disease has been the subject of considerable debate. Either technique is relatively specific and relatively sensitive in the diagnosis of pancreatic cancer (Table 3). Either can demonstrate hepatic metastasis and biliary dilatation, but CT scanning is superior in demonstrating large lymph node metastases and other extrapancreatic spread of disease. The two studies can be complementary to some extent. Ultrasound studies are confounded by obesity, ascites, and intestinal gas, whereas CT visualization of pancreatic masses is actually enhanced by the presence of natural radiographic contrast material (intestinal

Table 3. Approximate accuracy of diagnostic tests for adenocarcinoma of pancreas

Test	% sensitivity	% specificity	Predictability (%) of	
			Positive test	Negative test
Ultrasound	85	85	80	85
CT	75	80	75	80
ERCP	95	95	99	80
Angiography	70	95	85	80
Percutaneous biopsy	85	95	98	75
Operative biopsy	95	99	99	90

gas, peripancreatic fat) adjacent to the pancreas. Conversely, CT images of the pancreas can be indistinct in patients with cachexia and a limited amount of peripancreatic fat. In such patients, ultrasound may be more useful in outlining a pancreatic tumor mass. ORMSON et al. (1986) have shown that a normal homogeneous ultrasound echotexture in the pancreas is very useful in excluding pancreatic pathology when the CT scan is equivocal. The major limitation of a positive pancreatic echo study is its inability to differentiate pancreatitis from pancreatic malignancy.

In an effort to improve ultrasonic imaging of the pancreas, a small ultrasound probe can be mounted on the distal insertion tube of a flexible fiberoptic endoscope (DIMAGNO et al. 1980). In fact, such an instrument is currently commercially available. Endoscopic access to the target organ obviates several problems with routine ultrasound such as those occasioned by intestinal gas and bony structures. This study can be done with the same type of sedation as routine upper gastrointestinal endoscopy and has been well tolerated in limited experience (STROHM 1984).

Early data suggest that this technique may be very sensitive and specific in localizing small tumors in the pancreas (YASUDA et al. 1984, TIO and TYTGAT 1986). There are a few instances in which small (<2 cm) tumors have been detected by this technique but not by standard modalities such as CT or ERCP (YASUDA et al. 1988). Further experience with endoscopic ultrasound will be required to place it in its proper perspective; however, there is reason to believe that with this technique clinicians may be able to diagnose pancreatic tumors in a much earlier stage than has been possible hereto fore.

Endoscopic retrograde cholangiopancreatography provides a means to visualize the biliary tract and the pancreatic ducts radiographically. The resulting pancreatographs are almost always abnormal in patients with pancreatic cancer. The disease may produce stenosis of the major pancreatic duct (with or without proximal dilatation), complete obstruction of the duct, encasement of the duct, focal destruction of the gland, and/or cyst formation. Even though ERCP is highly sensitive and specific for pancreatic cancer (Table 3), the technical failure rate may be as high as 20% even in the hands of a skilled endoscopist (HALL et al. 1978).

The double-duct sign (involvement of the pancreatic duct and the adjacent portion of the common bile duct) is highly suggestive of pancreatic cancer but is not pathognomonic. Blockage of the pancreatic duct is likewise not pathognomonic of pancreatic cancer. Malignant processes tend to produce abrupt changes in the duct while a slow tapering of the duct is more likely to be due to a benign process.

The utility of magnetic resonance imaging (MRI) in the diagnosis and staging of pancreatic cancer has yet to be completely defined. However, one study (TSCHOLAKOFF et al. 1987) comparing CT and MRI has shown that MRI has advantages over CT in detecting liver metastases or vascular involvement invisible on CT. The MRI also showed superiority in evaluating the pancreas postoperatively when CT images were degraded by surgical clip artifacts. Major disadvantages of MRI include its inability to identify pancreatic calcifications important in differentiating carcinoma from inflammation.

Radioisotope pancreatic scans are not useful in diagnosis or staging of pancreatic cancer.

7.4.5 Definitive Diagnostic Procedures

The diagnosis of pancreatic cancer can only be definitively established histologically from specimens obtained by either biopsy or autopsy. Until relatively recently, surgeons have been reluctant to biopsy pancreatic lesions for fear of complications caused by leakage of pancreatic juice (MOOSA et al. 1986;

SCHULTZ and SAUNDERS 1963). This surgical dictum is a holdover from the preantibiotic days. More recent experience has shown that biopsy of pancreatic tumors is relatively safe (FAINTUCH and LEVIN 1986; GEORGE et al. 1975; HO et al. 1977; HYLAND et al. 1981; ISSACSON et al. 1974; KLINE and NEAL 1978; LEE 1982; LIGHTWOOD et al. 1976; MOOSA et al. 1986; WAY 1987; WEISS et al. 1982). Biopsy material may be obtained percutaneously or intraoperatively.

7.4.5.1 Percutaneous Biopsy

Of course, the easiest approach to diagnosis in patients with hepatic metastasis is percutaneous liver biopsy. Pancreatic lesions can be approached with a 22-gauge needle under CT or ultrasonic guidance. A cytologic specimen may be obtained through the needle by aspiration as the needle is passed back and forth through the lesion several times. Interpretation of the specimen requires an experienced cytopathologist, but processing of the sample is rapid and a diagnosis can be generally obtained within 20–30 min. A positive result is virtually diagnostic for pancreatic cancer. False-postive results have not been reported. False-negative results are obtained in approximately 15% of patients (HO et al. 1977; KLINE an NEAL 1978). The risk of abdominal wall implantation of tumor is less than one-tenth of 1% with this technique (FERRUCCI et al. 1979; SMITH et al. 1980).

7.4.5.2 Laparoscopic Biopsy

Abdominal laparoscopy provides a suboptimal view of the pancreas, but the main abdominal cavity and the inferior surface of the liver can be inspected. By puncturing the gastrocolic omentum, the lesser sac can be opened, providing a limited view of the body and tail of the pancreas. Most pancreatic lesions are inaccessible to biopsy via laparoscope, but histologic specimens may be obtained from peritoneal implants of disease (CUSHIERI et al. 1978; ISHIDA 1983; ISHIDA et al. 1984).

7.4.5.3 Intraoperative Biopsy

If exploratory laparotomy documents extrapancreatic extension of disease to liver, lymph nodes, mesentery, or peritoneum, the biopsy should be obtained from that site. If the lesion appears to be confined to the pancreas, tissue may be obtained by (in order of preference) needle aspiration, a core sample, or a wedge biopsy of the lesion. If needle aspiration does not provide a diagnosis satisfactorily, a core sample may be obtained. In this case, the lesion is best approached through the duodenal wall so that if drainage from the pancreas subsequently occurs, that drainage is into the duodenum rather than into the peritoneal cavity. In general, it is important to confirm the clinical diagnosis of cancer histologically even if resection is out of the question. This is because it has now become apparent that radiation therapy with or without chemotherapy can provide significant palliative benefit in the management of this disease.

7.5 Staging Systems

Three staging systems commonly used for adenocarcinoma of the pancreas are compared in Table 4 (BEAHRS and MYERS 1983; CUBILLA et al. 1978a; DOBELBOWER et al. 1986). Although the same general principles of cancer staging are employed in all three systems, it should be noted that the stage groupings are *not* identical. A patient with minimal extension of disease beyond the pancreas (T_2) may have his disease categorized as stage I in the American Joint Committee on Cancer (AJCC) classification but stage II in the Medical College of Ohio and CUBILLA systems. A patient with contiguous extension of disease to the liver would fall into stage II in the CUBILLA classification and stage IV in the Medical College of Ohio system, whereas a patient with hematogenous hepatic matastasis would clearly fall into stage IV in either the AJCC or the Medical College of Ohio system, but would be classified as stage III in the CUBILLA system. Such inconsistencies can confound interpretation of reported clinical data.

7.6 Pathologic Classification

Malignant tumors of the pancreas may originate from duct cells, acinar cells, connective tissue elements, or lymphoid elements or may be of mixed cell type (Table 5). Approximately three-fourths of all malignant pancreatic tumors are ductal adenocarcinomas (Table 5). Variations of this histology make up nearly 90% of tumors of the exocrine pancreas. The majority of these tumors are mucus producing. Perineural invasion occurs in at least 90% of cases, and nearly all lesions show a desmoplastic

Table 4. Comparison of staging systems currently used for adenocarcinoma of the pancreas

CUBILLA et al. (1978a)	American Joint Committee on Cancer (BEAHRS and MYERS 1983)	Medical College of Ohio (DOBELBOWER et al. 1986)
Stage I	*Stage I, T1, T2, N0, M0*	*Stage I*
Tumor confined to the pancreas	T1 – No direct extension of primary tumor beyond pancreas T1a – Tumor 2 cm or less in greatest diameter T1b – Tumor more than 2 in greatest diameter	Cancer confined to the pancreas. No extension of cancer beyond the capsule of the pancreas. No involved lymph nodes. No hepatic or other distant metastasis.
Stage II		*Stage II*
Involvement of regional lymph nodes and/or contiguous organs	T2 – Tumor extension to duodenum, bile ducts, or peripancreatic tissues N0 – Regional nodes not involved M0 – No (known) distant metastasis	Cancer extending locally beyond the pancreas to involve contiguous structures. No involved lymph nodes. No hepatic or other distant metastasis.
	Stage II, T3, N0, M0	
	T3 – Direct extension to stomach, spleen, colon, or adjacent large vessels	
	Stage III, T1–3, N1, M0	*Stage III*
	N1 – Regional nodes involved	Cancer metastatic in regional lymph nodes. No hepatic or other distant metastases.
Stage III	*Stage IV, T1–3, N0–N1, M1*	*Stage IV*
Distant metastasis	M1 – Distant metastasis in liver or other sites	Hepatic metastasis (either direct extension or hematogenous spread) or other distant metastasis.

Table 5. Histologic classification of the common tumors of the exocrine panreas (modified from CUBILLA and FITZGERALD 1984)

Tumor type	Relative frequency (%)
Duct cell origin	88.8
Duct cell carcinoma	76.6
Giant cell carcinoma	4.4
Adenosquamous carcinoma	3.1
Microadenocarcinoma	2.5
Mucinous (colloid) carinoma	1.4
Cystadenocarcinoma (mucinous)	0.8
Acinar cell origin	1.2
Acinar cell carcinoma	
Mixed cell origin	0.2
Connective tissue origin (sarcomas)	0.6
Lymphoma	0.1
Uncertain histogenesis	9.1

response around and distal to the tumor (CUBILLA and FITZGERALD 1978, 1984).

Giant cell carcinoma accounts for approximately 5% of malignant pancreatic tumors. This disease entity is characterized by huge cells with giant nuclei frequently producing large hemorrhagic necrotic cysts. Patients with this histologic variant commonly expire within a few weeks to months after diagnosis.

Adenosquamous carcinoma has a prognosis roughly the same as ductal adenocarcinoma but histologically shows a combination of epidermoid carcinoma and adenocarcinoma. Mucinous (colloid) adenocarcinoma contains large amounts of mucin and may grossly resembly a goiter because of its gelatinous cut surface; the tumor is also called "colloid" or gelatinous carcinoma. The sex ratio is higher for mucinous carcinomas than for ductal carcinomas. It has been suggested that prognosis is roughly proportional to the amount of mucin in the lesion (CUBILLA and FITZGERALD 1984).

Cystadenocarcinoma (mucinous) occurs in females six times as often as in males. Cystadenocarcinomas are typically large multilocular tumors lined by columnar mucin-producing epithelium in the tail of the pancreas. Complete resection results in 5-year survival in over half of the patients (KLÖPPEL and FITZGERALD 1986).

Acinar cell carcinoma is associated with a particularly poor prognosis (CUBILLA and FITZGERALD 1980a, b, 1984; FITZGERALD 1981). Typically this histologic variant occurs in elderly patients, many of whom present with distant metastasis from an occult primary lesion (CUBILLA and FITZGERALD 1984; WEBB 1977). Occasionally, it is associated with a syndrome of subcutaneous fat necrosis, polyarthritis, and eosinophilia, presumably related to release of lipase and other enzymes by the tumor (BURNS et al. 1974; SCHREIBER and PROBST 1977; TANNENBAUM et al. 1975).

Primary lymphoma of the pancreas is a rare, but theoretically radiocurable/chemocurable lesion (ACKERMAN et al. 1976; BAYLOR and BERG 1973; OGUMA et al. 1983; RICHARDS et al. 1958).

7.7 Prognostic Factors Influencing Choice of Treatment

The overall survival rate of patients with pancreatic cancer remains less than 1% (BAUMEL and DEIXONNE 1986; BRAGANZA and HOWAT 1979; LEICHMAN 1985; SILVERBERG and LUBERA 1989). Survival time is related to treatment factors and especially to the stage of the disease. Untreated patients demonstrate a median survival time of approximately 3 months from diagnosis to death with no patients surviving 5 years. Patients with local disease treated by curative resection display a median survival time of approximately 12 months and a 5-year survival rate of 6.4% (JORDAN 1987). Data from the Gastrointestinal Tumor Study Group (GITSG) trial suggest that the addition of radiotherapy and chemotherapy as adjuvants to curative resection may double median survival time and 5-year survival rate (KALSER und ELLENBERG 1985). Less than 150 patients with cancer of the pancreas are known to have survived 5 years regardless of treatment (JORDAN 1987).

With few exceptions, only jaundiced patients with tumors located in the head of the pancreas have any prospect of definitive resection. Back pain is generally indicative of retroperitoneal extension of cancer involving the celiac plexus and forebodes unresectability and shortened survival. Weight loss of more than 25 lb usually indicates relatively advanced unresectable disease. Malignant ascites, supraclavicular lymphadenopathy, and skin metastases are, likewise, signs of advanced unresectable disease and are harbingers of impending demise.

Prognosis is clearly related to tumor site, size, stage, and grade. Lesions located in the pancreatic head have a better prognosis than lesions in the body or the tail. There are only four known 5-year survivors of cancer in these distal regions of gland (GORDON-TAYLOR 1934; NAKASE et al. 1977), two treated by resection (HANNA and HASTINGS 1968), one by radiation therapy (DOBELBOWER and MILLIGAN 1984), and one by biopsy and observation. The patient treated by radiation therapy died of hepatic metastases 69 months postdiagnosis.

For patients with cancer confined to the pancreas treated by "curative" resection, survival time is inversely correlated with the size of the lesion (Table 6). Regardless of type of therapy, survival is inversely correlated with disease stage (Table 7). Patients with disease confined to the pancreas display a median survival time of 11 months as compared to 5 months for those patients with metastases to regional lymph nodes, and 3 months for patients with distant metastases (CUBILLA et al 1978a).

Tumors of low histologic grade appear to present at a lower stage and show an improved median survival as compared to tumors of higher histologic grade (EDIS et al. 1980; KLÖPPEL and FITZGERALD 1986; POLLARD 1981).

Of course, patients in good general condition at the time of diagnosis and treatment tend to fare better than those in poor condition. As expected, patient survival time is positively correlated with Kar-

Table 6. Survival[a] by size of tumor of patients with stage I adenocarcinoma of the pancreas treated by Whipple resection (modified from CUBILLA and FITZGERALD 1978)

Tumor size (cm)	No. of patients	Median survival (months)
1-1.9	5	29
2-2.9	5	17
3-3.9	5	6
4-4.9	4	5
5-10	6	2

[a] Operative deaths excluded

Table 7. Survival of patients with adenocarcinoma of the pancreas by extent of disease (modified from CUBILLA et al. 1978a)

Extent of disease	No. of patients	Median survival (months)
Confined to pancreas	47	11
Lymph node metastasis	69	5
Cancer beyond pancreas and lymph nodes	212	3

nofsky performance score (KALSER and ELLENBERG 1985; KALSER et al. 1985).

7.8 General Management

Cancer of the pancreas is rarely diagnosed at an early stage; hence, fewer than 20% of patients are candidates for surgical resection at the time of diagnosis. Resection with an operative mortality below 10% is being achieved by experienced pancreatic surgeons today (DOBELBOWER and HOWARD 1985). Response rates to single agent chemotherapy are generally less than 15% for this disease in cooperative group trials, and aggressive combination chemotherapy regimens produce response rates that are little better. Responses to chemotherapy in this disease are characteristically short-lived. For years, pancreatic adenocarcinoma was regarded as radioresistant (BORGELT et al. 1978; DOBELBOWER 1979a). Only recently has this notion been challenged.

Clinical management strategies for pancreatic cancer vary according to the extent of the disease: resectable, locally unresectable (disease confined to the upper abdomen with no evidence of hematogenous dissemination), or metastatic (DOBELBOWER and HOWARD 1987).

7.8.1 Resectable Disease

For patients with resectable adenocarcinoma of the pancreas, the treatment of choice is radical surgical excision. The operation should be preceded by CT scan of the liver to delineate hepatic metastases and duodenoscopy to identify and obtain a biopsy of ampullary and duodenal cancers. Percutaneous transhepatic cholangiography with drainage may permit preoperative decompression of the bile ducts (ELLISON et al. 1984). Distant abdominal disease is a contraindication to extensive pancreatic resection.

After the peritoneal cavity is entered, the peritoneum and liver are carefully palpated and inspected. Dimpling or retraction at the base of the mesocolon may represent local extension of cancer. Dilated venous structures may imply obstruction by tumor. Entering the lesser sac permits inspection of the pancreas while a complete Kocher maneuver permits assessment of the distal common duct, the superior mesenteric artery, and the posterior aspect of the pancreatic head. These maneuvers permit detection of most contraindications to resection. Careful examination of the branches of the celiac axis, superior mesenteric vessels, portal vein, and porta hepatis often identifies suspicious lymph nodes, from which a biopsy should be obtained and sent to the laboratory for frozen section analysis.

If no tumor is encountered to this point, the common bile duct, hepatic artery, and portal vein are dissected and isolated with vascular tapes. The superior mesenteric vessels are identified and the overlying peritoneum is incised over the caudal margin of the pancreas. The anterior aspect of the mesenteric vein is usually free of tributaries, so that the surgeon can develop a plane between this structure and the overlying portion of the pancreas and tunnel underneath the gland if invasion of the vein has not occurred. A Penrose drain may be used to elevate the tumor und the gland and facilitate further retropancreatic dissection. If no distant metastasis or vascular involvement is found, the lesion can be considered resectable. To this point, the procedure may be terminated without resection, but further steps are generally irreversible.

The traditional definitive surgical approach is pancreaticoduodenectomy (Whipple resection) or total pancreatectomy. The pros and cons of these two operations are compared in Table 8. A pylorus-preserving modification of the traditional pancreatectomy has been introduced in an effort to reduce the incidence of postgastrectomy syndrome and marginal ulceration (TRAVERSO and LONGMIRE

Table 8. Comparison of Whipple resection and total pancreatectomy for adenocarcinoma of the pancreas

	Whipple resection	Total pancreatectomy
Splenectomy	No	Yes
Histologic diagnosis of cancer required before resection	When feasible	Yes
Pancreatic duct transection (possible spillage of tumor cells)	Yes	Seldom
Pancreaticojejunostomy (possible leakage)	Yes	No
Pancreatic remnant (possible residual cancer)	Yes	No
Average number of lymph nodes removed	33	41
Operative mortality (experienced pancreatic surgeon)	0%-25%	0%-25%
Long-term metabolic defects	±	+++

1978, 1980). Surgical resection of the pancreas often lies at the limits of a surgeon's skill and the patient's endurance. For these reasons, surgical resection of cancer of the pancreas should only be undertaken by individuals who frequently perform this type of surgery. Patients with suspected resectable pancreatic adenocarcinoma diagnosed by most general surgeons in community hospitals should be decompressed by tube cholecystostomy and referred to major centers for definitive management. Palliative gastrointestinal and biliary bypass procedures may be indicated at the time of the first operation, but these should be done with cognizance of subsequent surgical and radiotherapeutic procedures so that definitive treatment is not compromised. In general, the patient is better off if he is subjected only to a one-stage, definitive resection (with adjunctive radio/chemotherapy).

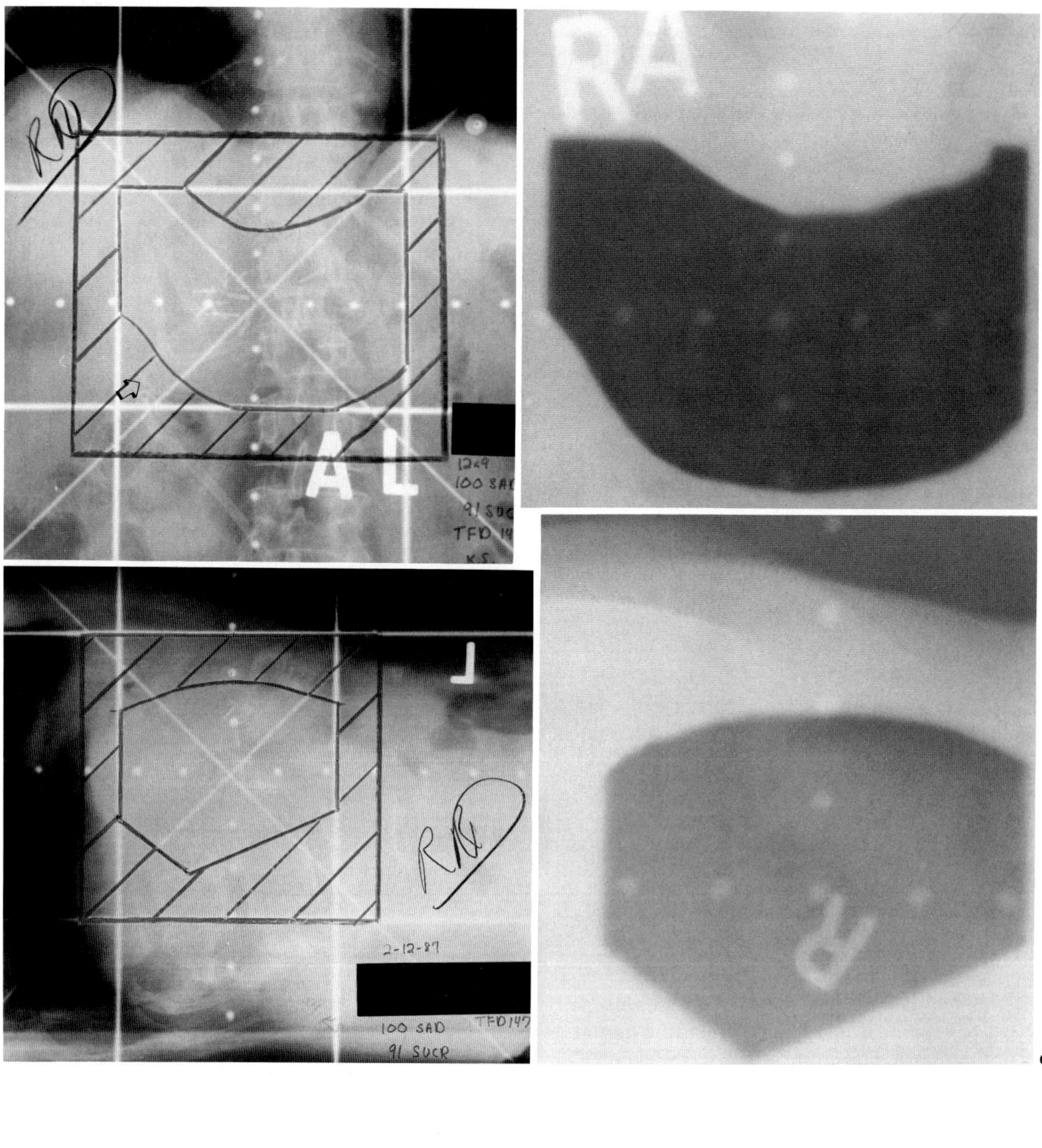

a b c d

Fig. 5. a Simulator film: anterior field, pancreas. Large clips outline gross tumor. Outline for custom beam shaping block is drawn on film. The biliary-enteric anastomosis *(arrow)* will be shielded. Note generous lateral margins. **b** Anterior beam film: pancreas. **c** Simulator film: lateral fields, pancreas. **d** Lateral beam film: pancreas

Following resection of the cancer, placement of radiopaque clips outlining the tumor bed will greatly facilitate delivery of postoperative external beam radiation therapy (Fig. 5). Intraoperative consultation with the radiation oncologist is also of value, allowing him or her to better appreciate the original position of the tumor, to determine more precisely the volume to be irradiated, and to visualize the anatomic locations of surgical anastomoses and of loops of bowel used to construct biliary and gastrointestinal diversions. Intravenous hyperalimentation, adequate ventilatory support, and good intensive nursing care after operation are essential elements of pancreatic surgery.

Data from the GITSG adjuvant study for cancer of the pancreas suggest that postresection radiation therapy in combination with 5-fluorouracil (5-FU) will enhance survival (Fig. 6) (KAISER and ELLENBERG 1985). Optimal radiation dosages and field arrangements have yet to be defined, as well as optimal chemotherapeutic agents and dosages. Available data suggest that a split course radiation dose as low as 30 Gy in 6 weeks may be of benefit when combined with 5-FU therapy. The radiation therapy fields should be closely tailored to the target volume (tumor bed), and loops of bowel used in construction of biliary or gastrointestinal bypasses should be excluded from the fields, if feasible. Likewise, the anastomoses, except for the pancreaticojejunal anastomosis, should be excluded from the radiation field (Fig. 5). Any pancreatic remnant should be included in the postoperative radiation fields.

When planning the target volume, the radiation oncologist must consider the location of the adjacent normal structures, particularly the spinal cord, kidneys, liver, and gut. A postoperative CT scan, with and without oral and intravenous contrast agents, usually suffices to define locations of most of these structures.

Radiation therapy should not be started until the patient has recovered from the surgical insult and any complications therefrom have been adequately managed. Development of pancreaticocutaneous fistula is not a contraindication to postoperative irradiation. A dose of 50 Gy in 1.8-2 Gy daily increments, five fractions weekly, is generally well tolerated after recovery from major upper abdominal surgery. Treatment should be interrupted or terminated if reaction to therapy is severe. Adjuvant 5-FU can be combined with these radiation doses on the first and last 3 days of radiation therapy in a dose of 500 mg/m^2. Alternatively, 5-FU can be administered as a continous 96 h infusion (1000 mg/m^2 - not to exceed 1.5 g/day) during the 1st and 5th weeks of irradiation. Weekly 5-FU should follow completion of external beam radiation therapy. The 5-FU should be delayed 1 week if the white blood count is less than 3000 or platelet count less than 75000.

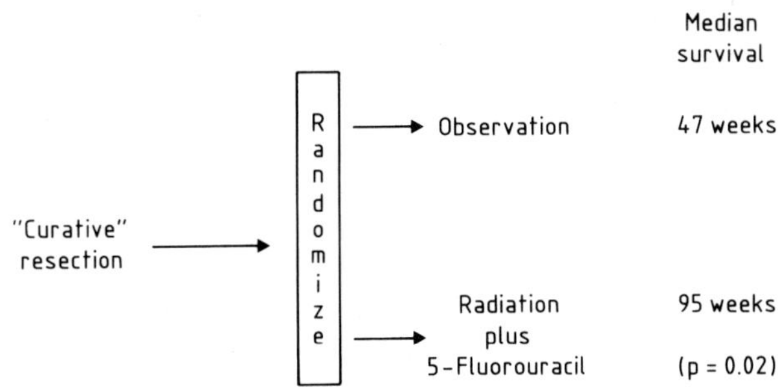

Fig. 6. Schema: Gastrointestinal Tumor Study Group Protocol 9173 (GITSG 9173 pancreas adjuvant study)

Nutritional support may be necessary during radiation or drug treatment. Patients should be followed closely by the radiation oncologist and medical oncologist (monthly intervals) to assess late toxicity. Resultant gastritis should be managed by the usual medical means.

7.8.2 Locally Unresectable Disease

The majority of patients with pancreatic cancer will present with unresectable disease. Approximately half of these will have no evidence of metastatic disease, but will be unresectable by virtue of local extension of disease into adjacent structures (e.g., portal vein, superior mesenteric vein or artery, stomach, colon, mesentery). Extensive surgical procedures in such patients may do more harm than good, but many can be treated with combined radiation therapy and chemotherapy. A few patients with such unfavorable disease may actually be salvaged with radiotherapeutic or combined modality approaches (DOBELBOWER and MILLIGAN 1984).

In light of the above information, it is important to clearly establish the diagnosis histologically at the time of first laparotomy and it is important to outline the gross tumor margins with radiopaque clips for purposes of subsequent radiation therapy. Intraoperative consultation with a radiation oncologist can be very helpful.

Many surgeons are reluctant to perform a biopsy of the pancreas because of the possibility of serious complication (pancreatic fistula). This surgical notion was conceived in the preantibiotic era, but persists even today. Recent reviews of experience with pancreatic cancer suggest that with current surgical and anesthetic techniques, and appropriate postoperative care, pancreatic biopsy is relatively accurate and safe (Faintuch and Levin 1986; Moosa et al. 1986; Way 1982; Weiss et al. 1982). Neoplasms that obstruct the pancreatic duct can produce proximal dilatation of the pancreatic duct system via increased intraductal pressure. For this reason, pancreatic biopsy should be limited to the neoplasm itself or be performed from the duodenal side of the pancreatic neoplasm. Percutaneous or intraoperative needle biopsies may be safer (albeit less reliable) than open biopsy. A transduodenal needle biopsy can conceivably avoid the possibility of pancreaticoperitoneal fistula. Obstruction of the pancreatic duct by tumor often produces peripheral pancreatitis, which may be difficult to clinically distinguish from neoplasm; however, careful palpation can sometimes make this determination easier, as can the "feel" of the tissues during biopsy. The radiation oncologist is often quite accomplished in palpation of neoplastic tissues; hence, his/her opinion regarding site of biopsy at the time of surgery may be helpful as it is for arranging clips to outline the gross limits of the neoplasm for subsequent external beam irradiation.

Radioisotope implanted directly into a pancreatic neoplasm is a mechanism for increasing the local radiation dose to the tumor, yet sparing the surrounding tissues (Fig. 7). The isotope most commonly employed for this purpose today is ^{125}I. Such isotope implants directly into the pancreas should only be performed by an experienced team under carefully controlled conditions because perioperative mortality rates as high as 25% have been reported from major centers (Table 9) (Al-Abdulla et al. 1981; Borgelt 1981; Dobelbower et al. 1986; Hilaris and Roussis 1975; Shipley et al. 1980; Syed et al. 1983; Whittington et al. 1981). Thorough exploration of the abdominopelvic cavity, especially the liver, peritoneum, and mesentery,

Table 9. Operative mortality: radioisotope pancreas implant

	No. of patients	Radioisotope	Operative mortality (%)
Thomas Jefferson University	52	^{125}I	23
Memorial Sloan Kettering	33	^{125}I	3
M. D. Anderson	25	^{198}Au	25
Syed	18	^{125}I	0
Medical College of Ohio	15	^{125}I	0
Shipley	12	^{125}I	0

Fig. 7. Cross-table intraoperative radiograph of needle placement for ^{125}I seed implant, pancreas

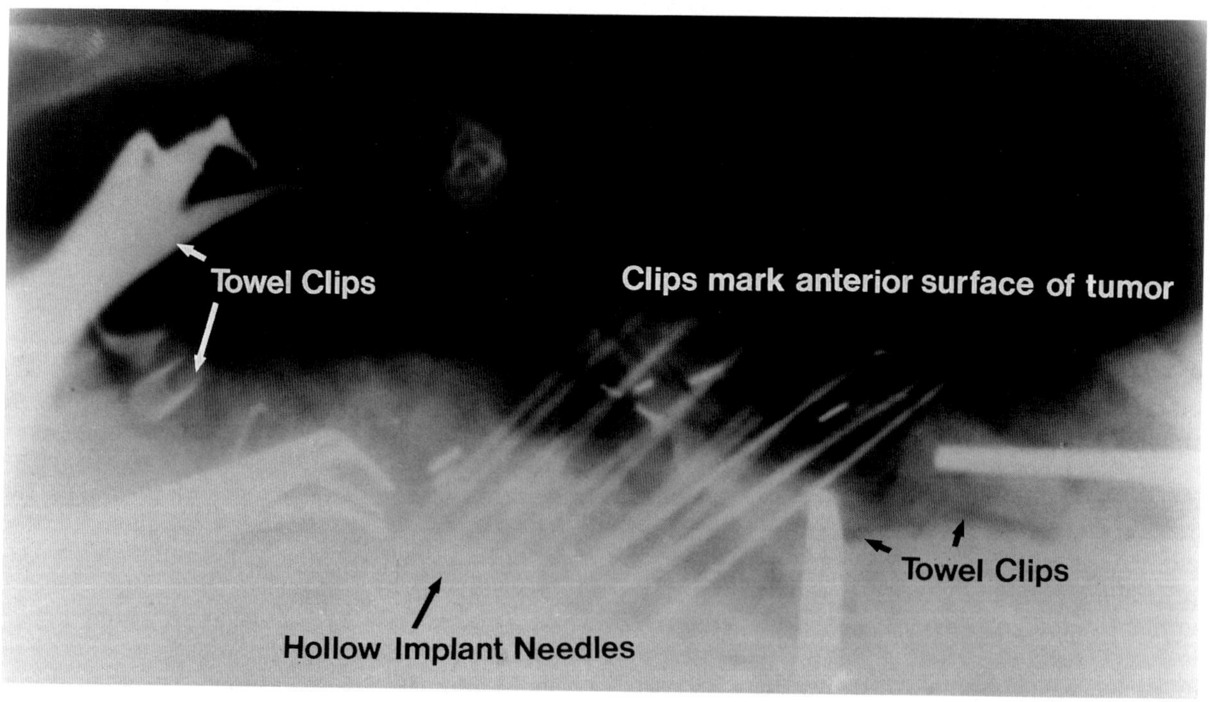

must be accomplished before a dicision is made to implant a pancreatic neoplasm. Adequate surgical exposure of the anterior surface of the tumor and excellent hemostasis are mandatory for pancreatic radioisotope implants. The radiation oncologist, who must have the final word regarding adequacy of exposure, placement of radiopaque clips, and type of radioisotope implant to be performed, must be thoroughly familiar with the local anatomy in each case (e.g., location of major vessels, biliary system, pancreatic duct). The tumor is measured and the radioisotope implant is mapped intraoperatively by the radiation oncologist with the assistance of a radiation physicist. An external beam radiation dose of 5 Gy may be administered preoperatively to minimize dissemination of tumor cells by the implant procedure (WHITTINGTON et al. 1981).

It is important to implant only the neoplasm with radioisotope and to avoid the surrounding uninvolved (or inflamed) pancreas. This will minimize complications from high-dose irradiation of normal pancreatic parenchyma and from violation of inflamed parenchyma by needles. The thickness of the neoplasm is judged by the operator placing the needles as there is usually a difference in consistency between tumor tissue and the surrounding normal structures. Occasionally, a needle penetrates a major blood vessel. Local pressure with a "peanut" usually achieves hemostasis. The operator should not use his/her finger for such purposes if radioisotope has already been implanted. A record of placement of radioactive sources through the needles at the time of withdrawal is made intraoperatively, but radiographs of the implant generally are not taken in the operating room. Postoperatively, stereoshift or (preferably) orthogonal radiographs can be obtained for dosimetric purposes.

Biliary and gastrointestinal bypass procedures are best accomplished after the radioisotope implant has been performed. Because of the low energy of ^{125}I-emitted gamma rays, the dose to the operator in constructing gastrointestinal and biliary bypasses, even in the face of a large pancreatic implant, is rather low. Both gastrointestinal and biliary diversions should be accomplished, even if "prophylactically," at the time of ^{125}I implant so as to minimize the necessity for subsequent surgical intervention. Choledochojejunostomy is preferred to choledochoduodenostomy or cholecystoduodenostomy because of the possibility of subsequent obstruction by tumor growth or radiation complication. If the anticipated life span is short, cholecystojejunal anastomosis is simpler. If the duodenum is not obstructed, a "circus movement" of food may follow a gastrojejunostomy and result in vomiting. Thus, a jejunojejunostomy, anastomosing the proximal and distal limbs of the gastrojejunostomy, may prevent this by preventing the reentry of food into the stomach.

Optimal radioisotope implant dosages have not been thoroughly worked out, but available evidence suggests that doses of 120–160 Gy over the period of complete isotope decay are well tolerated with ^{125}I. With other isotopes, the dosages must be modified.

Because of the occult local extesion of disease, radioisotope implant for pancreatic tumors should be combined with precision high dose (PHD) external beam therapy (DOBELBOWER et al. 1980a) and probably with chemotherapy as well (Fig. 8).

Pancreatic neoplasms require high local radiation doses to achieve control. These doses are in excess of the radiotolerance of surrounding normal tissues. High doses to the pancreas can be achieved with the use of PHD external beam radiation therapy techniques (Figs. 9, 10). This implies accurate three-dimensional localization of the tumor and adjacent normal structures. This is best accomplished with CT scanning in conjunction with radiopaque clips placed at the margins of the neoplasm at the time of surgery. The renal parenchyma may be further localized by anteroposterior and cross-table lateral views of the kidneys during the nephrogram phase of an intravenous pyelogram. The location of the spinal cord is established with the radiation therapy simulator. The location of the stomach and liver is best established by CT scan. The PHD irradiation techniques imply good patient immobilization, accurate reproduction of daily treatment set-up, careful simulation with attention to detail, the use of line lasers for reproducing patient set-up daily, and transposition of information from simulation to treatment machine set-up with especially high technologic standards and much attention to detail. Beams are angled, blocked, and often extensively shaped to conform the three-dimensional dose profile to the disease profile.

External beam radiation therapy should not commence until the patient is well recovered from the surgical procedure. The field should extend no more than 17 cm in greatest dimension. Doses of 1.8 Gy per fraction are well tolerated if the target volume is closely confined to the pancreatic mass. Only high-energy (over 4 meV) photons should be used for PHD external beam therapy. At the time of simulation for external beam radiation therapy, the excursion with respiration of the clipped pancreatic mass should be noted and the cephalic and caudal

Fig. 8

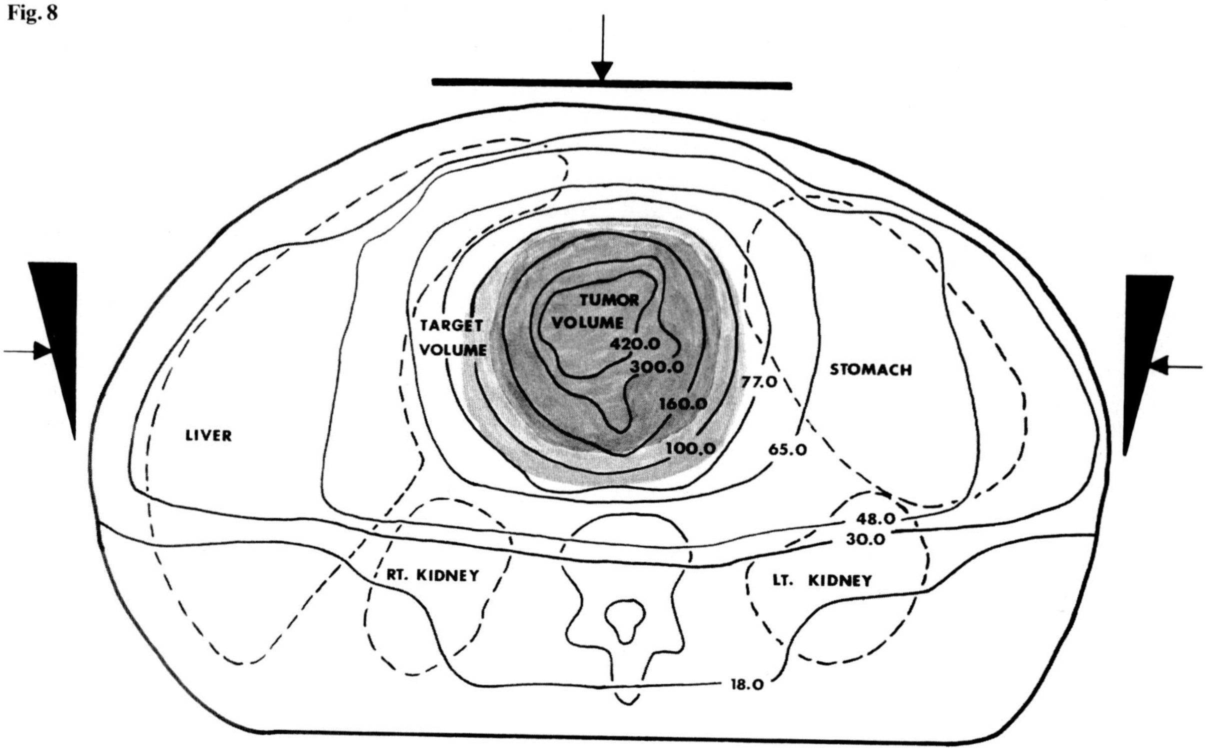

Fig. 9

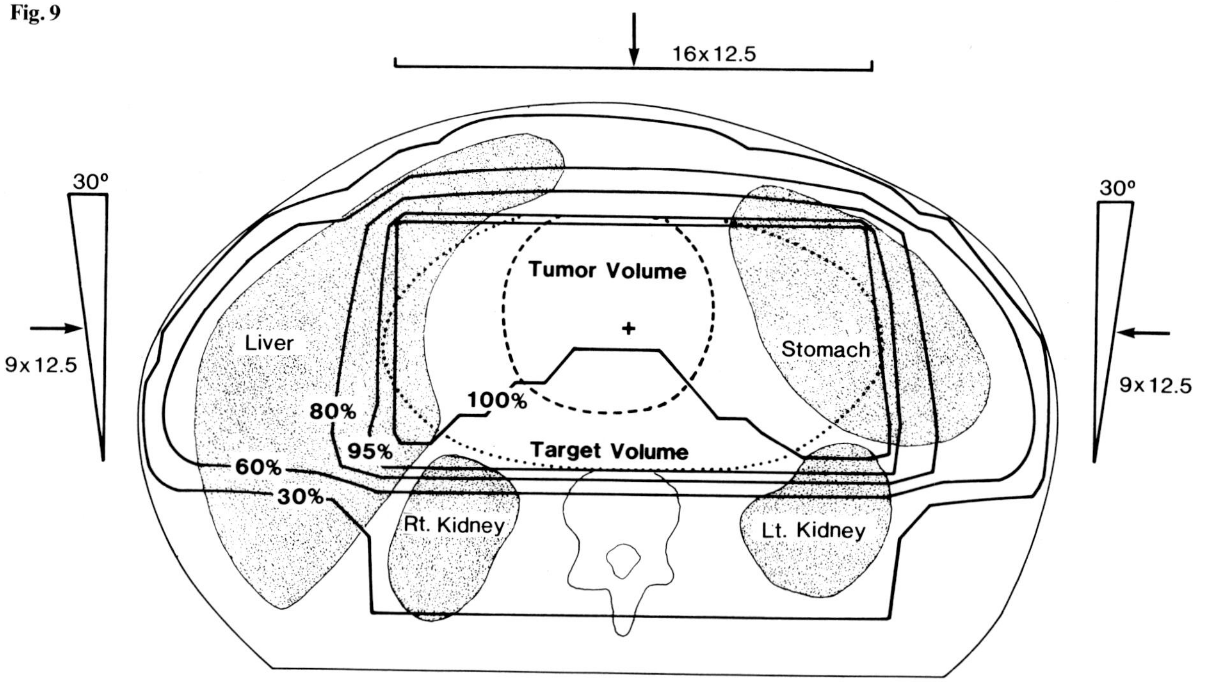

Fig. 8. Patient cross-section through pancreatic tumor *(darkly shaded)* with superimposed isodose distribution (Gy) from ^{125}I implant and shrinking external beam fields: custom-shaped anterior field and custom-shaped wedged lateral fields of 10 meVp photons. The 100 Gy isodose line surrounds the tumor volume. The target volume for external beam therapy is *lightly shaded*

Fig. 9. Isodose distribution: pancreas, 10 meVp X-rays, three shaped isocentric fields with wedged (30°) lateral fields

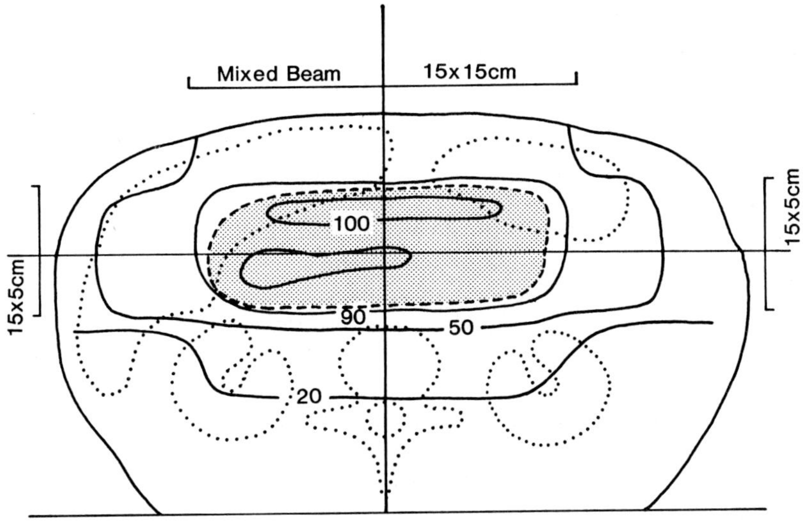

Fig. 10. Contour of patient with superimposed isodose distributions for pancreas treatment plan: opposed lateral 45 meVp photon fields and an anterior "mixed beam" field – 50% 45 meVp photons, 50% 20 meV electrons. The 90% isodose curve surrounds the target volume

radiation field borders adjusted accordingly. If the excursion is minimal, a margin of 2 cm should suffice. The right and left lateral field margins should be more generous, especially in the region of the common bile duct and the tail of the pancreas, as pancreatic tumors are known to spread along these routes in occult fashion (Fig. 5). The anterior field margin can be less generous because this is the border best defined by surgical clips as the approach is generally from the anterior. The posterior margin is often the most difficult to establish, but in no case should it be less than 1 cm. It is important to ascertain that at least 50% of the functioning renal parenchyma is outside the high-dose treatment volume (Figs. 9, 10). The dose to the remaining kidney tissue should be kept below 25 Gy, and the dose to the spinal cord should be kept below 45 Gy. After a dose of 45–55 Gy, the field may be reduced to irradiate just the neoplasm with "tight" margins, continuing to doses as high as 65 Gy. If external beam therapy is used alone without ^{125}I seed implant boost, we have carried the dose to 67 Gy in most instances.

5-Fluorouracil can be administered in conjunction with PHD external beam radiation therapy in a dose of 500 mg/m^2 daily for 3 days at the beginning and at the end of external beam radiation therapy or as a continuous 96 h infusion, 1000 mg/kg/day (not to exceed 1500 mg/day, during the 1st and 5th weeks of radiation therapy). Maintenance therapy can then continue with 5-FU in weekly doses of 500 mg/m^2. Drug administration should be delayed 1 week if the white count is less than 3000 or the platelet count is less than 75000.

It is thought that one of the reasons why pancreatic cancers are not better controlled by irradiation is that these bulky lesions contain a hypoxic core, the cells therein being relatively resistant to low linear energy transfer (LET) radiations. For high LET radiations (neutrons, protons, pi mesons, stripped nuclei, etc.), the radioprotective effect of hypoxia is obviated to some extent. For these radiobiologic reasons (among others), high LET radiation beams have been experimentally employed in the treatment of unresectable pancreatic cancer. The equipment necessary to produce such high LET beams is extremely sophisticated and costly. Hence, only a handful of centers in the world are equipped to conduct such studies.

Occasionally, the oncologist is presented with a patient with a clinical or surgical diagnosis of unresectable panreatic cancer. Under no circumstances should such patients be treated with PHD external beam radiation therapy, with radioisotope implant, or with chemotherapy in the absence of histologic confirmation of disease (or the appearance of metastases) because myriad disease processes, most notably pancreatitis, mimic pancreatic cancer. The risks of chemotherapy and/or high-dose irradiation are not warranted in such situations. Lower doses of radiation therapy may preclude subsequent definitive treatment. In such instances, attempts at histologic confirmation of the disease should be made.

7.8.3 Metastatic Disease

For patients presenting with metastatic pancreatic cancer, cure is currently impossible. The goal for such patients should be palliation of symptoms. This may require a combined surgical/radiotherapeutic/chemotherapeutic approach. Radiation therapy is extremely effective for palliation of local symptoms of pancreatic cancer except for gastrointestinal or biliary obstruction. These problems should be addressed surgically. Operation should include biopsy and, usually, cholecystojejunostomy, gastrojejunostomy, and, as already described, jejunojejunostomy to prevent reentry of food into the stomach. Careful attention to blood volume, serum albumin levels, and nutrition (intravenous hyperalimentation) is necessary to minimize operative mortality. Painful bone metastases regularly respond to a dose of 40 Gy administered in 15 equal increments over a period of 3 weeks. Likewise, upper abdominal pain and sometimes even nausea, vomiting, and anorexia are palliated by local upper abdominal irradiation, with or without adjuvant chemotherapy. Many studies have suggested that the palliative benefits of radiation in such instances are dose related, and it appears that one must achieve at least the radiobiologic equivalent of 45 Gy administered in equal 1.8 Gy increments over the course of 5 weeks in order to achieve a good response.

Obstruction of the gastrointestinal tract is best palliated surgically, whereas biliary obstruction can be treated by either surgical bypass or, in selected instances, with percutaneous transhepatic catheter drainage, with or without radioisotope placed in the catheter. Palliative Whipple procedures and palliative total pancreatectomies are not recommended because the surgical morbidity and mortality outweigh the potential benefit in patients destined to die shortly of the disease.

Five Fluorouracil has been used extensively for pancreatic cancer, with response rates reported as high as 67% in retrospective reports. In prospective controlled cooperative group trials, response rates of less than 20% have generally been observed. The combination of streptozotocin, mitomycin-C, and 5-FU is reported to have produced partial response in 43% of 23 evaluable patients so treated (WIGGANS et al. 1978). Median duration of response was 5 months and median duration of survival in responders was 9.5 months.

The patient who is asymptomatic after histologic proof of pancreatic cancer and appropriate palliative bypass surgery is best treated expectantly. Quality of life is important. Liberal use of analgesic is the best treatment for selected patients with symptomatic refractory disease. Nutritional support and comfort measures may be the most important tools available to the clinician in such instances. When present, diabetes or exocrine pancreatic insufficiency should be treated.

7.9 Results

7.9.1 Surgical Treatment

The overall results of treatment of pancreatic cancer are at best discouraging, and, in most cases, dismal. As of 1987, only 127 patients were known to have survived 5 years following pancreaticoduodenectomy for cancer of the pancreas (JORDAN 1987). Many reports continue to record no long-term survivals. (HERMRECK et al. 1974) reported 15.4% 5-year survival from a group of 26 patients treated with pancreaticoduodenectomy. Pancreatic pain is not relieved by bypass surgery, but chemical splanch nectomy (injection of 6% phenol or 50% alcohol around the celiac plexus) relieves pancreatic cancer pain in as many as 80% of cases (BRIDENBAUGH et al. 1964; FLANIGAN and KRAFT 1978). In many instances, the pain relief is permanent.

7.9.2 External Beam Therapy

For patients with locally unresectable disease, PHD external beam radiation therapy (preferably with adjuvant chemotherapy) yields 5% 5-year survival (DOBELBOWER and MILLIGAN 1984). Survival of the first 40 patients so treated is shown in Fig. 11. Appetite improved in 8 or 15 anorectic patients and relief of pain occurred in 22 of 32 patients. Sixteen patients lost 5 lb or more during therapy, while the weight of the remaining 24 patients either remained stable or increased (DOBELBOWER et al. 1980a).

Two patients survived 5 years postdiagnosis. One of the 5-year survivors presented with a large adenocarcinoma of the body of the pancreas. Median survival was 12 months. Survival of the patients treated with PHD radiotherapy seemed remarkably similar to that of a group of 31 patients treated with radical

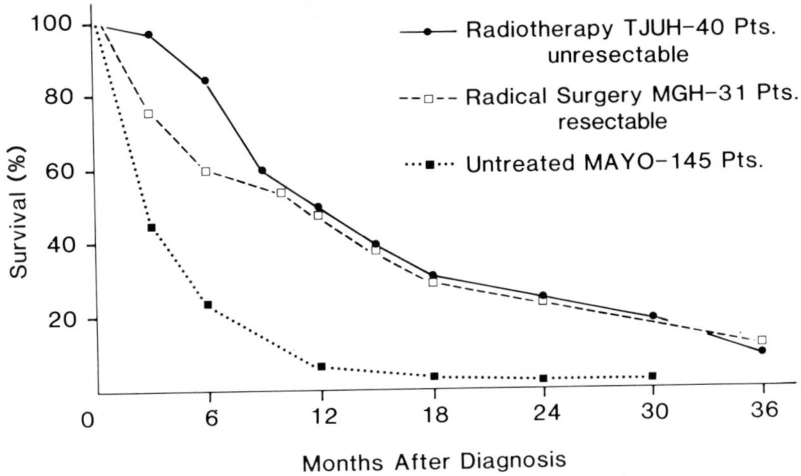

Fig. 11. Survival of patients with cancer of the pancreas treated with PHD external beam irradiation with or without 5-FU (●), or radical surgery (□) compared with survival of unselected untreated patients with pancreatic cancer (■)

surgery at Massachusetts General Hospital (Fig. 11) (DOBELBOWER et al. 1980b).

The available data suggest that PHD external beam therapy extends survival of patients with unresectable pancreatic cancer and may actually be curative in the occasional patient (DOBELBOWER and MILLIGAN 1984). Relief of pain and palliation of other symptoms seem to be the rule rather than the exception with PHD external beam therapy for unresectable pancreatic cancer (DOBELBOWER et al. 1980a).

7.9.3 Radiation Therapy as a Surgical Adjuvant

At many anatomic sites it has been found that malignant diseases are better treated by combining radiation therapy and surgery than by using either modality alone. This concept has been applied to the treatment of pancreatic cancer but it has been inadequately tested.

7.9.3.1 Postoperative Irradiation

Postresection adjuvant therapy by irradiation was tested by the GITSG (KALSER and ELLENBERG 1985). After "curative" resection (Whipple procedure or total pancreatectomy), patients were randomized to receive conventional moderate dose (40 Gy) split course external beam therapy plus adjuvant 5-FU or no adjuvant treatment. This study accessed only 49 patients over the course of 10 years. There appears to be a statistically significant survival benefit for the treatment group: 22 patients randomized to the observation arm displayed a median survival of 47 weeks, as compared to 95 weeks for 21 patients randomized to receive radiation therapy plus 5-FU (Fig. 6). Only one of the control group patients survived 5 years as compared to three patients in the adjuvant therapy group.

7.9.3.2 Preoperative Irradiation

To date, only one study of preoperative radiation therapy for pancreatic cancer has been reported (PILEPICH and MILLER 1980). This is understandable in view of the general unwillingness of radiation oncologists to irradiate patients without a histologic diagnosis. Also, many surgeons are reluctant to perform surgery in heavily irradiated fields because of the increased risk of complications.

PILEPICH and MILLER (1980) treated 17 patients with surgically documented carcinoma of the pancreas and with unresectable or borderline resectable disease with Cobalt 60 teletherapy or 46 meVp photons to doses of 40–50 Gy by anterior and posterior opposed fields that ranged from 11×6 cm to 15×16 cm. The dose was conventionally fractionated. Three-quarters of the patients received doses between 44 and 46 Gy. Six weeks postirradiation, patients were restudied for metastatic disease. Six of the 17 patients were not reoperated because five showed disease progression and one sustained a myocardial infarction in the interim. Of the 11 patients surgically explored, five had no resection because of surgically documented metastatic disease or unresectability. Six patients were treated by Whipple procedure and two of these survived 5 years. Treatment was implicated as a cause of death in only one patient. Certainly, these data suggest that preoperative radiation therapy for pancreatic cancer deserves further study.

Review of studies that have addressed the issue clearly suggests that therapy by irradiation is an effective adjuvant to surgery either preoperatively or postoperatively. It seems clear that radical pancreatic resection is feasible following irradiation to moderate doses in standard fractionation schemata. Combinations of preoperative and postoperative irradiation have not been studied, nor have optimal dose-time fractionation schemes been determined for either preoperative or postoperative irradiation. Certainly, much work remains to be done in these areas.

7.9.4 Radioisotope Implantation

HILARIS and ROUSSIS (1975) developed techniques for ^{125}I seed implantation of the pancreas. They reported only one perioperative death in 33 patients. One additional patient developed a pancreatic fistula after multiple pancreatic biopsies. Three of the six other major complications occurred in patients having concomitant bypass procedures and were not attributed to the implant. Median survival was 8 months, with the longest survivor alive with no evidence of disease at 33 months. In spite of the fact that nine of the implant patients had hepatic metastases at the time of surgery, the authors observed no difference in survival between the group of patients treated by ^{125}I seed implantation and a group of 39 patients treated by resection.

SHIPLEY et al. (1980) compared the survival of 12 patients treated with ^{125}I implant with that of ten patients treated by Whipple procedure at the same institution during the same time period. The patients treated with ^{125}I seed implantation all had biopsy-proven, localized pancreatic cancers less than 7 cm in diameter that were unsuitable for resection. In addition to 160 Gy by implant, patients received postoperative external beam therapy, and three were treated with adjuvant chemotherapy. Further confounding interpretation of the data, three of the resection group received adjuvant radiotherapy by ^{125}I implant of the pancreatic remnant. Postoperative complications were seen in five of the 12 implant patients: two developed pancreatic fistulae, one developed duodenal obstruction, two developed gastric hemorrhage, and four developed pancreatic exocrine insufficiency. Several patients suffered more than one complication. Seven of nine patients developed hematogenous metastases. Excellent palliation of back pain was noted in three of three patients. Survival of the patients treated by ^{125}I seed implantion seemed comparable so that of the ten patients treated by Whipple procedure.

WHITTINGTON et al. (1981) claimed 93% local control of pancreatic cancer in patients treated with ^{125}I implant and postoperative PHD external beam therapy with adjuvant chemotherapy. Patient survival did not appear to be better than that of patients treated with PHD external beam therapy alone, however. In seven patients a histologic diagnosis of cancer was available prior to isotope implantation, and these patients received a single dose of 5 Gy preoperatively by opposed anterior and posterior coplanar fields. Only one of these seven patients developed diffuse peritoneal seeding or incisional metastasis in contrast to five of seven patients not given preoperative single-dose radiation prior to implantation.

SYED et al. (1983) reported data from 18 patients treated with biliary bypass surgery, ^{125}I implant, and conventional external beam therapy. For ten patients, the ^{125}I implant was "sandwiched" between two courses of external beam therapy. Typically, patients received a dose of 30 Gy following biopsy and biliary bypass by 15 × 10 cm opposed anterior and posterior fields shielding the left kidney. The ^{125}I seed implant was performed 2 weeks after cessation of external beam therapy, and 3–4 weeks later an additional 15–20 Gy was administered, again with conventional external beam therapy. The implant dose was 100–150 Gy. The tumors of eight patients unsuitable for reexploration were implanted with ^{125}I seeds at the time of initial operation. These patients received 30–50 Gy external beam therapy 3–6 weeks postoperatively. Patients in the latter group were said to have a poorer prognosis as compared with that of the ten patients treated with the "sandwich" technique. Two patients had prolonged (11 weeks, 6 months) wound drainage. Three patients developed insulin-dependent diabetes. Two additional patients developed other complications of implant.

At the Medical College of Ohio, 13 patients were treated with biliary and gastrointestinal bypass, ^{125}I implant (140 Gy), and postoperative PHD external beam therapy (60 Gy) (DOBELBOWER et al. 1986). Half the group received adjuvant 5-FU. Operative complications were observed in five patients (pancreatic fistulae, upper gastrointestinal hemorrhage, pulmonary embolus, cholangitis, and intrahepatic cyst). The external beam portion of therapy was well tolerated. Patients lost an average of only 5½ Ib during the PHD external beam therapy. No patient lost more than 10 lb. In ten patients at risk 5 months or more, three long-term effects of radiation were noted in two patients: recurrent cholangitis and intrahepatic cyst (3 months) and gastric ulcer (6 months) in a patient treated for gastritis and ulcer prior to developing pancreatic cancer.

Pain was relieved in six of ten patients, and three of 11 patients with weight loss showed a reversal of that trend. No operative mortality was observed. Patient survival seemed comparable to that of patients treated at Jefferson with PHD external beam therapy with and without chemotherapy (Fig. 8).

7.9.5 Adjuvant Chemotherapy

Nearly every study that has addressed the issue has found benefit from systemic chemotherapy as an adjuvant to radiation treatment for unresectable cancer of the pancreas. The mechanism of such action is not entirely clear: radiosensitization of tumor by the systemic agent or direct cytotoxic effect on subclinical metastases.

Childs et al. (1965) pioneered the study of 5-FU as an adjuvant to radiation therapy. In a prospective double-blind study, he randomly assigned ambulatory patients with locally unresectable cancer to receive either radiation plus 5-FU or radiation plus injections of placebo. If the tumor was too large to be encompassed with anterior and posterior 20×20 cm beams of ^{60}Co or 6 meVp photons, patients were excluded from the study. A dose of 35–40 Gy was administered at 9–12 Gy per week. The 5-FU was given by bolus intravenous injection on the first 3 days of radiation therapy in a dose of 15 mg/kg of actual or ideal body weight (whichever was less). With 32 patients in each arm of the study, mean survival for the combined-modality group was 10.4 months as compared with 6.3 months for the single-modality group ($P < 0.05$).

At least 11 nonrandomized retrospective analyses tend to support the hypothesis that chemotherapy plus irradiation is more effective than irradiation alone in the treatment of this disease (Dobelbower 1979b). Mikhailichenko and Tamarkin (1972), however, reported a mean survival time of 28.1 months for patients treated with radiation alone after palliative surgery, 12.9 months for patients treated with 5-FU alone, and 8.3 months for patients treated with 5-FU and half the usual radiation dose.

A large-scale prospective randomized study of the adjuvant value of 5-FU was conducted by the GITSG (Gastrointestinal Tumor Study Group 1979). Patients were assigned to one of three treatment groups: (a) radiation therapy alone, 60 Gy, (b) 60 Gy plus concurrent and subsequent 5-FU; or (c) 40 Gy plus 5-FU. The 5-FU was administered as a bolus injection of 500 mg/m^2 on the first 3 days of each course of radiation therapy. The radiation treatment was given in 20 Gy courses over two or three 2-week periods, each followed by a 2-week rest. The initial 40 Gy was delivered to the entire pancreas, but if the patient was randomized to one of the 60 Gy arms, a boost dose of 20 Gy was delivered to just the tumor mass.

Survival data suggested a significantly shorter survival time for patients receiving radiation alone (17 weeks) than for patients in either of the combined modality arms (28 and 35 weeks for 60 and 40 Gy, respectively). For this reason, the radiation alone arm was dropped from study after 34 patients had been randomized to the radiation alone arm.

These data suggest that 5-FU is of benefit as an adjuvant to external beam radiation therapy for cancer of the pancreas. Scrutiny of the data shows that patients assigned to the radiation alone arm tended to have poorer initial performance status, a higher incidence of diabetes mellitus (suggesting greater pancreatic tumor involvement and greater tumor burden), more involvement of multiple sites (head, body, and tail of the pancreas), more lymph node involvement, and more frequent bile duct involvement than patients in the combined modality arms. It should also be noted that survival at 2 years is rather poor in all arms of this study as compared with survival of patients treated with PHD external beam therapy. Nonetheless, the bulk of evidence from this study and others does support the hypothesis that 5-FU enhances survival for patients with cancer of the pancreas treated with external beam radiation therapy. In the Jefferson PHD external beam pilot study (Dobelbower et al. 1980a), 12 patients received adjuvant chemotherapy. This was not by random allocation. The combined modality group displayed a median survival time of 15 months compared with 12 months for the remaining 28 patients treated with PHD external beam therapy alone. Again, this suggests that chemotherapy is a useful adjuvant to external beam therapy in treatment of cancer of the pancreas. It should be noted that neither of the 5-year survivors in the Jefferson study received adjuvant chemotherapy.

The preponderance of available data suggests that chemotherapy as an adjuvant to external beam irradiation may enhance short-term survival.

7.9.6 Chemotherapy Alone

SMITH et al. (1983) reported an 8-month median survival in a group of 17 patients with locally advanced pancreas cancer treated with 5-FU, doxorubicin, and mitomycin-C. He compared survival of these patients with that of patients treated with 60 Gy photons in the GITSG study and found them roughly equivalent. Again, it should be noted that survival of patients in the GITSG study is poor compared with that of patients treated with PHD external beam therapy, perhaps at least partly as a result of the relatively primitive techniques of irradiation dictated by GITSG protocol.

7.9.7 Intraoperative Radiation Therapy

Cancer of the pancreas is the disease that has been most often treated with intraoperative radiation therapy (IORT), with at least 720 cases documented in the scientific literature (DOBELBOWER 1987). The IORT doses employed range from 15 to 50 Gy in conjunction with preoperative doses ranging from 0 to 50 Gy and postoperative doses up to 50 Gy. MATSUDA (1982) reported data from 12 patients with locally unresectable pancreatic cancer treated with 18–30 Gy IORT followed by 9–41 Gy delivered by "conformation" external beam therapy. Median patient survival time was 12.5 months. ABE and TAKAHASHI (1981) reported data from 100 patients treated at 14 Japanese facilities. Thirty-nine percent of patients received IORT alone plus large field external beam radiotherapy. Only five patients survived more than 1 year. The average survival time was 5.8 months. The IORT dose ranged from 15 to 40 Gy. Eighty percent of patients with severe abdominal pain reported relief of same within 1 week after an IORT dose of more than 20 Gy. Thirty percent of patients developed diarrhea, 20% bloody stool. Two gastric ulcers and one duodenal ulcer were documented 2–40 weeks after IORT.

SHIPLEY et al. (1984a, b) and WOOD et al. (1982) reported clinical data from 29 patients with locally unresectable cancer of the pancreas treated with IORT doses escalated from 15 to 20 Gy while breathing 100% oxygen. Since 1982 misonidazole was administered prior to the IORT dose with no apparent improvement in survival. In addition, patients received 10–20 Gy fractionated external beam therapy directed to the primary tumor and the adjacent node-bearing areas prior to IORT and 27 patients received an additional 30–40 Gy fractionated external beam therapy with a four-field technique after IORT. Twenty patients received 5-FU (500 mg/m^3) on the first 3 days of postoperative irradiation and 15 patients received maintenance chemotherapy with 5-FU, doxorubicin, and mitomycin-C. Three significant operative complications were observed: a suture line leak at the gastric antrum, a candida pancreatic abscess, and delayed gastric emptying that required 5 weeks to resolve. Seventeen delayed complications were observed: four injuries of the pylorus or duodenum included in the IORT field (one obstruction, three hemorrhage), three cases of retroperitoneal fibrosis with obstruction, and ten cases of pancreatic insufficiency. Pain was relieved or significantly improved in all 16 patients presenting with same. Median survival time was reported as 16.5 months, but as the data matured, this decreased to 13.5 months.

SINDELAR and KINSELLA (1986a) conducted a prospective randomized trial of IORT in the management of patients with unresectable stage III (locally infiltrating tumor with nodal involvement) or stage IV (visceral or peritoneal metastasis) adenocarcinoma of the pancreas. During 1984 and 1985, 37 patients were evaluated for this study and 27 were found to be eligible. Some refused protocol treatment, and 22 patients were randomly allocated to receive experimental therapy, consisting of surgical biliary and gastric diversion, IORT (25 Gy with 18–22 meV electrons), and postoperative external beam irradiation (15 Gy with 6–8 meVp photons in 1.5–1.75 Gy increments over 5–6 weeks), or conventional treatment consisting of biliary and gastric bypass and postoperative external beam radiation therapy to a dose of 60 Gy in double-split course fashion (20 Gy over 2 weeks × 3). Patients in both IORT and control groups received 5-FU (500 mg/m^2) IV daily × 3 concomitant with the external beam radiotherapy and repeated in cycles every 4 weeks. Ten patients entered the experimental arm and 12 entered the control arm of the study.

Hepatic metastases were observed in ten of the IORT patients and eight of the control group. One early death from respiratory failure occurred in the IORT group. Significant complications of treatment were seen in approximately 40% of patients in each treatment group. The IORT patients had no acute toxicity, but three developed late (more than 6 months) duodenal hemorrhage. Dose-limiting acute radiation enteritis occurred in five patients, late enteritis in three patients. Median survival was 8.7 months in the IORT group, as compared to 8.1 months in the control group. All patients in the control group died within 18 months and in the IORT group within 24 months. The time to disease progression was longer in the IORT group. For patients with local disease only (stage III) at the beginning of treatment, the time to disease progression and the survival were superior in the IORT group compared to the control group.

One common thread that runs through reports of IORT for unresectable pancreatic cancer is pain relief. This has been observed in 50%-92% of patients presenting with pain and treated with IORT.

Intraoperative radiation therapy is presently being studied as an adjunt to surgical resection for cancer of the pancreas. Two of 26 pancreatic cancer patients reported by GUNDERSON et al. (1984) were treated for gross residual disease after resection. SHIPLEY et al. (1984a, b) treated four pancreatic cancer patients with IORT after radical resection. HIROAKA et al. (1984) treated the tumor beds (celiac axis, superior mesenteric artery, portal vein, inferior vena cava, aorta, etc.) of 12 patients to 30 Gy using 8 meV electrons immediately after pancreaticoduodenectomy. They compared data from this group of 12 patients to those of a comparable group of patients treated with pancreaticoduodenectomy alone. At 1 year, survival seemed improved in the IORT group, but a 2 years there was no appreciable difference.

SINDELAR and KINSELLA (1986b) conducted the first randomized, prospective, controlled trial of IORT used as an adjunct to surgery in the treatment of resectable cancer of the pancreas. They evaluated 132 patients referred for protocol treatment and found 63 eligible. Seven patients refused protocol therapy, and 56 were randomly allocated to receive surgical resection plus 20 Gy IORT with 9-12 meV electeons, or surgical resection alone (for disease confined to the panreas) plus postoperative external beam radiation therapy (50 Gy at 1.5-1.75 Gy per fraction) for lesions extending beyond the pancreatic capsule or with nodal involvement. Sixteen of 29 patients randomized to receive IORT were disqualified because of metastatic disease found at surgery, as were 15 of 27 patients allocated to receive routine therapy. Thus, 13 patients were treated with resection and IORT and compared to 12 patients treated with routine treatment.

Five of the 13 patients on the experimental arm (38%) died postoperatively compared to two of 12 (17%) patients treated conventionally. Significant complications were observed in approximately half of each group of patients. Between the two groups, no difference was observed in disease-free survival or time to recurrence. When operative deaths were excluded from analysis, the disease-free interval was increased in patients treated with IORT (18.5 months) as compared to the control group (12 months). Survival of the IORT patients tended to be longer than control patients, although statistical significance was not achieved. The local disease control rate was significantly superior in the IORT group. All control group patients failed locally within 12 months compared to 80% local control at 12 months in the IORT group.

It remains to be seen whether or not IORT will be an effective adjuvant to surgical resection for cancer of the pancreas.

7.9.8 Particle Beam Irradiation

Early results of the high LET treatment programs for cancer of the pancreas have been reviewed (DOBELBOWER 1986). In general, patient survival seems no better than for those treated by PHD external beam therapy (Fig. 8), and the complication rate seems higher (Table 10).

Table 10. Morbidity of particle beam therapy for adenocarcinoma of the pancreas

Group	Beam(s)	Dose ^{60}Co eq	Complications
M. D. Anderson	50 meV neutrons ±32 meVp X-ray	60 Gy, 10 weeks double split	None
Fermilab	66 meV neutrons	58.5 Gy, 6-7 weeks	Hemorrhage 2/31 Ulcer 3/31 Pancreatic insufficiency 4/31 Obstruction 1/31
Cleveland Clinic	25 meV neutrons +10 meVp X-rays	60 Gy, 6-8 weeks	Hemorrhage 4/30 Gangrene bowel 1/30
Berkeley	Helium, carbon, or neon ions	50-70 Gy, 6-8 weeks	Hemorrhage 6/58 Obstruction 2/55
MANTA	14 meV neutrons ±5-FU	52 Gy, 6 weeks	Hemorrhage 6/18 Hepatopathy 10/20 Colitis 2/20 Myelitis 1/25

The most encouraging early report from the high LET treatment programs comes from M. D. Anderson, where 15 patients were treated with 22 meV neutrons (Al-Abdulla et al. 1981). Of necessity, patients were treated with a fixed horizontal beam either sitting, standing, or kneeling. This circumstance complicated shaping of the radiation fields and resulted in increased thickness of the transit tissues at the level of the pancreas. This also complicated portal localization, as the pancreas may move caudally as much as 8 cm when a patient moves from a supine to a standing position. Thirteen patients in the neutron group were treated with neutrons combined with high-energy photons. Two neutron and three photon treatments were administered weekly. Two patients were treated with neutrons alone twice weekly. A double-split course schedule was employed similar to that used by Haslam et al. (1973). Treatment was planned to deliver an equivalent tumor dose of 60 Gy over 6–7 weeks for six patients and over 10 weeks in double-split course fashion for nine patients.

The majority of the neutron-treated patients experienced significant palliation of symptoms. Pain was relieved in seven of eight patients. No complications were observed. Three of the neutron-treated patients lived 16 months or more. Survival of the group seemed comparable to that of patients with unresectable pancreatic cancer treated with PHD external beam therapy of Jefferson.

Although the initial results from the high LET treatment programs may seem discouraging for pancreatic cancer, it must be borne in mind that these are emerging treatment modalities and that the complexity of the equipment required to generate such beams is several orders of magnitude beyond that of conventional radiotherapy equipment. Also, the radiobiologic effects of high LET particles are still under study. The theoretical advantages of high LET beams, particularly those capable of depositing greater dose at depth rather than at the surface or in the transit tissue by virtue of the Bragg peak phenomenon (Dobelbower and Milligan 1987), would appear to justify further clinical investigation at this time. Optimal time-dose fractionation schemes have yet to be elucidated. The combination of low LET and high LET radiations in the treatment of pancreatic cancer seems a fertile field for study.

7.10 Complications

Both surgical and radiotherapeutic approaches to cancer of the pancreas are fraught with early and late complications.

7.10.1 Surgery

The average mortality among 1544 patients treated surgically for pancreatic malignancy over the past 20 years is 20.7% (Jordan 1987). There is strong evidence to suggest that operative mortality has decreased sharply in recent decades and also that operative mortality is inversely proportional to the experience of the operating surgeon (Jordan 1987).

The incidence of early postoperative complications following pancreaticoduodenal resection for pancreatic cancer is quite high. Sepsis accounts for 13% of deaths after resection, the most common cause being peritonitis, intraperitoneal abscess, and wound infections (Jordan 1987). Pancreatic fistula is one of the most common and most dreaded postoperative complications. This also accounts for as many as half of postoperative deaths (Balasegaram 1976; Papachristou and Fortner 1981a, b). The incidence of pancreatic fistula, as well as the mortality therefrom, decreases with the experience of the operating surgeon. Biliary fistulae are less frequent (5%) and are associated with a much lower mortality than pancreatic fistulae. Cardiopulmonary complications occur in approximately 9% of patients (Jordan 1987). Renal faulure is a frequent cause of postoperative death. Hemorrhage from portal vein or splenic, hepatic, or superior mesenteric vessels can occur intraoperatively or during the first few postoperative days. It is the most common cause of postoperative death. Delayed gastric emptying is also a frequent postoperative complication. The incidence seems to vary from surgeon to surgeon. Mesenteric thrombosis accounted for 19% of deaths in the past, but now is an infrequent postoperative complication.

Delayed postoperative complications include mucosal ulceration at the surgical anastomoses. In the past it occurred in as many as 50% of patients but now is noted as a complication in 9%–12% of patients. Other delayed surgical complications include stricture of the biliary intestinal anastomoses, diabetes mellitus, and intestinal obstruction.

7.10.2 Precision High-Dose Radiation Therapy

Precision high-dose external beam radiation therapy directed to the panreas is relatively well tolerated.

The first 40 patients treated with PHD radiation therapy for cancer of the pancreas had locally aggressive, extensive, unresectable adenocarcinoma of the pancreas with no known distant metastases at the time of treatment. Only five of the 40 failed to complete treatment as planned. Treatment was well tolerated by the remaining patients. Only six of the 40 reported significant nausea; one of these was a patient with sepsis and one was a patient with rapid local progression of disease. Four of these six experienced episodes of vomiting during radiation therapy. Three patients experienced significant diarrhea, but all had diarrhea prior to commencement of external beam radiation therapy. One patient reported profound anorexia (in additon to nausea and vomiting), but this was present prior to treatment (DOBELBOWER et al. 1980a).

Seven of 27 patients at risk for 6 months or more developed symptoms that could conceivably be attributed to radiation therapy: five patients developed gastrointestinal bleeding and six, symptoms of gastritis. In all but two cases, the gastritis and/or bleeding coincided in time with local recurrence of cancer and might have actually represented manifestations of local tumor activity rather than radiation complications. Seven patients developed mild pancreatic insufficiency 2–10 months after beginning PHD radiation therapy. In every case, the symptoms were controlled by oral enzyme replacement. Two patients developed elevated fasting blood sugar levels subsequent to PHD irradiation (DOBELBOWER et al. 1980a).

No patient died of radiation complication. There were no cases of radiation myelitis or clinically significant radiation hepatitis or radiation nephritis.

7.10.3 Radioisotope Implant

Implantation of radioactive isotopes into pancreatic tumors is associated with operative mortality in as many as 25% of cases (Table 9) (BORGELT 1980; DOBELBOWER et al. 1986; WHITTINGTON et al. 1981). Other major complications include gastric ulcer, pulmonary embolus, sepsis, fistula, hemorrhage, cholangitis, pancreatic insufficiency, diabetes mellitus, and intestinal obstruction. A total of 68 complications have been reported in 101 patients treated with surgical implantation of radioactive sources into the pancreas (DOBELBOWER et al. 1986).

7.10.4 Particle Beam Therapy

Patients treated with high LET beams also seem to show a higher complication rate than those treated with more conventional radiation sources (Table 9) (BORGELT 1980; BUSH and KLIGERMAN 1980; CASTRO et al. 1980, 1982; GAHBAUER et al. 1980; KAUL et al. 1981; MANSELL et al. 1980; QUIVEY et al. 1980).

7.10.5 Intraoperative Radiation Therapy

Patients with pancreatic cancer treated by IORT are not free of complications. Seven treatment-related deaths have been reported in the periodic literature and serious morbidity (ulceration, hemorrhage, fistula formation, wound dehiscence, etc.) has been observed in as many as 30% of patients (DOBELBOWER 1987).

Delayed gastric emptying appears to be a common sequel of IORT for pancreatic cancer. GOLDSON (1979 personal communication) was the first to observe this. In a group of 23 pancreatic cancer patients receiving 10–20 Gy via IORT in addition to 45–50 Gy external beam radiation therapy, GUNDERSON et al. (1984) observed nine complications: two cases of delayed gastric emptying, two hemorrhage, two symptomatic fibrosis, and three severe nutritional problems. An analysis performed with 52 patients on study showed patient survival no better than that of patients treated with PHD external beam therapy alone, but local in-field failure was reduced to 7%.

Given the wide range of IORT doses and the varying combinations of preoperative and postoperative irradiation, it is not surprising that the results of treatment are quite varied.

Interpretation of the rather sketchy available data is confounded by the circumstance that the IORT has been delivered with adjuvant misonidazole, oxygen, 5-FU, and other chemotherapeutic agents. The preoperative and postoperative radiation therapy has been also occasionally combined with chemotherapy and even intraperitoneal ^{32}P installation.

7.11 Summary

In spite of continued advances in pre- and postoperative patient care, revolutionary new diagnostic modalities (CT, MRI, ERCP, ultrasound, tumor markers), development of extended surgical procedures, and improvements in radiation therapy dose delivery techniques, as well as development of high LET radiation beams, cancer of the pancreas still has a dismal prognosis. The basic problem is lack of early diagnosis. Partly, this is due to the location of the hermit organ and the biology of the disease; there are no early signs of pancreatic cancer, except for lesions fortuitously located in proximity to the common bile duct. Partly, it is due to our inability to achieve local control of disease. The GITSG showed an 83% incidence of local recurrence of disease in patients treated with "curative" surgery (Kalser and Ellenberg 1985). As our ability to achieve local control improves, our inability to deal with distant metastasis becomes evident. Whittington et al. (1981) claimed a local control rate of 93% using ^{125}I implantation of radioisotope in conjunction with PHD external beam therapy and multiagent chemotherapy. Local control did not translate into improved survival. In order to improve survival in this dread disease, we must be able to develop methods to diagnose the tumor at an earlier stage, to improve local control of gross disease in and around the pancreas, and to deal with systemic metastastic disease.

For patients with resectable disease, the preferred approached is, of course, surgical resection. The GITSG data suggest that the addition of radiation therapy and chemotherapy as adjuvants will improve survival (Kalser and Ellenberg 1985). The drugs and drug combinations as well as the radiation doses and techniques and combinations of radiation therapy and chemotherapy remain to be refined.

The current preferred approach for unresectable disease is PHD external beam radiation therapy with adjuvant chemotherapy. Here again, optimal combinations of the two modalities have yet to be determined.

For patients with metastatic pancreatic cancer, cure is currently impossible. The goal for such patients should be palliation of symptoms. This may require a combined surgical/radiotherapeutic/chemotherapeutic approach. Radiation therapy is extremely effective for palliation of local symptoms of pancreatic cancer except for gastrointestinal or biliary obstruction. These problems should be treated surgically. Operation should include biopsy and, usually, choledochojejunostomy, gastrojejunostomy, and jejunojejunostomy to prevent reentry of food into the stomach. Careful attention to blood volume, serum albumin levels, and nutrition (IV hyperalimentation) is necessary to minimize operative mortality. Painful bone metastases regularly respond to a dose of 40 Gy administered in 15 equal increments over a period of 3 weeks. Likewise, upper abdominal pain and sometimes even nausea, vomiting, and anorexia are palliated by local upper abdominal irradiation, with or without adjuvant chemotherapy. Many studies have suggested that the palliative benefits of radiation in such instances are dose related, and it appears that one must achieve at least the radiobiologic equivalent of 45 Gy administered in equal 1.8 Gy increments over the course of 5 weeks in order to achieve a good response.

Obstruction of the gastrointestinal tract is best palliated surgically, whereas biliary obstruction can be treated either by surgical bypass or, in selected cases, with percutaneous transhepatic catheter drainage, with or without radioisotope placed in the catheter. Palliative Whipple procedures and palliative total pancreatectomies are not recommended because the surgical morbidity and mortality outweigh the potential benefit in patients destined to die shortly of the disease.

Five-Fluorouracil has been used extensively for pancreatic cancer, with response rates reported as high as 67% in retrospective reports. In prospective controlled cooperative group trials, response rates of less than 20% have generally been observed (Litka and Schein 1986). The combination of streptozotocin, mitomycin-C, and 5-FU is reported to have produced partial response in 43% of 23 evaluable patients so treated (Smith et al. 1983). Median duration of response was 5 months and median duration of survival in responders was 9.5 months.

The patient who is asymptomatic after histologic proof of pancreatic cancer and appropriate palliative bypass surgery is best treated expectantly. Quality of life is important. Liberal use of analgesic is the best treatment for selected patients with symptomatic refractory disease. Nutritional support and comfort measures may be the most important tools available to the clinician in such instances. When present, diabetes or exocrine pancreatic insufficiency should be treated.

7.11.1 New Approaches

Improving the therapeutic ratio with the use of radiation sensitizers is currently being evaluated prospectively by the Eastern Cooperative Oncology Group Study 8282 testing PHD radiation therapy alone versus PHD radiation therapy plus 5-FU and mitomycin-C in radiosensitizing doses (Fig. 12). At the University of Innsbruck, Glaser and Frommhold (1986, personal communication) are investi-

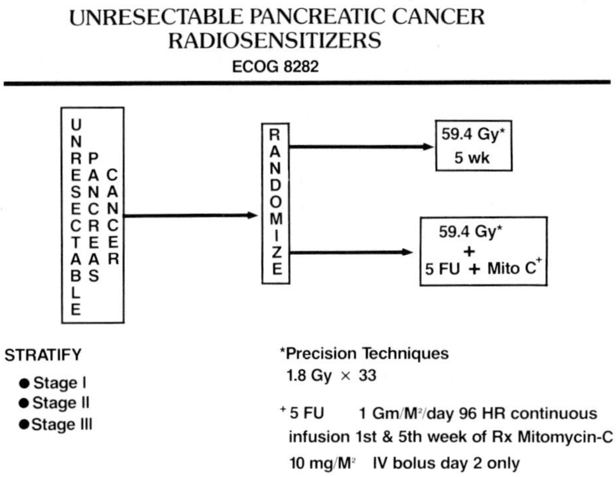

Fig. 12. Schema: Eastern Cooperative Oncology Group Study 8282

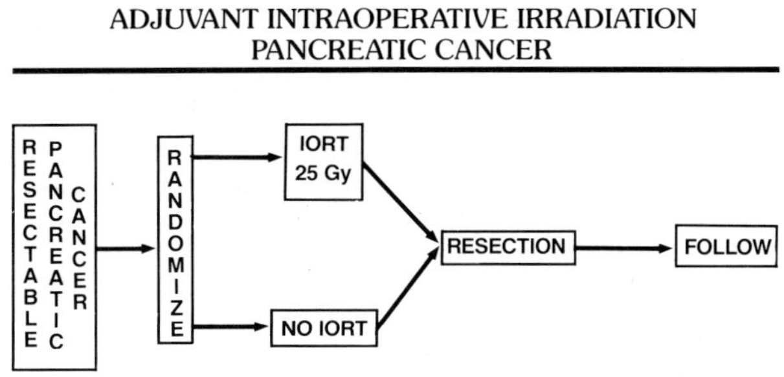

Fig. 13. Schema: Proposed study of adjuvant IORT for resectable pancreatic cancer

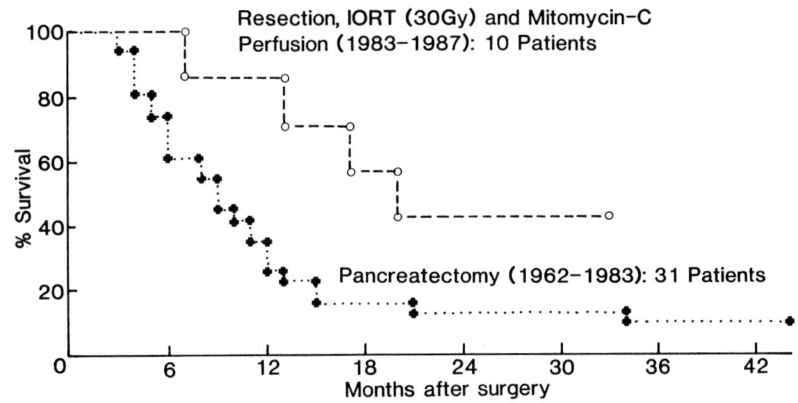

Fig. 14. Survival of patients with cancer of the pancreas treated with radical pancreatectomy (...+...) or resection, IORT, and mitomycin-C perfusion (--○--). (Ozaki et al. 1987)

gating the use of preresection IORT (Fig. 13). This approach is based on the premise that a single massive dose of radiation to the bulk of disease given immediately before surgical resection will render inactive any tumor cells disseminated during surgical manipulation. As the liver is the most frequent site of metastatic disease, prophylactic hepatic radiation after resection or in conjunction with local PHD radiation therapy for unresectable lesions might be a logical course of investigation. The use of IORT combined with chemotherapy as an adjuvant to surgical resection is being tested in Japan and the initial results are some of the best reported in the management of resectable pancreatic cancer (Fig. 14) (Ozaki et al. 1987).

In summary, it is evident that both improved diagnostic and therapeutic modalities are needed to enhance our ability to deal with adenocarcinoma of the panreas. Active research is underway on many fronts to deal with this problem. Much remains to be learned.

Acknowledgments. The authors thank Miss Sandra K. Price for the clerical preparation of this manuscript, Mr. Roy Schneider for his assistance in the preparation of Fig. 1, and Mrs. Faye Keen for her assistance in the preparation of Fig. 5.

References

Abe M, Takahashi M (1981) Intraoperative radiotherapy: The Japanese experience. Int J Radiat Oncol Biol Phys 7: 863–868

Ackerman NB, Aust JC, Bredenberg CE, Hanson VA Jr, Rogers LS (1976) Problems in differentiating between pancreatic lymphoma and anaplastic carcinoma and their management. Ann Surg 184: 705–708

Al-Abdulla ASM, Hussey DH, Olson MH, Wright AE (1981) Experience with fast neutron therapy for unresectable carcinoma of the pancreas. Int J Radiat Oncol Biol Phys 7: 165–172

Aoki K, Ogawa H (1978) Cancer of the pancreas, international mortality trends. World Health Stat Rep 31 (1): 2

Balasegaram M (1976) Carcinoma of the periampullary region. A review of a personal series of 87 patients. Br J Surg 63: 532–537

Banwo O, Versey J, Hobbs JR (1974) New oncofetal antigen for human pancreas. Lancet 1: 643–645
Baumel H, Deixonne B (1986) Results of therapy. In: Baumel H, Deixonne B (eds) Exocrine pancreatic cancer. Springer-Verlag, Berlin Heidelberg New York, pp 156–169
Baylor SM, Berg JW (1973) Cross-classification and survival characteristics of 5000 cases of cancer of the pancreas. J Surg Oncol 5: 335–358
Beahrs OH, Hensen DE, Hutter RUP, Myers MH (1988) Manual for staging of cancer. Lippincott, Philadelphia pp 109–110
Bell ET (1957) Carcinoma of the pancreas. I. A clinical and pathological study of 609 necropsied cases. II. The relation of carcinoma of the pancreas to diabetes mellitus. Am J Pathol 33: 499–523
Berg JW, Connelly RR (1979) Updating the epidemiologic data on pancreatic cancer. Semin Oncol 6 (3) 275–283
Best EW (1966) A Canadian study of smoking and health. Department of National Health and Welfare, Ottawa
Borgelt BB (1980) Radiation therapy with either gold grain implant or neutron beam for unresectable adenocarcinoma of the pancreas. In: Cohn I (ed) Pancreatic cancer: new directions in therapeutic management. Masson, New York, pp 55–62
Borgelt BB, Dobelbower Jr RR, Strubler KA (1978) Betatron therapy for unresectable pancreatic cancer: a preliminary report. Am J Surg 135: 76–80
Braganza JM, Howat HT (1979) Tumors of the exocrine pancreas. In: Howat HT, Sarles H (eds) The exocrine pancreas. Saunders, Philadelphia, pp 484–519
Bridenbaugh LD, Moore DC, Campbell DD (1964) Management of upper abdominal cancer pain: Treatment with celiac plexus block with alcohol. JAMA 190: 377–381
Burns WA, Matthews MJ, Hamosh M, VanderWeide G, Blum R, Johnson FB (1974) Lipase secreting acinar cell carcinoma of the pancreas with polyarthropathy – a light and electron microscopic, histochemical and biochemical study. Cancer 33: 1002–1009
Bush SE, Kligerman MM (1980) Pi meson radiotherapy for carcinoma of the pancreas. In: Cohn I Jr (ed) Pancreatic cancer: new directions in therapeutic management. Masson, New York, pp 89–96
Busnardo AC, DiDio LJA, Thomford NR (1988) Anatomicosurgical segments of the human pancreas. Surg Radiol Anat 10: 77–82
Castleman B, Scully R, McNeely BU (1972) Case records of the Massachusetts General Hospital, case 25–1972. N Engl J Med 286: 1353–1356
Castro JR, Quivey JM, Lyman JT, Chen GTY, Phillips TL, Tobias CA, Alpen EL (1980) Current status of clinical particle radiotherapy at Lawrence Berkeley Laboratory. Cancer 46: 633–641
Castro JR, Saunders WM, Quivey JM, Chen GT, Collier JM, Woodruff KH, Lyman JT, et al. (1982) Clinical problems in radiotherapy of carcinoma of the pancreas. Am J Clin Oncol 5: 579–587
Cederlof R, Friberg L, Hrubec Z, Lorich U (1975) The relationship of smoking and some social covariables to mortality and cancer morbidity. A ten-year follow-up in a probability sample of 55000 Swedish subjects age 18–69, Part 1/2. Karolinska Institute, Stockholm
Childs DS, Moertel DG, Holbrook MA, Reitemeir RJ, Colby MY (1965) Treatment of malignant neoplasms of the gastrointestinal tract with a combination of 5-fluorouracil and radiation. Radiology 84: 143–204
Cubilla AL, Fitzgerald PJ (1978) Pancreas cancer. I. Duct adenocarcinoma. Pathol Annu 1: 241–289
Cubilla AL, Fitzgerald PJ (1980a) Cancer (nonendocrine) of the pancreas. In: Fitzgerald PJ, Morrison AB (eds) The pancreas. Williams and Wilkins, Baltimore, pp 82–110 (International Academy of Pathology monograph)
Cubilla AL, Fitzgerald PJ (1980b) Surgical pathology of tumors of the exocrine pancreas. In: Moosa AB (ed) Tumors of the pancreas. Williams and Wilkins, Baltimore
Cubilla AL, Fitzgerald PJ (1984) Tumors of the exocrine pancreas. Atlas of tumor pathology. Second Series, Fascicle 19. Armed Forces Institute of Pathology, Washington, DC
Cubilla AL, Fitzgerald PJ, Fortner JG (1978a) Pancreas cancer – duct cell. Adenocarcinoma: Survival in relation to site, size, stage and type of therapy. J Surg Oncol 10: 465–482
Cubilla AL, Fortner JG, Fitzgerald PJ (1978b) Lymph node involvement in carcinoma of the head of the pancreas area. Cancer 41: 880–887
Cushieri A, Hall AW, Clark J (1978) Value of laparoscopy in the diagnosis of management of pancreatic carcinoma. Gut 19: 672–677
Dat NM, Sontag SJ (1982) Pancreatic carcinoma in brothers. Ann Intern Med 97 (2): 282
Del Favero G, Farini R, Fabris C, Bonvicini P, Piccoli A, Venturini R, Panucci A, Naccarato R (1984) CA 19-9 in the differential diagnosis of pancreatic cancer. Gastroenterology 86: 1059
Del Villano BC, Brennan S, Brock P, Bucher C, Liu V, McClure M, Rake B, et al. (1983) Radioimmunometric assay for a monoclonal antibody-defined tumor marker, CA 19-9. Clin Chem 29: 549–552
DiDio LJA (1948) Pars tecta pancreatis in man. Rev Sudam Morfol 6: 34–139
DiDio LJA (1956) Quelques données d'application de l'anatomie du pancréas. Rev Int Hepatol 6: 1093–1096
DiDio LJA, Anderson MC (1968) The sphincters of the digestive system. Anatomical, functional and surgical considerations. Williams and Wilkins, Baltimore
DiDio LJA (1970) Synopsis of anatomy. Mosby, St Louis
DiMagno EP, Buxton JL, Regan PT, Hattery RR, Wilson DA, Suarez JR, Green PS (1980) Ultrasonic endoscope. Lancet 1: 629–631
Dobelbower RR Jr (1979a) Cancer of the pancreas – radiation therapy. In: Thacher N (ed) Advances in medical oncology, research and education, vol IX. Digestive cancer. Pergamon, New York, pp 177–185
Dobelbower RR Jr (1979b) The radiotherapy of pancreatic cancer. Semin Oncol 6 (3): 1
Dobelbower RR Jr (1986) Therapy by irradiation. In: Go VLW; Gardner JD, Brooks FP, Lebenthal E, DiMagno E, Scheele GA (eds) The exocrine pancreas: biology, pathobiology, and diseases. Raven, New York, pp 699–711
Dobelbower RR Jr (1987) Intraoperative radiotherapy. Rev Bras Cancerol 33 (3): 207–226
Dobelbower RR Jr, Howard JM (1985) Pancreatic cancer. In: Brain MC, Carbone PP (eds) Current therapy in hematology-oncology 1985–1986. Decker, Philadelphia, pp 181–186
Dobelbower RR Jr, Howard JM (1987) Pancreatic cancer. In: Bayless TM, Brain MC, Cherniack RM (eds) Current therapy in internal medicine – 2. Decker, Philadelphia, pp 402–407
Dobelbower RR Jr, Milligan AJ (1987) Radiotherapeutic approaches to treatment of pancreatic cancer. In: Howard JM, Jordan GL Jr, Reber HA (eds) Surgical diseases of the pancreas. Lea and Febiger, Philadelphia, pp 734–747

Dobelbower RR Jr, Milligan AJ (1984) Treatment of pancreatic cancer by radiation therapy. World J Surg 8: 919-928

Dobelbower RR Jr, Borgelt BB, Strubler KA, Kutcher GJ, Suntharalingam N (1980a) Precision radiotherapy for cancer of the pancreas: technique and results. Int J Radiat Oncol Biol Phys 6: 1127-1133

Dobelbower RR Jr, Strubler KA, Vaisman I (1980b) Clinical applications of high energy electron beams: the pancreas, pleura and spine. In: Zuppinger A, Bataini JP, Irigaray JM, Chu F (eds) high energy electrons in radiation therapy. Springer-Verlag, Heidelberg pp 91-97

Dobelbower RR Jr, Merrick HW III, Ahuja RK, Skeel RT (1986) ^{125}I interstitial implant, precision high-dose external beam therapy and 5-FU for unresectable adenocarcinoma of pancreas and extrahepatic biliary tree. Cancer 58: 2185-2195

Doll R, Petro R (1976) Mortality in relation to smoking: 20 years' observation of male British doctors. Br Med J 2: 1525-1536

Edis AJ, Kiernan PD, Taylor WF (1980) Attempted curative resection of ductal carcinoma of the pancreas. Mayo Clin Proc 55: 531-536

Ellison EC, van Aman ME, Carey LC (1984) Preoperative transhepatic biliary decompression in pancreatic and periampullary cancer. World J Surg 8: 862-871

Faintuch J, Levin B (1986) Clinical presentation and diagnosis of exocrine tumors of the pancreas. In: Go VLW, Gardner JD, Brooks FP, Lebenthal E, DiMagno EP, Scheele GA (eds) The exocrine pancreas: biology, pathobiology and diseases. Raven, New York, pp 675-687

Ferrucci HF Jr, Wittenberg J, Margolies MN, Carry RW (1979) Malignant seeding of the tract after thin-needle aspiration biopsy. Radiology 130: 345-356

Fitzgerald PJ (1981) Pathology (nonendocrine). In: Cohn I Jr, Hastings PR (eds) Pancreatic cancer, report no 12. International Union Against Cancer, Geneva

Flanigan DP, Kraft RO (1978) Continuing experience with palliative chemical splanchnicectomy. Arch Surg 113: 509-511

Fontham E, Delayo C, Cohn I Jr (1987) Epidemiology of cancer of the pancreas. In: Howard JM, Jordan GL, Reber HA (eds) Surgical diseases of the pancreas. Lea and Febiger, Philadelphia, pp 613-626

Friedman JM, Fialkow PJ (1976) Carcinoma of the pancreas in four brothers. Birth Defects 1 (12): 145-150

Gahbauer R, Koh KY, Rodriquez-Antunez A, Jelden GL, Turco RF, Horton J, Blue J, et al. (1980) Preliminary results of fast neutron treatments in carcinoma of the pancreas. In: Cohn I Jr (ed) Pancreatic cancer: new directions in therapeutic management. Masson, New York, pp 63-65

Gastrointestinal Tumor Study Group (1979) Comparative therapeutic trial of radiation with or without chemotherapy in pancreatic carcinoma. Int J Radiat Oncol Biol Phys 5: 1643-1647

Gelder FB, Reese CJ, Moosa AR, Hall T, Hunter R (1978a) Purification, partial characterization and clinical evaluation of a pancreatic oncofetal antigen. Cancer Res 38: 313-324

Gelder FB, Reese CJ, Moosa AR, Hall T, Hunter R (1978b) Studies on an oncofetal antigen, POA. Cancer 42: 1635-1645

George P, Brown C, Gilchrist J (1975) Operative biopsy of the pancreas. Br J Surg 62: 280-283

Go VL, DiMagno EP (1977) The pancreas: pancreatic exocrine adenocarcinoma. Br J Hosp Med 18 (6): 567-571

Gordis L, Gold EB (1986) Epidemiology of pancreatic cancer. In: Go VLW, Gardner JD, Brooks FP, Lebenthal E, DiMagno EP, Scheele GA (eds) The exocrine pancreas: biology, pathobiology and diseases. Raven, New York, pp 621-636

Gordon-Taylor G (1934) The radical surgery of cancer of the pancreas. Ann Surg 100: 206-214

Grieve DC (1973) Adenocarcinoma of the pancreas (a review of 100 cases). J R Coll Surg Edinb 18: 221-226

Gunderson LL, Martin JK, Earle JD, Byer DE, Voss M, Fieck JM, Kvois LK, et al. (1984) Intraoperative and external beam irradiation with or without resection: Mayo pilot experience. Mayo Clin Proc 59: 691-699

Hall TJ, Blackstone MO, Coope MJ, Hughes RG, Moossa AR (1978) Prospective evaluation of endoscopic retrograde cholangiopancreatography in the diagnosis of periampullary cancers. Ann Surg 187: 131-137

Hammond EC (1966) Smoking in relation to the death rates of one million men and women. NCI Monogr 19: 126

Hanna CB, Hastings WD Jr (1968) Carcinoma of the pancreas: survival without resection. A case report. J SC Med Assoc 64: 8-10

Haslam JB, Cavanaugh PJ, Stroup SL (1973) Radiation therapy in the treatment of irresectable adenocarcinoma of the pancreas. Cancer 32: 1341-1345

Hermreck AS, Thomas CY IV, Friesen SR (1974) Importance of pathologic staging in the surgical managment of adenocarcinoma of the exocrine pancreas. Am J Surg 127: 653-657

Hilaris BS, Roussis K (1975) Cancer of the pancreas. In: Hilaris BS (ed) Handbook of interstitial brachytherapy. Publishing Science Group, Acton pp 251-262

Hirayama T (1972) Smoking in relation to the death rates of 265, 118 men and women in Japan. A report on five years of follow-up. 14th Science Writers Seminar of the American Cancer Society, Clearwater Beach

Hiroaka T, Watanabe E, Mochinaga M, Tashiro S, Miyauchi Y, Nakamura I, Yokoyama I (1984) Intraoperative irradiation combined with radical resection for cancer of the head of the pancreas. World J Surg 8: 766-771

Ho CS, McLoughlin MJ, McHattie JD, Tao LC (1977) Percutaneous fine needle aspiration biopsy of the pancreas following endoscopic retrograde cholangiopancreatography. Radiology 125: 351-353

Holyoke ED, Douglass HO Jr, Goldrosen MH, Chu TM (1979) Tumor markers in pancreatic cancer. Semin Oncol 6: 347-356

Hutchinson GB, MacMahon B, Jablon S, Land CE (1979) Review of report by Mancuso, Stewart and Kneale of radiation exposure of Hanford workers. Health Phys 37 (2): 207-220

Hyland C, Kheir SM, Kashlan MB (1981) Frozen section diagnosis of pancreatic carcinoma. A prospective study of 64 biopsies. Am J Surg Pathol 5: 179

Ishida M (1983) Peritoneoscopy and pancreas biopsy in the diagnosis of pancreatic diseases. Gastrointest Endosc 29: 211-218

Ishida H, Domzono T, Furukawa Y (1984) Laparoscopy and biopsy in the diagnosis of malignant intra-abdominal tumors. Endoscopy 16: 140-142

Issacson R, Weiland LH, McIlrath DC (1974) Biopsy of the pancreas. Arch Surg 109: 227-230

Jordan GL Jr (1987) Pancreatic resection for pancreatic cancer. In: Howard JM, Jordan GL, Reber HA (eds) Surgical diseases of the pancreas. Lea and Febiger, Philadelphia, pp 666-714

Kahn HA (1966) The Dorn study of smoking and mortality among U.S. veterans: Report on 8½ years of observation. NCI Monogr 19: 1

Kalser MH, Ellenberg SS (1985) Pancreatic cancer adjuvant combined radiation and chemotherapy following curative resection. Arch Surg 120: 899-903

Kalser MH, Barkin J, MacIntyre JM (1985) Pancreatic cancer: Assessment of prognosis by clinical presentation. Cancer 56: 397-402

Kaul R, Cohen L, Hendrickson F, Awschalom M, Hrejsa AF, Rosenberg I (1981) Pancreatic carcinoma: Results with fast neutron therapy. Int J Radiat Oncol Biol Phys 7 (2): 173-178

Kessler II (1970) Cancer mortality among diabetics. JNCI 44: 673-686

Klavins JV (1981) Tumor markers of pancreatic carcinoma. Cancer 47: 1597-1601

Kline TS, Neal MS (1978) Needle aspiration biopsy: A critical appraisal - eight years and 3267 specimens later. JAMA 239: 36-39

Klöppel G (1984) Pancreatic, non-endocrine tumors. In: Klöppel G, Heitz PU (eds) Pancreatic pathology. Livingston, Edinburgh, pp 79-113

Klöppel G, Fitzgerald PJ (1986) Pathology of nonendocrine pancreatic tumors. In: Go VLW, Gardner JD, Brooks FP, Lebenthal E, DiMagno EP, Scheele GA (eds) The exocrine pancreas: biology pathobiology and diseases. Raven, New York, pp 649-674

Lee YTN (1982) Tissue diagnosis for carcinoma of the pancreas and periampullary structures. Cancer 49: 1035-1039

Leichman LP (1985) Pancreatic cancer: medical aspects. In: Toledo-Pereyra LH (ed) The pancreas - principles of medical and surgical practice. Wiley, New York, pp 285-306

Lightwood R, Reber HA, Way LW (1976) The risk and accuracy of pancreatic biopsy. Am J Surg 132: 189-194

Lin RS, Kessler II (1981) A multifactorial model for pancreatic cancer in man. Epidemiologic evidence. JAMA 245: 147-152

Litka PA, Schein PS (1986) Chemotherapy of pancreatic cancer. In: Go VLW, Gardner JD, Brooks FP, Lebenthal E, DiMagno EP, Scheele GA (eds) The exocrine pancreas: Biology, pathobiology and diseases. Raven, New York, pp 689-697

MacDermott RP, Kramer P (1973) Adenocarcinoma of the pancreas in four siblings. Gastroenterology 65: 137-139

Mancuso TF, El-Attar AA (1967) Cohort study of workers exposed to betanaphthylamine and benzidine. J Occup Med 9: 277-285

Mancuso TF, Stewart A, Kneale G (1977) Radiation exposure of Hanfort workers dying from cancer and other causes. Health Phys 33: 369-385

Mansell J, Cohen L, Hendrickson F, Kaul R (1980) Preliminary report of the Fermilab experience using neutron irradiation for the treatment of pancreatic cancer. In: Cohn I Jr (ed) Pancreatic cancer: new directions in therapeutic management. Masson, New York, pp 67-76

Matsuda T (1982) Radiotherapy for pancreatic carcinoma combined with intraoperative radiotherapy and conformation radiotherapy. Seminar for high LET particle irradiation and other approaches to increasing effectiveness of radiation therapy for cancer, Oct 2-5, Kyoto University Club House, Kyoto

McIntire KR, Waldmann TA, Moertel GG, Go VLW (1975) Serum fetoprotein in patients with neoplasm of the gastrointestinal tract. Cancer Res 35: 991-996

McMichael AJ, McCall MJ, Hartshorne JM, Woodings TL (1980) Patterns of gastrointestinal cancer in European migrants to Australia: the role of dietary change. Int J Cancer 25 (4): 431-437

Metzgar RS, Asch HL (1988) Conference Report: "Antigens of human pancreatic adenocarcinomas: their role in diagnosis and therapy." December 7-8, 1987, Rockville, MD. Pancreas 3: 352-371

Mikal S, Campbell AJA (1950) Carcinoma of the pancreas. Diagnostic and operative criteria based on one hundred consecutive autopsies. Surgery 28: 963-969

Mikhailichenko VA, Tamarkin MA (1972) Combined treatment of cancer of the pancreas. Khirurgiia (Mosk) 48: 18-21

Moldow RE, Connelly RR (1968) Epidemiology of pancreatic cancer in Connecticut. Gastroenterology 55: 677-686

Moosa AR, Levin B (1981) The diagnosis of "early" pancreatic cancer: The University of Chicago experience. Cancer 47: 1688-1697

Moosa AR, Gadd M, Lavelle-Jones M (1986) Surgical treatment of exocrine pancreatic cancer. In: Go VLW, Gardner JD, Brooks FP, Lebenthal E, DiMagno EP, Scheele GA (eds) The exocrine pancreas: biology, pathobiology, and diseases. Raven, New York, pp 713-725

Moynan RW, Neerhout RC, Johnson TS (1964) Pancreatic carcinoma in children: Case report and review. J Pediatr 65: 711-720

Nakase A, Matsumoto Y, Uchida K, Honjo I (1977) Surgical treatment of cancer of the pancreas and the periampullary region - Cumulative results in 57 institutiones in Japan. Ann Surg 185: 52-57

Oguchi H, Homma T, Kawa S, Nagata A, Furuta S, Fukui M (1984) A pancreatic oncofetal antigen (POA): its characterization and application for enzyme immunoassay. Cancer Detect Prev 7: 51-58

Oguma S, Nippa K, Katsueki T, Gakka-Sasshi S (1983) Plasmacytoma of the head of the pancreas, 5 cases. J Clin Pathol 36: 147-152

Ormson MJ, Charboneau WJ, Stephens DH (1986) Sonography in patients with a possible pancreatic mass shown on CT. AJR 148: 551-555

Ozaki H, Kinoshiz T, Egawa S, Kishi K (1987) Combined treatment for resectable pancreatic carcinoma. In: Sugahara K (ed) New trends in gastroenterology, 1987. Japan Society of Gastroenterology, Kofu, pp 215-221

Papachristou DN, Fortner JC (1981a) Management of the pancreatic remnant in pancreatoduodenectomy. J Surg Oncol 18: 1-7

Papachristou DN, Fortner JG (1981b) Pancreatic fistula complicating pancreatectomy for malignant disease. Br J Surg 68: 238-240

Paul RE, Miller KH, Kahn PC, Callow AD, Edwards TL Jr, Patters JF (1965) Pancreatic angiography with application of subselective angiography of the celiac or superior mesenteric artery to the diagnosis of carcinoma of the pancreas. N Engl J Med 272: 283-287

Pilepich MV, Miller HH (1980) Preoperative irradiation in carcinoma of the pancreas. Cancer 46: 1945-1949

Pollard HM (1981) Staging of cancer of the pancreas. Cancer of the pancreas task force. Cancer 47: 1631-1637

Quivey JM, Castro JR, Chen GTY, Lyman JT, Tobias CA (1980) Helium ion radiotherapy in the treatment of pancreatic carcinoma: a preliminary analysis. In: Cohn I Jr (ed) Pancreatic cancer: new directions in therapeutic management. Masson, New York, pp 77-87

Richards WG, Katamann FS, Coleman FC (1958) Extramedullary plasmacytoma arising in the head of the pancreas: Report of a case. Cancer 11: 649–652

Rosch J, Keller FS (1981) Pancreatic arteriography, transhepatic pancreatic venography and pancreatic venous sampling in diagnosis of pancreatic cancer. Cancer 47: 1679–1684

Schmiegel WH, Eberl W, Kreiker C, Arndt R, Classen M, Creten H, Jessen K, et al. (1984) Differential diagnosis of pancreatic diseases use of tumor marker (CA 19-9, CEA, POA800, AFP) determinations in pancreatic juices and sera. Gastroenterology 86: 1236

Schreiber D, Probst HJ (1977) Sekretorisch aktives Karzinom des exokrinen Pankreas. Fallbericht und Literaturübersicht. Zentralbl Allg Pathol 121: 114–121

Schultz NJ, Saunders RJ (1963) Evaluation of pancreatic biopsy. Ann Surg 158: 1053–1057

Selikoff IJ, Seidman H (1981) Cancer of the pancreas among asbestos insulation workers. Cancer 47 (6): 1469–1473

Shipley WU, Nardi GL, Cohen AM, Ling CC (1980) Iodine-125 implant and external beam irradiation in patients with localized pancreatic carcinoma: A comparative study to surgical resection. Cancer 45: 709–714

Shipley WU, Tepper JE, Warshaw AL, Orlow EL (1984a) Intraoperative radiation therapy for patients with pancreatic carcinoma. World J Surg 8: 929–934

Shipley WU, Wood WC, Tepper JE, Warshaw AL, Orlow EL, Kaufman SD, Battit GE, Nardi GL (1984b) Intraoperative electron beam irradiation for patients with unresectable pancreatic carcinoma. Ann Surg 200: 289–296

Silverberg E, Lubera JA (1989) Cancer statistics, 1989. CA 39: 3–20

Silverberg E, Lubera JA (1988) Cancer statistics. CA 38: 5–22

Sindelar WF, Kinsella WT (1986a) Randomized trial of intraoperative radiotherapy in unresectable carcinoma of the pancreas. Int J Radiat Oncol Biol Phys [Suppl 1] 12: 148–149

Sindelar WF, Kinsella TJ (1986b) Randomized trial of intraoperative radiotherapy in resected carcinoma of the pancreas. Int J Radiat Oncol Biol Phys [Suppl 1] 12: 148

Smith FP, MacDonald JS, Schein PS, Ornitz RD (1980) Continuous seeding of pancreatic cancer by skinny-needle aspiration biopsy. Arch Intern Med 140: 855

Smith FP, Stablein DM, Korsmeyer S, Neefe J, Chun BK, Wolley PV, Schein PS (1983) Combination chemotherapy for locally advanced pancreatic cancer: Equivalence to external beam irradiation and implication for future management. J Clin Oncol 1: 413–415

Smith PE, Krementz ET, Reid RJ, Bufkin WJ (1967) An analysis of 600 patients with carcinoma of the pancreas. Surg Gynecol Obstet 124: 1288–1290

Steinberg WM, Gelfand R, Anderson KK, Glenn J, Sindelar W, Kurtzman S, Toskes PP (1984) Sensitivity and specificity of a new assay to diagnose cancer of the pancreas. Gastroenterology 86: 1266

Strohm WD (1984) Limits of ultrasound tomography and features of endoscopic ultrasound. Scand J Gastroenterol 19 (94): 7–12

Syed AMN, Puthawala AA, Neblett DL (1983) Interstitial iodine-125 implant in the management of unresectable pancreatic carcinoma. Cancer 52: 808–813

Tannenbaum H, Anderson LG, Schur PH (1975) Association of polyarthritis, subcutaneous nodules, and pancreas disease. J Rheumatol 2: 15–20

Taxy JB (1976) Adenocarcinoma of the pancreas in childhood. Cancer 37: 1508–1518

Tio TL, Tytgat GNI (1986) Endoscopic ultrasonography in staging local resectability of pancreatic and periampullary malignancy. Scand J Gastroenterol 21 (123): 135–142

Traverso WL, Longmire WP Jr (1978) Preservation of the pylorus in pancreatico-duodenectomy. Surg Gynecol Obstet 146: 959–962

Traverso WL, Longmire WP Jr (1980) Preservation of the pylorus in pancreatico-duodenectomy: a follow-up evaluation. Ann Surg 192: 306–310

Tryka AF, Brooks JR (1979) Histopathology in the evaluation of total pancreatectomy for ductal carcinoma. Ann Surg 190: 373–381

Tscholakoff D, Hricak H, Thoeni R, Winkler ML, Margulis (1987) Magnetic resonance imaging in the diagnosis of pancreatic disease. AJR 148: 703–708

Tsukimoto I, Watanabe K, Lin JB, Najajima T (1973) Pancreatic carcinoma in children in Japan. Cancer 31: 1203–1207

Waterhouse J, Muir C, Correa P, Powell J (1976) Cancer incidence in five continents, vol 3 IARC, Lyon (IARC scientific publication no 15)

Way LW (1987) Diagnosis of pancreatic and other periampullary cancers. In: Howard JM, Jordan GL, Reber HA (eds) Surgical diseases of the pancreas. Lea and Febiger, Philadelphia, pp 641–656

Webb JN (1977) Acinar cell neoplasms of the exocrine pancreas. J Clin Pathol 30: 103–112

Weiss SM, Skibber J, Dobelbower RR Jr, Whittington R, Rosato FE (1982) Operative pancreatic biopsy: ten-year review of accuracy and complications. Am Surg 48: 214–216

Weiss SM, Skibber JM, Mohiuddin M, Rosato FE (1985) Rapid intraabdominal spread of pancreatic cancer: influence of multiple operative biopsy. Arch Surg 120: 415–416

Whittington R, Dobelbower RR Jr, Mohiuddin M, Rosato FE, Weiss SM (1981) Radiotherapy of unresectable pancreatic carcinomas: a six-year experience with 104 patients. Int J Radiat Oncol Biol Phys 7: 1639–1644

Wiggans G, Woolley PV, MacDonald JS, Smythe T, Ueno W, Schein PS (1978) Phase II trial of streptozotocin mitomycin-C and 5-fluorouracil (SMF) in the treatment of advanced pancreatic cancer. Cancer 41: 387–391

Wood WC, Shipley WU, Gunderson LL, Cohen AM, Nardi GL (1982) Intraoperative irradiation for unresectable pancreatic carcinoma. Cancer 49: 1272–1275

Wynder EL, Mabuchi K, Maruchi N, Fortner JG (1973) Epidemiology of cancer of the pancreas. JNCI 50: 645–667

Yasuda K, Tanaka Y, Fujimoto S, Nakajima M, Kawai K (1984) Use of ultrasonography in small pancreatic cancer. Scand J Gastroenterol 19 (102): 9–17

Yasuda K, Mukai H, Fujimoto S, Nakajima M, Kawai K (1988) The diagnosis of pancreatic cancer by endoscopic ultrasonography. Gastrointest Endosc 34: 1–8

Zamcheck N, Martin EW (1981) Factors controlling the circulating CEA levels in pancreatic cancer. Cancer 47: 1620–1627

8 Biliary Cancer

J. H. MEERWALDT

CONTENTS

8.1 Introduction

Malignancies of the biliary tree are rare. These tumors cause a distinct combination of symptoms. According to this presentation and the possible treatment they can be divided by site as follows:

J. H. MEERWALDT, M.D., Ph.D., The Dr. Daniel den Hoed Cancer Center and Rotterdam Radio-Therapeutic Institute, Department of Radiotherapy, P.O. Box 5201, 3008 AE Rotterdam, The Netherlands

1. Lesser intrahepatic bile ducts
2. Right and left hepatic ducts
3. The hilar region and common hepatic duct
4. Gallbladder and cystic duct
5. Common bile duct
6. Region of the ampulla of Vater

Bile duct tumors tend to present relatively early because they occlude the lumen and thus cause jaundice. Although there is a slight difference according to site, the prognosis for these patients is generally poor.

Surgery has always been of great importance in the treatment of these tumors, whether it be curative or palliative, and offers the only possibility for cure. Until recently the role of radiation therapy was thought to be limited.

8.2 Topographical Anatomy

8.2.1 Gallbladder

The gallbladder is located at the medial lower border of the right lobe of the liver. Figure 1 shows the anatomical relationship to the surrounding organs. The neck of the gallbladder points to the hilum of the liver. Here the cystic duct, common hepatic duct, common bile duct, hepatic artery, and portal vein lie in close approximation.

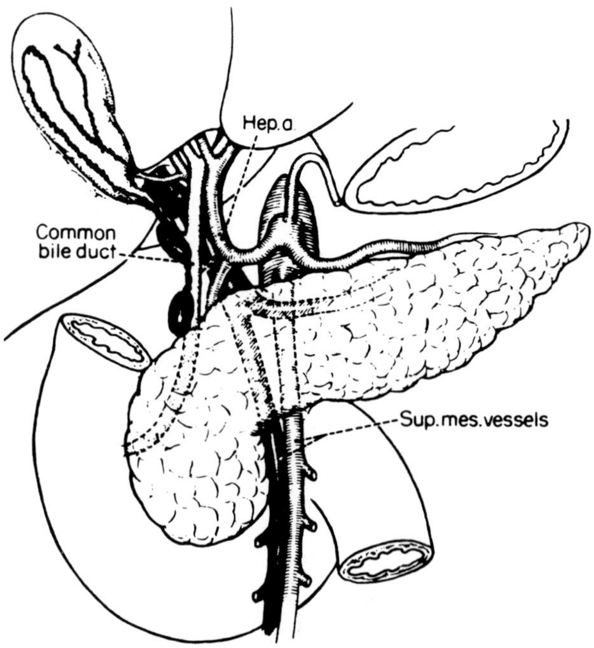

Fig. 1. Topographic anatomy of the gallbladder and bile ducts

There is a series of venules that cross the border between the gallbladder and the liver and in this special way a route for hematogenous liver metastases ist created.

8.2.2 Bile Ducts

Bile ducts arise in the liver where they collect the bile formed by the hepatocytes. The bile ducts lie in close proximity to the branches of the portal vein and the hepatic artery, but the flow is in the opposite direction. Within the liver these vessels lie together in the portal pedicle, their branches following each other closely.

Within the hilum of the liver we can identify a left and a right hepatic duct and right and left branches of the portal vein and hepatic artery. For surgical reasons a few anatomical variations as described by LONGMIRE et al. (1973) may be of importance.

Within the hepatoduodenal ligament, the hepatic duct – common bile duct and portal vein – and hepatic artery remain alongside. The last part of the common bile duct is situated within the pancreas. Based on these anatomical relationships, Longmire suggested dividing the bile duct into an upper third, hilar region – cystic duct; middle third, cystic duct – pancreas; and lower third, intrapancreatic part – ampulla.

8.2.3 Lymph Drainage

In the portal pedicle the lymph vessel runs alongside the bile ducts. Within the outer wall of the main hepatic duct and common bile duct an extensive lymph vessel system is present. Lymph nodes are situated in the hilum of the liver and in the hepatoduodenal ligament. Regional lymph nodes are the upper pancreaticoduodenal nodes. The lymph vessels of the gallbladder are situated within the gallbladder wall and follow the cystic duct until it reaches the common bile duct, where a common bile duct lymph node is situated. The next lymph node station is the upper pancreatico-duodenal lymph node group. No such thing as a hilar lymph node of the gallbladder exists according to FAHIM et al. (1962).

8.3 Natural History and Background of the Disease

8.3.1 Pattern of Spread

Tumors of the biliary tree not only tend to grow through the wall of the gallbladder or bile duct but also tend to extend along the wall in both directions.

8.3.1.1 Tumors of the Gallbladder

According to LITWIN (1967), the most frequent sites of metastases are liver 63%, regional lymph nodes 42%, colon 22%, porta hepatis 21%, omentum and duodenum 20%, and peritoneum 15%. TASAKA et al. (1975) found that well differentiated tumors tend to spread to the liver and poorly differentiated ones to the peritoneal surface.

8.3.1.2 Tumors of the Bile Ducts

Tumors of the bile ducts tend to invade surrounding structures. Proximal tumors invade the liver by local extension; in the hepaticoduodenal ligament the portal vein and the hepatic artery may be invaded; the distal tumors invade the pancreas and are often difficult to differentiate from pancreatic carcinomas. Bile duct tumors furthermore extend along the duct itself in both directions.

8.3.2 Natural History

Because of their primarily local extension into surrounding organs and structures complete resection is often not possible. Palliative procedures to relieve obstruction of the bile flow, either surgical of by interventional radiology, may be temporarily successful. But in the course of the disease recurrent obstruction and cholangitis will lead to hepatic failure and septic death (KLATSKIN 1965; OTTOW et a. 1985).

8.4 Epidemiology and Predisposing Factors

8.4.1 Incidence

The incidence of tumors of the gallbladder and extrahepatic bile ducts is strongly dependent on the way it is studied.

8.4.1.1 Gallbladder Cancer

In 56000 autopsies 203 gallbladder carcinomas were found, or 0.36% EDMONDSON (1967). This is about the same as ARMINSKI'S findings in 1949 (0.44%). On the total number of cholecystectomies STRAUCH (1960) found 1.5% malignancies, while BERGDAHL (1980) noted only 0.3%. The annual death rate for gallbladder carcinomas in the United States was about 6500 a year (1973), and in England and Wales 1100 a year (1966-1970), which accounts for 46 per million. In the United States gallbladder cancer accounts for 0.5% of all malignancies.

8.4.1.2 Bile Duct Cancer

At autopsy EDMONDSON (1967) reported 53 bile duct tumors in 56000 autopsies. WHELTON et al. (1969) and INGIS and FARMER (1975) consider statistics misleading because, even during laparotomy, the diagnosis is missed if not considered. LONGMIRE (1976) reported the incidence of new cases to be around 4500 a year in the United States.

8.4.2 Race

Both gallbladder and bile duct tumors are seen most frequently in Japanese (HAENSZEL and KURIHARA 1968; SEGI et al. 1969; INOUYNE and WHELAN 1978). They are also more common in American Indians, Mexican Americans, and Israelis (WATERHOUSE et al. 1976; KRAIN 1972; BLACK et al. 1977). They are regularly seen in the Chinese population of Hong Kong. KOO et al. (1981) and GUPTA et al. (1980) reported that these tumors account for 2.9% of all cancer cases in India. The Japanese retain the high incidence when they emigrate to Hawaii (INOUYNE and WHELAN 1978).

8.4.3 Age and Sex

Tumors of the biliary tree predominantly occur in older age. From 85% to 95% of the patients are over 50 years of age (STRAUCH 1960; LITWIN 1967). The male to female ratios are not the same throughout the world. LITWIN (1967) noted a 5 to 1 ratio in favor of women; GUPTA et al. (1980) found in India a ratio of 2.6 to 1. In England and Wales a ratio of 1.5 to 1 was noted, while INOUYNE and WHELAN (1978) reported an almost 2 to 1 ratio in favor of males in Hawaii.

8.4.4 Predisposing Factors

8.4.4.1 Gallbladder Cancer

In carcinoma of the gallbladder the association with gallstones is stressed. In 72%–95.9% of the cases of carcinoma stones were present (STRAUCH 1906; NEILL and de NEESE 1973). KOO et al. (1981) reported an association of only 23% (in a Chinese population).

The incidence of carcinoma of the gallbladder in 34 242 biliary operations, however, was low (1.41% or 482) (STRAUCH 1960). This incidence depends on race and sex (LOWENFELS et al. 1985). The duration of symptoms might also be of importance; 54% had symptoms of gallstones for longer than 4 years, while 26% had symptoms for over 20 years.

Chronic irritation might in this way be an contributing factor to the development of gallbladder cancer. Other contributing factors suggested are infection and chemical carcinogenesis. Obstruction by stones and infection, which in approximately two-thirds of the patients accompanies stones, favors conversion of bile acids to more carcinogenic substances (LOWENFELS 1978). Also in the presence of choledochal cysts and ulcerative colitis an increased risk of gallbladder cancer is reported (RITCHIE et al. 1974).

FORTNER (1955) reported induction of gallbladder cancer in the cat following implantation of a pellet of methylcholanthrene. Whether methylcholanthrene is also produced in vivo by degradation of cholic acid, deoxycholic acid, and cholesterol is questionable (FORTNER 1958).

8.4.4.2 Bile Duct Cancer

The relation between stones and bile duct cancer is not as clear as in gallbladder cancer. The frequency of stones in association with this cancer is not more than 30% (ROSS et al. 1973; HART et al. 1971). Only 1 in 60 patients with bile duct tumors underwent a cholecystectomy prior to the disease (GEORGE et al. 1981).

A role of chemical carcinogenic substances is postulated; this includes metabolic changes of bile acids or some compounds of rubber and wood-finishing industries (FORTNER 1958; KRAIN 1972).

In patients with ulcerative colitis a tenfold increase of incidence was reported by RITCHIE et al. (1974). The age at which the tumor is seen is on average 10 years earlier than in the general population.

The incidence of bile duct cancer in choledochal cysts is about 20 times greater (FLANIGAN 1977).

Sclerosing cholangitis is suggested to be a possible premalignant condition (WHELTON et al. 1969). In around 10% a cholangiocarcinoma is found at some time in the course of the disease. Oriental cholangiohepatitis is a disease almost exclusively of the Chinese. Cholangiocarcinoma can be a late complication (BELMARIC 1973).

8.5 Diagnostic Workup

8.5.1 History and Physical Findings

8.5.1.1 Gallbladder Cancer

There are no clear signs indicating a malignancy of the gallbladder. At an early stage symptoms may mimic those of cholecystitis of gallstones. Fat intolerance and frank colic pains are most frequently observed. A second presenting form might be obstructive jaundice; differential diagnosis is between benign obstruction or obstruction by an pancreatic tumor. A mass may be palpable in the right upper abdomen. A third presentation might be that of nonspecific gastrointestinal disorder, often accompanied by weight loss and anorexia.

8.5.1.2 Bile Duct Cancer

Inherent to their anatomical location, tumors of the bile duct will cause obstructive jaundice. In 34/50 cases this was the first presenting symptom. Sometimes pruritis or abdominal pain precedes the jaundice. Weight loss, nausea, and vomiting often accompany these symptoms (GEORGE 1983).

On clinical examination often no other specific signs are to be found. Sometimes the liver is enlarged but feels smooth.

8.5.2 Laboratory Data

Exhaustive laboratory investigations are not necessary to confirm the suspicion of obstructive jaundice. Liver and renal function tests are mandatory, as are serum alkaline phosphatase, bilirubin, conjugated and unconjugated, and levels of transferases. Renal functions should be measured to warn for possible hepatorenal syndrome. In the case of unsuspected symptomless gallbladder carcinoma no laboratory abnormalities will be found.

8.5.3 Diagnostic Procedures

It has been stressed that the diagnostic procedures, once a malignancy of the biliary tree is suspected, strongly depend on the discipline of the responsible physician and the therapeutic strategy he considers following.

8.5.3.1 Gallbladder Cancer

In tumors of the gallbladder *a plain abdominal X-ray* might show calculi. A "porcelain gallbladder" could also be revealed. According to POLK (1966) this was associated with carcinoma in 22%. With *oral cholecystograms* and *intravenous cholangiograms* a nonvisualization of the gallbladder is found in the majority of the patients.

Occasionally, filling defects might be observed. The bile ducts can be demonstrated by a percutaneous transhepatic cholangiogram (PTC) or an endoscopic retrograde cholangiopancreaticogram (ERCP). In 1600 ERCP cannulations OGOSHI and NIWA (1977) found 60 cases of gallbladder carcinoma. Failure to demonstrate the carcinoma was due to duodenal infiltration in 11 cases; 4 false positive and 4 false negative cases were reported.

Ultrasound can diagnose gallstones and cholecystitis and is reliable for detection of extension to the liver and for liver metastases. The accuracy of diagnosis is about 85% (PALLA et al. 1980; YEH 1979). Ultrasound and *CT scan* correlate well (YEH 1979). *In summary*, LITWIN (1967) reported a correct preoperative diagnosis in 19%; KOO et al. (1981), however, stated that the investigations were not helpful. The most important factor is to consider the diagnosis.

8.5.3.2 Bile Duct Cancer

When a tumor of the bile duct is suspected, the patient generally presents with obstructive jaundice. Plain abdominal X-rays, oral cholecystograms, and intravenous cholangiograms and barium meals are of little diagnostic value.

Ultrasound makes it possible to diagnose dilated bile ducts and to delineate the level of obstruction. The sensitivity of the investigation is around 90%, the specificity around 99% (RICHTER et al. 1983). The etiology of obstructive jaundice can often be suggested (LEE et al. 1977). Furthermore, neighboring structures can be identified, such as the liver, metastases or infiltration, and obstruction of the portal vein.

CT Scan. Valuable pictures of the dilated bile ducts and, if present, a mass are obtained.

PTC. The success rate in demonstrating dilated bile ducts is reported to be between 90% and 100% (ELIAS 1976; JAIN et al. 1977; OKUDA et al. 1974). Complications are reported in occur in 5%-7% (OKUDA et al. 1974).

According to MCPHERSON et al. (1982) the majority of patients complain of abdominal pain (30/37) following percutaneous transhepatic biliary drainage (PTBD). In about 10% intraperitoneal bile leakage occurred. In about 30% of the patients bacteriemia resulted an 2/37 patients developed hepatic abscesses. With ultrasound-guided PTBD, puncture related complications and leakage are less likely to occur (LAMERIS et al. 1985). Infectious conditions remain the main problem of any intervention with an obstructive bile duct system.

ERCP. Demonstration of the duct system is possible. Depending on the extent of the obstruction it will be possible to demonstrate the strictured area and the dilated ducts proximal to the obstruction. Especially if contrast is instilled under pressure and the system is not drained afterwards, the possibility of cholangitis and sepsis is present (ELIAS 1976). During ERCP it is also possible to biopsy the lesion.

8.5.3.3 Which Diagnostic Procedure to Choose

Which diagnostic procedure to choose in the case of a patient with obstructive jaundice is strongly dependent on physician expertise and availability of equipment. Individual diagnostic tests were compared by different authors. ELIAS (1976) directly compared ERCP and PTC. The two procedures were similar but PTC was more readily available. MATZEN et al. (1981) and ERTAN et al. (1981) showed similar results with recent technical advantages.

VAS and SALEM (1981) compared ultrasound, PTC, and the combination of both and demonstrated the latter to be superior. LAMERIS et al. (1985) demonstrated the safety of ultrasound-guided PTC. RICHTER et al. (1983) analyzed the different diagnostic approaches for sensitivity, specificity, and costs.

In patients with a low probability of extrahepatic obstruction the optimal strategy, ultrasound followed by PTC if dilated ducts are present, has a sensitivity of 92% and a specificity of 99%. In patients with a higher probability of extrahepatic obstructive jaundice, the optimal procedure, PTC, has a sensitivity of 97% and a specificity of 98%. However, some surgeons feel that a peroperative cholangiogram is sufficient and condemn preoperative cholangiograms, both PTC and ERCP, because of the possible bacterial contamination, which hampers postoperative recovery.

For a definitive diagnosis, a histological or cytological specimen is necessary. Which procedure to choose will again depend on the overall strategy. If one is going to intervene surgically anyway, there is no need to try to biopsy the tumor preoperatively. However, if a preoperative histological diagnosis is desirable, biopsies can be taken either during an ERCP or via the percutaneous transhepatic route. Ultrasound-guided cytological biopsy can also give diagnostic information. According to EVANDER et al. (1980) in 42%–53% of the patients a preoperative cytological diagnosis can be made.

If neither biopsy nor cytology reveals the true identity of the disease, the only possibility left of confirming malignancy is a diagnostic laparotomy. However, the majority of these patients are of old age, and may not be suitable for explorative surgery. One should also keep in mind that especially with the scirrhous type adenocarcinomas, even generous biopsies taken at laparotomy may fail to provide a definite histological diagnosis.

8.6 Pathology

Carcinomas of the biliary tree are in general adenocarcinomas (60%–98%); undifferentiated carcinomas (6%–10%) and squamous cell carcinomas (3%–6%) are less frequently observed (STRAUCH 1960; LITWIN 1967). The adenocarcinomas are mainly of the scirrhous type, but papillary carcinomas are also observed. TODOROKI et al. (1980a) describes four types of adenocarcinoma: polypoid, nodular, scirrhous, and diffusely infiltrating. A better survival for patients with polypoid adenocarcinomas was suggested. This was confirmed by KOZUKA et al. (1984), who also described the relation between the histological type and the extent of infiltration of the tumor. Contrary to these findings are the results of GEORGE (1983), who suggested that the more abundant the fibrous tissue in the scirrhous type, the better the prognosis.

8.7 Prognostic Factors

It is generally stated that cure for tumors of the biliary tree is only achieved when radical resection is possible. Whether this type of surgery is possible depends on factors related to the patient and factors related to the disease. Patient-related factors are concomitant diseases which preclude surgery. As 85%–95% of the patients are over 50 years of age, this preclusion is not exceptional.

8.7.1 Gallbladder Cancer

Curative resection of gallbladder carcinoma is generally limited to those cases in which the gallbladder is resected for apparently benign disease, and the diagnosis of carcinoma is made by chance on histological examination (NEVIN et al. 1976; BERGDAHL 1980). A correlation between survival and depth of tumor invasion and spread of the disease was suggested by NEVIN et al. (1976). Disease confined to the mucosa or mucosa and muscularis had a better prognosis. Based on the infiltration of the tumor, NEVIN et al. (1976) proposed a staging system, stages I–V: stage I, the tumor being intramucosal only: stage II, mucosa and muscularis involvement; stage III, involvement of all three layers of the wall; stage IV, involvement of all three layers and the cystic lymph node; and stage V, involvement of the liver or distant metastases. This staging system appeared to work well when applied to 399 cases taken from the literature.

The importance of an extensive surgical procedure was stressed by FAHIM et al. (1962). As confirmed by KOGA (1985), more aggressive surgery based on the depth of invasion will possibly extend the postoperative survival time.

8.7.2 Bile Duct Cancer

In carcinomas of the bile duct long-term survival is also related to the resectability for cure. Contrary to carcinomas of the gallbladder, bile duct tumors are seldom found by accident as the patient generally presents with symptoms of obstruction. The resectability depends on whether the tumor infiltrates into the surrounding structures (portal vein and hepatic artery mainly) and on the extent of the obstruction. If the disease is not localized, and lymph node metastases or distant metastases are present, resection for cure is not possible.

The location of the obstruction is suggested to be of influence on the prognosis. Lesions of the middle and distal thirds of the bile ducts would have a better outlook than lesions of the proximal third.

8.8 General Management

8.8.1 Carcinoma of the Gallbladder

8.8.1.1 Surgery

Cholecystectomy is the treatment of choice of gallbladder cancer. However, curative resection is generally limited to those where the malignancy is only discovered by chance on histological examination (BERGDAHL 1980; KOGA 1985). Due to the lack of specific signs and symptoms, 70%–90% of the tumors are beyond the resectable stage at laparotomy (COLLIER et al. 1984). When at laparotomy the diagnosis of gallbladder cancer is made, more extensive surgery has been advocated (VAITTINEN 1970; PIEHLER and CRICHBOW 1978), e.g., excision of the gallbladder bed and enbloc dissection of the lymphatic drainage along the common bile duct. If a curative resection is not possible, surgery is directed to relieve obstruction of the gastrointestinal tract or bile duct obstruction. KRAIN (1972) analyzed 1808 cases: no treatment 39%, surgery 35%, surgery plus radiotherapy 22%, and radiotherapy only 2%.

8.8.1.2 Radiotherapy

Postoperative adjuvant radiotherapy was analyzed by VAITTINEN (1970). Following curative resection only, the median survival was 29 months; following resection plus radiotherapy, 63 months. No details about the radiotherapy were reported.

In our own small series of five patients treated with postoperative radiotherapy because of microscopic disease left behind following resection, four are doing well, with no evidence of disease 4–20 months following radiotherapy.

8.8.1.3 Chemotherapy

Only small series of patients have been treated with different schedules, mainly including 5-FU (OSWALT and CRUZ 1977), sometimes combined with radiotherapy. No survival advantage was observed. Mitomycin C is suggested as an active drug in advanced disease (von EYBEN et al. 1980). Combined modality treatment of radiotherapy and chemotherapy followed by maintenance chemotherapy was described by KOPELSON et al. (1977). A small advantage in median survival was suggested.

8.8.2 Carcinoma of the Bile Ducts

8.8.2.1 Surgery

Resection of the tumor with histologically free margins of the surgical specimen is the only hope for cure up to now. However, due to tumor extension and the general condition of the patient, resection for cure is only possible in around 20% of the patients (BLUMGART 1982). The treatment of bile duct tumors also depends on the location. According to LONGMIRE et al. (1973) the bile duct system can be divided into a proximal third, a middle third, and a distal third comprising the intrapancreatic part of the bile duct and the periampullary region.

8.8.2.1.1 Proximal Third of the Bile Duct. General surgical management of these tumors consists of resection of the tumor with histologically tumor-free margins. Due to the extension of the tumor, resection for cure is possible in around 20% of the patients. Following preoperative screening, 60% of the patients thought to be resectable could be resected with hope for cure (BLUMGART et al. 1984; LAUNOIS et al. 1979). If resection for cure is not possible, palliative resection or surgical bypass with a hepaticoduodenostomy can be performed. Another possibility is stenting of the duct through the region of the tumor to relieve the obstruction. In the past this was the only possible treatment.

8.8.2.1.2 Middle Third of the Bile Duct. Surgical treatment of tumors of the middle third is in principle not different from that of tumors of the proximal third. Because of its location, however, resection is more easily performed. If resection is not possible, a choledochojejunostomy can be performed.

8.8.2.1.3 Distal Third of the Bile Duct. Surgical treatment of tumors of the distal third is generally the same as the treatment of tumors of the pancreas. If feasible, a pancreaticoduodenectomy is performed. If this procedure cannot be performed, a bypass operation is indicated.

8.8.2.2 Adjuvant Radiotherapy

Postoperative radiotherapy is advocated by several authors (TERBLANCHE et al. 1972; PIPELICH and LAMBERT 1978). In a small series of our own group five patients were treated with external radiotherapy, mainly because the margins of the surgical specimen were not tumor-free. These five patients are still alive with a follow-up period between 6 and 36 months. It is necessary to study the pattern of relapse of patients following a specific kind of surgical treatment. In this way it might be possible to define a group of patients who might benefit from either adjuvant radiotherapy or adjuvant chemotherapy.

8.8.2.3 Chemotherapy

Systemic chemotherapy has as yet not been fully evaluated. Case histories of single agents, 5-FU, BCNU, and mitomycin C have been reported. Mitomycin C is suggested as giving the best results (Haskell 1980).

8.8.2.4 Palliative Interventional Radiology Procedures

Apart from the surgically placed stents and tubes, it is also possible to place a tube by the transhepatic percutaneous route. With ultrasound-guided PTC it is easy to identify a major intrahepatic duct and place a catheter. If it is possible to pass the guide wire through the obstructed area, internal drainage either by a multiple side holes catheter or endoprosthesis can be achieved (Figs. 2, 3). If not, the catheter is left proximal of the stenosis to provide external drainage. Repeated attempts to pass the stenosis after a few days of external drainage are often successful.

Another possibility is to place an endoprosthesis by way of an ERCP. Following a papillotomy a guide wire is introduced. If the obstructed area is passed, an endoprosthesis is placed over the guide wire. Both methods, the PTBD- and the ERCP-placed endoprosthesis, are complementary. For proximal bile duct tumors a PTC-placed endopros-

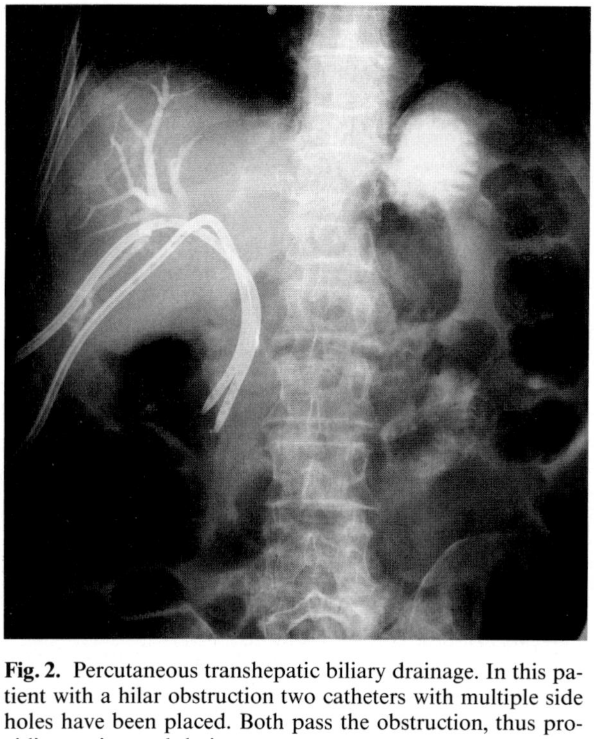

Fig. 2. Percutaneous transhepatic biliary drainage. In this patient with a hilar obstruction two catheters with multiple side holes have been placed. Both pass the obstruction, thus providing an internal drainage

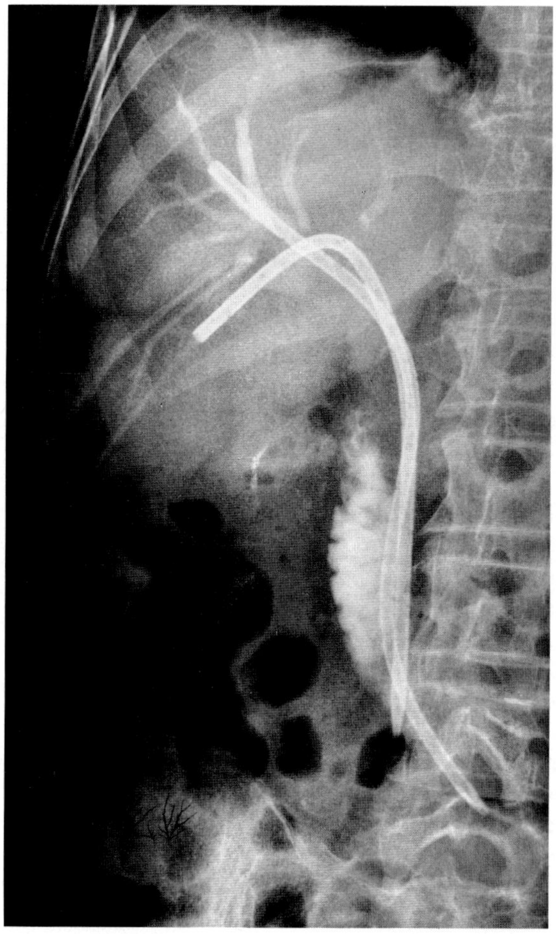

Fig. 3. Percutaneously placed endoprostheses. Following the ▷ intraluminal irradiation the external drains were removed and endoprostheses were placed by the percutaneous route (same patient as in Fig. 2)

thesis has advantages above the ERCP-placed one, less failures and less complications, while for middle and distal tumors the ERCP method is easier (HUIBREGTSE 1984, and our own experience).

8.9 Radiotherapy

8.9.1 Introduction

Radiotherapy of tumors of the biliary tree can be given with different intentions:

1. As the only treatment with hope of cure
2. As a combined treatment, adjuvant to surgery
3. As a palliative treatment

8.9.1.1 Radiotherapy as the Only Treatment with Hope of Cure

For gallbladder carcinoma no data are available to indicate that this option is a reality. Tumors are seldom diagnosed at an early stage and in this case surgery is the treatment of choice. For carcinoma of the bile duct a number of long-term survivals following radiotherapy only have been reported. Whether it is already possible to speak of cure remains to be seen, although a disease-free survival of 5-6 years seems to point that way (TERBLANCHE et al. 1972; KARANI et al. 1985).

8.9.1.2 Radiotherapy as a Combined Treatment

In patients with a carcinoma of the gallbladder the local recurrence rate following "curative" surgery is 86% in patients not surviving 5 years, and in those surviving 5 years it is 48% (KOPELSON et al. 1977). This indicates a place for another local therapy, i.e., radiotherapy. According to KRAIN (1972) 22% of the patients with carcinoma of the gallbladder are treated with surgery and radiotherapy. VAITTINEN (1907) reported a median survival of 29 months following "curative" surgery and of 63 months following surgery and radiotherapy. However, the patient groups are probably not comparable and no data of randomized prospective studies are available. For patients with a carcinoma of the bile duct few data are available on the pattern of failure following "curative" surgery. In a review of ten series of 70 patients who survived the postoperative period after "curative" surgery, 18 (26%) are known to have developed a local recurrence (KOPELSON et al. 1977). This might indicate that for subsets of patients, adjuvant radiotherapy might be worthwhile.

8.9.1.3 Radiotherapy as a Palliative Treatment

A palliative treatment is aimed at relieving the symptoms of the disease. The main symptom of tumors of the biliary tree is jaundice. Although elimination of jaundice following radiotherapy is reported, percutaneous transhepatic bile drainage or endoscopically inserted endoprosthesis offers an alternative which takes less time out of the patient's remaining life span.

Comparison of survival data for drainage alone or combined with radiotherapy indicates a longer survival for the combination (TERBLANCHE 1976). In addition, radiotherapy is a good palliative treatment for pain caused by infiltration or an expanding mass. KOPELSON et al. (1977) reported an overall palliative effect in 92% of the patients following radiotherapy and 51% following palliative surgery.

8.9.2 Radiation Therapy Techniques

Radiotherapy can be delivered in different ways:

1. External radiotherapy
2. Intraluminal radiotherapy
3. Intraoperative radiotherapy
4. Combinations of 1, 2, and 3

8.9.2.1 External Radiotherapy

When describing external radiotherapy, several factors are of importance:

1. The target volume
2. The dose and fractionation
3. The radiation therapy technique

8.9.2.1.1 Target Volume. The target volume of external radiotherapy has not been very well outlined by the different authors. In most reported series vague terms like "the tumor and the regional lymph nodes" are quoted without defining these areas. HOPFAN and WATSON (1978) described a standard radiation portal for porta hepatis radiotherapy. This was based on extensive celiac angiographic findings and consideration of the lymphatic drainage. This standard portal encompassed the porta hepatis and the lymph drainage along the common bile duct and the celiac axis.

According to MITTAL et al. (1985) the following areas should be considered for irradiation, as indicated by the pattern of failure: tumor bed, liver, porta hepatis and regional nodes, abdominal scar,

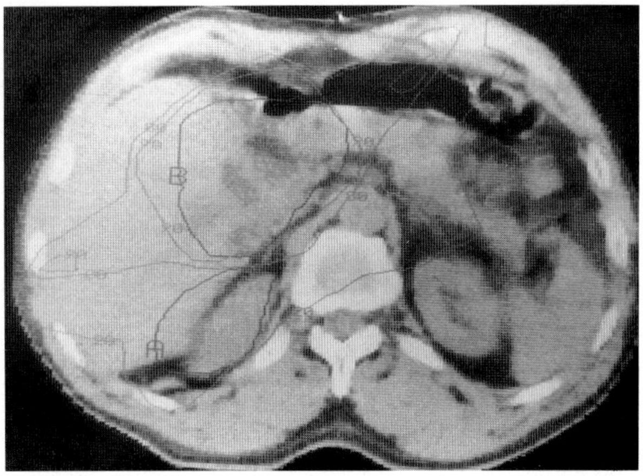

Fig. 4. Computer tomography assisted treatment planning. The target volume and organs at risk are outlined on a central reference image. Treatment plans are then individualized and calculated by computer and superimposed on the reference image

and the peritoneal cavity. However, elective irradiation of the whole liver and of the whole peritoneal cavity of a sufficient dose would not be feasible.

If a CT scan-based planning system is used, the considerations mentioned by HOPFAN and WATSON (1978) and MITTAL et al. (1985) should be kept in mind. The tumor with adequate margins or the area where the tumor was prior to surgery should be included. The whole length of the common bile duct should be in the target volume; the reason for this is the lymph drainage within the wall of the duct and the extension of the tumor along the duct wall. An example of such a CT scan-based planning is shown in Fig. 4.

The superior pancreaticoduodenal nodes and the nodes along the celiac axis should also be included in the target volume.

8.9.2.1.2 Dose. As in radiotherapy of other sites of the abdomen, the total dose is not so much determined by what is necessary but by what is tolerated.

A total dose of 45–50 Gy is suggested as being a minimal requirement (SMORON 1977; MITTAL et al. 1985).

In our own experience a dose of 30–40 Gy to a group of patients with microscopic disease only, might be sufficient. Therefore, we would suggest a dose of 40 Gy to areas with microscopic or no detectable disease. With a shrinking field technique, a boost can be delivered to the tumor and to grossly involved lymph nodes.

Some authors report total doses of up to 60–70 Gy (KOPELSON et al. 1977; BUSKIRK et al. 1984; MITTAL et al. 1985). A dose-effect relationship is suggested by SMORON (1977), better survival for doses above 45 Gy, and by MITTAL et al. (1985), better survival for TDF values of more than 70. However, one should keep in mind that there could be a bias, as these are not randomized studies and the patients receiving less treatment were probably in a poor general condition (SMORON 1977).

Little or no data are available on fractionation of the dose. In most series the fractionation schedule is not reported or differed within wide ranges as the patients were treated over a longer period.

Recently, some patients have been treated with multiple daily fractions, and the tolerance was good. Survival is comparable to that achieved with conventional fractionated radiotherapy.

8.9.2.1.3 Technique of Irradiation. The technique used by the different authors have not been reported in detail. Only HOPFAN and WATSON (1978) described his anterior-posterior (AP) and posterior-anterior (PA) plan parallel opposing fields technique. Other authors simply state the use of two-, three-, or four-field techniques.
Our group has been using CT scan-planned radiotherapy (SIDOS system). By carefully delineating the target volume and the critical normal tissues, it is possible to choose the technique with delivers a given dose of radiation with at the same time a maximum sparing of normal tissue (Fig. 4). This means that we have no standard technique but that we individualize techniques in each patient.

8.9.2.2 Intraluminal Radiotherapy

Intraluminal radiotherapy is the best name for this treatment; other names are used like interstitial radiotherapy or brachytherapy. Interstitial means placed in the tissue and brachytherapy means therapy from close range. But intraluminal therapy is exactly what is done. The radiation source is brought into the lumen of the bile duct. This type of treatment has advantages and disadvantages. Firstly, it is only possible if we have a route into the bile duct system, so we need a drainage of some kind. Secondly, we have to bring the source(s) into the system or into the drain. This handling of the drains can give rise to the most feared complication, cholangitis.

An advantage is that it is possible to deliver a reasonable dose of radiation without causing damage to surrounding normal tissues. This is explained by the rapid fall-off of all radioactive sources; this means that the dose decreases rapidly with increasing distance. At the same time this is a disadvantage; if a tumor measures 2–3 cm, there is quite a difference in dose absorbed by the inner and the outer margins of the tumor – this can easily be a four- to fivefold difference.

8.9.2.2.1 Target Volume. The volume should encompass the tumor, and as the lymph nodes and the lymph drainage are situated along the common bile duct, the whole length of the common bile duct should also be encompassed. Adequate margins, especially at the proximal site, should be taken to avoid marginal recurrences.

8.9.2.2.2 Dose. The dose of radiation delivered is generally specified at a depth of 5 mm; this is supposed to be the wall thickness of the bile duct. Doses fo 10–60 Gy have been given. Apart from the group of FLETCHER et al. (1983) and KARANI et al. (1985) only incidental reports are given of intraluminal radiotherapy only. Little is known about the tolerance of the bile duct to irradiation. For this reason the dose was at first delivered in two fractions. Later on the same dose was given as one fraction. In general the radiation is delivered at a low dose rate, e.g., 44 Gy in 55 h (FLETCHER et al. 1983), or even lower, 26 Gy in 78 h (MOLT et al. 1986).

ITAMI et al. (1986) reported a method of delivering the dose (30 Gy) in about 30 min, repeated 1 week later. This high dose rate program is easier for the patient and was associated with less complications.

8.9.2.2.3 Technique of Irradiation. As mentioned above, it is necessary to have a route for applying the radiation source. In most reports the authors described different percutaneous drainage procedures, U-tubes, T-tubes, surgically placed transtumoral tubes, or percutaneous transhepatic drainage procedures. A catheter with either preplaced sources or an afterloading catheter is placed in the drain. For tumors of the hilar region it is advisable to drain and implant both sides. The length of the radioactive material necessary is determined and placed in the catheter. Figure 5 shows an X-ray of the iridium wires and isodose curves.

Another possible route is using a nasobiliary drain placed by gastroduodenoscopy. PHILIP et al. (1984) reported on this method, the advantage being no percutaneous source of infection. Only three patients have been treated so far and no follow-up data are available.

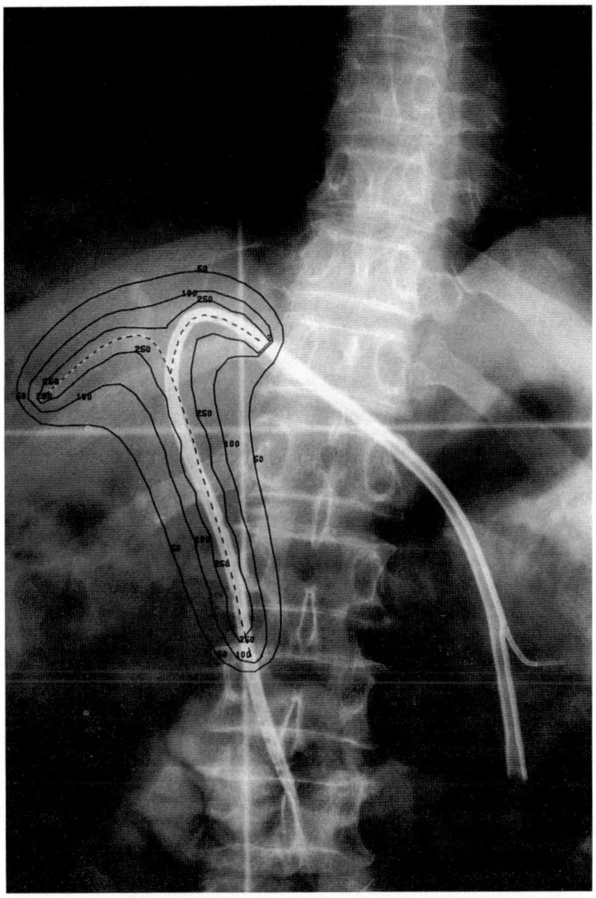

Fig. 5. Intraluminal irradiation. Both sides of the biliary tree (left and right) are drained by PTBD (only the drain on the left side is a radiopaque one). An iridium wire (---) is visible inside both drains. The isodose curves drawn around the wires indicate the rapid fall off (100% = ±1 cm of the axis of the wire; the other isodoses are 250% and 50%)

8.9.2.3 Intraoperative Radiotherapy

Intraoperative radiotherapy is another possibilty to avoid normal tissue damage. During surgery it is possible to visualize the lesion and remove as far as possible the normal tissues from the area to be irradiated. Unfortunately, it is only possible to deliver one fraction.

TODOROKI et al. (1980b) reported on 11 patients with advanced unresectable tumors of the biliary tree who were treated this way. The radiation modality used was electrons. Depending on their energy they have a rapid fall-off and the depth of penetration can be chosen this way. It is suggested to debulk as much of the tumor as possible prior to irradiation.

8.9.2.4 Combinations of External and Intraluminal or Intraoperative Radiotherapy

To deliver as high a dose of radiation as possible in the target area it is necessary to combine the different radiotherapy techniques. External radiotherapy only cannot be delivered to a sufficient dose without causing adverse side-effects. Intraluminal radiotherapy does not penetrate enough to treat large tumors and the lymph node regions. Intraoperative radiotherapy can only be delivered in one fraction, which is probably not enough. From these points of view it is understandable that combinations of the three methods are generally used. Whether this will result in better survival and what is the tolerance of combined methods has yet to be determined.

8.10 Results of Standard Techniques

At the present time there is no standard technique on which the results can be reported. Therefore, we report here on the results of the different techniques mentioned.

8.10.1 Surgery

8.10.1.1 Gallbladder Cancer

Surgical resection, if possible, is the treatment of choice; however, it is only possible in 10%–20% of the patients. The operative mortality is a high as 25%, increasing to 40% if hepatic resection is included.

When related to stage, BELTZ and CONDON (1974) and PIEHLER and CRICHBOW (1978) reported a mean survival of 16.7–21.5 months for stage A (confined to the mucosa and diagnosed only histologically postoperatively) and of 8.5 weeks to 1.7 months for stage D (widespread disease) (see Table 1).

If staged according to NEVIN et al. (1976) a 5-year survival for stage I was 6/10 patients, while for stage V, 0/21 patients survived (see Table 2).

BERGDAHL (1980) reported data from 32 patients where the diagnosis was made on histological examination only. In 21 the tumor invaded all layers of the wall and all died within 2½ years. Median survival was less then 1 year. However, of the 11 patients with infiltration of the mucosa or submucosa only, a 5-year survival of 64% and a 10-year survival of 44% was reported.

Table 1. Survival of patients with gallbladder carcinoma according to stage (BELTZ and CONDON 1974)

Stage	Mean survival	No. of patients
A	16.7 months	14
B	24.7 months	4
C	15.4 weeks	28
D	8.5 weeks	71

Table 2. Comparison between stage, grade, and survival in patients with carcinoma of the gallbladder

	No. of patients surviving 5 years	Total no. of patients
Stage		
I	6	10
II	5	7
III	2	23
IV	1	5
V	0	21
Grade		
I	10	16
II	3	26
III	1	24

SHIEH et al. (1981) also reported a 5-year survival of 66%. However, MOERTEL (1982) had doubts as to whether these "early cancers" of the gallbladder were truly malignant tumors.

8.10.1.2 Bile Duct Cancer

Surgery is the treatment of choice if possible, but resection with hope of cure is only feasible in 20% of the patients. The resectability depends on the surgeon's attitude and aggressiveness.

Radical resection in this subset of patients resulted in a mean survival of 14–24 months (LAUNOIS et al. 1979; BLUMGART 1982). The mortality of this procedure is determined by whether or not hepatic resection is included in the operation, and if so, can increase to as much as 35%. If no "curative" resection is possible, and a palliative procedure is performed, a mean survival of 7–16 months is reported (EVANDER et al. 1980; CAHOW 1979). In a review of a collective series of 483 patients, BROE and CAMERON (1981) reported the mean survival data for treatment by T-tubes to be 9.9 months, for intrahepatic cholangiojejunostomy 11.0 months, for palliative transhepatic stenting 18.6 months, and for tumor resection 22.2 months.

OTTOW et al. (1985) also reviewed data of different surgical and nonsurgical treatments.

8.10.2 Radiotherapy

8.10.2.1 Gallbladder Cancer

Results of radiotherapy of tumors of the gallbladder are scant, although according to KRAIN (1972) 22% of the patients are treated by a combination of surgery and radiotherapy. Most reported patients had obstruction of the bile ducts and were treated as patients with bile duct tumors.

Of a compilative group of data of four studies survival ranged from 3.5 to 15.8 months; only one patient out of 16 survived more than 1 year (BUSKIRK et al. 1984; KOPELSON et al. 1977; MITTAL et al. 1985). However, some long-term survivors have been reported.

ACKERMAN and DEL REGATO (1970) reported a patient who survived 5 years following radiotherapy and VAITTINEN (1970) reported a median survival following "curative" surgery of 29 months and following surgery and radiotherapy of 63 months. However, no data on the radiotherapy technique are reported. On the basis of these case reports and small series it is difficult to validate this treatment. Larger uniformly treated series of patients are needed.

8.10.2.2 Bile Duct Cancer

In reporting results of radiotherapy of bile duct tumors there are several difficulties. Firstly, in general, numbers of patients are small; in the reported series there is a mean number of 15.5 patients (range 5–34). If then divided by different treatment modalities the numbers of patients become even smaller. Secondly, as some of the treatments took place during a long period, different doses and fractionations or combinations were used. Thirdly, survival as the only objective evaluation point is not always calculated from the same moment, i.e., laparotomy or time of diagnosis or start of radiotherapy or completion of radiotherapy. This can make a considerable difference, e.g., JOHNSON reported a mean survival of 8.3 months following completion of radiotherapy, while it was 16.1 months following laparotomy. KOPELSON reported the same: 8.0 vs 16.1 months. This might be an explanation for some of the differences observed.

The following results are published:

- External radiotherapy only: 8.0–12.5 months (BUSKIRK et al. 1984; FOGEL and WEISSBERG 1984; KOPELSON et al. 1977; MITTAL et al. 1985)
- Intraluminal radiotherapy only: 7.6–16.8 months (FLETCHER et al. 1983; KARANI et al. 1985; KUTZNER et al. 1985; MORNER et al. 1984)
- Intraoperative radiotherapy only: 10.9 months (TODOROKI et al. 1980b)
- Combinations of external and intraluminal radiotherapy: 4.5–18 months (BUSKIRK et al. 1984; HERSKOVIC et al. 1985; JOHNSON et al. 1985; MOLT et al. 1986; MEERWALDT, unpublished)

These data are the collective results of 13 studies, divided by treatment technique, reporting their mean survival data. Of more importance would be to know how many patients were still alive after, say, 6 months or 1 year. Only a few authors published these data, as, in general, follow-up is short.

FLETCHER et al. (1983) reported 16/19 patients who have lived for more than 6 months and 9/19 who have lived for more than 1 year. In a later report of this same group, KARANI et al. (1985) reported 21/30 patients who have lived for more than 1 year and 5/30 who have lived for more than 2 years. HERSKOVIC et al. (1985) reported 7/16 patients who lived for more than 6 months and 5/16 who lived for more than 1 year.

Furthermore, a number of long-term survivors of 5–6 years have been reported (TERBLANCHE et al. 1972; CAMERON et al. 1982; KARANI et al. 1985). Especially noticeable is a series of five patients by TERBLANCHE et al. (1972), who reported three patients alive with NED at 2.5, 3.75, and 3.75 years' follow-up; one had a recurrance at 2.5 years and one had died at 5 years' follow-up. These patients were treated by external radiotherapy, 60 Gy, following palliative surgery.

In conclusion, one could say that radiotherapy can produce long-term survivors; whether these patients are indeed cured remains to be seen.

8.10.3 Interventional Radiology Drainage Techniques

To relieve the symptoms of obstruction a percutaneous biliary drainage procedure can be performed. This can be either an external or an internal drainage, or an endoprosthesis can be inserted.

PASSARIELLO et al. (1985) reported on a multicentric investigation of 731 cases, of which 536 were patients with a neoplastic obstruction. The median survival of the group only treated by percutaneous drainage was 2.4 months. Patients who were subjected to biliary drainage followed by curative surgery had a median survival of 5 months and those with palliative surgery 2.9 months.

At 2 years there was virtually no survival difference between the three groups.

Within our group we have expertise in both techniques of interventional radiology available, the endoprosthesis placed by the percutaneous transhepatic route and the one placed by the endoscopic route. During 1983 and 1984, 144 patients were treated for malignant obstruction. The results of this procedure strongly depend on the location of the obstruction. For strictures of the common bile duct the success rate was 82%, resulting in relief of obstruction in 77% and only 10% complications. For hilar strictures the success rate was 70%; resulting in relief of obstruction in only 55% while 41% experienced complications. Therefore the endoscopic route is less suitable for hilar strictures. The transhepatic route, however, resulted in less complications and a higher success rate for hilar strictures compared to strictures of the common bile duct (Lameris 1986). Problems arising following interventional procedures are infections and clotting of the catheter. As long as the bile flow is not obstructed, contamination of the bile will not result in infection and fever. However, manipulations of the drain, especially if the intrahepatic pressure increases, will lead to sepsis.

According to Fletcher et al. (1983) the mean life span of a drain before it has to be exchanged is 4.2 months.

8.11 Patterns of Failure

The pattern of failure following radiotherapy tells us something about the efficacy of the treatment. However, one should keep in mind that not all patients in a given reported series were treated uniformly.

Fogel and Weissberg (1984) reported the autopsy findings in ten patients: nine failed locally, two failed in the lymph nodes, five had liver metastases, two lung metastases, two skin metastases, and two metastases to other abdominal organs or peritoneal metastases.

Herskovic et al. (1985) reported in 12 patients only three local (marginal) failures, one lymph node failure, four liver failures, two distant metastases, and two peritoneal metastases.

Mittal et al. (1985) reported a local failure rate of 10/18 (56%), liver metastases in 12/18 (67%), and peritoneal metastases in 5/18 (28%).

The only conclusion one can draw from these data is that despite considerable treatment, the local failure rate is still between 25% and 90%. In the series of Herskovic et al. (1985) the 3/12 local failures were marginal recurrences, due to a target volume which was chosen too small. Patients in the other two series were only treated by external radiotherapy and the exact target volume is not known.

8.12 Complications of Treatment

In this section we will only discuss complications of radiotherapy.

8.12.1 Acute Side-effects Due to External Beam Radiotherapy

In general only mild side-effects, nausea and weight loss in about half of the patients, are reported. Some patients cannot complete their course of radiotherapy, but this is mainly due to deterioration of an already poor general condition or septic complications.

8.12.2 Acute and Long-term Side-effects Due to Intraluminal Radiotherapy

Cholangitis and other infectious complications are a serious threat to these patients. From 30% to 74% of the patients experience multiple periods of fever (Buskirk et al. 1984, Fletcher et al. 1983; Herskovic et al. 1985; Mittal et al. 1985), resulting in 10%–20% septic death (Fletcher et al. 1983; Herskovic et al. 1985). In most cases these episodes of fever could be traced back to manipulation of the bile drains. No local acute side-effects of the intraluminal radiotherapy could be detected.

The occurrence of fibrotic changes as a late effect has been suggested but has not been substantiated.

8.12.3 Late Effects of External Beam Radiotherapy

Major late complications mainly due to external beam radiotherapy are upper gastrointestinal bleedings reported to occur in 10%–20% of the patients. On endoscopy or laparotomy a friable duodenal mucosa or frank ulceration was found in 10%–20% (Buskirk et al. 1984; Johnson et al. 1985; Molt et al. 1986; Mittal et al. 1985).

Two out of nine patients receiving more than 50 Gy developed duodenal ulceration (Mittal et al. 1985). Whether intraluminal radiotherapy might also have contributed to these complications is not certain. In general, the duodenum is not included in the volume of the high dose levels.

8.13 Summary

8.13.1 The Main Problems in Treatment

8.13.1.1 The Late Stage at Presentation

The first problem ist the late stage of the disease when it is diagnosed. The majority of the patients with carcinoma of the gallbladder or bile ducts (80%-90%) have unresectable lesions at the time of diagnosis (INOUYNE and WHELAN 1978; BLUMGART et al. 1984). As the disease at an early stage is not distinguishable from benign conditions, such as cholecystitis or gallstones, the most important point is to consider the diagnosis if symptoms change.

8.13.1.2 Surgical Problems

For the group of patients in whom resection with hope of cure is possible, this is the treatment of choice. However, with an increasingly aggressive attitude of the surgeon, the mortality varies between 11% and 28%, increasing to 35% if hepatic resections are included (EVANDER et al. 1980).

In collected reviews of 1060 patients only 3% survived more than 5 years and approximately 60% developed a local-regional recurrence (BUSKIRK et al. 1984).

8.13.1.3 Radiotherapy Problems

Although some textbooks still say that there is no place for radiotherapy in the treatment of tumors of the biliary tree, the results mentioned in this chapter indicate the opposite. Radiotherapy is a good palliative treatment with a success rate of 92% (KOPELSON et al. 1977). Radiotherapy can prolong the disease-free survival (KOPELSON et al. 1977; TERBLANCHE et al. 1972; LEES et al. 1980). Long-term survivors have been reported, but whether these patients are cured by radiotherapy has still to be determined. The problem is, what is the optimal way to give radiotherapy? There is no proven difference in mean survival for the different techniques. Larger groups of uniformly treated patients reported in similar fashion will be needed. Another problem is the place of adjuvant radiotherapy. Case histories and small series indicate a better survival of patients who, following "curative" resection or resection with microscopic tumor left behind, are treated with radiotherapy.

8.13.2 Comparative Studies

No randomized trials have been done for tumors of the biliary tree. Maybe the time has come to try and perform such study trials or, if necessary, to start sound pilot studies. These studies should be based on patterns of failure and known dose-effect results. Three different groups of patients should be defined:

1. The group where adjuvant radiotherapy following surgery might have a place
2. The group where radiotherapy with hope of cure is indicated
3. The group where only palliative treatment is indicated

Much is still open for discussion, e.g.: What kind of radiotherapy (external, intraluminal, intraoperative, or in combination)? What dose of radiotherapy? What target volume? Should liver and peritoneal cavity also be included to a low dose?

Only close cooperation between the surgical oncologist, the interventional radiologist, the gastroenterologist, the radiation oncologist, and the medical oncologist, preferably on a multicentric or even international basis, will be able to solve this problem.

8.13.3 Preferred Therapy

8.13.3.1 Gallbladder Cancer

Radical surgery, if possible, is the treatment of choice. Surgeons should be aware of the possibility of malignancy and, when such is suspected, take the necessary action.

The indications for postoperative adjuvant radiotherapy have yet to be defined.

If at laparotomy resection is not possible, the surgeon should map the extent of the disease as accurately as possible in three dimensions. Within the team the best approach has to be discussed. "Curative" or palliative, and if palliative, what treatment offers the best palliation?

8.13.3.2 Bile Duct Cancer

Radical surgery, if possible, is the preferred therapy for biliary cancer. The role of postoperative adjuvant radiotherapy has yet to be defined. If at laparotomy resection for cure is not possible, the surgeon should not perform partial resections which preclude radiotherapy with hope of cure. Within the team the best approach has to be discussed, "curative" or palliative. In either case a biliary drainage will be necessary. Radiotherapy with hope of cure might be a combination of external radiotherapy of at least 40 Gy and intraluminal radiotherapy of at least 40 Gy at 5 mm.

Palliative radiotherapy can either be given by external beam radiotherapy or by intraluminal radiotherapy, or by a combination. A problem still to be determined is whether for individual patients percutaneous bile drainage plus radiotherapy offers better palliation than percutaneous bile drainage alone.

Intraoperative radiotherapy might prove to have a place in the primary treatment of these tumors. However, the availability of appropriate facilities might be the primary problem for performing this kind of treatment.

References

Ackermann LV, Del Regato JA (1970) Cancer, diagnosis, treatment and prognosis. Mosby, St Louis

Arminski T (1949) Primary carcinoma of the gall bladder. Cancer 2: 379

Belmaric J (1973) Intrahepatic bile duct carcinoma and *C. sinensis* infection in Hong Kong. Cancer 31: 468-473

Beltz WR, Condon RE (1974) Primary carcinoma of the gall bladder. Ann Surg 180: 180-184

Bergdahl L (1980) Gallbladder carcinoma first diagnosed at microscopic examination of gallbladder removed for presumed benign disease. Ann Surg 191: 19-22

Black WC, Key CR, Carmany TB, Herman D (1977) Carcinoma of the gall bladder in a population of Southwestern American Indians. Cancer 39: 1267-1279

Blumgart LH (1982) New perspectives and old problems in biliary surgery. Surg News 13: 1-4

Blumgart LH, Benjamin IS, Hadjis NS, Bearley R (1984) Surgical approaches to cholangiocarcinoma at the confluence of hepatic ducts. Lancet 1: 66-70

Broe PJ, Cameron JL (1981) The management of proximal biliary tract tumors. Adv Surg 15: 47-91

Buskirk SJ, Gunderson LL, Adson MA, Martinez A, May GR, McIlrath DC, Nagorney DM (1984) Analysis of failure following curative irradiation of gall bladder and extrahepatic bile duct carcinoma. Int J Radiat Biol Oncol Phys 10: 2013-2023

Cahow CE (1979) Intrahepatic cholangiojejunostomy: a new simplified approach. Am J Surg 137: 443-448

Cameron JL, Boe P, Zuidema GD (1982) Potential bile duct tumours: surgical management with Silastic biliary stents. Ann Surg 196: 412-419

Collier NA, Carr D, Hemmingway A, Blumgart LH (1984) Preoperative diagnosis and its effect on the treatment of carcinoma of the gall bladder. Surg Gynecol Obstet 159: 465-470

Edmondson HA (1967) Tumors of the gall bladder and extrahepatic bile ducts. In: Atlas of tumour pathology, sect 7. Armed Forces Institute of Pathology, Washington DC, p 26

Elias E (1976) A randomized trial of percutaneous transhepatic cholangiography with Chiba needle versus endoscopic retrograde cholangiography for bile duct visualization in jaundice. Gastroenterology 71: 439-443

Ertan A, Kandilci U, Danisoglu V (1981) A comparison of percutaneous transhepatic cholangiographic and endoscopic retrograde cholangiopancreatography in postcholecystectomy jaundice. J Clin Gastroenterol 3: 67-72

Evander A, Fredlund P, Hoevels J (1980) Evaluation of aggressive surgery for carcinoma of the extrahepatic bile ducts. Ann Surg 191: 23-29

Fahim RB, McDonald JR, Richards JC, Ferris DO (1962) Carcinoma of the gall bladder: a study of modes of spread. Ann Surg 156: 114-124

Flanigan DP (1977) Biliary carcinoma associated with biliary cysts. Cancer 40: 880-883

Fletcher MS, Brinkley D, Dawson JL, Nunnerley H, Williams R (1983) Treatment of hilar carcinoma by bile drainage combined with internal radiotherapy using 192 iridium wire. Br J Surg 70: 733-735

Fogel TD, Weissberg JB (1984) The role of radiation therapy in carcinoma of the extrahepatic bile ducts. Int J Radiat Oncol Biol Phys 10: 2251-2258

Fortner JG (1955) The experimental induction of primary carcinoma of the gall bladder. Cancer 8: 689-700

Fortner JG (1958) An appraisal of the pathogenesis of primary carcinoma of the extrahepatic biliary tract. Surgery 43: 563-571

George PA (1983) Cancer of the bile ducts. In: Gazet JC (ed) Carcinoma of the liver, biliary tract and pancreas. Arnold, London, pp 104-151

George PA, Brown C, Foley RTE (1981) Carcinoma of the hepatic duct function. Br J Surg 68: 14-18

Gupta S, Udupa KN, Gupta S (1980) Primary carcinoma of the gall bladder: A review of 328 cases. J Surg Oncol 14: 35-44

Haenszel W, Kurihara M (1968) Studies of Japanese migrants. I. Mortality from cancer and other diseases among Japanese in the United States. INCI 40: 43-68

Hart J, Modan B, Shani M (1971) Cholelithiasis in the aetiology of gall bladder neoplasms. Lancet 1: 1151-1153

Haskell CM (1980) Cancer of the liver: In: Haskell CM (ed) Cancer treatment. Saunders, Philadelphia, pp 319-357

Herskovic AM, Engler MJ, Noell KT (1985) Radical radiotherapy for bile duct carcinoma. Endocuriether Hyperthermia Oncol 1: 119-124

Hopfan S, Watson R (1978) Porta hepatis irradiation. Int J Radiat Oncol Biol Phys 4: 333-336

Huibregtse K (1984) Endoscopic insertion of stents through malignant and benign biliary strictures. In: Yap SM, v. d. Sluis SF, Lamers CBHW (eds.) Recent advances in diagnostic and therapeutic endoscopy. Mur Kostverloren, Aalsmeer, pp 57-63

Ingis DA, Farmer RG (1975) Adenocarcinoma of the bile ducts. Relationship of anatomic location to clinical features. Am J Dig Dis 20: 253-261

Inouye AA, Whelan TJ (1978) Carcinoma of the extrahepatic bile ducts. Am J Surg 136: 90-95

Itami J, Saegusa K, Tsuchiya Y, Mamiya T, Miyoshi T, Ohto M, Arimizu N (1986) Intrakavitäre High-dose-rate-Afterloading-Bestrahlung beim inoperablen malignen Gallengangsverschluß. Strahlenther Onkol 162: 105-110

Jain S, Long RG, Scott J, Dick R, Sherlock S (1977) Percutaneous transhepatic cholangiography using the "Chiba" needle - 80 cases. Br J Radiol 50: 175-180

Johnson DW, Safai C, Goffinet DR (1985) Malignant obstructive jaundice: treatment with external-beam and intracavitary radiotherapy. Int J Radiat Oncol Biol Phys 11: 411-416

Karani J, Fletcher M, Brinkley D, Dawson JL, Williams R, Nunnerley H (1985) Internal biliary drainage and local radiotherapy with iridium-192 wire in treatment of hilar cholangiocarcinoma. Clin Radiol 36: 603-606

Klatskin G (1965) Adenocarcinoma of the hepatic duct at its bifurcation within the porta hepatis. Am J Med 38: 241-256

Koga A, Yamauchi S, Izumi Y, Hamanaka N (1985) Ultrasonigraphic detection of early and curable carcinoma of the gallbladder. Br J Surg 72: 728-730

Koo J, Wong J, Cheng FC, Ong GB (1981) Carcinoma of the gall bladder. Br J Surg 68: 161-165

Kopelson G, Harisiadis L, Tretter P, Chang CH (1977) The role of radiation therapy in cancer of the extra-hepatic biliary system. Int J Radiat Oncol Biol Phys 2: 883-894

Kozuka S, Tsubone M, Hachisuka K (1984) Evolution of carcinoma in the extrahepatic bile ducts. Cancer 54: 65-72

Krain LS (1972) Gall bladder and extrahepatic bile duct carcinoma. Geriatrics 27: 111-117

Kutzner J, Klose K, Keller E (1985) Palliative Gallengangsbestrahlung. Strahlentherapie 161: 669-672

Laméris JS, Obertop H, Jeekel J (1985) Biliary drainage by ultrasound guided puncture of the left hepatic duct. Clin Radiol 36: 269-274

Laméris JS, Stoker J, Dees J, Nix GAJJ, v. Blankenstein M, Jeekel J (1987) Non-surgical palliative treatment of patients with malignant biliary obstruction - The place of endoscopic and percutaneous drainage. Clin Radiol 38: 603-608

Launois B, Campion JP, Brissot P (1979) Carcinoma of the hepatic hilus. Ann Surg 190: 151-157

Lee TG, Henderson SC, Ehrlich R (1977) Ultrasound diagnosis of common bile duct dilatation. Radiology 124: 793-797

Lees CD, Zapolanski A, Cooperman AM (1980) Carcinoma of the bile ducts. Surg Gynecol Obstet 151: 193-198

Litwin MS (1967) Primary carcinoma of the gall bladder; review of 28 cases. Arch Surg 95: 236-240

Longmire WP (1976) Tumors of the extrahepatic biliary radicals. In: Hickey RC (ed) Current problems in cancer. Year Book Medical Publishers, Chicago, p 1

Longmire WP, McArthur MS, Bastounis EA (1973) Carcinoma of the extrahepatic biliary tract. Ann Surg 178: 333-345

Lowenfels AB (1978) Does bile promote extra-colonic cancer? Lancet 2: 239-241

Lowenfels AB, Lindström CG, Conway MJ, Hasting PR (1985) Gall stones and risk of gall bladder cancer. INCI 75: 77-80

Matzen P, Haubek A, Holst-Christensen J, Lejerstofte J, Juhl E (1981) Accuracy of direct cholangiography by endoscopic or transhepatic route in jaundice - a prospective study. Gastroenterology 81: 237-241

McPherson GAD, Benjamin IS, Habib NA, Bowley NB, Blumgart LH (1982) Percutaneous transhepatic drainage in obstructive jaundice: advantages and problems. Br J Surg 69: 261-264

Mittal B, Deutsch M, Iwatsuki S (1985) Primary cancers of extrahepatic biliary passages. Int J Radiat Oncol Biol Phys 11: 849-854

Moertel CG (1982) The gall bladder: extrahepatic bile ducts. In: Holland JF, Fre E III (eds) Cancer medicine. Lea and Febiger, Philadelphia, pp 1785-1790

Molt P, Hopfan S, Watson RC, Botet JF, Brennan MF (1986) Intraluminal radiation therapy in the management of malignant biliary obstruction. Cancer 57: 536-544

Morner F, Ardiet JM, Bret P, Gerard JP (1984) Radiotherapy of high bile duct carcinoma using intracatheter iridium 192 wire. Cancer 54: 2069-2073

Neill RH, de Weese RS (1973) Primary carcinoma of the gall bladder. Am J Surg 125: 726-729

Nevin JE, Moran TJ, Kay S, King R (1976) Carcinoma of the gall bladder. Cancer 37: 141-148

Ogoshi K, Niwa M (1977) The diagnostic evaluation of ERCP in pancreatic and biliary carcinoma. Gastroenterol Jpn 12: 218-223

Okuda K, Tanikawa K, Emura T, Kuratomi S, Jinnouchi S, Urabe K, Sumikoshi T, et al. (1974) Nonsurgical, percutaneous transhepatic cholangiography - diagnostic significance in medical problems of the liver. Dig Dis 19: 21-26

Oswalt CE, Cruz AB (1977) Effectiveness of chemotherapy in addition to surgery in treating carcinoma of the gall bladder. Rev Surg 34: 436-438

Ottow RT, August DA, Sugarbaker PH (1985) Treatment of proximal biliary tract carcinoma: An overview of techniques and results. Surgery 97: 251-262

Palla L, Rizzatto G, Pozzi-Mucelli RS, Bazzocchi M (1980) Grey-scale ultrasonography in the evaluation of carcinoma of the gall bladder. Br J Radiol 53: 662-667

Passariello R, Pavone P, Simonetti G, Modini C, Lasagni RP, Mannella P, Gazzaniga GM, et al. (1985) Percutaneous biliary drainage in neoplastic jaundice. Acta Radiol [Diagn] (Stockh) 26: 681-688

Philip J, Hagenmüller F, Manegold K, Szepesi S, Classen M (1984) Endoskopische, intraduktuale Strahlentherapie hochsitzender Gallengangskarzinome. Dtsch Med Wochenschr 109: 422-426

Piehler JM, Crichbow RW (1978) Primary carcinoma of the gall bladder (collective reviews). Surg Gynecol Obstet 147: 929-942

Pipelich MV, Lambert PM (1978) Radiotherapy of carcinomas of the biliary extrahepatic system. Radiology 127: 767-770

Polk HC (1966) Carcinoma of the calcified gall bladder. Gastroenterology 50: 582-585

Richter JM, Silverstein MD, Shapiro R (1983) Suspected obstructive jaundice: A decision analysis of diagnostic strategies. Ann Intern Med 99: 46-51

Ritchie JR, Allan RN, Macartney J (1974) Biliary tract carcinoma associated with ulcerative colitis. Q J Med 170: 263-279

Ross AP, Bruasch JW, Warren KW (1973) Carcinoma of the proximal bile ducts. Surg Gynecol Obstet 136: 923-928

Segi M (1969) Cancer mortality for selected sites in 24 countries, no 5. Department of Public Health, Tohuku University School of Medicine, Sendas

Shieh CJ, Dunn E, Standard JE (1981) Primary carcinoma of the gall bladder. Cancer 47: 996-1004

Smoron GL (1977) Radiation therapy of gall bladder and biliary tract. Cancer 40: 1422-1424

Strauch GO (1960) Primary carcinoma of the gall bladder. Presentation of seventy cases from the Rode Island Hospital and a cumulative review of the last ten years of the American literature. Surgery 47: 368-383

Tasaka K, Watanabe H, Enjoji M (1975) Carcinomas of the pancreas, gall bladder, extrahepatic bile ducts and duodenal papilla. A statistical observation of 137 autopsy cases. Fukuoka Acta Med 66: 486-499

Terblanche J (1976) Is carcinoma of the main hepatic duct junction an indication for liver transplantation or palliative surgery? Surgery 79: 127-128

Terblanche J, Saunders SJ, Louw JH (1972) Prolonged palliation in carcinoma of the main hepatic duct junction. Surgery 71: 720-731

Todoroki T, Okamura T, Fukuo K, Nistimura A, Otsu H, Soto H, Iwasaki Y (1980a) Gross appearance of main hepatic duct carcinoma and its prognosis. Surg Gynecol Obstet 150: 33-41

Todoroki T, Iwasaki Y, Okamura T, Nagoshi K, Asakura H, Nakano M, Inada T, et al. (1980b) Intraoperative radiotherapy for advanced carcinoma of the biliary systems. Cancer 46: 2179-2184

Vaittinen E (1970) Carcinoma of the gall bladder, a study of 390 cases diagnosed in Finland 1953-1967. Ann Chir Gynaecol [suppl 59] 168: 7-81

Vas W, Salem S (1981) Accuracy of sonography and transhepatic cholangiography in obstructive jaundice. J Can Assoc Radiol 32: 111-113

von Eyben F, Hellekant C, Mattson W, Ljungquist U, Jonsson K (1980) Mitomycin C in advanced gall bladder carcinoma. Acta Radiol Oncol 19: 81-84

Waterhouse J, Muir C, Correa P, Powell J (eds) (1976) Cancer incidence in five continents, vol 3. IARC, Lyon

Whelton MJ, Petrelli W, George P (1969) Carcinoma at the junction of the main hepatic ducts. Q J Med 150: 211-230

Yeh HC (1979) Ultrasonography and computed tomography of carcinoma of the gall bladder. Radiology 133: 167-173

9 Colon Cancer

MOHAMMED MOHIUDDIN and LYDIA KOMARNICKY

CONTENTS

The annual incidence of new cancer cases in the United States is 930000, with large bowel cancer accounting for 45000 new cases in males, and 53000 in females. The annual estimated number of deaths from colon cancer is 51000 (24800 male, 27000 female) (SILVERBERG et al. 1986). There has been an increase in the incidence of cancer of the colon over the past several decades. In 1947 the age adjusted incidence rates were 25.1/100000 in males and 26.8/100000 for females; in 1973-77 the incidence was 36.9/100000 in males and 31.3/100000 in females. At the same time there has been a decline in the incidence of carcinoma of the rectum (DEVESA and SILVERMAN 1978). Over the age of 50, the incidence of large bowel cancer rises dramatically. In the 80-year age group, the male to female ratio is approximately 2.5 to 1. The incidence in the United States for colon cancer is higher than in most other countries (DEVESA and SILVERMAN 1978). In Japan, colon cancer is rarely seen, but the incidence rises sharply in immigrants from Japan, especially those people who change their dietary habits to those of the United States population (HAENSZEL and KURIHARA 1968). Several clinical studies (ARMSTRONG and DOLL 1985; REDDY 1978; SHERLOCK 1986) indicate that environmental factors and diet are very important in the development of colon cancer. There is an increased risk with intake of foods containing refined carbohydrates, low fiber, animal fat, and protein. Genetic factors also are linked to an increased risk of the development of colon cancer. In familial polyposis (ANDERSON 1975), a hereditary disease associated with a high occurrence of colon cancer, tumor development is seen as early as the second decade of life. Nearly two-thirds of patients who present with familial polyposis and symptoms are likely to have a concurrent cancer when first seen. It is estimated that half of the patients with this disease will develop their tumor by age 30 and almost all will develop a cancer by age 50. Other genetic disorders associated with the development of colon cancer include Gardner's syndrome, Crohn's disease, and chronic ulcerative colitis, where the development of cancer is 5-10 times greater than in the general population. There is also a reported increased risk in asbestos workers, as well as those patients with a history of colon or rectal cancer (incidence of a second primary in the order of 2%-5%).

MOHAMMED MOHIUDDIN, M.D., Professor of Radiation Therapy and Nuclear Medicine, LYDIA KOMARNICKY, M.D., Department of Radiation Therapy and Nuclear Medicine, Thomas Jefferson University Hospital, Eleventh and Walnut Street, Philadelphia, PA 19107, USA

9.1 Anatomy of the Large Intestine

The large intestine measures 5 ft in length and extends from the termination of the ileum to the anus, comprising one-fifth of the whole extent of the intestinal canal. The various segments of the large bowel exhibit differences in degree of mobility, peritoneal covering, and lymphovascular drainage.

The cecum lies at the ileocolic junction and forms a large blind pouch from which the appendix arises on its medial aspect. As a rule it is entirely enveloped by peritoneum, but in a small percentage of cases (5%) the peritoneal covering is not complete, and the posterior surface is uncovered. The appendix is also completely covered by peritoneum and is attached to the mesentery of the small intestine via the mesoappendix. The arterial and venous supply is by way of the ileocolic branches of the superior mesenteric vessels. The lymphatic drainage is to the superior mesenteric lymph nodes.

The ascending colon is continuous with the cecum and lies in the right iliac region. It extends upward to the inferior surface of the liver where it then becomes the hepatic flexure, and is continuous with the transverse colon. The peritoneum covers the anterior and lateral walls of the ascending colon, binding it to the posterior abdominal wall. It is noted to be a relatively immobile structure. The arterial and venous blood supply is from the ileocolic and right colic branches of the superior mesenteric vessels. The lymphatics drain into lymph nodes along the colic vessels to ultimately reach the superior mesenteric nodes.

The transverse colon extends across the abdomen and is completely invested by peritoneum. It is the most moveable part of the colon and is suspended by the transverse mesocolon. The arterial supply of the proximal two-thirds of the transverse colon is via a branch of the superior mesenteric and the middle colic artery. The distal third is supplied by a branch of the inferior mesenteric (left colic artery). The lymphatics drain to the superior mesenteric lymph nodes from the proximal two-thirds of the transverse colon, while the distal third drains to the inferior mesenteric lymph nodes. The transverse colon ascends to become the splenic flexure, which is inferior to the spleen and lacks a posterior peritoneal cover.

The descending colon lies in the left iliac area and extends to the pelvic brim. The peritoneum covers the anterior and lateral aspects, binding it to the posterior abdominal wall. The arterial and venous supply is from the superior and inferior left colic branches of the inferior mesenteric vessels. The lymphatics drain to the inferior mesenteric lymph nodes.

The sigmoid colon is situated in the left iliac fossa and extends from the margin of the iliac crest to the brim of the true pelvis opposite the left sacroiliac joint. The first portion of the sigmoid is covered by peritoneum on the anterior and lateral aspects and is fixed to the retroperitoneal tissues. The sigmoid loop, the midsegment of the colon, is entirely surrounded by peritoneum and retained in place by the sigmoid mesocolon. This segment is freely mobile and often lies in the true pelvis. The arterial and venous supply is from the inferior left colic branches of the inferior mesenteric vessels. The lymphatics drain into pericolic nodes along the course of the arteries and eventually to the inferior mesenteric lymph nodes.

The partial or complete peritoneal covering of the different segments of the colon often determine the mobility of the structure, the access of tumors to the peritoneal cavity, and pathways of local-regional failure following surgery.

The wall of the intestine comprises four layers: the mucosa, submucosa, muscularis propria, and adventitia. The *mucosa* contains no villi and consists of long tubular glands which extend through its entire thickness. The surface comprises columnar cells having striated borders, with interspersed goblet cells. At the base of the glands (crypts of Lieberkühn) there are rapidly dividing cells which replenish this surface mucosa as necessary. Between the glands lies the *lamina propria,* a vascular connective tissue with interspersed lymphoid nodules which may extend into the submucosa. The muscularis mucosa is considered part of the mucosa and consists of a thin layer of smooth muscle. The next layer is the *submucosa* which is a fibrous, highly cellular, connective tissue layer with accumulations of lymphoid tissue. It also contains large blood vessels that send finer vessels into the other layers. The *muscularis propria* consists of two layers of smooth muscle. The fibers of the inner layer run in a circular fashion; those of the outer layer run lengthwise. The longitudinal muscle fibers gather into three longitudinal bands called taeniae. The *adventitia* comprises several layers of loose connective tissue and is variably covered by peritoneum as discussed earlier.

Throughout the colon there is a rich supply of lymphatic vessels which arise as blind channels in the mucosa, coalesce and form larger lymphatic vessels in the submucosa, and send branches into the muscle layers, eventually passing into the mesentery and draining into paracolonic lymph nodes.

9.2 Pathology

The precancerous nature of colonic polyps is one of considerable controversy. There is some evidence to indicate that colon cancer arises in pre-existing polyps rather than de novo. There are several types of polyp: (1) hyperplastic polyps, (2) tubular adenomas, (3) villous adenomas, (4) tubulovillous adenomas. Hyperplastic polyps represent the majority and do not become malignant. Tubular adenomas of the colon are common after the age of 70, but only about 5%–10% of these adenomas develop invasive cancers. Villous adenomas can become malignant and do so in a high proportion of cases (30%–70%). Tubulovillous adenomas develop invasive malignancies in approximately 20% of cases. When a cancer arises from polypoid lesions, it is more likely to arise from the villous component.

The likelihood of invasive cancer in adenoma rises with increasing size, especially if larger than 2 cm (MUTO et al. 1975).

Most of the carcinomas of the colon are well-to-moderately differentiated adenocarcinomas. Some show intracellular mucin production and are called signet ring carcinomas which tend to be very aggressive, and grow circumferentially through the wall of the colon, usually with regional lymph node metastasis (GILCHRIST and DAVID 1947). Other tumors have excessive amounts of extracellular mucin production and are called mucinous adenocarcinomas. These tumors have a similar natural history to that of signet ring adenocarcinomas (SYMONDS and VICKERY 1976). Anaplastic or undifferentiated tumors can also be found in the colon. These tend to be locally extensive with early lymphatic invasion with a very poor prognosis. GRINELL (1939) has demonstrated that poorly differentiated tumors are associated with increased spread through the bowel wall and lymph node involvement. Tumors generally exhibit the poorest differentiation at their deep margins and these areas should be used for grading.

9.3 Prognostic Factors

Several prognostic factors affect survival in cancer of the colon (Table 1). These include duration and degree of symptoms, age (HOERNER 1958), fistulization (SANFELIPPO and BEAHRS 1972), and perforation (SPRATT and SPJUT 1967). Tumors with circumferential involvement of the bowel seem to have a worse prognosis than tumors with partial involvement (LOEFLER and HAFNER 1967). If there is a component of obstruction from the circumferential disease, survival appears to be further compromised (LOEFLER and HAFNER 1967). The poor prognosis of patients may be due more to a higher incidence of lymph node metastasis, rather than the configuration of the tumors (COLLER et al. 1941).

Fungating tumors that project into the bowel lumen usually have less penetration of bowel wall as compared to ulcerating tumors (GRINNELL 1964). Exophytic tumors of the colon have a high survival (83% at 5 years), compared to ulcerating tumors – 28% (GRINNELL 1939). Ulcerating tumors invariably have tumor extension through the entire thickness of the wall and are associated with a high incidence of lymph node metastasis (TAYLOR 1962). Similarly, tumor fixation is a major prognostic indicator, with bowel penetration directly correlating with the efficacy of treatment.

Table 1. Factors adversely affecting prognosis

Clinical	Pathological
Age <30 years	Anaplastic
Obstruction	Signet ring
Perforation	Mucoid
Fistula formation	Pericolonic extension
Circumferential	Lymph node involvement
Ulcerated	

9.4 Staging

DUKES (1944) initially introduced a histopathological staging system for cancer of the rectum. It was based on depth of penetration into the bowel wall by tumor with the presence or absence of lymph node metastasis. Since then, Dukes' classification has undergone several modifications that have resulted in considerable confusion, especially in the designation of subcategories, in stages B and C. The staging system most commonly utilized is the Astler-Coller modification of Dukes' classification (ASTLER and COLLER 1954). Further modifications are utilized to define gross involvement of adjacent organs. The prognostic significance of site and number of lymph node involvement is considered to be very critical in some staging systems. These subclassifications are shown in Table 2.

Table 2. Various subclassifications of stage C

DUKES	C – any node
	C1 – paracolonic nodes
	C2 – nodes at apex of mesentery
ASTLER-COLLER	C1 – stage B1 with any node
	C2 – stage B2 with any node
GITSG	C1 – <4 nodes
	C2 >5 nodes

9.5 Treatment

9.5.1 Surgery

The goal of surgical resection of colon cancer is to remove the affected segment of large bowel with its attached mesentery and lymph nodes. Generally when possible, wide margins of normal colon are also removed adjacent to the tumor. During the procedure, it is important not to dissect tissue planes adjacent to the tumor or traumatize the tumor in any way. It is recommended that large ligatures be placed around vessels and soft tissue near the tumor to prevent leakage of malignant cells from lymphatics into the operative field.

For cancers arising in the *right colon* (cecum, asending colon, hepatic flexure), a one stage, right hemicolectomy is standard therapy. The proximal margin is formed by dividing the ileum, approximately 3-6 in. from the ileocecal valve, while the distal margin is formed by dividing the transverse colon proximal to the midcolic artery. Cancers arising in the *transverse colon* are treated by removing the entire length of the transverse colon along with the middle colic artery. For lesions of the *splenic flexure* and *descending colon*, the standard therapy consists of a left hemicolectomy, in which the proximal margin is proximal to the mid-transverse colon and the distal margin is at the junction of the sigmoid and rectum. Tumors located in the *sigmoid colon* are treated with a left hemicolectomy; however, in these cases the proximal margin is the mid-descending colon, while the distal margin is at the rectosigmoid junction. Following resection an end-to-end anastomosis is performed with closure of the mesentery. Copious irrigation of the operative site with saline followed by abdominal irrigation with antibiotic solutions is recommended before closure of the abdomen.

Newer surgical techniques for resection of colon cancers have been devised in an attempt to improve postoperative survival. TURNBULL et al. (1967) have recommended the so-called no touch technique which includes minimal manipulation of the primary tumor, ligation of vascular trunks prior to dissection, and ligation of colon proximal and distal to the tumor (KOPELSON 1983a). These techniques have been developed in an attempt to decrease the likelihood of dissemination of tumor cells intraluminally, intraperitoneally, and systemically during surgery. Others have recommended the use of colonic lavage as well as iodized sutures, both in an attempt to prevent an anastomotic recurrence. At the time of surgical intervention, the determination of tumor fixation, peritoneal implantation, and/or liver metastasis is very important.

9.5.1.1 Patterns of Failure

While surgery continues to remain the primary mode of treatment for cancer of the colon, survival depends on the stage of the disease, and ranges from about 35% for stage C cancers to a high of 80% for stage A and B1 tumors. The site and location of the tumor in the colon, the presence or absence of peritoneal covering, and the proximity of adjacent tissues and organs have a significant bearing on the spread of disease. Direct extension of disease into adjacent organs occurs especially with tumors of the hepatic and splenic flexure, and large tumors can often grow to involve the abdominal wall. The rich interconnecting lymphatics that are present in the wall of the colon drain tumor cells into lymphatic channels that follow the vascular pedicles to the pericolonic and mesenteric lymph nodes. Hematogenous dissemination of tumors occurs via the portal circulation and is primarily to the liver, and occasionally metastasis occurs in the lung. Once tumors infiltrate through to the peritoneal surface, it is not uncommon to find diffuse peritoneal implantation and seeding throughout the peritoneal cavity (GUNDERSON et al. 1985b). The pattern of true relapse of cancers of the colon following surgical resection has still not been fully documented. Clinical impressions suggest liver metastasis as the dominant failure pathway (50%-80%). However, recurrences in the tumor bed region and involvement of other intra-abdominal structures are vastly underestimated because they occur in silent areas. Autopsy series (RUSSELL et al. 1984) indicate that less than 10% of patients relapse with metastasis to the liver alone without other areas of involvement. Recent studies specifically looking for patterns of relapse from colon cancer indicate that local regional relapse of tumor may be a significantly greater problem than previously appreciated. The use of planned, "second look" operations following curative resection has shed new light in determining the true pattern of relapse.

GUNDERSON et al. (1985b) reported their observations on 91 patients with colon primary tumors outside the true pelvis who underwent second look laparotomies following curative resection (Table 3). Recurrence or metastatic cancer was found in 58 of the 91 patients (64%). Twenty (22%) of the total group had a local tumor bed failure alone, and this made up 34% of patients whose cancers relapsed; 24 (26%) additional patients had both local and distant failure so that local failure as a component of the overall failure pattern occurred in 44 patients, making up 76% of the total group of patients relapsing with disease. Distant failure alone was uncommon and was seen in only seven patients (7%), or 12% of the total group of patients who relapsed. When analyzed by site of disease, the cecum had the highest (11/37 or 30%) rate of local failure, followed by the splenic flexure and descending colon - 18 patients (19%); the lowest incidence of relapse occurred in the transverse and ascending co-

Table 3. Pattern of relapse at planned second look (GUNDERSON et al. 1985b)

	No. of patients	Percent showing relapses ($n = 58$ pts)	Percent of total ($n = 91$ pts)
Local only	20	34%	22%
Local and distant	24	42%	48%
Distant only	7	12%	7%
Peritoneal seeding[a]	19	33%	21%

[a] Peritoneal seeding as only site occurred in 4 (4%) of patients

lon. Some of the differences may have been due to the presence of lower stage lesions in the ascending and transverse colon.

RUSSELL et al. (1984), in a study of 550 patients with colon cancer, found that recurrences developed in 186 patients (34%). Of these patients 64 underwent an exploratory laparotomy at the time of the cancer recurrence. Local recurrences were observed in 29 of 61 patients (47.5%) in whom the initial operative site was re-explored (Table 4). This was the only site of recurrence in 10 of the 64 patients (15.5%). Liver metastasis were found in 23 of 54 patients (42.5%); however, liver metastasis alone without a component of other failure was observed in only three (4.5%) patients. Lymph node involvement was observed in 13 of 38 (44%) patients, but in only three patients was it the sole site of involvement. In general, nodal involvement was most commonly observed at the mesenteric roots. Peritoneal seeding was observed in 28 of 64 patients (44%), but was the only site of metastasis in 2 of 64 (3%) patients. Of 64 patients evaluated 62 had intra-abdominal metastasis.

WILLETT et al. (1983), in a retrospective review of 533 patients with colon cancer undergoing curative resection and followed a minimum of 5 years, found that 370 patients remained free of disease. Sixty-one (11%) patients developed distant metastasis only, 32 (6%) developed local failure alone, and 70 (13%) developed both local and distant failure. The overall local failure rate was 19% (102/533) or 62.5% of the total group of failures. Failure was documented by clinical and radiographic examination or re-exploration and autopsy information.

In a similar study of 280 patients, CASS et al. (1976) found that 105 (37%) patients subsequently developed recurrent cancer; 60% (63/105) presented with local recurrence alone, 14% (15/105) with concomitant local recurrence and distant metastasis, and 26% (27/105) with distant metastasis

Table 4. Pattern of relapse at symptomatic second look in 61 patients[a] (WILLETT et al. 1983)

	Only site (%)	Total relapses (%)
Local	10 (15.5)	29/61 (47.5)
Liver	3 (4.5)	23/54 (42.5)
Nodes	3 (4.5)	13/39 (34)
Peritoneal seeding	2 (3)	28/64 (44)
Extra-abdominal	-	2/64 (3)

[a] Denominator represents only areas explored and not all patients

Table 5. Pattern of relapse following curative resection

	No. of patients	No. of recurrences	LR	LR+DM	DM
WILLET et al. (1983)	533	163	32	70	61
CASS et al. (1976)	280	105	63	15	27
OLSEN et al. (1980)	214	49	14	10	25
Total	1027	317	109	95	113
Percent of total		31%	11%	9%	11%
Percent of recurrences			34%	30%	36%

LR, local recurrence; DM, distant metastasis

alone. Similar results have been reported in other studies as well (Table 5).

Discrepancies in the rate of local failure alone in these various series may be due to the method of assessment of local failure in the course of the natural history of the disease. A higher rate of local failure alone is observed when planned second look surgical procedures are undertaken (34%), as compared to "symptomatic" second look procedures (15.5%). When recurrences have had time to mature, clinical assessment tends to document disease only in accessible areas, i.e., liver, and underestimates the site of primary relapse. It is clear from these studies, however, that local recurrence alone or as a part of the total component of failure occurs in up to two-thirds of the patients who fail following curative resection of colon tumors. The key prognostic variables that affect the pattern of failure appear to be the depth of penetration of the tumor through the wall of the colon, the location of the tumor at different segments of the colon, and the presence or absence of involved lymph nodes.

The impact on the pattern of local recurrence is dominated by the stage of the disease, as seen in

Table 6. Impact of stage on pattern of local recurrence

Stages	Cass et al. (1976)	Willet et al. (1983)	Olsen et al. (1980)	Total (%)
A	0/19	1/29	1/26	2/74 (3)
B1	3/20	2/89	7/116	12/225 (5)
B2	9/48	18/124	13/60	40/232 (17)
B3	20/58	25/52	–	45/110 (41)
C1	4/19	0/15	7/37	11/71 (15)
C2	33/64	32/49	9/42	74/155 (48)
C3	–	24/49	–	24/49 (49)

Table 7. Pattern of local relapse in cancer of the cecum

Stage	Tong et al. (1959)	Simonovsky and Feldman (1976)	Willet et al. (1983)	Total (%)
A	0/6	0/1	0/3	0/10 (–)
B1	1/23	0/5	2/26	3/54 (6)
B2	8/32	7/54	0/23	15/109 (14)
B3	3/4	3/12	5/31	11/47 (23)
C1	0/2	0/1	0/4	0/7 (–)
C2	11/18	8/26	7/15	26/57 (46)
C3	5/5	4/9	7/15	16/29 (55)

Table 6. Patients with early tumors confined to the muscularis propria (stages A, B1) have a low incidence of local recurrence. Extension through the bowel wall (stage B2 or greater) have a considerably higher incidence of local recurrence (greater than 30%), especially when there are also positive lymph nodes (stage C2) or adjacent organ involvement (stages B3, C3).

The location of the tumors in the different segments of the colon can also affect the pattern of local failure and is often a function of the degree of peritoneal covering of that segment of the colon. Generally, the more fixed the area of the colon, the greater is the likelihood of local failure. The cecum, ascending colon, descending colon, and the proximal and distal portions of the sigmoid together with splenic and hepatic flexures remain fixed or partially fixed. The transverse and midsigmoid colon are freely mobile segments of the colon.

Tong et al. (1959) found that 31% of 90 patients with cancer of the cecum, resected for cure, demonstrated local recurrence of tumor. Local recurrence was defined as recurrence at the anastomotic site or at the tumor bed. No local recurrences were observed in stage A tumors, but in patients with B1, B2, and B3 tumors, the rate of local recurrence was 4%, 25%, and 75% respectively. In patients with stage C2 or C3 tumors, the local recurrence was 61%. The majority of recurrences were also associated with failure in the abdominal cavity, retroperitoneal tissues, or liver, and in only 18% was the local recurrence the sole site of failure. Similar results have been reported by others (Table 7). In studies involving second look procedures, the cecum was one of the highest sites for local recurrence in the colon. Gunderson et al. (1985b) reported 11 of 37 patients to have isolated local recurrences, and 11 of 37 having both local and distant failure, with only six patients showing distant metastasis.

Table 8. Pattern of local relapse in cancer of the ascending colon

Stage	Willet et al. (1983)	Russell et al. (1959)	Total (%)
A	–	0/3	0/3
B1	0/12	0/6	0/18
B2	1/21	6/18	7/39 (18)
B3	0/4	0/1	0/5
C1	0/1	2/3	2/4 (50)
C2	2/10	9/16	11/26 (42)
C3	3/4	3/3	6/7 (86)

Tumors of the ascending colon have been reported by several authors to have a moderate local failure rate. Cass et al. (1976) reported local failures in 0 of 15, and Olson et al. (1980) had 7 of 95 patients failing locally. Gunderson et al. (1985b) reported that only 2 of 17 patients developed local failure alone, whereas 7 of 17 patients developed both local and distant failure. No patients with ascending colon lesions developed distant metastasis alone. The pattern of failure by stage of disease in other series is shown in Table 8.

Tumors of the transverse colon have been reported generally to have a very low incidence of failure. Cass et al. (1976) did find 4 of 15 patients with disease recurring locally, whereas Olson et al. (1980) had no local failures in ten patients. Gunderson et al. (1985b) found that three of eight transverse colon lesions developed a local recurrence and Willett et al. (1983) reported that only 3 of 47 patients developed local failure alone, whereas six additional patients developed a component of local failure and distant metastasis. Distant metastases alone were observed in six patients. The local failures were all in patients with stage B3, C2, or C3 cancers.

Tumors of the descending colon in some series have been reported as having a slightly higher inci-

Table 9. Pattern of local relapse in cancer of the descending colon

Stage	WILLET et al. (1983)	RUSSELL et al. (1959)	Total (%)
A	0/2	0/5	0/7 (-)
B1	0/7	1/11	1/18 (6)
B2	1/16	5/18	6/34 (18)
B3	5/10	0/1	5/11 (45)
C1	0/1	3/5	3/6 (50)
C2	3/11	1/3	4/14 (29)
C3	2/4	1/1	3/5 (60)

dence of local recurrence. CASS et al. (1976) found local recurrence in 18 of 71 patients (25%), and GUNDERSON et al. (1985b) found 15 of 16 patients developing local recurrence either alone or as a component of local and distant failure. In series reported by WILLETT et al. (1983) and RUSSELL et al. (1959), local failure again appeared to be a function of stage of disease, with high rates of failure occurring in patients having extracolonic extension of disease or positive regional nodes (Table 9).

Tumors of the sigmoid are often considered together, but the proximal and distal regions of the sigmoid tend to be relatively immobile as compared to the midsigmoid, which is surrounded by peritoneum and is mobile with a mesenteric attachment. OLSON et al. (1980) reported an 11% (9/85) rate of local recurrence, and an 11% rate of distant metastasis (9/85), with an additional 10% (8/85) of patients developing a local and regional recurrence. WILLETT et al. (1983) found tumors of the midsigmoid and tumors of the high and low sigmoid had similar but negligible failure rates for stages A, B1, and C1 (0 of 26 vs 1 of 39). The rate of local relapse was high but similar for stages B3, C2, and C3: 11 of 33 (33%) to the mid sigmoid versus 22 of 51 (44%) for the high and low sigmoid colon. The patterns of failure were very different for stage B2 tumors. Midsigmoid lesions had a local failure rate of 1 of 24 (4%), as compared to 9 of 38 (24%) for high and low sigmoid lesions.

9.5.2 Adjuvant Therapy

Over the past 20 years, the major focus of adjuvant therapy programs in an effort to improve curability and survival of patients with carcinoma of the colon has been on the use of adjuvant chemotherapy. There are, however, few agents with significant cytotoxic activity against colon cancer. The most effective group of drugs appears to be the fluorinated pyrimidines. The overall response even with these drugs is only about 20% (ANSFIELD 1975; CARTER and FRIEDMAN 1987; MCGLONE et al. 1982; MOERTEL 1979). 5-Fluorouracil (5-FU) continues to remain the mainstay of most therapeutic programs in this disease. Despite 40 years of experience with this drug, the best method of administration of 5-FU has not been established. Intravenous administration appears to be more effective than oral administration (BATEMAN et al. 1975; HAHN et al. 1975), and there appears to be greater efficacy with prolonged infusion than with bolus administration over short periods of time (GRAGE and MOSS 1981; SEIFERT et al. 1975). There appears to be a decreased hematological toxicity with prolonged infusions and a dose-response relationship also appears to exist, with increased frequency of responses when the drug is pushed to toxicity. Other agents of limited but known activity include the nitrosoureas, mitomycin-C, immune stimulants such as MER-BCG, and levamisole (Table 10). Multidrug regimens have also been used but have not proven to be any more effective than single agent therapy with 5-FU (Table 11). The most promising of the combination appeared to be 5-FU and methyl-CCNU, but a randomized study with this combination has shown that not only is there no survival advantage with this drug combination (BAKER et al. 1975), but there may be an enhanced risk of a leukemogenic effect with the use of methyl-CCNU (GASTROINTESTINAL TUMOR STUDY GROUP 1984).

The initial large scale randomized studies in testing the efficacy of adjuvant chemotherapy in colon

Table 10. Single agent chemotherapy for colon cancer

Author	Agent	Response rate
ANSFIELD (1975)	5-FU	13%-33%
MOERTEL (1979)	FUDR	22%
LAVIN et al. (1980)	Methyl-CCNU	15%
CARTER and FRIEDMAN (1987)	Mitomycin-C	16%

FUDR, Fluorodeoxyuridine

Table 11. Multiagent chemotherapy for colon cancers

Author	Agents	Response rate
POSEY and MORGAN (1977)	FU + methyl-CCNU	30%
MOERTEL (1979)	MOF	44%
BUROKER et al. (1978)	FU + mitomycin-C	37%
KEMENY et al. (1980)	MOF + streptozotocin	32%

MOF, combination of 5-FU, methyl-CCNU, and vincristine

cancer were conducted by the Veterans Administration Surgical Oncology Group utilizing a short course thio-TEPA regimen (DWIGHT et al. 1973). There was no survival advantage observed. In the subsequent study (HIGGENS et al. 1978) of adjuvant 5-FU, no advantage for either short course treatment or a prolonged adjuvant program with 5-FU over a full year following surgery was found. The Central Oncology group (GRAGE and MOSS 1981) used a higher dose of 5-FU, treating patients to toxicity and continuing over a 1-year period. There appeared to be a small increase in the disease-free interval and survival only in patients with stage C rectal cancer, but no overall statistical significant advantage for the group as a whole.

More recent trials of adjuvant therapy have utilized combinations of chemotherapy (MOERTEL et al. 1975) and/or immunotherapy (HOOVER et al. 1985). The GASTROINTESTINAL TUMOR STUDY GROUP (1984) initiated a trial in 1975 for Dukes' B2 and C colon cancers comparing surgery alone versus adjuvant chemotherapy with 5-FU and methyl-CCNU, adjuvant immunotherapy with MER (methanol extract with residue of BCG), and both adjuvant modalities combined. A total of 621 patients have been entered into the trial and after a median follow-up of 5½ years, no significant differences were noted either in recurrence or survival rates among the four treatment programs. Leukemia developed in seven patients, all of whom had received 5-FU and methyl-CCNU. The conclusions of the study did not support the use of chemotherapy with 5-FU and methyl-CCNU, MER-BCG, or the combination as an adjuvant treatment program for colon carcinoma. Studies of adjuvant therapy undertaken by the Eastern Cooperative Oncology (ENGSTROM et al. 1982; KEMENY et al. 1980) and Southwest Oncology Groups (BUROKER et al. 1978), utilizing single and multiagent chemotherapy, to date have not shown any advantage over surgery alone.

Since a significant number of patients with colon cancer develop liver metastasis, investigators have tried to develop adjuvant programs to deal specifically with the prevention of liver disease. TAYLOR et al. (1979) reported a novel approach to adjuvant chemotherapy by inserting a catheter into the portal vein either directly or via the umbilical vein at the time of surgical resection of the colon primary and then infusing the patients with 5-FU over a 7-day postoperative period. There was an impressive advantage reported for the treated group. With a mean follow-up of 25 months, 23 of 47 control patients had died as compared to only 7 of 43 patients receiving adjuvant therapy. Liver metastases were reported in only two patients receiving the 5-FU. These results need to be confirmed in a randomized study which is currently being undertaken by the EORTC and the Mayo Clinic.

9.5.3 Radiation Therapy

Primary or adjuvant radiation therapy in the treatment of colon cancer has had very little application. Adjuvant irradiation in the treatment of rectal cancer has proven to be very effective in reducing the rate of local recurrence in the pelvis and the perineal tissues (ROMSDAHL and WITHERS 1978). The recent recognition of the high rate of local failure following curative resections of colon cancers has created a renewed interest in the application of adjuvant radiation therapy for colon tumors. To date, no major randomized studies have been undertaken, but small institutional pilot studies (Table 12) appear to show early promise (BRENNER et al. 1983; GHOSSEIN et al. 1981; TURNER et al. 1977; REDDY et al., personal communication). Most adjuvant programs have utilized postoperative irradiation, as this allows careful patient selection based on histopathological staging. Adequate dose levels for carcinoma of the rectum require delivery of 45–50 Gy in 5–6 weeks. Achieving such doses to the major segments of the extrapelvic colon is hampered by the increased volume of small bowel that necessarily has to be encompassed by the radiation field, and also the limited radiation tolerance of other critical viscera, i.e., the kidney, liver, stomach, and other organs. In an effort to minimize acute and chronic complications, meticulous precision and innovative treatment techniques need to be utilized.

One of the early series of postoperative irradiation for colon cancer was reported by GHOSSEIN et

Table 12. Five-year survival vs stage for cancer of the colon

	Stage				
	B2	B3	C1	C2	C3
Surgery					
WILLETT et al. (1983)	65%	51%	50%	39%	29%
MCGLONE et al. (1982)	51%	67%	65%	15%	22%
BRENNER et al. (1983)	–	–	–	36%	–
Surgery + XRT					
DUTTENHAVER et al. (1986)	60%	78%	100%	57%	49%
BRENNER et al. (1983)	–	–	–	–	65%
WHONG et al. (1985)	–	83%	–	53%	–

al. (1981); 47 colon tumors were treated with a moving strip technique to the whole abdomen using 4-cm strips and delivering 21 Gy in six fractions over 8 days. Patients with residual disease received a "cone down" field for an additional 10–15 Gy. In patients with curative resection and no residual disease, only 14% of patients failed locally. Complications were observed in 3 of 47 patients; all three of these were due to small bowel damage. REDDY et al. (personal communication) used an approach of combined 5-FU infusion and concomitant total abdominal irradiation for Dukes' B2 and C colon cancers. 5-FU was given in a dose of 300–500 mg/m^2 on the first 5 days of radiation treatment and repeated on days 28–32. Radiation was delivered to an open abdominal field using 1–1.5 Gy per fraction for a total of 25 Gy and an additional boost of 15 Gy to the tumor bed area. Eighteen patients were treated in the program, two of whom developed significant gastrointestinal toxicity and required hospitalization. Six patients had mild gastrointestinal toxicity requiring symptomatic treatment. With a median follow-up of 18 months, 15 patients were alive with no evidence of disease.

In a prospective study at MGH (DUTTENHAVER 1986), patients at high risk for local recurrence after curative resection of adenocarcinoma of the colon were offered a course of postoperative irradiation. Eighty patients with stage B2, B3, C2, or C3 tumors were entered in the study and given a total dose of 45 Gy at 1.8 Gy per fraction to the tumor bed with a 4–5 cm margin around the initial tumor. With a median follow-up of 39 months, 26 (33%) patients had disease that relapsed: 19 patients had the tumor bed area restaged either surgically or by abdominal CAT scan, and the remaining seven patients had advanced metastatic disease with a clinically negative abdominal examination. Thirteen of the 19 patients failed locally, but only one patient had a solitary local recurrence. Eleven of the 13 local recurrences were in patients with stage C2 or C3 tumors. Actuarial survival at 3 years for stages B2, B3, C2, and C3 disease was 84%, 84%, 73%, and 49%, respectively. Morbidity was carefully studied. None of the patients developed renal dysfunction; several patients experienced transient elevation of liver function tests after partial hepatic irradiation; however, no patient had clinical radiation hepatitis. Nausea and diarrhea were observed during the course of treatment in the majority of patients with 13 patients requiring a treatment break. Late small bowel morbidity was observed in two patients requiring operative intervention. The local failure rate was 5% and 25% for stages B3 and C2, respectively, with postoperative irradiation, as compared to an expected failure rate of 30% for these stages with surgery alone. There was no improvement in local control for stage C3 tumors, and this may be due to an inadequate dose of radiation for treatment of bulk disease.

In a similar study undertaken at the Princess Margaret Hospital (TAYLOR et al. 1979), 82 patients with cancer of the colon above the peritoneal reflection received postoperative local abdominal irradiation. Forty-eight patients were treated in an adjuvant fashion, and 34 patients were treated for gross residual disease following surgery. The 5-year actuarial survival was 67% in the patients treated as an adjuvant and local relapse was observed in 3 of 20 patients with stage B2 disease, and 9 of 28 patients with stage C disease. Patients with gross residual disease did not fare well, with 32 of 34 patients developing progressive disease, and 28 of these patients having progression locally. Mild intestinal toxicity was observed during the treatment, but was controlled by conservative measures. Major acute and delayed morbidity was seen predominantly in patients treated for residual disease, and only 4 of 48 treated patients in the adjuvant group developed small bowel toxicity.

In a small group of 17 patients with tumors of the sigmoid colon resected for cure and given postoperative radiation therapy, KOPELSON et al. (1983a) found that improved regional disease control was obtained as compared to patients treated with surgery alone (91% versus 70%), and an improved 5-year survival was seen in patients with stage B2 or B3 tumors who received postoperative irradiation as compared to unirradiated controls (100% versus 64%, $P<0.05$%), but no survival advantage was seen in patients with stage C2 or C3 tumors.

In a similar study of 15 patients with more proximal tumors of the colon, KOPELSON et al. (1983b) also observed that none of eight patients given postoperative radiation therapy developed a local recurrence, and four of seven patients treated with surgery alone had disease that recurred in the tumor bed/abdominal wall area. SHEHATA et al. (1984) treated 21 patients with adenocarcinoma of the cecum with postoperative irradiation following a right hemicolectomy and compared the results to a matched group of patients treated with surgery alone, and found a reduced incidence of local failure in the treated group (5% vs 19%, $P<0.02$) but no significant differences in median survival or incidence of distant metastasis. One patient developed ileitis that resolved with conservative management.

These preliminary data indicate that postoperative irradiation certainly has efficacy for improving local control of disease in the tumor bed area and may also have a potential for improving survival in selected groups of patients.

Further randomized studies need to be undertaken to establish the full role of adjuvant radiation therapy, either alone or in combination with systemic chemotherapy, and to better define the optimum radiation dose and treatment volume. Preoperative irradiation has not been utilized in the management of colon cancer primarily because of the lack of a clinical staging system and an inability to identify high risk patients who are likely to benefit from such treatment. However, initially unresectable colon carcinomas can benefit from high dose preoperative irradiation (45–50 Gy) in an effort to improve the resectable rate. EMAMI et al. (1982) reported treating 44 carcinomas of the colon and rectum that were considered unresectable with high dose preoperative irradiation. In 33 patients resection was attempted, of whom 26 were able to undergo a complete resection; seven additional patients had an incomplete resection. The mean survival of patients not explored was 8 months; the mean survival of the seven patients undergoing incomplete resection was 17 months, and mean survival of patients having a complete resection was 29 months, with 9 of 28 patients alive for more than 5 years. It therefore appears that in patients who have initially unresectable disease, an attempt at high dose preoperative irradiation followed by radical surgical resection may allow salvage of a significant number of patients.

In planning adjuvant radiation therapy, every effort to minimize both acute and chronic morbidity should be made. Identification and localization of intra-abdominal viscera by contrast enhancement is essential. It is important to assess not only the position of the kidneys, but to determine if both kidneys are functioning satisfactorily, and if only one kidney is functioning, to minimize the radiation dose to the functional kidney. Small bowel radiographs should be used in designing the radiation therapy ports and in minimizing the amount of small bowel included in the treatment fields. Patient positioning can also minimize the volume of small bowel in the treatment field and is especially important where boost doses beyond 45 Gy are given for the attempted control of residual disease (GUNDERSON et al. 1985a). A minimum tumor dose of 45 Gy in 5 weeks should be planned using 1.8–2 Gy per fraction, 5 days a week. When microscopic residual disease is present, an additional 15–20 Gy should be delivered using small volume irradiation. Total doses should not exceed 65–70 Gy. Tumor volumes should encompass the tumor bed area with a generous 4–5 cm margin, and in stage C tumors, an attempt should be made to encompass the nodal drainage areas. Nodal recurrences occur primarily at the root of the mesentery, and therefore, for right colon lesions, the root of the superior mesenteric vessels and the celiac nodes should be included in the treatment portal. For left-sided lesions, resection of the inferior mesenteric vessels is often more complete, and therefore, treatment fields may be individualized based on the extent of nodal involvement. The task of radiotherapy treatment planning is made easier if the tumor bed area is marked with radiopaque clips at the time of surgical resection.

References

Anderson DE (1975) Familial susceptibility. In: Fraumeni JF Jr (ed) Persons at high risk of cancer. Academic, New York

Ansfield F (1975) A randomized phase III study of four dosage regimens of 5-FU, – a preliminary report. Proc Am Assoc Clin Oncol 67: 224

Armstrong B, Doll R (1985) Environmental factors and cancer incidence and mortality in different countries with special reference to dietary practices. Int J Cancer 15: 617–631

Astler VS, Coller FA (1954) The prognostic significance of direct extension of carcinoma of the colon and rectum. Ann Surg 139: 846–851

Baker LH, Matter R, Talley R, Vaitkevicius V (1975) 5-FU versus 5-FU and methyl-CCNU in gastrointestinal cancer. Proc Am Soc Clin Oncol 16: 229

Bateman J, Irwin L, Pugh R et al. (1975) Comparison of intravenous and oral administration of 5-fluorouracil for colorectal carcinoma. Proc Am Assoc Cancer Res 16: 242

Brenner HJ, F.R.C.R., Bibi C, Chaitchik S (1983) Adjuvant therapy for Dukes C adenocarcinoma of colon. Int J Radiat Oncol Biol Phys 9: 1789–1792

Buroker T, Kim PN, Gropper C et al. (1978) 5-Fluorouracil infusion with mitomycin-C versus 5-fluorouracil infusion with methyl-CCNU in the treatment of advanced colon cancer: a SWOG study. Cancer 42: 1228–1233

Carter SK, Friedman M (1987) Integration of chemotherapy into combined modality treatment of solid tumors II. Large bowel carcinoma. Cancer Treat Rev 1: 111–129

Cass AW, Million RR, Pfaff WW (1976) Patterns of recurrence following surgery alone for adenocarcinoma of the colon and rectum. Cancer 37: 2861–2865

Coller FA, Kay EB, MacIntyre RS (1941) Regional lymphatic metastasis in carcinoma of the colon. Ann Surg 114: 56–63

Devesa SS, Silverman DT (1978) Cancer incidence and mortality trends in the United States: 1935–1974. JNCI 60: 545–571

Dukes CE (1944) The surgical pathology of rectal cancer. Proc R Soc Lond 37: 131–144

Duttenhaver J, Hoskins RB, Gunderson LL, Tepper JE (1986) Adjuvant postoperative radiation therapy in the management of adenocarcinoma of the colon. Cancer 57: 955–963

Dwight RW, Humphrey EE, Higgens GA et al. (1973) FUDR as an adjuvant to surgery in cancer of the large bowel. J Surg Oncol 5: 243-349

Emami B, Pilepich M, Willett C, Munzenrider J, Miller H (1982) Effect of preoperative irradiation on resectability of colorectal carcinomas. Int J Radiat Oncol Biol Phys 8: 1295-1299

Engstrom P, MacIntyre J, Douglass H, Muggia F, Mittelman A (1982) Combination chemotherapy of advanced colorectal cancer utilizing 5-fluorouracil, semustine, dacarbazine, vincristine, and hydroxyurea. Cancer 49: 1555-1560

Gastrointestinal Tumor Study Group (1984) Adjuvant therapy of colon cancer: Results of a prospectively randomized trial. N Engl J Med 310: 737-743

Ghossein NA, Samala EL, Alpert S et al. (1981) Elective postoperative radiotherapy after incomplete resection of a colorectal cancer. Dis Colon Rectum 24: 252-256

Gilchrist RK, David VC (1947) A consideration of pathological factors influencing five-year survival in radical resection of the large bowel and rectum for carcinoma. Ann Surg 126: 421-438

Grage TB, Moss SE (1981) Adjuvant chemotherapy in cancer of the colon and rectum: Demonstration of effectiveness of prolonged 5FU chemotherapy in a prospectively controlled randomized trial. Surg Clin North Am 61: 1321-1329

Grinnell RS (1939) The grading and prognosis of carcinoma of the colon and rectum. Ann Surg 109: 500-533

Grinnell RS (1964) The chance of cancer and lymphatic metastasis in small colon tumors discovered on x-ray examination. Ann Surg 159: 132-138

Gunderson L, Russell A, Llewellyn H, Doppke K, Tepper J (1985a) Treatment planning for colorectal cancer: Radiation and surgical techniques and value of small-bowel films. Int J Radiat Oncol Biol Phys 11: 1379-1393

Gunderson L, Sosin H, Levitt S (1985b) Extrapelvic colon - areas of failure in a reoperation series: Implications for adjuvant therapy. Int J Radiat Oncol Biol Phys 11: 731-741

Haenszel WM, Kurihara M (1968) Studies of Japanese migrants: I. Mortality from cancer and other diseases among Japanese in the United States. JNCI 40: 43

Hahn RG, Moertel CG, Schmitt AJ (1975) A double blind comparison of intensive course 5-fluorouracil by oral versus intravenous route in the treatment of colorectal carcinoma. Cancer 35: 1031-1035

Higgens GA, Lee LE, Dwight RW, Keehn RJ (1978) The case for adjuvant 5-fluorouracil in colorectal cancer. Cancer Clin Trials 1: 35-41

Hoerner MT (1958) Carcinoma of the colon and rectum in persons under twenty years of age. Am J Surg 96: 47-53

Hoover H, Surdyke MG, Dangel R, Leona C, Hanna M (1985) Prospectively randomized trial of adjuvant active-specific immunotherapy for human colorectal cancer. Cancer 55: 1236-1243

Kemeny N, Yagoda A, Braun D, Golbey R (1980) Therapy for metastatic colorectal cancer with a combination of methyl-CCNU, 5-fluorouracil, vincristine and streptozotocin (MOF-Strep). Cancer 45: 876-881

Kopelson G (1983a) Adjuvant postoperative radiation therapy for colorectal carcinoma above the peritoneal reflection - sigmoid colon. Cancer 51: 1593-1598

Kopelson G (1983b) Adjuvant postoperative radiation therapy for colorectal carcinoma above the peritoneal reflection - antimesenteric wall, ascending and descending colon and cecum. Cancer 52: 633-636

Lavin P, Mittelman A, Douglass H et al. (1980) Survival and response to chemotherapy for advanced colorectal adenocarcinoma: an ECOG report. Cancer 46: 1536-1541

Loefler I, Hafner CD (1964) Survival rate in obstructing carcinoma of colon. Arch Surg 89: 716-718

McGlone P, Bernie W, Elliott D (1982) Survival following extended operations in extracolonic invasion by colon cancer. Arch Surg 117: 595-599

Moertel CG (1979) Large bowel. In: Holland J, Frei LE (eds) Cancer medicine. Lea and Febeger, Philadelphia, pp 1597-16727

Moertel CG, Schmitt AJ, Hahn RG, Reitemeier RJ (1975) Therapy of advanced gastrointestinal cancer with a combination of 5-FU, methyl-CCNU and vincristine. JNCI 54: 69-71

Muto T, Bussey HJR, Morson BC (1975) The evolution of cancer of the colon and rectum. Cancer 36: 2251-2270

Olson RM, Perencevich NP, Malcolm AW, Chaffey JT, Wilson R (1980) Patterns of recurrence following curative resection of adenocarcinoma of the colon and rectum. Cancer 45: 2969-2974

Posey LE, Morgan LR (1977) Methyl-CCNU versus methyl-CCNU plus 5-fluorouracil in carcinoma of the large bowel. Cancer Treat Rep 61: 1453-1458

Reddy BS (1978) Carcinogenesis of the colon and rectum. In: Enker WE (ed) Carcinoma of the colon and rectum. Year Book Medical, Chicago, pp 326-343

Romsdahl M, Withers HR (1978) Radiotherapy combined with curative surgery. Arch Surg 113: 446-453

Russell AH, Tong D, Dawson LE, Wisbeck WM, Griffin TW, Laramore GE, Luk KH (1959) Adenocarcinoma of the retroperitoneal ascending and descending colon: Sites of initial dissemination and clinical patterns of recurrence following surgery alone. Int J Radiat Oncol Biol Phys 9: 361-365

Russell AH, Tong D, Dawson LE, Wisbeck W (1984) Adenocarcinoma of the proximal colon. Sites of initial dissemination and patterns of recurrence following surgery alone. Cancer 53: 360-367

Sanfelippo PM, Beahrs OH (1972) Factors in the prognosis of adenocarcinoma of the colon and rectum. Arch Surg 104: 401-406

Seifert P, Baker L, Reed M, Vaitkevicius VK (1975) Comparison of conntinuously infused 5-fluorouracil with bolus injection in the treatment of patients with colorectal adenocarcinoma. Cancer 36: 123-138

Shehata WM, Meyer RL, Krause RJ, Jazy FK, Cormier WF (1984) Postoperative adjuvant irradiation and 5-FU for adenocarcinoma of the cecum. Cancer 54: 2850-2853

Sherlock P (1986) Etiology of colorectal cancer. In: Bean DH, Higgins GA, Weinstein JJ (eds) Colorectal tumors. Lippincott, Philadelphia

Silverberg E, Lubera J (1986) Cancer statistics, 1986. CA 57: 26

Simanovsky M, Feldman M (1986) Adenocarcinoma of the cecum. Analysis of 106 cases. Cancer 58: 1766-1769

Spratt JS, Spjut HJ (1967) Prevalence and prognosis of individual clinical and pathologic variables associated with colorectal carcinoma. Cancer 20: 1976-1985

Symonds DA, Vickery AL (1976) Mucinous carcinoma of the colon and rectum. Cancer 37: 1891-1900

Taylor FW (1962) Cancer of the colon and rectum: A study of routes of metastases and death. Surgery 52: 305-308

Taylor I, Rowling J, West C (1979) Adjuvant cytotoxic liver perfusion for colorectal hospital. Br J Surg 66: 833-834

Tong D, Russell AH, Dawson LE, Wisbeck WM, Griffin TW, Laramore GE, Luk KH (1959) Adenocarcinoma of the cecum: Natural history and clinical patterns of recurrence following radical surgery. Int J Radiat Oncol Biol Phys 9: 357-360

Turnbull RB, Kyle K, Watson FR, Spratt J (1967) Cancer of the colon. Influence of the no-touch technic on survival rates. Am Surg 166: 420-425

Turner S, Vierira EF, Ager PJ, Alpert S, Efron G, Ragins H, Weil P, Ghossein NA (1977) Elective postoperative radiotherapy for locally advanced colorectal cancer. Cancer 40: 105-108

Willett C, Tepper JE, Cohen A, Orlow E, Welch C, Donaldson G (1983) Local failure following curative resection of colonic adenocarcinoma. Int J Radiat Oncol Biol Phys 10: 645-651

Wong CS, Harwood AR, Cummings BJ, Keane TJ, Thomas GM, Rider WD (1985) Postoperative local abdominal irradiation for cancer of the colon above the peritoneal reflection. Int J Radiat Oncol Biol Phys 11: 2067-2071

10 Rectal Cancer

B. J. CUMMINGS

CONTENTS

B. J. CUMMINGS, M.B., Ch.B., FRCPC, F.R.C.R., F.R.A.C.R., Radiation Oncologist, Professor, Department of Radiology, University of Toronto, The Princess Margaret Hospital, 500 Sherbourne Street, Toronto, Ontario, M4X 1K9, Canada

10.1 Introduction

The treatment of rectal malignancies has changed considerably during the past two decades. There has been greater emphasis on the conservation of anorectal function, and there has been increasing use of radiation therapy as a supplement to or as an alternative to resection. The preceding chapter discussed cancers which arose in the large intestine above the peritoneal reflection. This demarcation point is not an arbitrary one, although the reasons for distinguishing treatment principles according to the relationship of a tumor to the peritoneum have changed. In 1876 CRIPPS emphasized the risk of fatal peritonitis following the excision of any growth above the level of the peritoneal reflection (MORGAN 1965). Today the peritoneal reflection is important as a landmark below which a cancer can spread extensively without entering the peritoneal cavity.

In this chapter, the role of radiation therapy in the management of adenocarcinoma of the rectum will be stressed, for tumors of this type make up more than 90% of all rectal malignancies. Detailed accounts of the natural history and of the principles of surgical management are included because the effects of radiation therapy must be evaluated against the background of established surgical practice, and the design of any radiotherapy program must consider the patterns of spread of the cancer. Brief mention is also made of the use of radiation for uncommon conditions such as carcinoid tumors, smooth muscle tumors, lymphomas, and squamous cell carcinomas of the rectum.

10.2 Topographical Anatomy

As a result of the embryonic development of the intestinal tract and peritoneal cavity, the rectum comes to lie largely behind and inferior to the peritoneal cavity. The rectum is usually considered to be the distal 12–15 cm of the large bowel above the pectinate line, but since the junction of the rectum and sigmoid colon is indefinite, several different conventions have arisen in descriptive anatomy. To many the term "rectosigmoid" represents the last 5–8 cm of the sigmoid colon, which is intraperitoneal, and the uppermost 5 cm of the rectum (GOLIGHER 1984). Reports from St. Mark's Hospital include the rectosigmoid junction in the "upper third of the rectum" (MORSON and BUSSEY 1967). Such lack of agreement on the definition of the various anatomical regions can cause difficulties in comparing different series.

The rectum descends along the curve of the sacrum from about the level of the third sacral vertebra to the level of the muscles of the pelvic diaphragm below the tip of the coccyx. Superiorly, the rectum is the same width as the sigmoid colon, and inferiorly it bulges into the distensible rectal ampulla, which is normally empty, with its walls collapsed, except just before and during defecation. The rectum has three lateral flexures, the upper and lower convex to the right, and the middle convex to the left. At each of these flexures the mucous membrane of the rectal wall folds into the rectal lumen as the crescent-shaped valves of the rectum. The valves incompletely divide the rectum into three segments, each about 4–5 cm in length.

The upper third of the rectum is covered completely by peritoneum except posteriorly where the peritoneum is reflected away to the lateral pelvic walls. The middle third is covered on its anterior surface only, and the lower third has no direct relationship to the peritoneum. The middle rectal valve approximates to the level of the anterior peritoneal reflection, but there is considerable individual and sexual variation in the level at which the rectum loses any peritoneal covering (GOLIGHER 1984). In men, this anterior peritoneal reflection is usually about 7–9 cm from the anal verge, and in women it lies more inferiorly some 5–8 cm above the anal verge. Although most anatomists consider the rectum has no mesentery, the term "mesorectum" is often used in the surgical literature to describe the uppermost short thick posterolateral peritoneal and fibrofatty region (GOLIGHER 1984), and is even extended by some to include all of the connective tissues posterior to the rectum down to the level of the pelvic floor (HEALD et al. 1982).

The pararectal fossae are created by the reflection of the peritoneum laterally from the rectal to the pelvic walls. Anteriorly in the male the peritoneal reflection creates the rectovesical pouch, and in the female the rectouterine or rectovaginal pouch. These fossae and pouches are usually occupied by loops of small intestine or sigmoid colon. In the male, the extraperitoneal rectum is related anteriorly to the prostate, seminal vesicles, ureters, and bladder wall. In the female, the vaginal wall is the only usual extraperitoneal relation.

Between the peritoneal level above and the levator ani muscles below, the rectum is surrounded and supported by various segments of the pelvic fascial attachments and fatty areolar connective tissue. The posterior aspect of the extraperitoneal rectum is loosely attached to the front of the sacrum by connective tissue, and a thin layer of fascia, the fascia propria, covers the vessels and lymphatics on the back of the rectum. The parietal pelvic fascia is thickened over the anterior aspects of the sacrum and coccyx, and an avascular layer of fibrous tissue attaches this presacral fascia to the posterior aspect of the rectum at the anorectal junction. Release of this fascial attachment, and division of the lateral fascial and vascular connections of the middle and lower rectum, allows some straightening and apparent lengthening of the rectum, so that distances measured from the anal verge with the rectum in its anatomical position may be shorter than those measured in the mobilized rectum. Anteriorly the extraperitoneal rectum is covered by relatively thin segments of the visceral pelvic fascia which extend from the peritoneal reflection to the urogenital diaphragm, and separate the rectum from other pelvic organs.

The main arterial supply is provided by the superior hemorrhoidal artery, a branch of the inferior mesenteric artery, and there are important additional contributions from the middle and inferior hemorrhoidal branches of the internal iliac arteries. Venous blood from all levels of the rectum drains to the portal system through the superior hemorrhoidal and inferior mesenteric veins, and the middle and inferior hemorrhoidal veins of the internal iliac system offer pathways to the systemic venous system predominantly from the middle and lower rectum.

The major vascular and lymphatic vessels of the rectum follow similar courses. An intramural lymphatic plexus in the submucous and subserous layers of the rectum drains into the extramural lymphatics. The extramural lymph flow follows superior, lateral, and inferior pathways, although for all levels of the rectum the initial lymph node stations are in the immediate pararectal tissues. Although the superior pathway is principally through the superior hemorrhoidal and inferior mesenteric vessels, there may occasionally be some drainage to the paracolic and epicolic glands of the sigmoid colon, or to the paraportal lymphatics (SUGARBAKER and CORLEW 1982). Lymph flows laterally along the middle hemorrhoidal vessels to the internal iliac glands on the lateral pelvic wall. Inferior flow may also communicate with the anal lymphatic plexuses which run to the inguinal, and internal and external iliac glands. Injection techniques suggest that lymph flow from the intraperitoneal rectum is predominantly upwards, but that from the extraperitoneal rectum may follow all three routes. Vital dye injection studies have also demonstrated apparent lymphatic communications between the lower third of the rectum and the posterior vaginal wall, uterus, broad ligaments, tubes, and ovaries, and from the middle third of the rectum to the broad ligaments (ENQUIST and BLOCK 1966).

10.3 Adenocarcinoma of the Rectum

10.3.1 Epidemiology and Risk Factors

There are important differences in many aspects of the epidemiology of carcinomas of the colon, and of the rectum, although the two sites may often be considered together (ZIEGLER et al. 1986).

Large bowel cancer in general is more common in developed countries. The highest incidence rates are reported in North America, and Australia and New Zealand, with intermediate rates in Western Europe and Scandinavia, and somewhat lower rates in Eastern Europe. The lowest rates are found in Asia, Africa, and Latin America (CORREA and HAENSZEL 1978). Typical of high risk countries is the United States, where the age-adjusted incidence rates per 100000 for 1973-1977 for white males were 36.7 for colon cancer and 19.1 for rectal cancer, and those for white females were 31.0 for colon cancer and 11.5 for rectal cancer (YOUNG et al. 1981). There are divergent trends over time in high and intermediate risk countries, where colon cancer rates have tended to rise while those for rectal cancer have remained stationary or declined (CORREA and HAENSZEL 1978).

Within low risk populations there is a fairly even incidence of cancer in each segment of the large bowel, but in high and intermediate risk populations the rectum and rectosigmoid have the highest incidence (DEJONG et al. 1972), although the distal few centimeters of the rectum appear to be relatively unaffected by such variations (CORREA and HAENSZEL 1978). It is of interest that in familial multiple polyposis the distal 5 cm of the rectum is also relatively spared of adenomas compared with the more cephalad 10 cm (DECOSSE and BOYLE 1985).

The male to female ratio is usually about 1.4:1 for rectal cancer in the high risk countries, and is nearer to 1:1 for colon cancer. The sex ratio is more variable in Africa, Asia, and Latin America (CORREA and HAENSZEL 1978).

The divergent trends in incidence between countries, and within larger countries, support the notion of multifactorial etiology and susceptibility (LOGAN 1976), and the rapid increase in risk after migration from a low incidence to a higher incidence area, even in the first generation, argues strongly for the importance of external environmental factors (ZARIDZE 1983). Although there is strong evidence that nutritional factors play a major etiological role in large bowel cancer (REDDY et al. 1980), the evidence for many of the postulated relationships, such as those to high fat and low fiber diets, is not unequivocal (STUBBS 1983; ZARIDZE 1983). The concomitance in geographical distribution of risk with endocrine-dependent tumors (breast and endometrium) and arteriosclerotic heart disease is more clearly defined for colon cancer than for rectal cancer (CORREA and HAENSZEL 1978). There are no specific occupational factors associated with rectal cancer (SPIEGELMAN and WEGMAN 1985).

The development of cancer of the rectum, and indeed of cancers throughout the large bowel, is considered on the basis of animal and human studies to be a multistage process. Two major theories are advocated. The first, and the more popular, postulates that carcinomas of the colorectum evolve through the formation of adenomatous polyps which subsequently display phases of dysplasia and superficial noninvasive carcinomatous changes. The several arguments in favor of this theory have been reviewed recently (MORSON et al. 1983; SUGARBAKER et al. 1985). VOGELSTEIN et al. (1988) correlated the presence of oncogene mutations (*ras*-gene) and allelic deletions in several chromosomes of genes which normally suppress tumorigenesis with the finding of colorectal adenomas and carcinomas. The molecular alterations studied accumulated in parallel with the postulated clinical progression from adenomas to invasive cancers. The second theory argues that dysplastic and subsequently malignant changes can be observed in the mucosa without any intervening adenomatous change (SHAMSUDDIN et al. 1981). It is likely that both theories are correct, with the former being the more common pathway. In North American and European communities, 5%-10% of the general population have one or more adenomas of the large intestine (WINAWER et al. 1984). MORSON and BUS-

SEY (1985) demonstrated that after the removal of an adenomatous polyp, the risk over the subsequent 15 years of developing a further polyp is 1 in 2, and that of developing a colorectal carcinoma, 1 in 15.

Recent studies suggest that an inherited susceptibility to colonic adenomatous polyps and colorectal cancer is responsible for the majority of colonic neoplasms observed clinically, even when the polyp or cancer is apparently isolated (CANNON-ALBRIGHT et al. 1988). Uncommon, but better recognized, autosomal dominant conditions which give rise to colorectal carcinomas include the multiple polyposis syndrome and its variants. Inflammatory bowel disease, especially ulcerative colitis, and to a lesser extent granulomatous colitis, in either their sporadic or familial forms, are associated with an increased risk of colorectal carcinoma. First-degree relatives of patients with colorectal cancer have about three times the risk of the normal population of developing the disease, which could be due to genetic or to environmental factors (STUBBS 1983).

While a detailed discussion of the role of population screening is beyond the scope of this review, in a recent symposium HARDCASTLE (1986) noted that population screening for symptoms of colorectal cancer is not profitable, that the only practical methods of population screening are tests to detect blood in the feces but that these have many limitations (SIMON 1985), that endoscopic methods of screening which have a higher yield are practical only in high risk groups, and that in the population as a whole the only risk factor that is helpful in selection is the age of the patient, so that screening a population below the age of about 45 is unlikely to be cost-effective. A variety of schedules for screening high risk groups has been proposed (WINAWER et al. 1984).

10.3.2 Natural History

Carcinomas of the rectum spread by local infiltration of the tissues adjacent to their point of origin, and by metastasizing through lymphatic and venous channels, and through natural anatomical passages such as the peritoneal cavity and the bowel lumen. The complex processes of tumor spread have been reviewed by FIDLER and HART (1985).

10.3.2.1 Local Spread

Spread within the rectal wall occurs more rapidly and extensively in the circumferential axis than in the longitudinal axis (GOLIGHER 1984). The extent of subclinical spread beyond the visible and palpable limits of the primary tumor is of particular importance in determining the margin of tissue to be treated, and the distal extent of spread especially has attracted great interest since if affects the safety of restorative anastomoses. Extensive intramural spread is uncommon. WILLIAMS (1984) has catalogued eight series totaling 449 cases. Distal intramural spread was reported in 33 (7%), and in only 11 (3%) did tumor extend more than 2 cm beyond the gross tumor. When spread beyond 2 cm did occur it was usually associated with advanced lesions and long-term survival was rare. Data on proximal intramural spread are more difficult to find, but there is no evidence to suggest that it is any greater than in the distal direction (BLACK and WAUGH 1948).

Radial spread through the rectal wall allows the cancer to penetrate the perirectal fat and adjacent organs. In many series, the tumor has extended completely through the muscularis propria of the rectal wall in 85% of patients or more by the time rectal cancer is diagnosed (DUKES and BUSSEY 1958). The dense presacral fascia provides some resistance to infiltration but the sacral plexus and sacrum may be invaded. Anterior extension by an extraperitoneal rectal cancer leads to early adherence to and invasion of local viscera since there is relatively little perirectal fat in this area and fascial layers are thin. Anterior or lateral radial extension by a carcinoma above the peritoneal reflection may lead to transperitoneal metastases, or invasion of adjacent organs within the peritoneal cavity. In one series of 496 patients who underwent complete resection, tumor adherence to adjacent structures and difficulty in dissection were described in 75 (15%), and in an additional 22 (4%) there was invasion of adjacent structures (WITHERS et al. 1981). As the extent of radial spread increases, the risk of both lymph node metastases (DUKES and BUSSEY 1958) and hematogenous metastases (DIONNE 1965) increases.

Several authors have endeavored to measure the growth rate of colorectal carcinomas. Rectal cancers have not been identified separately in these studies, which usually include only small numbers of cases, and measurements of serial barium studies must inevitably lack precision. Median volume doubling times of 195 days (BOLIN et al. 1983) and 620 days (WELIN et al. 1963), within an overall range of 80–2400 days have been reported. Linear growth rates showed a ten fold difference in median values between series (BOLIN et al. 1983).

10.3.2.2 Lymphatic Metastases

Lymph node metastases are found in from 30% to 50% of patients who undergo radical resection (MORSON and BUSSEY 1967; PAHLMAN and GLIMELIUS 1984; PILIPSHEN et al. 1984). They are relatively uncommon until the cancer has extended through the muscularis propria. HERMANEK and GALL (1986) reviewed early invasive carcinomas in which the tumor had not penetrated beyond the submucosa. Of 130 patients who had classical resections, metastatic lymph nodes were found in only four (3%). They found that 75% of these early tumors were 3 cm or smaller in diameter, 92% were polypoid, 51% were grade 1 differentiation, and 47% grade 2 differentiation. They also reviewed 19 other series of early colorectal carcinomas treated by classical surgery and found that the rates of lymph node metastases reported ranged from 0% to 19%, with an average of 12%. The risk of nodal metastases when the cancer has infiltrated into, but not completely through, the muscularis propria varies widely, from 10% (4 of 40) (RICH et al. 1983) to 38% (31 of 82) (HOJO et al. 1982). When the tumor has penetrated completely through the muscularis propria, the proportion of patients with nodal metastases typically ranges between about 30% (42 of 132) (PAHLMAN and GLIMELIUS 1984) and 55% (156 of 279) (HOJO et al. 1982).

In general, metastases are said to be found most commonly in the pararectal node groups and along the upward pathways following the superior hemorrhoidal vessels. The role of the lateral pathways to the pelvic side walls is frequently disputed. However, in series which have included pelvic wall dissection for extraperitoneal rectal cancers, lateral pelvic wall metastases were reported in from 9% (11 of 122) (STEARNS and DEDDISH 1959) to 16% (38 of 241) (HOJO et al. 1982). The risk of pelvic wall metastases from upper rectal carcinomas was lower (10 of 148, 7%) (HOJO et al. 1982).

Although most authors consider that the progression of lymph node metastases follows a steady sequential pattern, initially to the pararectal nodes and thence to the superior hemorrhoidal and/or the internal iliac chains, apparent "skip metastases" have been reported, usually in fewer than 5% of cases (GABRIEL et al. 1935), although JINNAI (cited by GOLIGHER 1984) claimed such irregular patterns in just over 30% of his specimens of rectal cancer. Distal extramural spread through the subserosal and perirectal lymphatics is also uncommon. In three series in which a total of 1650 specimens were examined, 104 (6%) had lymph node metastases lying distal to the primary lesion, but in only 30 (2%) were these nodes more than 2 cm below the primary tumor, and most of these patients also had extensive metastases in other node groups (WILLIAMS 1984).

The discrepancies between the incidence of node metastases in the various series cited reflect not only the heterogeneity of the different patient populations and variations in the extent of surgical resection, but also differences in the techniques of preparing and examining the tissues excised.

10.3.2.3 Hematogenous Metastases

The relatively high survival rates in patients who do not have lymph node metastases are considered evidence that hematogenous metastases are uncommon unless lymphatic invasion is also present, but the relationship between lymphatic and hematogenous metastasis is complex (FIDLER and HART 1985). The venous drainage of the rectum communicates with both the portal and systemic venous systems. The organs most frequently involved are the liver and the lungs, and metastases to other organs in the absence of liver or lung involvement were found at autopsy in only 4% (3 of 70) (BROWN and WARREN 1938). An increased risk of systemic metastases has been correlated with the extent of cancer invasion into and through the rectal wall (BROWN and WARREN 1938; DIONNE 1965), with the presence of lymph node metastases (BROWN and WARREN 1938; DIONNE 1965), with high tumor grade (DIONNE 1965), with ulcerating tumors compared to exophytic cancers (DIONNE 1965), and with carcinomas from the upper rather than the lower rectum (BROWN and WARREN 1938; DIONNE 1965). As would be anticipated, the risk of systemic metastases can also be correlated with the identification of venous invasion by tumor. In one series, 35% (128 of 365) of those in whom venous invasion was identified developed liver metastases, compared to 14% (48 of 338) of those without identifiable venous infiltration (TALBOT et al. 1980). Attempts to correlate patterns of metastases with the finding of tumor cells in venous blood have been less successful (GOLIGHER 1984; GRIFFITHS et al. 1973).

It has been suggested that metastases grow more rapidly than primary rectal cancers. The median doubling time of pulmonary metastases from colorectal cancer has been variously measured at 116 days (COLLINS et al. 1956) and 109 days (WELIN et al. 1963), and the mean doubling time of liver metastases at 70 days (HAVELAAR et al. 1984) and 112 days (FINLAY et al. 1982). However, since the rate of growth of the primary tumors is not known in those patients in whom the growth rate of metastases was measured, it is possible that the aggressive biological characteristics of those tumors which metastasize might also be reflected in relatively rapid growth by the primary tumor.

10.3.2.4 Implantation

Although there are conflicting data on the viability of exfoliated tumor cells (ROSENBERG et al. 1978; UMPLEBY et al. 1984), these discrepancies may reflect variations in experimental technique. UMPLEBY et al. (1984) demonstrated apparently viable tumor cells in the bowel lumen from colorectal lavage or resection margin irrigation specimens in 70% (52 of 74 studies). The number of viable tumor cells did not correlate with the stage, differentiation, size, or fixity of the tumor. Attempts to identify malignant cells in peritoneal washings from patients with colorectal cancer have had limited success, and have not correlated with tumor stage or prognosis (URDANETA et al. 1983).

10.3.2.5 Untreated Rectal Cancer

The natural history of cancer of the rectum when it is left untreated has been described by several authors. The median survival of such patients from the first symptom of rectal cancer was from 14 to 21 months (DALAND et al. 1936; PESTANA et al. 1964; REPORTS on PUBLIC HEALTH and MEDICAL SUBJECTS 1927) with a range of from 1 to 49 months (DALAND et al. 1936). DALAND et al. (1936) compared the survival of their untreated patients with that of patients whose only treatment was a colostomy, and concluded that a colostomy sometimes improved patient comfort but did not prolong life. Despite the general belief that operability rates have risen, and that the value of palliative resections and of radiation is now better appreciated (MORGAN 1965), even recently in some centers a considerable number of patients were not considered suitable for any treatment. For example, of 5800 patients recorded by the Birmingham Cancer Registry between 1950 and 1961, 1282 (22%) received no treatment and had a crude 5-year survival rate of only 1.8% (SLANEY 1971).

10.3.2.6 Survival Patterns After Treatment

The major risk of tumor recurrence occurs during the first 5 years following treatment, and 70% of recurrences are detected during the first 2 years in most series (BERGE et al. 1973; CARLSSON et al. 1987). Selection factors make it difficult to compare different series, but trends in survival patterns are generally similar. For example, the corrected 5- and 10-year survival rates for all 5800 patients in the Birmingham Registry were 29% and 25% respectively (SLANEY 1971). The comparable rates for 2216 patients who had curative resections at a specialized center, St. Mark's Hospital, London, between 1928 and 1952 were 57% and 52% (MORGAN 1965). Thus, there are relatively few deaths due to cancer between 5 and 10 years, and MORGAN (1965) found no appreciable decrease in survival due to tumor recurrence after 10 years.

10.3.2.7 Failure Patterns After Major Surgery

The patterns of failure after apparently complete surgical resection are of particular interest. Where such failures are well localized it is conceivable that adjuvant radiation therapy might be beneficial, and both local and distant failure sites should be amenable to treatment by systemic adjuvant therapy. That local recurrence within the pelvis is a major problem is demonstrated by clinical studies, elective reoperation series, and autopsy series. While general trends are consistent, there is wide variation in the incidence of recurrence in the various series, reflecting such factors as different criteria for treatment, the recording of only the site of first recurrence rather than recurrence at any time over the patient's lifetime, and the rigor of proof of recurrence required.

The rates of pelvic recurrence after curative resection in selected representative series are shown in Table 1. When the carcinoma has not yet penetrated the muscularis propria the risk of recurrence is negligible, but when tumor is within the muscularis the range of recurrence rates reported is from as low as 1% to nearly 30%. Similarly, extension into the perirectal tissues carries a risk of from 5% to over 40%, and can also be related to the degree of extrarectal extension (GUNDERSON and SOSIN 1974; MORSON and BUSSEY 1967). The greatest risk occurs when lymph node metastases are present, the rates then ranging from about 15% to 50%. Most series suggest than the level of risk is increased if the cancer has both extended through the rectal wall and

Table 1. Pelvic recurrence after curative resection (recurrence/number treated)

		Stage							Total
ASTLER and COLLER (1954)		A		B_1	B_2	C_1		C_2	
DUKES and BUSSEY (1958)			A		B		C		
Clinical series	*Site*								
MORSON and BUSSEY (1967)	R, RS[a]		2/247 (1)[b]		31/583 (5)		122/762 (16)		155/1592 (10)
SLANETZ et al. (1972)[c]	R, RS		4/92 (4)		46/189 (24)		77/215 (36)		127/496 (26)
RICH et al. (1983)	R	0/3 (0)		3/36 (8)	18/59 (31)	2/4 (50)		20/40 (50)	43/142 (30)
PAHLMAN and GLIMELIUS (1984)	R, RS	1/13 (8)		11/40 (28)	39/90 (43)	5/12 (42)		18/42 (43)	74/197 (38)
PILIPSHEN et al. (1984)	F	0/5 (0)		18/128 (14)	32/111 (29)	11/49 (22)		44/89 (49)	105/382 (28)
Reoperation series									
GUNDERSON and SOSIN (1974)	R, RS			0/1 (0)	4/6 (67)	4/17 (24)		33/40 (83)	41/64 (64)

[a] R, rectum; RS, rectosigmoid; [b] ( ), percentage; [c] Series includes cancers 8–18 cm from anal verge.

involved regional nodes. This is consistent with the theory that local recurrences arise from tumor cells spilled from unrecognized direct extension within intercellular spaces, or from transected tumor-bearing lymphatic, vascular, or perineural channels. It has recently been demonstrated that extension of cancer to the lateral resection margins identified only on serial sectioning may be a major cause of local recurrence (QUIRKE et al. 1986). The rates of pelvic recurrence described in the retrospective studies in Table 1 are reasonably in keeping with the rates reported from the surgical control groups in prospective adjuvant therapy trials, in which the assessment protocols were presumably more consistent. For example, the EORTC reported a 35% 5-year actuarial local recurrence rate (GERARD et al. 1985), and Memorial Hospital noted 17% (42 of 251) pelvic failures (STEARNS 1980). Analyses of some large series of surgically treated cases, however, have shown local recurrence rates of only 10% (MORSON and BUSSEY 1967) to 15% (PHILLIPS et al. 1984), and individual surgeons have described even lower rates (HEALD et al. 1982; HEALD and RYALL 1986). The importance of these wide variations in the interpretation of studies of adjuvant therapy is self-evident.

The location of the recurrences within the pelvis has varied somewhat according to the nature of the preceding surgical procedure. LOFGREN et al. (1957) found that recurrences after anterior resection were usually close to the anastomosis area and primary tumor site, consistent with regrowth from transected vessels, rather than from unrecognized nodal metastases which would have been more remote. The sites of recurrence in the Minnesota reoperation series (GUNDERSON and SOSIN 1974), in which most patients had had an abdominoperineal resection, were predominantly in the anterior or posterior pelvis, with relatively few laterally, and perineal recurrence was usually accompanied by failure elsewhere in the pelvis. However, in the Memorial Hospital analysis, most recurrences after either low anterior resection or abdominoperineal resection showed some degree of fixation to the sacrum or pelvic side walls (PILIPSHEN et al. 1984).

Many pelvic recurrences are diagnosed within 2 years of surgery, and in the United Kingdom Large Bowel Cancer Project prospective survey the mean time to local recurrence was 20 months (ALDRIDGE et al. 1986).

A substantial number of patients who develop pelvic recurrence have concomitant or subsequent systemic metastases also. For example, in one EORTC trial (GERARD et al. 1985), the sites of first recurrence tabulated among 166 patients followed for an average of 3 years were in the pelvis alone in 21 (13%), at distant sites alone in 16 (10%), and both local recurrence and distant metastases were found in 21 (13%). Somewhat different patterns were reported in 58 patients who had had curative resection of Astler-Coller stage B_2 or stage C cancers, where the first recurrence was found in the

Table 2. Sites of cancer at autopsy (percent)

Tumor site	BERGE et al. (1973) 134 cases	SHINDO (1974) 1197 cases	WELCH and DONALDSON (1979) 56 cases		BROWN and WARREN (1938) 70 cases	
			Plus other sites	Only site	Plus other sites	Only site
Pelvic soft tissues	45	56	41	12	NS	NS
Lymph nodes	NS	38	59	–	NS	NS
Liver	62	47	48	2	49	33
Lung	64	45	52	5	46	9
Peritoneal	25	NS	25	–	NS	–
Bone	19	12	16	–	16	3
Brain	8	4	5	–	–	1
Spinal cord	NS	4	4	–	NS	–
Adrenals	19	NS	18	–	27	–

NS, not stated

pelvis only in 12 (21%), at distant sites in 18 (31%), and both local and distant failure was found in only two (3%) (GASTROINTESTINAL TUMOR STUDY GROUP 1985). In the reoperative series, although 64% had pelvic failure as a component of their disease recurrence, 36% in the series also developed distant metastases (GUNDERSON and SOSIN 1974). While these variations emphasize the difficulty in comparing different series, the conclusion that distant metastases are at least as great a problem as local recurrence in determining survival is inescapable. However, the symptoms from local recurrence tend to dominate whenever it occurs (GILBERT 1978).

Autopsy series give a further perspective on the patterns of tumor distribution and the findings in several series are shown in Table 2. The high risk of pelvic recurrence is apparent, although at the time of death metastases in other sites were usually present also. The predeliction for metastases to localize in the liver and lungs is clear. Despite the relatively high rate of peritoneal metastases noted in two autopsy series (BERGE et al. 1973; WELCH and DONALDSON 1979), peritoneal seeding was very rarely identified in reoperative (GUNDERSON and SOSIN 1974) or clinical series (PILIPSHEN et al. 1984).

Although local recurrence is common after apparently complete primary resection, it was considered the cause of death in only about 20% of cases, the remainder of cancer-related deaths being due to distant metastases (DIONNE 1965; WELCH and DONALDSON 1979). The mode of death from local recurrence was uremia, bowel obstruction, or local sepsis (WELCH and DONALDSON 1979). Local recurrence was much more likely to contribute to the cause of death when only a palliative resection or no resection at all had been possible (TAYLOR 1962).

10.3.3 Diagnostic Workup

10.3.3.1 History and Physical Findings

Although about 2% of patients are asymptomatic when their rectal cancer is found, the great majority have one or more symptoms. The most common symptoms recorded in a study of 258 patients with cancer of the rectum or rectosigmoid were melena (present in 85%), constipation (46%), diarrhea (30%), tenesmus (30%), and abdominal pain (26%) (POSTLETHWAIT 1949). A feeling of incomplete evacuation or of a rectal mass is also fairly common. Pain or interference with urinary or sexual function may occur due to involvement of the presacral nerve plexuses. FEINSTEIN et al. (1975a) suggested that there is merit in separating symptoms according to whether they arise from the primary tumor, from metastases, or from systemic effects of the tumor (e.g., weight loss), and in their series of 318 patients recorded primary symptoms in 93%, systemic symptoms in 45%, and metastatic symptoms in 9%. Physical examination is directed to identifying the location of the primary tumor, its general morphological characteristics, and in particular whether the tumor and rectal wall are mobile, or tethered or fixed to extrarectal tissues (YORK MASON 1976). In general, lymph node metastases are not accessible to clinical evaluation, although pararectal node involvement may be detectable by digital examination (NICHOLLS et al. 1982; PAPILLON 1982). A careful search for signs of metastases or for systemic effects of cancer is also necessary.

Digital examination will generally enable the physician to evaluate the lower rectum up to about 8 cm from the anal verge. Visualization of the tumor and biopsies to establish the diagnosis require proctosigmoidoscopy.

10.3.3.2 Imaging in the Pelvis

Although a barium enema is not generally necessary to evaluate low rectal cancers, the remainder of the colon must be assessed to identify other possible colorectal pathology. Barium enema and colonoscopy each has advantages and these tests may be complementary (NOLAN 1982). The reported incidence of synchronous primary cancers in the colorectum ranges from 2% to 8%, and benign polyps have been found in from 12% to 62% of patients with single cancers, and in from 38% to 86% of patients with synchronous cancers (LANGEVIN and NIVATVONGS 1984). Computerized tomography demonstrates tumor spread more than about 2 cm in extent but is less reliable in demonstrating lesser degrees of spread, which can usually be assessed more accurately by digital examination (DIXON et al. 1981; WILLIAMS and HUSBAND 1987). Early experience with magnetic resonance imaging suggests that it too provides useful information about extension into perirectal tissues and adjacent organs, but is not able to assess the extent of limited bowel wall infiltration (BUTCH et al. 1986). Some authors have found that intrarectal ultrasound can give reliable information about limited tumor infiltration, and the probe can also be used to evaluate tumors beyond the reach of digital examination (HILDEBRANDT and FEIFEL 1985).

Unfortunately, although the presence of lymph node metastases is such a significant prognostic factor, none of the available imaging techniques is able to identify involved nodes accurately. Magnetic resonance imaging and computerized tomography may identify enlarged nodes, yet not all node enlargement is due to metastases (GABRIEL et al. 1935), and tumor in normal-sized nodes cannot be visualized (BUTCH et al. 1986; DIXON et al. 1981). Pelvic lymphoscintigraphy has also proven disappointing (EGE and CUMMINGS 1980). The possible role of labeled monoclonal antibodies in identifying both lymphatic and hematogenous metastases is being investigated (KIM et al. 1980; LEYDEN et al. 1986).

10.3.3.3 Identification of Metastases

The detection of liver metastases may be achieved with varying degrees of accuracy by biochemical and imaging techniques. KEMENY et al. (1982) evaluated γ-glutamyl transpeptidase, serum glutamic-pyruvic transaminase, lactic dehydrogenase, serum glutamic-oxaloacetic transaminase, leucine aminopeptidase, and 5-nucleotidase. The accuracy of these tests ranged from 53% to 65%, with no significant difference between tests, and no apparent advantage from any combination of tests.

A study of various forms of hepatic imaging showed that the accuracy of liver scintiscans, ultrasound, and computerized tomography (CT) (with or without water-soluble contrast enhancement) ranged between 80% and 84%, with similar specificity and sensitivity ranges (SMITH et al. 1982). However, while most lesions less than 2 cm in diameter were not detected by any technique, and all methods were able to identify most lesions 4 cm or more in size, CT was the most efficient at detecting 3 cm diameter metastases. Possible physical reasons for the failure to identify small metastases, and the value of agents which enhance tumor recognition, have been discussed by BARTRAM (1985). The limitations of inspection and palpation of the liver at laparotomy, without additional investigation, were demonstrated by the finding of liver metastases in 5 of 31 autopsies performed on patients who died shortly after operation, having been adjudged free of liver metastases at surgery (GOLIGHER 1941).

Pulmonary metastases are generally identified by a conventional chest X-ray. More precise definition of the number and location of metastases may be achieved by whole lung tomography or computerized tomography of the chest. However, lung tomography is not necessary in the routine preoperative evaluation of the patient with rectal cancer (GIANOLA et al. 1984).

The risk of bone or central nervous system metastases is so low that no specific investigations of these systems are indicated in the asymptomatic patient.

Serum markers, including carcinoembryonic antigen (CEA), tissue polypeptide antigen, and a monoclonal-antibody-defined carcinoma associated carbohydrate antigen (CA-50), and others, have been evaluated alone and in combination (COOPER and O'QUIGLEY 1982; LAURENCE and NEVILLE 1983; STAHLE et al. 1988). While there was some correlation with both tumor stage and prognosis, the role of such markers in determining treatment is not yet clear.

10.3.3.4 Pretreatment Assessment

On the basis of the above considerations, pretreatment assessment should include barium enema and/or colonoscopy, a chest X-ray, an abdominal CT scan to assess both the liver and the pelvis, conventional liver biochemistry such as alkaline phosphatase, and a blood count. Despite such screening, about 10% of patients will be found to have metastases beyond the pelvis at laparotomy (CEDERMARK et al. 1985; STOCKHOLM RECTAL CANCER STUDY GROUP 1987). Tests such as the CEA may have value as a baseline for the later detection of possible recurrent or residual disease.

10.3.3.5 Follow-up

There is considerable debate about the merits of any systematic evaluation and investigation of patients after their initial treatment. KRONBORG (1986) reviewed the various arguments for and against systematic follow-up, and suggested that few metastases or recurrences are curable when they are detected, that a new metachronous colorectal cancer may have the same prognosis whether diagnosed at planned follow-up examination or when it causes symptoms, and that follow-up programs are expensive and associated with significant psychological stress for the patients. He argued that resources would be better redirected to prospective trials, including those evaluating follow-up programs. AUGUST et al. (1984) considered that there is evidence that systematic surveillance can detect recurrent disease in some patients at a time when it is possibly curable, and that some serious clinical complications from tumor recurrence might be avoided. The capabilities of the various investigations discussed above also apply to their use in follow-up. In one prospective study, the most helpful investigations in the detection of metastases after resection of the initial cancer were CEA levels and abdominal CT, together with endoscopy for possible anastomotic recurrence (GIANOLA et al. 1984).

The area of greatest controversy currently is the value of serial CEA estimations. When elective reoperations were performed on asymptomatic patients at high risk of recurrence, but prior to the development of the CEA assay, salvage was achieved in only 10% of patients with recurrent rectal cancer (GRIFFEN et al. 1969). Several groups have suggested that regular postoperative CEA evaluation can identify an increased number of patients with local recurrence or metastases potentially resectable for cure, and that long-term survival rates are increased in those in whom resection is possible (AUGUST et al. 1984; MARTIN et al. 1985). Unfortunately, even with the 4- to 5-month lead time which elevation of the CEA appears to give over symptomatic recurrence, the identification of pelvic recurrence is difficult and pelvic reoperation is associated with considerable morbidity (MARTIN et al. 1985). It is to be hoped that the role of CEA in follow-up will be clarified by the current prospective randomized trial being conducted in the United Kingdom, in which patients undergo monthly CEA assay, and if a significant rise is recorded, the patient is randomly assigned either to investigation and treatment prompted by the elevation in CEA, or to a control group in which the surgeon is not informed of the CEA result and bases management on conventional clinical indications (NORTHOVER 1985).

10.3.4 Prognostic Factors

Many factors have been claimed to exert prognostic influence in patients with rectal carcinomas, and have been evaluated by multivariate or, more commonly, univariate statistical techniques. The local extent of the cancer in relation to the rectal wall and the presence of lymph node metastases and distant metastases have long been recognized as the strongest prognostic factors.

In efforts to refine these categories, the possible influence of a number of other factors has been evaluated. In general, only one representative reference from the voluminous literature in this field is given.

Suggested prognostic factors include tumor descriptive characteristics such as the distance from the anal verge (FREEDMAN et al. 1984), tumor configuration (TALBOT 1982), tumor size (WOLMARK et al. 1983), mobility or fixation (WORKING PARTY MRC, 1982), and circumferential involvement of the rectum (WORKING PARTY MRC, 1982). General characteristics considered include the age of the patient at diagnosis (BLOCK and ENKER 1971; RECIO and BUSSEY 1965); the duration of symptoms and the presence of other gastrointestinal pathology (FEINSTEIN et al. 1975a); systemic tumor-related symptoms (ZORZITTO et al. 1982); and presentation with anemia (ZORZITTO et al. 1982), obstruction (WOLMARK et al. 1983), or perforation (PHILLIPS et al. 1984). The prognostic significance of a large number of histopathological features has also been discussed, including tumor grade (DUKES and BUSSEY 1958); the nature of the advancing edge of the tumor (SPRATT and SPJUT 1967); the histological subtype of the carcinoma, especially mucoid (SYMONDS and VICKERY 1976) or scirrhous (WOOLAM et al. 1965); the cellular DNA content of the cancer (ARMITAGE et al. 1985); the thymidine labeling index (TROTTER et al. 1985); the degree of local extension beyond the muscularis propria and whether it was detectable on gross inspection or on microscopy (GUNDERSON and SOSIN 1974); local inflammatory reaction (SPRATT and SPJUT 1967); and the presence of tumor at resection margins (FREEDMAN et al. 1984; QUIRKE et al. 1986). In addition to lymph node involvement, the number and location of the metastatic nodes (DUKES and BUSSEY 1958; HOJO et al. 1982) have been studied, and the immunomorphologic characteristics of non-tumor-bearing regional nodes have been thought important (PIHL et al. 1977). Invasion of veins (TALBOT et al. 1980) and of perineural spaces (SEEFELD and BARGEN 1943) have also been studied. The assay of oncogene products may eventually provide useful prognostic information (MICHELASSI et al. 1988).

10.3.5 Staging Systems

How is this plethora of data to be incorporated into a staging system? Two preliminary points must be made. Firstly, the objective of staging is to describe the extent of the tumor in order to make decisions about treatment, and to identify groups of patients with similar prognostic expectations. The latter is fundamental to the evaluation and comparison of treatments. Most currently used staging systems for rectal cancer are based on histopathological details which can be determined only after surgical treatment has been carried out and can play no part in shaping decisions about what treatment to apply to the primary tumor. Because no generally accepted method exists for ensuring the prognostic comparability of patients before surgical therapy is applied, comparisons cannot be made readily between patients treated by wide surgical resection and those treated by other techniques, and any preoperative therapy which alters histopathological features confounds efforts to compare patients treated with such methods to those undergoing surgical resection alone (ZORZITTO et al. 1982). The second point is that the development of new diagnostic tests which demonstrate metastases which were formerly unrecognized, and the frequent subdivisions of previously established histopathological staging systems so that only subgroups are compared rather than overall patient groups, may produce misleading optimism over results, due to what has been called "stage migration" (FEINSTEIN et al. 1985).

The cornerstones of histopathological staging systems remain the extent of primary tumor progression through the rectal wall and the presence or absence of regional lymph node metastases, although other histopathological features may be equally important (JASS and MORSON 1987). The confusion which can arise through misinterpretation of the many eponymous variations of the system originally propounded by DUKES (1932) has been summarized by KYRIAKOS (1985). The correlations of a new staging system proposed by the American Joint Committee on Cancer (AJCC) and the Union Internationale Contre le Cancer (UICC) with the two most commonly used alternatives are shown in Table 3. The proven reproducibility of staging by histopathological criteria ensures that it will retain a major role in the management of rectal carcinoma.

Efforts to develop staging systems based on features other than histopathological criteria alone have taken two forms. Clinical examination and imaging techniques have been used to try to gather the information normally considered in histopathological staging systems. The alternative approach has been to examine new staging systems based on

Table 3. A comparison of stage groupings

Depth of cancer in rectal wall	DUKES (1932)	ASTLER and COLLER (1954)	UICC/AJCC (HUTTER and SOBIN 1986)	
			Stage	Grouping
Into submucosa	A	A	I	$T_1N_0M_0$
Into muscularis propria	A	B1	I	$T_2N_0M_0$
Into subserosa or into unperitonealized perirectal tissues	B	B2	II	$T_3N_0M_0$
Into visceral peritoneum or directly into other organs or tissues	B	B2	II	$T_4N_0M_0$
Regional nodes involved	C	(C)[a]	III	Any T, N_1, M_0
	–	–	IV	Any T, any N, M_1

[a] Astler-Coller stage C1 = cancer has not extended beyond muscularis propria but regional node metastases are present; Astler-Coller stage C2 = cancer has extended completely through muscularis propria and regional node metastases are present.

tumor-related items in the patient's history and physical examination, and on imaging techniques.

NICHOLLS et al. (1982) reported concordance between preoperative clinical examination and pathological assessment in about 80% of patients for the degree of local tumor extension, and in about 65% for the presence or absence of pararectal lymph node metastases. The level of penetration through the bowel wall was underestimated in up to 15%. Similar studies have been undertaken by NETRI et al. (1985). Although NICHOLLS and his colleagues (1982) felt that the clinical classification they proposed was sufficiently reliable to be worthy of further study, others have emphasized that it does not reproduce exactly the stage distribution achieved by histopathological systems. Such criticism is valid, but irrelevant if the system proves reproducible and achieves the objectives of staging described previously. Similarly, although the limitations of imaging techniques in reproducing the detail achieved by histopathological examination have been commented on earlier, there appear to have been no attempts to develop staging systems based on imaging independent of histopathological correlation. Exponents of intrarectal irradiation techniques such as PAPILLON (1982) have demonstrated that patients can be selected appropriately for this form of treatment by techniques other than histopathological evaluation of the whole rectal tumor and its regional lymph nodes.

Alternative staging systems based on features in the history and physical examination, supplemented by imaging and laboratory techniques, have been developed (FEINSTEIN et al. 1975a, b; ZORZITTO et al. 1982). In these systems, clinical staging was as effective as anatomical staging in identifying groups of patients with similar prognosis. Also, the prognostic effects of histopathological staging systems and the clinical stage groups were shown to be independent of each other, although some of the prognostic impact of each system of classification was present in the other (ZORZITTO et al. 1982). The clinical groups provided a gradient within each of the histopathological groups and vice versa, so that the systems could be complementary and improve the segregation of prognostic groups and decision making for individual patients (FEINSTEIN et al. 1975a, b; ZORZITTO et al. 1982). A staging system based on biochemical analyses has also been proposed (DEVESA 1984).

Further development of staging systems based on features other than histopathological criteria will be needed for studies of preoperative adjuvant therapy and of techniques which conserve anorectal function through procedures more limited than conventional radical resection.

10.3.6 Surgery

The growing volume of literature on the role of chemotherapy and radiation therapy should not obscure the fact that at the present time surgical resection is the principal treatment for rectal cancer. The technical aspects of the various surgical procedures have been described by GOLIGHER (1984), and the historical evolution of these operations was summarized by ROTHENBERGER and WONG (1985) and BALLANTYNE (1988).

10.3.6.1 Major Resections

In specialized referral centers, resectability rates increased from about 40% in 1920–1930 to about 90% by 1950–1960 (GOLIGHER 1981; LOCKHART MUMMERY et al. 1976). This reflected not only advances in surgical technique and supportive care, but also a greater willingness to resect the pelvic tumor for palliation even when the presence of distant metastases made cure unlikely. The crude 5-year survival rates reported by major surgical centers are typically about 45%–50%, and rise to 50%–70% when corrected for perioperative deaths and deaths due to intercurrent disease (GOLIGHER 1981; LOCKHART MUMMERY et al. 1976). The results from cancer registries, which reflect total community experience, are sometimes less satisfactory, with resection rates of 50%–70%, and corrected 5-year survival rates of only about 30%–35% (EISENBERG et al. 1967; SLANEY 1971). The crude 5-year survival rates after curative resection in major centers average about 85% for Dukes stage A cancers, 60% for stage B, and 30%–40% for stage C (ROTHENBERGER and WONG 1985).

Operative mortality rates have steadily improved so that during the past 20 years the risk from the major pelvic resections is generally less than 5%, although some centers still report rates as high as 10% (GOLIGHER 1981; ROTHENBERGER and WONG 1985). Major morbidity, especially cardiopulmonary complications and urinary tract dysfunction, may occur in up to 50% following major pelvic surgery (ROTHENBERGER and WONG 1985); pelvic dissection leads to autonomic nerve damage with sexual dysfunction in 50% or more of men (WILLIAMS and JOHNSTON 1983), and a permanent colostomy is frequently accompanied by both psychosocial and physical morbidity (WILLIAMS and JOHNSTON 1983).

Over the past 20 years there has been an increased preference for restorative resections over procedures which sacrifice the anal sphincters, and it has been estimated that abdominoperineal resection (APR) or similar procedures (collectively considered here as APR) are now used for no more than 10%–25% of cancers of the middle third of the rectum, and for only about one-half of those arising in the lower third (GOLIGHER 1984; ROTHENBERGER and WONG 1985). In a review of the rationale for the preservation of the anal sphincter with carcinomas of the middle and distal rectum, WILLIAMS (1984) noted several influential factors. These included reappraisal of the patterns of tumor spread, especially with regard to the distal margin of clearance required at resection, technical advances including stapled anastomoses, greater appreciation of the long-term morbidity associated with a permanent colostomy, and progress in the understanding of the physiological mechanisms of anal continence. For cancer of the upper third of the rectum, resection and anastomosis has been accepted treatment for many years. In the middle third of the rectum, comparison of APR and restorative resections in patients with similar stage tumors, and with tumors at a similar distance from the anal verge, shows no significant advantage in favor of either procedure. There are only a few studies so far reported which include tumors of the distal third of the rectum, and there is still considerable debate on whether there is a higher risk of local recurrence following restorative procedures at this level (PILIPSHEN et al. 1984; WILLIAMS 1984). The rationale for reducing the margin of normal rectum resected distal to the tumor from the long-held guideline of 5 cm to 2 cm has been presented earlier (Sect. 10.3.2), where both intramural and extramural distal extension beyond 2 cm were shown to be infrequent. Studies of anorectal function (WILLIAMS 1984; WILLIAMS et al. 1980) have demonstrated that the reflexes which mediate fecal continence are preserved even if the rectum is excised completely and are probably situated in the extrarectal tissues, especially the muscles of the pelvic floor. Even with very low anastomoses nearly all patients achieve continence for solid feces although some have less certain control of flatus and liquid feces. Normal function is not regained after surgery for from 6 to 18 months, a period of time which presumably allows anal tone to recover and rectal capacity to increase, so that control is improved and the frequency of bowel action decreased.

Although in principle radical surgery for rectal carcinoma requires removal of all the major regional lymphatic pathways, in practice this is rarely done and most conventional resections include only the pararectal and superior hemorrhoidal lymph node groups. The results of series in which more extensive lymphatic dissections have been performed suggest that benefit is minimal, and morbidity significant. The superior level of ligation of the lymphovascular vessels and nodal dissection is commonly at the origin of the superior hemorrhoidal artery. Although some authors have suggested that the risk of node involvement between the origin of the inferior mesenteric artery and the branching of the left colic and superior hemorrhoidal artery is between 5% and 20% in those with nodal metastases at a lower level (ROTHENBERGER and WONG 1985), no consistent advantage has been found from ligation and dissection to the level of origin of the inferior mesenteric artery (GRINNELL 1966; NICHOLLS 1982) or from para-aortic dissection above this level (STEARNS and DEDDISH 1959). Debate on the need for removal of the pelvic wall nodes with carcinomas situated below the peritoneal reflection has been rekindled by the publications of ENKER et al. (1986) and HOJO et al. (1982). Although ENKER et al. (1986) reported improved survival rates in all disease stages following en bloc pelvic lymphadenectomy compared with nonrandomized controls treated by conventional resection, they did not provide data on the location of the involved lymph nodes found. Other reports have suggested that when pelvic side wall lymph nodes are involved, survival rates are low and are not markedly improved by the extended dissection (HOJO et al. 1982; STEARNS and DEDDISH 1959), and all have noted an increase in morbidity from the procedure. Contrary to the experience of STEARNS and DEDDISH (1959), both ENKER et al. (1986) and HOJO and KOYAMA (1982) found that extended pelvic lymphadenectomy appeared to reduce the risk of pelvic recurrence. Since HOJO and KOYAMA (1982) observed fewer pelvic recurrences even in patients who did not have nodal metastases, it may be that the important technical factor is the width of the dissection beyond the primary cancer, reducing the risk of transecting tumor, rather than removal of the nodes themselves. The majority of surgeons at present consider that extended lymph node dissections may disclose more involved nodes but do not improve survival greatly. This is consistent with observations that the first nodes involved are those in the perirectal area and that survival decreases as the number of nodal metastases increases (DUKES and BUSSEY 1958; HOJO et al. 1982).

Adherence to adjacent organs is found at laparotomy in from 5% to 20% of cases. A review by SUGARBAKER and CORLEW (1982) of six series totaling 288 patients who had adherence between the primary colorectal cancer and adjacent structures showed that in 62% (range 38% to 80%) the adhesions were due to infiltrative cancer and not to inflammation. Although in these series complete en bloc resection resulted in 5-year survival rates not greatly different from those of patients with cancers of similar stage in whom no adhesions were found, there must presumably have been considerable selection in determining which patients could undergo resection, and other investigators have reported higher local recurrence rates and poorer survival rates when adjacent organs were invaded by cancer or when resection was technically difficult (WITHERS et al. 1981). ENQUIST and BLOCK (1966) suggested elective removal of the posterior vagina, uterus, ovaries, fallopian tubes, broad ligaments, and adjacent peritoneum because of the demonstration of lymphatic communications between these organs and the lower and mid rectum. In a retrospective review, VEAZEY and MCBRIDE (1979) found that pelvic recurrence occurred in 4% (1 of 22) of women who had had a previous or concomitant hysterectomy and APR, and in 36% (15 of 41) of those who had not had a hysterectomy. Although systematic removal of all of the tissues suggested by ENQUIST and BLOCK (1966) has not found general favor, more interest has been expressed in the role of prophylactic oophorectomy to remove a site of possible occult metastases (GRAFFNER et al. 1983, MACKEIGAN and FERGUSON 1979). However, the benefits of this procedure are also unproven.

10.3.6.2 Local Surgical Procedures

The use of treatments which focus only on the local primary tumor, rather than conventional restorative excisions or APR, has excited considerable discussion, particularly when such treatment would avoid the need for a permanent colostomy (BAKER 1980; ROTHENBERGER and WONG 1985). Local treatment is appropriate only when the primary tumor has not involved adjacent organs or regional lymph nodes, so that extended resection would be necessary for cure, and when the local modality can be applied safely. Experience has shown that careful patient selection can result in high local tumor control and cure rates, with low morbidity. The tumor characteristics which identify patients with a probability of regional node metastases of less than about 10% are free mobility, an exophytic profile, a size of less than 3–5 cm, and low or intermediate histological grade on biopsy (BAKER 1980; HERMANEK and GALL 1986; PAPILLON 1982). It is usual also to consider for local treatment only those tumors less than 10 cm from the anal verge to reduce the risk of intraperitoneal perforation and even then pelvic abscess is a potential hazard. Even in large centers, the number of patients selected for local treatment is small, and over a 25-year period only 4% of 3972 patients were treated by local excision at St. Mark's Hospital (MORSON et al. 1977). Complete local excision resulted in a crude 5-year survival rate of 82%, and other series have also reported 5-year survival rates of about 85%, and local recurrence rates of about 10% (DEDDISH 1974; HAGER et al. 1983). Electrocoagulation has also been used as definitive local therapy, and with rather broader selection criteria than those suggested above, MADDEN and KANDALAFT (1973) achieved a 5-year survival rate of 63% and a local recurrence rate of 13%, with only two episodes of perforation in 204 patients, but 16% experienced significant bleeding. Many authors favor the use of excision if localized treatment is selected, arguing that only this method provides complete histopathological details, enabling the physician to make a decision on the need for additional treatment immediately (MORSON 1985).

10.3.6.3 Pelvic Recurrence and Metastases

The surgical management of pelvic recurrence after previous major surgical resections, or of metastases, may result in useful palliation and can sometimes be curative.

Anastomotic recurrences are frequently also fixed to the pelvic side walls so that complete excision, generally by APR, is not usually possible (ENKER et al. 1986), and SLANETZ et al. (1972) reported only one 5-year survivor among 22 patients who developed anastomotic recurrence after anterior resection. Even elective reoperation on high risk asymptomatic patients (GRIFFEN et al. 1969) or surgery on asymptomatic patients with rising CEA titers (MARTIN et al. 1985) infrequently leads to cure when cancer has recurred in the pelvis and is often associated with significant morbidity. While incomplete excision has sometimes resulted in pain relief, a perineal incision often fails to heal and adds persistent discharge to the patient's discomfort (POLK and SPRATT 1971). Intestinal obstruction may require laparotomy for relief, but ureteric obstruction can rarely be managed except by nephrostomy or other diversionary surgery, and life expectancy is usually brief despite such surgery (Brin et al. 1975).

The role of surgery for the management of hepatic and pulmonary metastases has been reviewed by August et al. (1984). These authors suggested that as many as 10%-15% of those with hepatic metastases from a colorectal primary tumor might be candidates for potentially curative hepatic resection. They surveyed ten series in which operative mortality ranged from 0% to 12%, and 5-year survival varied from 18% to 52%, with most around 40%. Although there were some differences between series, the patients most likely to benefit seemed to be those with either solitary or a few metastases localized to a limited area of the liver. A similar survey of the literature relating to pulmonary resection for colorectal cancer metastases disclosed an operative mortality of 0%-4%, and a 5-year survival rate of generally about 25%-30%. Most patients who underwent pulmonary resection had solitary nodules. A solitary pulmonary nodule in a patient who has had colorectal cancer may also be due to a new primary lung cancer (Cahan et al. 1974) or to benign pulmonary disease (Gianola et al. 1984).

10.3.7 Radiation Therapy

Although the palliative effects of radiation therapy for recurrent or metastatic rectal cancer have long been accepted, a significant role for radiation in the management of primary rectal cancer has been developed only in the past 15 years. The efforts of early radiation therapists were reviewed by Scherer (1972) in the previous edition of this text. The development and general availability of megavoltage apparatus, the realization that rectal adenocarcinomas are not radioresistant, and concerted efforts to improve survival rates and to conserve anorectal function have all contributed to the still growing interest in the irradiation of rectal adenocarcinomas. At the present time, the role of radiation therapy may be considered conveniently as adjuvant to surgical resection, as primary therapy, and as treatment for recurrent or metastatic carcinoma.

10.3.7.1 Adjuvant Radiation Therapy

Carcinomas which recur in the pelvis after resection are rarely curable, and considerable efforts have been devoted to preventing such recurrences through modifications of surgical technique and trials of adjuvant therapy. The intentions of adjuvant radiation therapy can be summarized as follows:

1. Reduction of the viability of tumor cells within the irradiated volume, leading to a reduction in local recurrence, or in distant metastases if the irradiated cells are subsequently released from the pelvis
2. Improvement in survival rates due to reduction in the rates of local recurrence and metastases
3. Achievement of the previous objectives without undue morbidity

Most of the treatment programs studied have been developed empirically, although animal models have been used to study some aspects of adjuvant irradiation (Agostino and Nickson 1960; Powers and Tolmach 1964; Whiteley 1967).

Many of the studies of adjuvant radiation therapy were designed primarily to identify improvements in survival rather than to study patterns of failure. However, the realization that a reduction in pelvic recurrence rates might be accompanied by a decrease in morbidity, but that distant metastases might still determine survival rates, has led to some reappraisal of the objectives for adjuvant irradiation, and to greater interest in programs which combine pelvic regional and systemic adjuvant therapy. Because of the differences in the way patients are selected for treatment, in the criteria for determining tumor recurrence patterns, and in the duration over which patients are followed, it is difficult to compare one study with another, and considerable emphasis must be placed on the results of randomized trials, although even these may contain flaws in design and interpretation (Cummings 1984, 1986; Buyse et al. 1988).

10.3.7.1.1 Preoperative Radiation Therapy. The principal reasons given for favoring preoperative irradiation have been the intent of reducing the rate of distant metastases as well as the rate of pelvic recurrence from cells released during surgical resection, reduction of the size of the primary tumor and of regional node involvement to facilitate resection, and the lower likelihood of late radiation enteritis due to the lesser volumes of small bowel generally present in the pelvis when the rectum and sigmoid colon have not been removed. The major potential disadvantages of preoperative irradiation are the difficulties in selecting only those patients at high risk of pelvic recurrence, and the delay to definitive surgical resection which might permit further tumor growth.

Table 4. Preoperative radiation therapy for primary resectable rectal cancer

	Patients	Radiation dose (total dose, Gy/fractions/days)	Volume irradiated	Interval between radiation and surgery (days)	Overall 5-year survival (%)		Pelvic recurrence rate (%)		Comment
					RTS	S	RTS	S	
Randomized series									
RIDER et al. (1977)	125	5/1/1	Pelvis, 15 × 18 cm	Same day	36	38	NS	NS	
STEARNS (1980)	790	20/8/10	Pelvis, 16 × 18 cm	2–42	57	58	11	17	
HIGGINS et al. (1975)	700	20/10/14 plus 5/10/14 perineal boost	Pelvis, 20 × 20 cm	14 (average)	49	39	NS	NS	
Second Report of an MRC WORKING PARTY (1984)	850	5/1/1 or 20/10/14	Pelvis, 15 × 18 cm	<7	42 40	38	45 47	43	Actuarial rates at 5 years
HIGGINS et al. (1986)	361	31.5/18/24	Pelvis plus para-aortics to L2	<5	43	47	NS	NS	
GERARD et al. (1985, 1988)	410	34.5/15/19	Pelvis plus para-aortics to L2	4–15	62	60	15	35	$P<0.05$ for pelvic recurrence
STOCKHOLM RECTAL CANCER STUDY GROUP (1987)	545	25/5/5–7	Pelvis plus para-aortics to L2	1–7	45	45	9	20	Median follow-up 3 years. $P<0.01$ for pelvic recurrence
KLIGERMAN et al. (1972)	31	46/23/31	Pelvis plus para-aortics to L2	28	41	25	8	0	
Nonrandomized series									
HERZOG et al. (1986)	68	16/4/3 12-h intervals	Pelvis, 14 × 14 cm	Same day	66	–	7	–	
RODRIGUEZ-ANTUNEZ et al. (1973)	70	24/3/3	Posterior pelvis, 8 × 15 cm	10–15	58	–	7	–	
PUJOL et al. (1978)	116	40/18/24	Pelvis, 13 × 13 cm	2–5	59	44	10	11	Historical controls, same center
STEVENS et al. (1976)	57	50–60/24–30/40–56	Primary tumor, 10 × 10 cm	28–49	55	38	0	11	Historical controls, same center

RTS, radiation therapy followed by surgery; S, surgery; NS, not stated

The radiation doses given prior to surgery have ranged from single doses of 5 Gy to fractionated courses of 50–60 Gy (Table 4). The shorter courses have been based on the premise that only small numbers of well-oxygenated tumor cells will not be removed at surgery, and may be sterilized by such doses (NIAS 1967; POWERS and TOLMACH 1964). Also, histopathological staging and prognostic attributes are probably not altered if only a few days elapse between the beginning of radiation therapy and subsequent surgery. The more prolonged courses of radiation therapy allow potential sterilization of greater numbers of tumor cells, and more closely approximate the radiation doses and schedules shown to be effective for the treatment of subclinical disease in other areas of the body.

The volume irradiated has also varied considerably and has ranged from small fields of about 10 × 10 cm designed to include only the primary tumor and immediately adjacent tissues to volumes extending from the perineum to the origin of the superior hemorrhoidal artery and intended to encompass all of the lymphatic-bearing tissue to be resected.

Both gross and microscopic changes have been observed in the primary tumor and regional nodes in many of the studies, although such changes are difficult to quantitate. Failure to identify any histological change does not necessarily mean radiation therapy has been ineffective, since biologically significant alteration in cell viability is generally not apparent until the cell enters mitosis one or more cell cycles after being irradiated. As discussed elsewhere, gross regression of rectal carcinomas after irradiation may take several months (CUMMINGS et al. 1983) whereas only a few days or weeks usually separate radiation and surgery in preoperative adjuvant therapy studies. Even so, reductions in the size of the primary tumor and in the number of both normal and involved lymph nodes have been observed within a few days after radiation doses as low as 20 Gy in 2 weeks (HIGGINS et al. 1975; SECOND REPORT of an MRC WORKING PARTY, 1984). Reduction in the size of the primary tumor and nodes as a result of irradiation may alter the histopathological stage when surgery is eventually performed, and hence alterations in survival rates or in patterns of failure must be sought only in the overall patient groups, or in subgroups matched by features other than the histopathological stage.

There is little evidence from the randomized trials that doses up to 35 Gy in 3 weeks improve survival significantly (Table 4). A meta-analysis of the randomized trials published in English up till December 1986, none of which had used a dose of more than 34.5 Gy in 3 weeks, found that 5-year survival was improved by only 2.9% in the patients who had received preoperative radiation (BUYSE et al. 1988). Higher doses show more promise, although such results require confirmation in larger randomized studies.

Even if survival rates were not markedly improved, a reduction in the morbidity associated with pelvic recurrence would be regarded as a successful outcome of adjuvant irradiation. The patterns of tumor recurrence also do not appear to have been affected significantly by low dose radiation therapy. However, the EORTC trial has demonstrated a significant reduction in pelvic recurrence rates from 35% to 15% following 34.5 Gy in 3 weeks (GERARD et al. 1985), and higher dose nonrandomized studies (PUJOL et al. 1978; STEVENS et al. 1979) point in the same direction, with none of the 44 patients in the Oregon series who received 50 Gy or more in 5 weeks suffering local recurrence (STEVENS et al. 1979). MOHIUDDIN and MARKS (1987) have demonstrated that restorative anastomoses may be performed after resection of very low lying rectal cancers with a reduced risk of local recurrence following doses of 45 Gy in 5 weeks. There is no evidence to suggest that partial tumor regression as a result of preoperative radiation therapy has led to inadequate resection margins and higher local recurrence rates in any series.

The limited data available on the rates of distant metastases after preoperative irradiation do not suggest any improvement as judged by the randomized studies. The low metastatic rates reported in high dose nonrandomized studies remain to be confirmed (PUJOL et al. 1978; STEVENS et al. 1979).

Postoperative mortality within 30 days of surgery has ranged from about 1% to 12%, and in none of the randomized studies apart from that in Stockholm (STOCKHOLM RECTAL CANCER STUDY GROUP 1987) has the difference between irradiated and nonirradiated patients been significant. In general, acute toxicity from the radiation therapy and from the subsequent surgery has not been severe. The increased mortality in the Stockholm study occurred mainly in patients over 75 years of age. The effects of radiation therapy on the healing of surgical incisions and restorative anastomoses are discussed in Sect. 10.3.7.4. There have been no reports of any increase in late radiation-induced bowel or bladder toxicity, perhaps because most studies have not involved high radiation doses.

The promise of preoperative irradiation is so far largely unrealized. Randomized trials of doses up

Table 5. Postoperative radiation therapy for completely resected primary rectal cancer

	Patients	Stages included (Astler-Coller)	Radiation dose (total dose, Gy/fractions/days)	Volume irradiated	Overall 5-year survival		Pelvic recurrence		Comment
					SRT	S	SRT	S	
Randomized series									
GASTROINTESTINAL TUMOR STUDY GROUP (1985)	227	B2C1C2	40–48/20–26/32–39	Whole pelvis to L5	50%	45%	20% (10/50)	24% (14/58)	First recurrences only recorded
			40–44/20–24/32–39 plus chemotherapy	Whole pelvis to L5	58%		11% (5/46)		
FISHER et al. (1988)	555	B2C1C2	46–47/26–27/35–39 plus 5–6/3–4/3–4	Whole pelvis to L5 Small volume boost	40%	43%	16% (30/184)	25% (45/184)	First failures only. Pelvic recurrence rates $P=0.06$
BALSLEV et al. (1986)	494	B2C1C2	50/25/49 split course	Posterior pelvis to L5	No significant difference expected		16% (38/244)	18% (44/250)	Interim analysis
Nonrandomized series									
FEIGEN et al. (1988)	97	B2C1C2	24/12/16	Posterior pelvis	41%		49%		Actuarial analysis
WITHERS et al. (1981)	73	B2C2	45/25/35 plus 6/3/3	Pelvis Small volume boost	No difference		9%	25%	Historical controls, actuarial analysis
TEPPER et al. (1987)	162	B2C1C2	45/25/35 plus 5.4–7.2/3–4/3–4	Posterior pelvis Small volume boost	57% (disease-free survival)	39%	19%	39%	Historical controls, actuarial analysis

SRT, surgery followed by radiation therapy; S, surgery

to 31 Gy in 3.5 weeks have failed to demonstrate improvements in recurrence rates or in survival. The initial report of the EORTC trial in which 34.5 Gy in 3 weeks produced a significant decrease in pelvic recurrence rates and an increase in survival in those undergoing complete tumor resection (GERARD et al. 1985, 1988), and the low recurrence rates associated with nonrandomized higher dose studies, suggest that further trials with radiation doses equivalent to 45 Gy in 4–5 weeks or more are needed.

10.3.7.1.2 Postoperative Radiation Therapy. If radiation therapy is deferred until after resection, patients can be selected on the basis of the correlations established between the histopathological stage and the risk of pelvic recurrence. Many investigators exclude patients in whom the cancer has not either penetrated completely through the rectal wall or involved the regional nodes. From 3% to 35% (GILBERT 1978; MOOSSA et al. 1975) of patients would be excluded by these criteria. However, when the cancer arises in the extraperitoneal rectum, or is poorly differentiated, the risks of local recurrence are relatively higher, and some investigators also treat these patients. Also, while broad categories of risk based on histopathological criteria have been defined, one effort to use operative and histopathological findings to predict the risk of local recurrence in individual patients produced results no better than random selection (MOOSSA et al. 1975). The potential disadvantages of postoperative radiation therapy are the risks associated with the irradiation of the increased amounts of small bowel usually present in the pelvis after removal of the rectum; the theoretical possibility of relative hypoxia, and therefore of relative radioresistance, of residual cancer cells due to vascular disruption during surgery; and the fact that any tumor cells released outside the pelvis during surgery will be beyond the limits of the radiation fields.

Most studies of postoperative adjuvant irradiation have used medium to high doses of 45- to 55-Gy megavoltage external beam therapy in 5–6 weeks, and the volume irradiated has generally been confined to the posterior pelvis. Novel approaches to adjuvant therapy have included the injection of radiocolloids into lymphatics (ADAMS et al. 1970) or interstitially (BRUSILOVSKY and KARLACHENKO 1981).

Several studies, both randomized and nonrandomized, have concluded that postoperative external beam irradiation can reduce the risk of pelvic recurrence to about 10%–20% (Table 5). However, at 24 months in the Danish national study local recurrence rates were significantly decreased only in those who had had nodal metastases, and these authors also have not so far found any reduction in local recurrence rates in those who had had fixed tumors (BALSLEV et al. 1986). Neither the randomized trial conducted by the GASTROINTESTINAL TUMOR STUDY GROUP (1985) nor the first reports from the Danish national study (BALSLEV et al. 1986) have yet shown a significant survival advantage. The National Surgical Adjuvant Breast and Bowel Project (NSABP) trial has shown a statistically significant ($P = 0.05$) increase in survival following treatment with adjuvant 5-fluorouracil (5-FU), methyl-CCNU, and vincristine, while pelvic radiation reduced the risk of local recurrence but did not improve the survival rate (FISHER et al. 1988).

The main toxicity reported has been late radiation enteritis requiring laparotomy. The rates have not been excessive, provided the radiation fields and doses have been restricted (see Sect. 10.3.7.4).

There are a large number of studies still in progress which should resolve the role of postoperative adjuvant radiation. A Swedish study comparing preoperative with postoperative radiation has been opened, but it has no surgical control arm and the radiation doses in the two study groups are not identical (PAHLMAN et al. 1985b). WINKLER (1985) has compared preoperative, postoperative, and combined radiotherapy programs but without a surgical control group.

10.3.7.1.3 Combined Preoperative and Postoperative Radiation Therapy. The concept of "sandwich" therapy has been studied by a few investigators who wished to exploit the theoretical benefits of low dose preoperative irradiation (NIAS 1967; POWERS and TOLMACH 1964), and to minimize interference with surgicopathological staging parameters so that high dose radiation therapy could be given only to those found to have poor prognostic features. It is difficult to draw any conclusions from these studies, most of which were not randomized (Table 6). The larger trials of low dose preoperative radiation (SECOND REPORT of an MRC WORKING PARTY 1984; RIDER et al. 1977) showed no benefits in overall survival or recurrence patterns, and the results obtained with the "sandwich" programs are consistent with studies of postoperative radiation therapy in general. When approximately equal doses of radiation have been given preoperatively and postoperatively, patients are said to have tolerated the split course therapy well (WINKLER 1985), but there is no strong suggestion that this approach offers any other benefits over alternative adjuvant regimens.

10.3.7.1.4 Combined Radiation and Chemotherapy. Because both pelvic recurrence and systemic metastases contribute to failure patterns in so many patients, there have been several studies in which radiation therapy has been combined with systemic adjuvant therapy. The results of trials of systemic adjuvant therapy alone are discussed in Chap. 12. The drug which has been studied most frequently in

Table 6. Combined preoperative and postoperative irradiation for primary resectable rectal cancer

	Preoperative radiation (total dose, Gy/fractions/days)	Postoperative radiation (total dose, Gy/fractions/days)	Local recurrence		Follow-up (months)
			Preoperative only AB1[a]	Preoperative plus postoperative B2C1C2[a]	
Randomized series					
WINKLER (1985)	25/10/12	25/10/12	–	4/97 (4%)	Interim analysis
	25/10/12	–	–	10/96 (10%)	
	–	50/20/28	–	5/86 (6%)	
Nonrandomized series					
MOHIUDDIN et al. (1984)	5/1/1	45/25/35	1/29 (3%)	2/31 (7%)	24–48
GUNDERSON et al. (1983b)	5/1/1 or 10/5/5	45–50/25–28/35–42	2/16 (13%)	0/14 (0)	26+
SHANK et al. (1987)	15/5/5	41.4/23/31	1/13 (8%)	0/22 (0)	17–40
BAYER et al. (1985)	30/10/12	30/15/19	2/29 (7%)	7/28 (25%)	3–52
BRENNER et al. (1980)	20/NS/10	26/NS/NS plus 5-FU and CCNU	–	0/26 (0)	14–42

NS, not stated; [a] Astler-Coller stages.

combination with radiation therapy is 5-FU, mainly because it is relatively more effective against colorectal cancer than most other agents, and in early studies the combination appeared more effective than irradiation alone for the treatment of advanced colorectal cancer (MOERTEL et al. 1969). In addition, some laboratory studies suggest that 5-FU and radiation may interact synergistically (BYFIELD et al. 1982), although this is not always the case (STEEL 1988).

There have as yet been no reports of any large randomized trials in which preoperative irradiation was combined with chemotherapy and compared with surgical resection alone. The EORTC did conduct a randomized trial in which preoperative irradiation (34.5 Gy in 15 fractions in 19 days) was compared with similar irradiation combined with 375 mg/m^2 5-FU by bolus intravenous injection on each of the first 4 days of irradiation (BOULIS-WASSIF et al. 1984). Acute toxicity and postoperative mortality rates were slightly higher in those who received both radiation and chemotherapy. The corrected survival rates were similar, as were the rates of local recurrence and metastasis. Nonrandomized studies of preoperative radiation therapy combined with 5-FU and mitomycin C showed reductions in the size of the primary tumor, but direct comparisons with preoperative radiation therapy alone, or with surgical resection alone, have not been reported (BUROKER et al. 1976; SISCHY et al. 1985).

Several randomized studies of postoperative combined therapy have been reported and others are in progress. The GASTROINTESTINAL TUMOR STUDY GROUP (1985) trial of postoperative adjuvant therapy included a treatment option of 40–44 Gy in 4.5 weeks with intravenous 5-FU 500 mg/m^2 on each of the first and last 3 days of radiation therapy, followed by 18 months intermittent 5-FU and methyl-CCNU. This schedule was associated with more acute gastrointestinal and hematological toxicity than either irradiation or chemotherapy alone, but only 5 of 46 (11%) patients developed local recurrence as a site of initial failure, compared to 14 of 58 (24%) in the surgical control arm and 10 of 50 (20%) treated with adjuvant radiation therapy alone. Overall survival at 7 years was significantly better ($P = 0.005$) for the combined treatment group compared to the surgical control arm (GASTROINTESTINAL TUMOR STUDY GROUP 1986). The preliminary report from another randomized trial has also suggested that a combination of postoperative pelvic irradiation, 50 Gy in 5 weeks, and 5-FU and methyl-CCNU can reduce both overall and local recurrence rates compared to irradiation alone (KROOK et al. 1986). A smaller trial from Israel suggested improved survival rates in patients who received postoperative irradiation, 5-FU, and MER-BCG compared with those who received only irradiation and 5-FU (ROBINSON et al. 1982).

Some studies of combinations of radiation and chemotherapy have been associated with increased toxicity, especially in the intestinal tract, and such toxicity was sufficiently serious to lead to one randomized trial being abandoned (SPARSO et al. 1984).

Since systemic metastases play such a major role in determining the outcome in cancer of the rectum, it is likely that there will be many more studies combining systemic and regional adjuvant therapy.

10.3.7.2 Primary Radiation Therapy

The role of radiation therapy as primary treatment for rectal cancer is relatively limited, but it has been demonstrated that both intrarectal and interstitial therapy, and external beam therapy can contribute to efforts to cure rectal cancer and to conserve normal anorectal anatomy and function.

10.3.7.2.1 Intrarectal and Interstitial Therapy Radiation therapists have worked in parallel with surgeons to identify patients in whom local procedures are appropriate. The history of the development of intrarectal contact X-ray therapy and interstitial radiation therapy has been reviewed by PAPILLON (1982). Both the natural history of rectal cancer and the physical characteristics of the available equipment determine the selection of patients.

The criteria for patient selection are similar to those used in choosing patients for local surgical procedures, since the objective is to identify patients at low risk of having nodal metastases. The most important predictors of lymphatic spread appear to be the depth of infiltration into the bowel wall and the tumor grade, together with the size and general configuration of the tumor (see Sect. 10.3.2.2). The extent of spread within the rectal wall can be assessed quite accurately by digital evaluation of mobility, as evidenced by the excellent results reported from several centers. The principal advantage of intracavitary irradiation in comparison with local excision is that the treatment can be given on an outpatient basis and without general anesthesia, and with minimal risk of perforation of the rectum, and the main disadvantage is that the whole specimen is not available for histological examination.

Table 7. Intrarectal irradiation

	Number	Failure		Surgical salvage	Deaths from rectal cancer	Survival	Comment
		Local	Nodal				
PAPILLON (1982)	158 polypoid	4%	4%	8/11	10%	73% 5-year crude	All patients followed at least 5 years
	49 ulcerated	10%	10%		12%	69% 5-year crude	
SISCHY (1985)	129	5%	NS	NS	2%		Follow-up 1–132 months
BASRUR and KNIGHT (1983)	23	13%	NS	2/3	4%		Follow-up 3–43 months
LAVERY et al. (1987)	62	18%	NS	8/11	10%		Follow-up 1–10 years
PARTURIER-ALBOT (1981)	1136 mucosa, submucosa	5%	NS	50/50 (80%)	NS	NS	All patients followed at least 5 years.
	532 muscularis propria	80%	NS	375/417 (79%)	NS	NS	Depth of cancer penetration based on clinical evaluation and post-RT biopsy excision specimens.
	377 serosa or beyond	99%	NS	241/318 (36%)	NS	NS	Bracketed percentages are proportions in whom surgery was sphincter saving.

NS, not stated

The total radiation dose given by contact X-ray therapy appears high, a total exposure of 10000–14000 R being delivered in four or five divided doses over 6–8 weeks. However the rapid falloff in dose from the 50-kVp tube, and the short FSD of about 4 cm, is such that the dose at 1 cm is approximately 25%, and at 2 cm only 10% of the applied dose (PAPILLON 1982). Tumors treated in this way usually shrink rapidly, so that the timing and size of doses can be adjusted individually. Because of the low penetration of the 50-kVp beam, and the difficulty in getting good apposition of the applicator to some tumors, PAPILLON (1982) recommended the use of interstitial therapy to supplement intracavitary therapy for indurated ulcerated cancers. Two iridium 192 wires, usually of 4 cm active length, were inserted into the base of the tumor under local anesthesia 1 month after the last intracavitary application to deliver a dose of 20–30 Gy in 24 h. Cancers up to about 3 × 5 cm in diameter, and within the distal 8–10 cm of the rectum have been treated successfully by these techniques. The treatment is well tolerated and PAPILLON (1982) reported temporary superficial necrosis in only 8% of cases. Telangiectasia at the site of treatment may later cause slight rectal bleeding (SISCHY 1985). The equipment is safe to operate provided simple precautions are taken (PURDY et al. 1985).

The most extensive experience with these techniques has been reported by PAPILLON (1982), although others have reported similar rates of success in smaller series (Table 7). The rate of local or regional node failure has been less than 10%, and salvage surgery has often been successful, so that corrected survival rates are 90% or better. The intensive radiation treatment used by PARTURIER-ALBOT (1981) differs from the more protracted approach favored by PAPILLON (1982). PARTURIER-ALBOT (1981) gave high doses of 10000–15000 R contact X-ray therapy in only one or two treatments separated by 8–15 days, and did not use interstitial therapy. Tumors which were considered to have penetrated the muscularis propria when first examined were removed by excision biopsy 2 or 3 months later. Contact irradiation alone was considered to have sterilized 95% (1075/1136) of tumors confined to the mucosa and submucosa, but only 12% (113/909) of those which extended more deeply. Surgical excision, using techniques which preserved the anal sphincter in 62%, resulted in cure in 85% of the remaining cases, so that the overall cure rate was 91% (1854/2045). Local excision in this series appears to have served a purpose similar to the interstitial therapy recommended by PAPILLON (1982).

PAPILLON (1984, 1987) has also reported a technique in which the indications for intracavitary irradiation may be extended for patients who are poor surgical risks, with distal rectal cancers too thick or large for primary intracavitary treatment. Initially, external beam radiation to a dose of 30 Gy (tumor maximum dose 39 Gy) in ten fractions in 12 days was given with cobalt 60 by a posterior arc technique. Eight weeks later, the patient was reevaluated, and if possible treated by conservative radiation therapy with intracavitary and/or interstitial therapy, and otherwise referred for radical surgery. With this approach 37 of 45 patients, followed a minimum of 3 years, had control of their rectal cancer by conservative radiation therapy and a further 50 who had received the initial external beam radiation underwent radical surgery. Variations on this approach have also been described (SISCHY 1985).

Techniques combining various combinations of intracavitary, interstitial, and external beam therapy have been applied successfully to selected adenocarcinomas which arise at the anorectal junction (PAPILLON 1982), and for carcinomas arising in villous adenomas or for large villous adenomas which cannot be resected (PAPILLON 1982; SISCHY 1985). All authors who have used intracavitary irradiation have reported that it can be used effectively for palliation also.

Interstitial therapy is used infrequently as the sole method of treating rectal cancer because anatomical factors make application relatively difficult, and because of the bulk of many rectal cancers and the risk of nodal metastases. However, techniques have been developed, particularly using templates to facilitate multiple needle implants whereby interstitial therapy was used to supplement intracavitary therapy (PAPILLON 1982), or external beam therapy (PUTHAWALA et al. 1982) or electrocoagulation (WASSINK 1956).

Conservative treatments, whether surgical or radiotherapeutic, which are directed solely to the primary rectal tumor do not offer any treatment to the regional lymph nodes. Through case selection, the risk of failure in the regional nodes in most reported series has been less than 5%. In an effort to reduce the risk of failure due to the progression of unrecognized lymph nodes, and to extend the indications for conservative local treatment of the primary cancer, particularly in younger patients, two approaches have been suggested. Elective pelvic lymphadenectomy has been proposed, and used either as elective prophylactic treatment or as therapy for patients with clinically abnormal pararectal lymph nodes (MAYER et al. 1982; WASSINK 1956). MAYER et al. (1982) have demonstrated that after intracavitary radiation therapy it is possible to remove the superior hemorrhoidal, pararectal, and middle hemorrhoidal lymph nodes without compromising the vascular supply of the rectum. An alternative approach is to combine conservative treatment of the primary rectal tumor by intrarectal irradiation (PAPILLON 1984) or local excision (RICH et al. 1985b) with external beam irradiation which covers the regional lymph nodes. Neither of these treatment programs has yet been tested widely.

10.3.7.2.2 External Beam Irradiation. There have been relatively few studies of the effects of radical external beam radiation therapy other than as adjuvant therapy or for the management of inoperable rectal cancers. However, there is evidence that some mobile cancers can be controlled by high dose external beam radiation therapy alone. Between 1958 and 1977, the Princess Margaret Hospital group (CUMMINGS et al. 1983; RIDER 1975) treated 61 patients who had clinically mobile rectal cancers and who were unable or unwilling to undergo major surgical procedures. The usual radiation dose was 50 Gy in 20 fractions in 4 weeks. Of the 61 patients, 24 (39%) had their cancer controlled by radiation therapy alone, and the 5-year uncorrected survival rate was 37%. One patient died of acute radiation enteritis. Other authors have described similar results in small groups of patients (AMALRIC et al. 1972; SKLAROFF 1973; TAYLOR et al. 1987). It is not possible to compare the cure rates in these series with those of surgically treated patients because of patient selection and the lack of a clinical staging system. Because the majority of the patients treated in Toronto were not eligible for resection, there was an opportunity to observe the natural history of the irradiated tumors. The regression rate was often slow so that complete regression took many months. Thus, 3 months after radiation therapy, complete regression had occurred in only 12 of 57 (21%) measurable tumors, and complete regression sometimes took as long as 6–7 months. These observations are compared with those from preoperative adjuvant radiation therapy series in Fig. 1, where it can be seen that the relatively smaller number of patients with complete tumor regression in these series is consistent with the shorter intervals between the end of radiation therapy and surgical resection. It is possible that this effect might be exploited in the conservative treatment of rectal cancer by extending the period of observation after high dose irradiation, and then treating residual rectal tumor by the techniques described by PAPILLON (1984).

10.3.7.2.3 Unresectable Primary Carcinomas. The results of irradiation alone in the 10% of patients who present with tumors which are unresectable by virtue of attachment of adjacent pelvic tissues are unsatisfactory, with 5-year survival rates of 15% or less (ARNOTT 1975; CUMMINGS et al. 1983; TIERIE 1978; WANG and SCHULTZ 1962). Consequently there has been greater interest in the possibility that radiation therapy might cause sufficient regression to render these tumors resectable. The interpretation of the literature is difficult because of the lack of generally accepted criteria for resectability. The problems in comparing series were well illustrated by NEWMAN and STEARNS (1975), who reported that

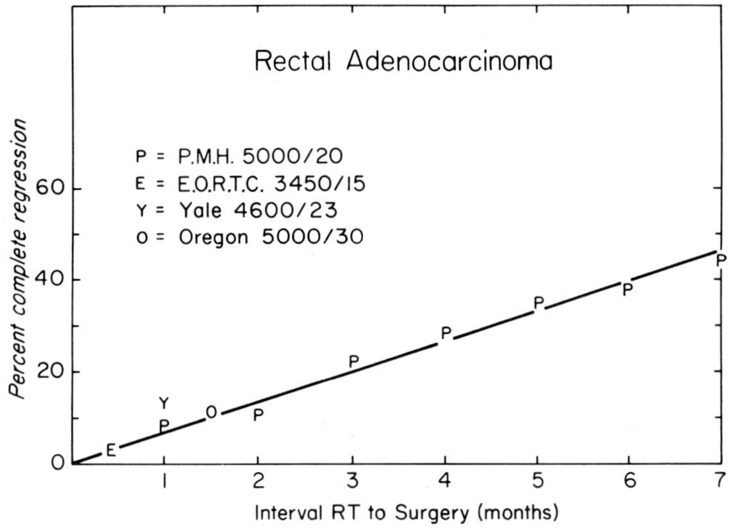

Fig. 1. Time course of complete regression of mobile primary rectal adenocarcinomas following external beam radiation. *P* = observations in 57 patients treated by radiation alone, 50 Gy in 20 fractions in 4 weeks. Complete regression occurred in 44% by 7 months after irradiation (CUMMINGS et al. 1983). In preoperative adjuvant radiation therapy series, no residual tumor was found at the time of planned surgery in 3 of 121 (3%) after 34.5 Gy in 15 fractions in 3 weeks, interval to surgery 2 weeks (BOULIS-WASSIF et al. 1984 = *E*); in 4 of 31 (13%) after 46 Gy in 23 fractions in 4.5 weeks, interval to surgery 4 weeks (KLIGERMAN et al. 1972 = *Y*); and in 5 of 50 (10%) after 50 Gy in 30 fractions in 6 weeks, interval to surgery about 6 weeks (STEVENS et al. 1976 = O)

Table 8. Preoperative radiation therapy for unresectable rectal cancer

Reference	Dose	Complete resections/ number irradiated	Postoperative mortality	Pelvic recurrence/ number resected	Survival
STEVENS et al. (1976)	50–62 Gy/5.5–7 wk	11/35 (31%)	0	2/11	12% 5 yr overall
KLIGERMAN and URDANETA-LAFEE (1974)	35–58 Gy/4–7 wk	9/15 (60%)	1/15	4/9	20% 5 yr overall
DOSORETZ et al. (1983)	40–52 Gy/4–6 wk	16/25 (64%)	3/19	5/13	28% 5 yr overall, 43% curative surgery
JAMES and SCHOFIELD (1985)	20 Gy/1 wk to 45 Gy/4 wk	18/42 (43%)	6/29	NS	NS
TIERIE (1978)					
Unresectable	64 Gy/6.5 wk	0/32 (0%)	–	–	0% 5 yr
Marginally resectable	35 Gy/3 wk	20/21 (95%)	1	3/20	50% 3 yr overall
TEPPER et al. (1986)	50.4 Gy/ 5.5 wk + 10–20 Gy intraoperative	18/29 (62%)	0	13% actuarial rate 8% curative surgery	60% 3 yr overall, 70% curative surgery
MELLA et al. (1984)	31.5 Gy/3 wk, reassess after 3 wks, further RT then or postoperatively	15/55 (27%)	2	NS	NS
PAHLMAN et al. (1985a)	46 Gy/4.5 wks, reassess after 3 wks, surgery or further RT + 5-fluorouracil	2/15 (13%)	1	NS	NS

NS, not stated

they were able to perform curative resections in about 60% of patients referred from other centers with a diagnosis of unresectable rectal cancer.

Some of the results obtained from resection following preoperative irradiation are shown in Table 8. Several of these series also include patients with sigmoid or rectosigmoid cancers. Although some 50%–100% of patients have undergone laparotomy subsequent to radiation therapy, there is a wide variation in the likelihood of resection and while many series suggest a resectability rate of 30%–65%, at least one center reported that not a single tumor became resectable (TIERIE 1978). This latter series illustrates the problems of differences in categorization since, if that center's series of marginally resectable tumors is combined with their unresectable cases, then the resectability rate after irradiation was 38%. Criteria for categorizing resectability have been suggested by DOSORETZ et al. (1983). Additional irradiation beyond the initial course has not increased resectability rates appreciably (BJERKESET and DAHL 1980; MELLA et al. 1984; PAHLMANN et al. 1985a). Postoperative mortality rates in these series were no greater than those for primary resection. Morbidity rates were moderate, with the most frequent problems being delays in healing, abscesses, and fistula formation especially between bowel and urinary tract (EMAMI et al. 1982; JAMES and SCHOFIELD 1985; STEVENS et al. 1976).

However, even though resectability rates appeared to be increased, recurrence occurred in the pelvis despite radiation and resection in up to 50% of the patients in these series (Table 8). In efforts to reduce this risk, some centers have evaluated intraoperative boost radiation therapy following preoperative external beam irradiation (GUNDERSON et al. 1983a, 1988; TEPPER et al. 1986) or have given additional postoperative external beam therapy (BJERKESET and DAHL 1980). In these series, when complete resection was possible the reported local recurrence rates were less than 15%, but because of the different selection criteria it is impossible to be certain that this improvement was due solely to the higher radiation dose. There seems to be little theoretical or real advantage to extending the radiation field beyond the pelvis, and JAMES and SCHOFIELD (1985) suggested that toxicity was greater when the whole rather than the posterior pelvis was irradiated.

When residual tumor remained in the pelvis after attempted primary resection, there is some evidence that microscopic residual disease was more likely to be controlled by high doses of 60–65 Gy in 6–7 weeks (ALLEE et al. 1981; GHOSSEIN et al. 1981; ROMINGER et al. 1985). However, there was little suggestion of improved response to higher doses when gross tumor remained, and the cancer eventually regrew in most patients.

10.3.7.2.4 Pelvic Recurrence After Radical Surgery
Many of the problems encountered by the patient who presents with an unresectable primary rectal carcinoma are similar to those due to pelvic recurrence after previous surgery. Cure by radiation therapy alone is rare, yet symptoms such as pelvic pain and tenesmus, bleeding, and discharge are not palliated by colostomy alone. Although most authors agree that irradiation is frequently effective therapy for symptomatic pelvic tumors, it is also clear that the duration of response is usually only about 3–6 months (JAMES et al. 1983; VILLALON and GREEN 1981; WILLIAMS et al. 1957) and the symptom-free interval is often only about one-third of the length of overall survival after recurrence is diagnosed (PACINI et al. 1986). Symptomatic response rates range from about 50% to 90%, with objective responses usually being incomplete and reported in from 25% to 60% of patients. The results obtained by ARNOTT (1975) are typical. In his series of 325 patients with primarily inoperable or locally recurrent cancer, two-thirds were adjudged suitable for palliative radiation only, 30–35 Gy in 10 fractions over 2 weeks, and the remainder received radical treatment, 55 Gy in 20 fractions over 4 weeks. Within the whole group, complete or partial symptomatic responses were found in 75% with pain and 60% with discharge, and the tumor itself regressed completely in 10% and partially in 55%. Amongst those who received the higher radiation dose, 32% of the primarily inoperable cancers and 50% of the recurrent tumors regressed completely. However, the 5-year survival rate was only 3% in recurrent cases, compared to 16% in those with primarily unresectable cancer.

The retrospective nature of many studies, and the inevitable case selection, make analysis of dose-response factors difficult. Although some have suggested that higher doses are more likely to lead to prolonged symptomatic relief and objective response (OVERGAARD et al. 1984; WANG and SCHULTZ 1962), others have not found any strong relationship between dose and symptomatic response (CIATTO and PACINI 1982; DOBROWSKY and SCHMID 1985; JAMES et al. 1983; PACINI et al. 1986). Efforts to deliver higher doses to localized areas of recurrence have included intraoperative boost radiation therapy and interstitial irradiation (GUNDERSON et al. 1983a). While high doses are necessary to produce the complete responses which are a prerequisite for possible cure (ARNOTT 1975; OVERGAARD et al. 1984), the majority of these patients show evidence of extrapelvic disease within about 2 years (OVERGAARD et al. 1984). Because of this, some authors have suggested that it is more appropriate to give short courses of low dose irradiation of about 20 Gy in 2 weeks, and to repeat this when necessary (WHITELEY et al. 1970). In an effort to identify those patients most likely to benefit from high dose irradiation, ARNOTT (1983) studied CEA levels and found that there seemed to be little advantage in terms of tumor response or survival from radical irradiation even when tumor was apparently confined to the pelvis if the CEA level was greater than 75 ng/ml. The variety of opinions on how to manage these patients is undoubtedly contributed to also by the heterogeneity of methods of assessing response rates and their duration.

The role of several experimental techniques has still to be resolved. As would be expected, these have so far been evaluated principally in patients with advanced tumors. Improvements in tumor response and control have been sought through efforts to overcome the radioresistance of the hypoxic tumor cells identifiable in rectal cancers (WENDLING et al. 1984), by irradiation under hyperbaric oxygen (DISCHE and SENANAYAKE 1972), hypoxic cell sensitizers (PHILLIPS et al. 1982), and neutron irradiation (BATTERMAN 1982; BRETEAU et al. 1986; DUNCAN et al. 1987); through efforts to deliver higher radiation doses with intraoperative boost therapy (GUNDERSON et al. 1983a; 1988; TEPPER et al. 1986) or with radiolabeled antibodies targetted to cancerous tissue (BARTH et al. 1986; BEGENT et al. 1987; HAMMERSMITH ONCOLOGY GROUP 1984); and through the combination of hyperthermia with radiation (BERDOV and MENTESHASHVILI 1984; MANNING et al. 1982; SAPOZINK et al. 1986). Varying degrees of tumor response have been reported and more detailed trials have been planned for some of these approaches.

There have been several attempts to improve the duration and quality of response in advanced cancers by combining irradiation with chemotherapy or immunotherapy. Although MOERTEL et al. (1969) reported a statistically significant improvement in the duration of symptomatic control and in mean survival in patients with locally unresectable large bowel cancer who received 35-40 Gy in 4 weeks, plus intravenous 5-FU 15 mg/kg daily on the first 3 days of radiation therapy, symptom control still lasted only about 17 months and the technique has not been used widely. Others have used higher doses of radiation and 5-FU in nonrandomized studies and also concluded that combined therapy was better (BOULIS-WASSIF 1983; VONGTAMA et al. 1975). However, negative results were reported from a trial in which patients with residual, inoperable, or locally recurrent carcinoma were randomized to receive either radiation therapy alone (45-51 Gy in 5-6 weeks, with a boost to a maximum of 70 Gy in 8 weeks when the position of the small bowel allowed) or similar radiation therapy (to a maximum of 60 Gy in 7 weeks) with intravenous bolus 5-FU 500 mg/m^2 on each of the first 3 days of radiation plus maintenance 5-FU and methyl-CCNU (ROMINGER et al. 1985). There were no statistically significant differences between treatments with respect to survival, complete remission rates, or local failure rates. A randomized pilot study conducted by the Eastern Cooperative Oncology Group (ECOG) comparing 55 Gy in 6 weeks with high dose split course irradiation (20 Gy in 2 weeks repeated three times in 10 weeks), in which each patient received a bolus intravenous injection of 5-FU 500 mg/m^2 on each of the first 3 days of every course of radiation therapy, produced quite severe acute toxicity in the continuous course program, and although late toxicity rates were similar in each group, 7 of 30 (23%) patients developed serious late toxicity (DANJOUX et al. 1985). Multiple drug chemotherapy (5-FU, methotrexate, and cyclophosphamide) has also been combined with irradiation but was not compared with a simpler regimen (VON DER MAASE et al. 1982). The addition of nonspecific immunotherapy to radiation therapy did not improve symptomatic response or survival (O'CONNELL et al. 1982).

The possible influence of scheduling on any interaction of cytotoxic drugs and radiotherapy has been largely unexplored as yet. Most studies with 5-FU have used bolus intravenous injections. Some laboratory studies (BYFIELD et al. 1982) have suggested that continuous infusions of 5-FU may be more effective, and preliminary clinical studies have been designed to address this issue (BUROKER et al. 1976; RICH et al. 1985a).

As is the case with adjuvant therapy, it seems likely that the natural history of advanced rectal carcinoma will encourage further studies of combinations of regional and systemic therapy.

10.3.7.2.5 Metastases. The most common sites of metastases from rectal cancer are the liver, lung, bone, and central nervous system. Treatment with radiation can frequently palliate symptoms, but the limited tolerance of the normal tissues and the disseminated nature of the metastases generally preclude efforts at curative radiation therapy.

KINSELLA (1983) has reviewed the role of radiation alone or combined with chemotherapy in the treatment of liver metastases. From his survey of the literature, he concluded that a total dose of 20-30 Gy in 2 to 3 Gy fractions in 2-3 weeks to the entire liver was within liver tolerance and resulted in considerable pain relief in 75%-90% of patients, and in a temporary decrease in liver size and improvement in liver function tests in up to 40%. While most experience has been with external beam therapy, there has been some limited investigation of the use of radioactive microspheres to deliver higher radiation doses in selected patients (GRADY 1979). Studies of combinations of external beam radiation therapy or intrahepatic arterial infusion of radioactive microspheres with infusional chemotherapy, usually 5-FU or FUDR, have reported response rates of 40%-70% with acceptable toxicity rates (KINSELLA 1983). The administration of intravenous misonidazole concomitantly with hepatic irradiation (LIEBEL et al. 1981) appeared to give results similar to those of radiation alone.

The indications for radiation therapy, and the response, for metastases to the lung, central nervous system, or bone are similar to those for metastases from any primary carcinoma.

10.3.7.3 Radiation Therapy Techniques

The role of radiation therapy in the treatment of rectal cancer is still evolving so that few firm principles have yet been established. It appears, however, that the most satisfactory results have been obtained by doses close to pelvic organ tolerance, except where palliation was the objective.

Efforts to identify the most effective radiation schedules have so far been empirical, but some laboratory and clinical data are now available which may guide future studies. The radiosensitivity of colonic mucosa (and its ability to repair sublethal damage) is similar to that of other epithelial cells (WITHERS and MASON 1974). The potential doubling time of carcinomas of the rectum in man has been reported to range between 2 and 45 days (DESCHNER and LIPKIN 1976; TROTT and KUMMERMEHR 1985), which is substantially faster than the range of doubling times observed radiologically (BOLIN et al. 1983), the difference presumably being accounted for, at least in part, by cell losses. TROTT and KUMMERMEHR (1985) speculated that colorectal cancers might respond favorably to accelerated fractionation if their potential doubling time is only about 5 days, and preliminary studies in this field have started (PERESLEGIN et al. 1985). With conventional single daily fractions, DERDEL et al. (1985) suggested, on the basis of a nonrandomized study and a review of the literature, that daily fractions of 2.5 Gy may be more effective against rectal carcinoma than lower doses. The developing field of predictive assays of tumor radiosensitivity and radiocurability was surveyed by PETERS et al. (1986), and efforts are being made to use these assays in the treatment of rectal cancer (STREFFER et al. 1986).

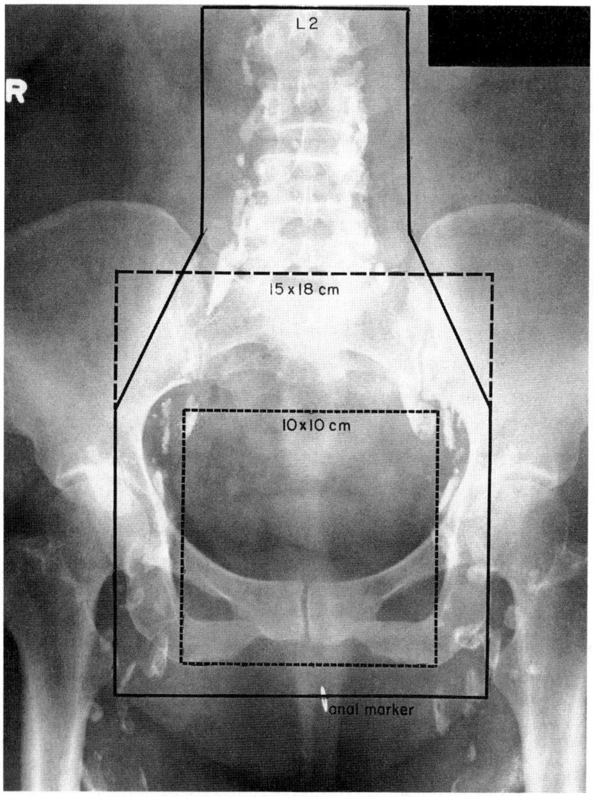

Fig. 2. The volume irradiated by various field sizes covering the primary rectal area only, 10 × 10 cm, the pelvis only, 15 × 18 cm, and the primary tumor and all regional nodes, including the pelvic and superior hemorrhoidal systems, perineum to second lumbar vertebra. (CUMMINGS 1984)

The general principles of radiation therapy planning have been reviewed by GUNDERSON and his colleagues (1985). Delivery of the higher doses usually recommended for inoperable or unresectable cancer, as opposed to the somewhat lower doses of 45–50 Gy in 5 weeks generally used for adjuvant therapy, may be possible only if areas at high risk can be localized by radiopaque clips and if sensitive small bowel can be excluded from the high dose radiation volume. A variety of techniques has been proposed to achieve this, including the interposition of omentum (RUSS et al. 1984) or artificial material (SUGARBAKER 1983) in the pelvis. The amount of small bowel in the pelvis can also be reduced by positioning the patient prone, and by using natural bladder distension to displace bowel (GUNDERSON et al. 1985). Since the risk of damage is proportional both to dose and the volume of small bowel irradiated (GALLAGHER et al. 1986; GREEN et al. 1975), small bowel series to localize the bowel and plan the radiation therapy are often useful, and shrinking field techniques have also been recommended (GUNDERSON et al. 1985). Patients who are thin, female, and in whom the small bowel is relatively immobile appear to be at particular risk of radiation enteritis (GALLAGHER et al. 1986; GREEN et al. 1975).

The use of high radiation doses generally precludes extension of the radiation fields up to the usual level of ligation of the superior hemorrhoidal vascular pedicle at around the second lumbar vertebra as has been done in several of the low and medium dose adjuvant studies (Fig. 2). It is still not clear whether any advantage is to be gained from extending fields beyond the primary rectal cancer and the immediate perarectal lymph nodes. However, the internal iliac nodes can readily be irradiated since they lie posteriorly in the pelvis. A few authors have suggested that the inguinal lymph nodes should also be treated when the primary cancer is in the distal rectum (MENDENHALL et al. 1985). The perineum is usually irradiated if an APR has been performed or is planned, although some suggest this may not be necessary (THOMAS et al. 1986b). Since recurrence of carcinoma in the perineum is usually accompanied by recurrence higher in the posterior pelvis (GUNDERSON and SOSIN 1974; VONGTAMA et al. 1975), irradiation of the perineum alone is not usually indicated. For these reasons, most radiation therapy plans are designed to encompass the posterior pelvis, and many possible field arrangements have been described, including multiple fields (GUNDERSON et al. 1985), arc rotation (PAPILLON 1984), and wedge techniques (DOBBS and BARRETT 1985). Computerized tomography may be useful for treatment volume localization (BATTISTA et al. 1980; DOBBS and BARRETT 1985). Studies of tissue density have shown that the net effect of tissue inhomogeneities in the pelvis is small, the differences between the predicted and measured radiation doses in the rectum being generally less than 2% (BATTISTA et al. 1980).

10.3.7.4 Toxicity of Radiation Therapy

General reviews of the morphological and functional aspects of tolerance of the intestine to radiation therapy may be found elsewhere (KINSELLA and BLOOMER 1980; RUBIN and CASARETT 1968). Concepts of the radiation dose levels which produce late radiation proctitis have generally been derived from the study of patients with genitourinary cancers, in whom only the anterolateral walls of the rectum usually receive the full radiation dose. Based on observations in such patients, a dose of 55 Gy in 5.5 weeks has been considered to carry a risk of less than 5% of serious injury at 5 years, while a risk of 25%-50% is associated with doses of about 80 Gy in 8 weeks (RUBIN and CASARETT 1968). Many other authors have suggested that the risk of late radiation proctitis rises steeply above doses of about 60-65 Gy in 6 weeks. Evaluation of the toxicity of treatment programs which include radiation therapy has become an important part of the assessment of all studies, to ensure that any benefits from such treatment are not offset by unacceptable levels of damage to normal tissues.

There are relatively few data on the tolerance of the irradiated rectum when part of the rectal wall has been involved by cancer, or on the tolerance of the rectum as it is reconstructed after restorative anastomosis. In one study of ten men with chronic radiation proctitis who had received 50 Gy/20 fractions/4 weeks for prostatic carcinoma, there were significant reductions in rectal volume at sensory threshold, in maximal tolerance, and in rectal compliance even in the absence of radiological abnormality, and histological studies suggested that smooth muscle hypertrophy and myenteric plexus damage contributed to the abnormal physiological findings (VARMA et al. 1985). When chronic radiation proctitis does occur and does not respond to medical management, a conservative surgical approach by diversion has usually been recommended (ANSELINE et al. 1981), although some suggest that resection and reanastomosis can be successful (MARKS and MOHIUDDIN 1983). The special problems of toxicity associated with high dose intraoperative radiation therapy in the pelvis have been discussed by TEPPER et al. (1984).

Considerable interest has been expressed in the influence of preoperative irradiation on the integrity of subsequent bowel anastomoses, and on tissue healing in general. Restorative anastomoses have been performed after varying doses of radiation (Table 9). It is not always clear whether the leaks reported in these series were of clinical importance, but it has been suggested that, after radiation doses equivalent to more than about 45 Gy in 4.5 weeks, some temporary protection of a restorative anastomosis is advisable. This protection may be provided by a temporary colostomy (STEVENS et al. 1978) or by short-term parenteral nutrition (GARY-BOBO et al. 1979). Several experiments in animals to test the reliability of colon anastomosis after irradiation have been performed (BUBRICK et al. 1984; DEGGES et al. 1983; MORGENSTERN et al. 1984), but they do not accurately reflect the clinical situation, and confirm only that anastomotic integrity decreases as the dose of prior radiation increases (CUMMINGS 1985). Several authors have noted some delay in the healing of the perineal incision following medium to high dose radiation (GERARD et al. 1985; GLIMELIUS et al. 1982). The EORTC reported a mean delay to healing of 60 days when APR was performed within 2 weeks of 34.5 Gy in 19 days, compared to 40 days in the control group (GERARD et al. 1985).

The risk of radiation-induced enteritis, as measured by the need for reoperation for small bowel obstruction, has been assessed principally in the postoperative adjuvant trials (Table 10). This risk has ranged from 4% to 10% compared to about 5% in the surgical control groups (BALSLEV et al. 1986; TEPPER et al. 1987). The risk of serious bowel toxicity was higher if the volume irradiated to high dose was extended above the pelvis to the level of the second lumbar vertebra, and if only anterior and posterior fields were used (WITHERS et al. 1981). As

Table 9. Colorectal anastomotic leaks following preoperative radiation therapy

	Radiation dose (total dose, Gy/fractions/days)	Interval radiation to surgery (days)	Leak rate		Comment
			RTS	S	
Randomized series					
WORKING PARTY MRC (1982)	5/1/1	Same day or up to 7	4/53 (8%)	21/70 (30%)	
	or 20/10/14	1–7	8/52 (16%)		
PORTER and NICHOLLS (1986)	15/3/3	<5	17/115 (15%)	15/116 (13%)	
STOCKHOLM RECTAL CANCER STUDY GROUP (1987)	25/5/5	1–7	6/126 (5%)	12/123 (10%)	
Nonrandomized series					
GARY-BOBO et al. (1979)	40/18/24	2–5	2/37 (5%)		Parenteral feeding in postoperative period
ROBERSON et al. (1985)	45/25/35	7–21	0/26 (0%)		No diverting colostomy
			0/14 (0%)		Diverting colostomy for usual surgical indications
STEVENS et al. (1978, 1979)	50–60/24–30/40–56	28–49	2/9 (22%)	1/79 (1%)	Contemporary unrandomized controls from same center

RTS, radiation therapy followed by surgery; S, surgery

Table 10. Laparotomy for small bowel enteritis following postoperative radiation therapy

	Radiation dose (total dose, Gy/fractions/days)	Radiation fields	Obstruction/perforation/ileus		Comment
			SRT	S	
Randomized series					
THOMAS et al. (1986a)	40–48/20–26/32–39	Anterior: posterior pelvis to L5	2/50 (4%)	NS	
	40–44/20–24/32–39 plus chemotherapy	Anterior: posterior pelvis to L5	3/46 (7%)	NS	2 fatal
BALSLEV et al. (1986)	50/25/49 split course	Posterior pelvis to L5	25/244 (10%)	9/250 (4%)	$P<0.001$
Nonrandomized series					
WITHERS et al. (1981), VIGLIOTTI et al. (1987)	45/25/35	Anterior: posterior pelvis; to L5-S1 (some to L2)	Fields above L5 4/16 (25%)	27/509 (5%)	Historical controls, same center
	plus 6/3/3	Lateral small field boost	Fields below L5 10/89 (11%)		
TEPPER et al. (1987)	45/25/35 plus 5.4–7.2/3–4/3–4	Four field posterior pelvis; boost to reduced volume	9/165 (5%)	7/142 (5%)	Historical controls, same center

SRT, surgery followed by radiation therapy; S, surgery; NS, not stated

might be expected, the risk of severe acute toxicity from concurrent chemotherapy and irradiation was greater in both the adjuvant trials (BOULIS-WASSIF et al. 1984; THOMAS et al. 1986a) and the advanced cancer treatment programs (ARNOTT 1975), and there is some evidence to suggest that late toxicity from these regimens is also greater (ROMINGER et al. 1985; SPARSO et al. 1984; THOMAS et al. 1986a).

The risk of induction of other malignancies by radiation used to treat rectal cancer appears to be low, although there are relatively few data as yet from large series with long-term follow-up (JAO et al. 1987).

10.3.8 Conclusion

Since the last edition of this volume, the role of radiation therapy in the treatment of rectal adenocarcinoma has become more firmly established. In the relatively small group of patients who present with superficial carcinomas, intrarectal irradiation offers one method of conservative management, and in selected cases with larger tumors external beam radiation therapy can also effect cure without the need for resections which sacrifice the anal sphincters. Adjuvant radiation therapy, either preoperative or postoperative, reduces the risk of pelvic recurrence, provided a sufficiently high dose is given, but has not yet improved survival rates significantly. The natural history of rectal cancer, with its high risk of extrapelvic metastases, and the failure of radiation therapy alone or combined with surgical resection to cure large numbers of those patients who present with extensive primary or recurrent pelvic carcinomas, suggest that the next major phase of development will come from studies which combine radiation and chemotherapy.

10.4 Uncommon Rectal Malignancies

Each of the conditions considered here is uncommon and makes up no more than 1% or 2% of all rectal malignancies. Most series are small, and assessment of the role of radiation therapy either alone or in combination with other modalities is based as much on the management of similar tumors which occur in other sites as on the treatment of those which have arisen primarily in the rectum.

10.4.1 Smooth Muscle Tumors

The distinction between leiomyoma and leiomyosarcoma on histological grounds is difficult and it has been suggested that even well differentiated tumors should be considered as of low grade malignancy. Direct local extension and hematogenous metastases are much more common than lymphatic spread. Five-year survival rates are typically about 25%. WALSH and MANN (1984) have suggested that the small polypoid and generally asymptomatic tumors which arise from the rectal muscularis mucosae can be treated adequately by local excision, but that local excision of the larger symptomatic lesions arising from the muscularis propria is followed by recurrence rates of 60%–80%. They therefore recommended radical excision for such lesions, since local recurrences were rarely salvaged. A comparison between smooth muscle tumors of the rectum and elsewhere in the gastrointestinal tract has been presented by LEE (1986). There does not appear to be any information on whether conservative excision coupled with high dose postoperative irradiation, in programs similar to those used for soft tissue sarcomas of the limbs, might be curative in this condition.

10.4.2 Carcinoids

Rectal carcinoids are found equally in men and women, and occur most frequently in the fifth and sixth decades. There is an increased risk of colonic adenocarcinoma in patients with rectal carcinoids (NAUNHEIM et al. 1983). The risk of metastases increases rapidly as the tumor grows beyond 1 cm in size. Five-year survival rates are about 90% when no metastases are present, about 45% in those with nodal metastases, and about 10% in those with distant metastases (GODWIN 1975). Excessive secretion of the agents which cause the various manifestations of the carcinoid syndromes is extremely rare from rectal carcinoids. The usual management is surgical resection. Radiation therapy may have some palliative value for symptomatic carcinoid tumor masses (KEANE et al. 1981).

10.4.3 Lymphomas

Lymphomas arising primarily in the rectum make up about 5% of all gastrointestinal lymphomas, but less than 1% of all primary rectal malignancies. In one series which separated the rectum as a distinct site, B cell follicular center cell lymphomas of the diffuse type were the most frequently identified subtype (VAN DER HEULE et al. 1982). Primary Hodgkin's disease is extremely rare in the rectum. Lymphomas which arise primarily in the rectum and cecum are said to have a poorer prognosis than lymphomas which develop primarily in other parts of the gastrointestinal tract. The general features of gastrointestinal lymphomas have been reviewed by PERREN and BLACKLEDGE (1986), who suggested a scheme of management which included excision, combined with radiation and/or chemotherapy for incompletely excised, disseminated, or high grade lymphomas. Primary anorectal lymphomas may occur more frequently as a manifestation of acquired immune deficiency syndrome (AIDS) (IOACHIM et al. 1987).

10.4.4 Squamous Cell Carcinomas

Only about 40 cases of this rare condition were found in the English literature by LAFRENIERE and

KETCHAM (1985). The tumor has been reported to occur with equal frequency in men and women, most commonly presenting in the fifth decade. Survival rates were difficult to assess from the scattered reports but, in general, the principles of treatment followed had been similar to those for rectal adenocarcinoma. Lymph node metastases were present in about 50% at diagnosis, and appeared to be an adverse prognostic feature. When restorative anastomosis is not possible, it would seem reasonable to treat squamous cell carcinomas which arise in the rectum similarly to epidermoid cancers of the anal canal (see Chap. 11), although there are as yet few data to support this opinion (LAFRENIERE and KETCHAM 1986).

Acknowledgements. The assistance of Miss E. Eisenreich and Miss G. Griffiths in preparing the manuscript and compiling the reference list is gratefully acknowledged.

References

Adams JT, Schwartz SI, Rubin P, Rob CG (1970) Intralymphatic 5-fluorouracil and radioactive gold as an adjuvant to surgical operation for colorectal carcinoma. Dis Colon Rectum 13: 201-206

Agostino D, Nickson JJ (1960) Preoperative X-ray therapy in a simulated colon carcinoma in the rat. Radiology 74: 816-818

Aldridge MC, Phillips RKS, Hittinger R, Fry JS, Fielding LP (1986) Influence of tumor site on presentation, management and subsequent outcome in large bowel cancer. Br J Surg 73: 663-670

Allee PE, Gunderson LL, Munzenrider JE (1981) Postoperative radiation therapy for residual colorectal carcinoma. Int J Radiat Oncol Biol Phys 7: 1208

Amalric R, Clement R, Juin P, Lipowsky G, Spitalier JM (1972) La radiotherapie des cancers du rectum. A propos de 100 cas. J Radiol Electrol Med Nucl 54: 613-616

Anseline PF, Lavery IC, Fazio VW, Jagelman DG, Weakley FL (1981) Radiation injury of the rectum. Ann Surg 194: 716-724

Armitage NC, Robins RA, Evans DF, Turner DR, Baldwin RW, Hardcastle JD (1985) The influence of tumor cell DNA abnormalities on survival in colorectal cancer. Br J Surg 72: 828-830

Arnott SJ (1975) The value of combined 5-fluorouracil and X-ray therapy in the palliation of locally recurrent and inoperable rectal carcinoma. Clin Radiol 26: 177-181

Arnott SJ (1982) Radiotherapy. Recent Results Cancer Res 83: 113-125

Arnott SJ (1983) Plasma carcinoembryonic antigen (CEA) as an indicator for radical or palliative radiotherapy in patients with rectal cancer. Cancer Detect Prev 6: 155-159

Astler VB, Coller FA (1954) The prognostic significance of direct extension of carcinoma of the colon and rectum. Ann Surg 139: 846-851

August DA, Ottow RT, Sugarbaker PH (1984) Clinical perspective of human colorectal cancer metastasis. Cancer Metastasis Rev 3: 303-324

Baker AR (1980) Local procedures in the management of rectal cancer. Semin Oncol 7: 385-391

Ballantyne GH (1988) Theories of carcinogenesis and their impact on surgical treatment of colorectal cancer. A historical review. Dis Colon Rectum 31: 513-517

Balslev I, Pederson M, Teglbjaerg PS, Hanberg-Soerensen F, Bone J, Jacobsen NO, Overgaard J, Sell A, Bertelsen K, Hage E, Fenger C, Kronborg O, Hansen L, Hoestrup H, Noergaard-Pedersen B (1986) Postoperative radiotherapy in Dukes B and C carcinoma of the rectum and rectosigmoid. Cancer 58: 22-28

Barth RF, Alan F, Soloway AH, Adams DM, Steplewski Z (1986) Boronated monoclonal antibody 17-1A for potential neutron captive therapy of colorectal cancer. Hybridoma 5, Supp 1: 543-550

Bartram CI (1985) Imaging. Br J Surg 72 [Suppl]: S49-S50

Basrur VR, Knight PR (1983) Intracavitary radiation for rectal carcinoma. J Can Assoc Radiol 34: 42-46

Batterman JJ (1982) Results of d+T fast neutron irradiation on advanced tumors of bladder and rectum. Int J Radiat Oncol Biol Phys 8: 2159-2164

Battista JJ, Rider WD, Van Dyk J (1980) Computed tomography for radiotherapy planning. Int J Radiat Oncol Biol Phys 6: 99-107

Bayer I, Turani H, Lurie H, Chamoff C (1985) The sandwich approach: Irradiation - surgery - irradiation in rectal cancer. Dis Colon Rectum 28: 222-224

Begent RHJ, Bagshawe KD, Pedley RB, Searle F, Ledermann JA, Green AJ, Keep PA, Chester KA, Glaser MG, Dale RG (1987) Use of second antibody in radioimmunotherapy. Natl Cancer Inst Monogr 3: 59-61

Berdov BA, Menteshashvili GZ (1984) Thermo-radiotherapy in local rectal cancer. Vopr Onkol 30: 87-91

Berge T, Ekelund G, Mellner C, Pihl B, Wenckert A (1973) Carcinoma of the colon and rectum in a defined population. Acta Chir Scand 139 [Suppl 438]: 1-86

Bjerkeset T, Dahl O (1980) Irradiation and surgery for primarily inoperable rectal adenocarcinoma. Dis Colon Rectum 23: 298-303

Black WA, Waugh JM (1948) The intramural extension of carcinoma of the descending colon, sigmoid and rectosigmoid. Surg Gynecol Obstet 87: 457-464

Block GE, Enker WE (1971) Survival after operations for rectal carcinoma in patients over 70 years of age. Ann Surg 174: 521-527

Bolin S, Nilsson E, Sjodahl R (1983) Carcinoma of the colon and rectum - growth rate. Ann Surg 198: 151-158

Boulis-Wassif S (1983) Ten years experience with a multimodality treatment of advanced stages of rectal cancer. Cancer 52: 2017-2024

Boulis-Wassif S, Gerard A, Loygue J, Camelot D, Buyse M, Duez N (1984) Final results of a randomized trial on the treatment of rectal cancer with preoperative radiotherapy alone or in combination with 5-fluorouracil, followed by radical surgery. Cancer 53: 1811-1818

Brenner S, Lantner B, Seligmann R (1980) Adjuvant therapy in treatment of rectal carcinoma. Int J Radiat Oncol Biol Phys 6: 1378-1379

Breteau N, Destembert B, Favre A, Sabattier R, Schlienger M (1986) An interim assessment of the experience of fast neutron boost in inoperable rectal carcinomas in Orleans. Bull Cancer (Paris) 5: 591-595

Brin EN, Schiff M, Weiss RM (1975) Palliative urinary diversion for pelvic malignancy. J Urol 113: 619-622

Brown CE, Warren S (1938) Visceral metastasis from rectal carcinoma. Surg Gynecol Obstet 66: 611-621

Brusilovsky MI, Karlachenko NI (1981) Interstitial radiotherapy with 198-Au after radical surgery for rectal cancer. Med Radiol 26: 15

Bubrick MP, Blake DP, Kochsiek GG, Feeney DA, Johnston GR, Strom RL, Hitchcock CR (1984) Low anterior anastomotic dehiscence following preoperative irradiation with 6000 rads. Dis Colon Rectum 27: 176-181

Buroker T, Nigro N, Correa J, Vaitkevicius VK, Samson M, Considine B (1976) Combination preoperative radiation and chemotherapy in adenocarcinoma of the rectum: preliminary report. Dis Colon Rectum 19: 660-663

Butch RJ, Stark DD, Wittenberg J, Tepper JE, Saini S, Simeone JF, Mueller PR, Ferrucci JT (1986) Staging rectal cancer by MR and CT. AJR 146: 1155-1160

Buyse M, Zeleniuch-Jacquotte A, Chalmers TC (1988) Adjuvant therapy of colorectal cancer. Why we still don't know. JAMA 259: 3571-3578

Byfield JE, Calabro-Jones P, Klisak R (1982) Pharmacological requirements for obtaining sensitization of human tumor cells in vitro to combined 5-FU or ftorafur and X-rays. Int J Radiat Oncol Biol Phys 8: 1923-1933

Cahan WG, Castro EB, Hajdu SI (1974) The significance of a solitary lung shadow in patients with colon carcinoma. Cancer 33: 414-421

Cannon-Albright LA, Skolnick MH, Bishop T, Lee RG, Burt RW (1988) Common inheritance of susceptibility to colonic adenomatous polyps and associated colorectal cancers. N Engl J Med 319: 533-537

Carlsson U, Lasson A, Ekelund G (1987) Recurrence rates after curative surgery for rectal carcinoma, with special reference to their accuracy. Dis Colon Rectum 30: 431-434

Cedermark B, Theire NO, Rieger A (1985) Preoperative short-term radiotherapy in rectal carcinoma. A preliminary report of a prospektive radnomized study. Cancer 55: 1182-1185

Ciatto S, Pacini P (1982) Radiation therapy of recurrences of carcinoma of the rectum and sigmoid after surgery. Acta Radiol Oncol 21: 105-109

Collins VP, Loeffer RK, Tivey H (1956) Observations on growth rates of human tumors. AJR 76: 988-1000

Cooper EH, O'Quigley J (1982) Biochemical markers. Recent Results Cancer Res 83: 67-76

Correa P, Haenszel W (1978) The epidemiology of large bowel cancer. Adv Cancer Res 26: 1-141

Cummings BJ (1984) Adjuvant radiation therapy for rectal adenocarcinoma. Dis Colon Rectum 27: 826-836

Cummings BJ (1985) The effect of preoperative irradiation on the healing of colorectal anastomoses. Am J Surg 149: 695-696

Cummings BJ (1986) A critical review of adjuvant preoperative radiation therapy for adenocarcinoma of the rectum. Br J Surg 73: 332-338

Cummings BJ, Rider WD, Harwood AR, Keane TJ, Thomas GM (1983) Radical external beam radiation therapy for adenocarcinoma of the rectum. Dis Colon Rectum 26: 30-36

Daland EM, Welch CE, Nathanson I (1936) One hundred untreated cancers of the rectum. N Engl J Med 214: 451-456

Danjoux CE, Gelber RD, Catton GE, Klaassen DJ (1985) Combination chemo-radiotherapy for residual recurrent or inoperable carcinoma of the rectum: E.C.O.G. Study (EST 3276). Int J Radiat Oncol Biol Phys 11: 765-771

DeCosse JJ, Boyle JC (1985) Overview of epidemiology and risk factors associated with colorectal cancer. In: Ingall JR, Mastromarino AJ (eds) Carcinoma of the large bowel and its precursors. Liss, New York, pp 1-12

Deddish MR (1974) Local excision. Surg Clin North Am 54: 877-880

Degges RD, Cannon DJ, Lang NP (1983) The effect of preoperative radiation on healing of rat colonic anastomoses. Dis Colon Rectum 26: 598-600

DeJong UW, Day NE, Muir CS, Barclay THC, Bras G, Foster FH, Jussawalla DJ, Kurihara M, Linden G, Martinez I, Payne PM, Pedersen E, Ringertz N, Shanmugaratnam T (1972) The distribution of cancer within the large bowel. Int J Cancer 10: 463-477

Derdel J, Mohuiddin M, Kramer S, Marks G (1985) Is dose/time fractionation important in treating rectal cancer. Int J Radiat Oncol Biol Phys 11: 579-582

Deschner EE, Lipkin M (1976) Cell proliferation in normal, preneoplastic and neoplastic gastrointestinal cells. Clin Gastroenterol 5: 543-561

Devesa J, Avedillo D, Morales V, Puertas AN (1984) Automatic preoperative classification of carcinoma of the colon and rectum. Surg Gynecol Obstet 158: 482-487

Dionne L (1965) The pattern of blood-borne metastasis from carcinoma of rectum. Cancer 18: 775-781

Dische S, Senanayake F (1982) Radiotherapy using hyperbaric oxygen in the palliation of carcinoma of the colon and rectum. Clin Radiol 23: 512-518

Dixon AK, Kelsey Fry I, Morson BC, Nicholls RJ, York Mason A (1981) Preoperative computed tomography of carcinoma of the rectum. Br J Radiol 54: 655-659

Dobbs J, Barrett A (1985) Practical radiotherapy planning. Arnold, London, p 162

Dobrowsky W, Schmid AP (1985) Radiotherapy of presacral recurrence following radical surgery for rectal carcinoma. Dis Colon Rectum 28: 917-919

Dosoretz DE, Gunderson LL, Hedberg S, Hoskins B, Blitzer PH, Shipley W, Cohen A (1983) Preoperative irradiation for unresectable rectal and rectosigmoid carcinomas. Cancer 52: 814-818

Dukes CE (1932) The classification of cancer of the rectum. J Pathol Bacteriol 35: 323-332

Dukes CE, Bussey HJR (1958) The spread of rectal cancer and its effect on prognosis. Br J Cancer 12: 309-320

Duncan W, Arnott SJ, Jack WJL, Orr JA, Kerr GR, Williams JR (1987) Results of two randomized clinical trials of neutron therapy in rectal adenocarcinoma. Radiother Oncol 8: 191-198

Ege GN, Cummings BJ (1980) Interstitial radiocolloid iliopelvic lymphoscintigraphy: technique, anatomy and clinical application. Int J Radiat Oncol Biol Phys 6: 1483-1490

Eisenberg H, Sullivan PD, Foote FM (1967) Trends in survival of digestive system cancer patients in Connecticut, 1935 to 1962. Gastroenterology 53: 528-546

Emami B, Pilepich M, Willett C, Munzenrider JE, Miller HH (1982) Effect of preoperative irradiation on resectability of colorectal carcinomas. Int J Radiat Oncol Biol Phys 8: 1295-1299

Enker WE, Heilweil ML, Hertz REL, Pilipshen SJ, Stearns MW, Sternberg SS, Janov AJ (1986) En bloc pelvic lymphadenectomy and sphincter preservation in the surgical management of rectal cancer. Ann Surg 203: 426-433

Enquist IF, Block IR (1966) Rectal cancer in the female: selection of proper operation based upon anatomic studies of rectal lymphatics. Prog Clin Cancer 2: 73-85

Feigen M, Cummings B, Hawkins N, Keane T, O'Sullivan B, Wong S (1988) Low dose postoperative adjuvant radiation therapy for rectal cancer is ineffective. Radiother Oncol 13: 181-186

Feinstein AR, Schimpff CR, Hull EW (1975a) A reappraisal of staging and therapy for patients with cancer of the rectum. I. Development of two new systems of staging. Arch Intern Med 135: 1441-1453
Feinstein AR, Schimpff CR, Hull EW (1975b) A reappraisal of staging and therapy for patients with cancer of the rectum. II. Patterns of presentation and outcome of treatment. Arch Intern Med 135: 1454-1462
Feinstein AR, Sosin DM, Wells CK (1985) The Will Rogers phenomenon. Stage migration and new diagnostic techniques as a source of misleading statistics for survival in cancer. N Engl J Med 312: 1604-1608
Fidler IJ, Hart IR (1985) Principles of cancer biology: cancer metastasis. In: deVita VT, Hellman S, Rosenberg SA (eds) Cancer. Principles and practice of oncology, 2nd edn. Lippincott, Philadelphia, pp 113-124
Finlay IG, Brunton CF, Meek D, McArdle CS (1982) Rate of growth of hepatic metastases in colorectal carcinoma. Br J Surg 69: 689
Fisher B, Wolmark N, Rockette H, Redmond C, other NSABP Investigators (1988) Postoperative adjuvant chemotherapy or radiation therapy for rectal cancer: results from NSABP Protocol R-01. J Natl Cancer Inst 80: 21-29
Freedman LS, Macaskill P, Smith AN (1984) Multivariate analysis of prognostic factors for operable rectal cancer. Lancet 2: 733-736
Gabriel WB, Dukes C, Bussey HJR (1935) Lymphatic spread in cancer of the rectum. Br J Surg 23: 395-413
Gallagher MJ, Brereton HD, Rostock RA, Zero JM, Zekoski DA, Poyss LF, Richter MP, Kligerman MM (1986) A prospective study of treatment techniques to minimize the volume of pelvic small bowel with reduction of acute and late effects associated with pelvic irradiation. Int J Radiat Oncol Biol Phys 12: 1565-1573
Gary-Bobo J, Pujol H, Solassol C, Broquerie JL, Nguyen M (1979) L'irradiation preoperataire du cancer rectal. Resultats a 5 ans de 116 cas. Bull Cancer 66: 491-496
Gastrointestinal Tumor Study Group (1985) Prolongation of the disease-free interval in surgically treated rectal carcinoma. N Engl J Med 312: 1465-1472
Gastrointestinal Tumor Study Group (1986) Survival after postoperative combination treatments of rectal cancer. N Engl J Med 315: 1294-1295
Gerard A, Berrod JL, Pene F, Loygue J, Langier A, Bruckner R, Camelot G, Arnand JP, Metzger U, Buyse M, Dalesio O, Duez N (1985) Interim analysis of a phase III study on preoperative radiation therapy in resectable rectal carcinoma. Cancer 55: 2373-2379
Gerard A, Berrod JL, Pene F, Loygue J, Langier A, Bruckner R, Camelot G, Arnand JP, Metzger U, Buyse M, Dalesio O, Duez N (1988) Preoperative radiotherapy and radical surgery as combined treatment in rectal cancer. Recent Results Cancer Res 110: 130-133
Ghossein NA, Samala EC, Alpert S, Deluca FR, Ragins H, Turner SS, Stacey P, Flax H (1981) Elective postoperative radiotherapy after incomplete resection of colorectal cancer. Dis Colon Rectum 24: 252-256
Gianola FJ, Dwyer A, Jones AE, Sugarbaker PH (1984) Prospective studies of laboratory and radiologic tests in the management of colon and rectal cancer patients. I. Selection of useful preoperative tests through an analysis of surgically occult metastases. Dis Colon Rectum 27: 811-818
Gilbert SG (1978) Symptomatic local tumor failure following abdominoperineal resection. Int J Radiat Oncol Biol Phys 4: 801-807
Glimelius B, Graffman S, Pahlman L, Rimsten A, Wilander E (1982) Preoperative irradiation with high-dose fractionation in adenocarcinoma of the rectum and rectosigmoid. Acta Radiol Oncol 21: 373-379
Godwin JD (1975) Carcinoid tumors. An analysis of 2837 cases. Cancer 36: 560-569
Goligher JC (1941) The operability of carcinoma of the rectum. Br Med J 2: 393-397
Goligher J (1981) Results of operations for large bowel cancer. In: DeCosse JJ (ed) Large bowel cancer. Churchill Livingstone, Edinburgh, pp 154-165
Goligher JC (1984) Diseases of the anus, rectum and colon, 5th edn. Bailliere Tindall, London
Grady ED (1979) Internal radiation therapy of hepatic cancer. Dis Colon Rectum 22: 371-375
Graffner HOL, Alm POA, Oscarson JEA (1983) Prophylactic oophorectomy in colorectal carcinoma. Am J Surg 146: 233-235
Green N, Iba G, Smith WR (1975) Measures to minimize small intestine injury in the irradiated pelvis. Cancer 35: 1633-1640
Griffen WO, Humphrey L, Sosin H (1969) The prognosis and management of recurrent abdominal malignancies. Curr Probl Surg 1-43
Griffiths JD, McKinna JA, Rowbotham HD, Tsdakidis P, Salsbury AJ (1973) Carcinoma of the colon and rectum: circulating malignant cells and five-year survival. Cancer 31: 226-236
Grinnell RS (1966) Results of ligation of inferior mesenteric arteries at the aorta in resections of carcinoma of the descending and sigmoid colon and rectum. Surg Gynecol Obstet 120: 1031-1036
Gunderson LL, Sosin H (1974) Areas of failure found at operation (second or symptomatic look) following "curative" surgery for adenocarcinoma of the rectum: Clinicopathologic correlation and implications for adjuvant therapy. Cancer 34: 1278-1292
Gunderson LL, Cohen AC, Dosoretz DD, Shipley WU, Hedberg SE, Wood WC, Rodkey GV, Suit HD (1983a) Residual, unresectable or recurrent colorectal cancer: external beam irradiation and intraoperative electron beam boost ± resection. Int J Radiat Oncol Biol Phys 9: 1597-1606
Gunderson LL, Dosoretz DE, Hedberg SE, Blitzer PH, Rodkey G, Hoskins B, Shipley WY, Cohen AC (1983b) Low dose preoperative irradiation, surgery, and elective postoperative radiation therapy for resectable rectum and rectosigmoid carcinoma. Cancer 52: 446-451
Gunderson LL, Martin JK, Beart RW, Nagorney DM, Fieck JM, Wieand HS, Martiny A, O'Connell MJ, Martenson JA, McIlrath DC (1988) Intraoperative and external beam irradiation for locally advanced colorectal cancer. Ann Surg 207: 52-60
Gunderson LL, Russell AH, Llewellyn HJ, Doppke KP, Tepper JE (1985) Treatment planning for colorectal cancer: radiation and surgical techniques and value of small-bowel films. Int J Radiat Oncol Biol Phys 11: 1379-1393
Hager TH, Gall FP, Hermanek P (1983) Local excision of cancer of the rectum. Dis Colon Rectum 26: 149-151
Hammersmith Oncology Group (1984) Antibody-guided irradiation of malignant lesions: three cases illustrating a new method of treatment. Lancet 1: 1441-1443
Hardcastle JD (1986) Symposium: Screening for colorectal cancer. Int J Colorect Dis 1: 63-78
Havelaar IJ, Sugarbaker PH, Vermess M, Miller DL (1984) Rate of growth of intraabdominal metastases from colorectal cancer. Cancer 54: 163-171

Heald RJ, Ryall RDH (1986) Recurrence and survival after total mesorectal excision for rectal cancer. Lancet 2: 1479-1482

Heald RJ, Husband EM, Ryall RD (1982) The mesorectum in rectal cancer surgery: the clue to pelvic recurrence? Br J Surg 69: 613-616

Hermanek P, Gall FP (1986) Early (microinvasive) colorectal carcinoma. Pathology, diagnosis, surgical treatment. Int J Colorect Dis 1: 79-84

Herzog J, Schmidt B, Fassbender T, Heitland W, Hubeuer KM (1986) Präoperative Kurzzeitbestrahlung bei Rectumkarzinomen. Strahlentherapie 163: 648-653

Higgins GA, Conn JH, Jordan PH, Humphrey EW, Roswit B, Keehn RJ (1985) Preoperative radiotherapy for rectal cancer. Ann Surg 181: 624-631

Higgins GA, Humphrey EW, Dwight RW, Roswit B, Lee LE, Keehn RJ (1986) Preoperative radiation and surgery for cancer of the rectum. Veterans Administration Surgical Oncology Group Trial II. Cancer 58: 352-359

Hildebrandt U, Feifel G (1985) Preoperative staging of rectal cancer by intrarectal ultrasound. Dis Colon Rectum 28: 42-46

Hojo K, Koyama Y (1982) The effectiveness of wide anatomical resection and radical lymphadenectomy for patients with rectal cancer. Jpn J Surg 12: 111-116

Hojo K, Koyama Y, Moriya Y (1982) Lymphatic spread and its prognostic value in patients with rectal cancer. Am J Surg 144: 350-354

Hutter RVP, Sobin LH (1986) A universal staging system for cancer of the colon and rectum. Arch Pathol Lab Med 110: 367-368

Ioachim HL, Weinstein MA, Robbins RD, Sohn N, Lugo PN (1987) Primary anorectal lymphoma. A new manifestation of the acquired immune deficiency syndrome (AIDS). Cancer 60: 1449-1453

James RD, Schofield PF (1985) Resection of 'inoperable' rectal cancer following radiotherapy. Br J Surg 72: 279-281

James RD, Johnson RJ, Eddleston B, Zheng GL, Jones JM (1983) Prognostic factors in locally recurrent rectal carcinoma treated by radiotherapy. Br J Surg 70: 469-472

Jao SW, Beart RW, Reiman HM, Gunderson LL, Ilstrup DM (1987) Colon and anorectal cancer after pelvic irradiation. Dis Colon Rectum 30: 953-958

Jass JR, Morson BC (1987) Reporting colorectal cancer. J Clin Path 40: 1016-1023

Keane TJ, Rider WD, Harwood AR, Thomas GM, Cummings BJ (1981) Whole abdominal radiation in the management of metastatic gastrointestinal carcinoid tumor. Int J Radiat Oncol Biol Phys 7: 1519-1521

Kemeny MM, Sugarbaker PH, Smith TJ, Edwards BK, Shawker T, Verness M, Jones AE (1982) A prospective analysis of laboratory tests and imaging studies to detect hepatic lesions. Ann Surg 195: 163-167

Kim EE, Deland FH, Casper S, Corgan RL, Primus FJ, Goldenberg DM (1980) Radioimmunodetection of colorectal cancer. Cancer 45: 1243-1247

Kinsella TJ (1983) The role of radiation therapy alone and combined with infusion chemotherapy for treating liver metastases. Semin Oncol 10: 215-222

Kinsella TJ, Bloomer WD (1980) Tolerance of the intestine to radiation therapy. Surg Gynecol Obstet 151: 273-283

Kligerman MM, Urdaneta-Lafee N (1974) Observations on fifteen inoperable/nonresectable cases of rectal cancer given preoperative irradiation. AJR 120: 624-626

Kligerman MM, Urdaneta N, Knowlton A, Vidone R, Hartman PR, Vera R (1972) Preoperative irradiation of rectosigmoid carcinoma including its regional lymph nodes. AJR 114: 498-503

Kronborg O (1986) Controversies in follow-up after colorectal carcinoma. Theor Surg 1: 40-46

Krook J, Moertel C, Wieand H, Collins R, Gunderson L, Kubista T, Beart R (1986) Radiation vs sequential chemotherapy-radiation-chemotherapy. Proc ASCO 5: 82

Kyriakos M (1985) The President's cancer, the Dukes classification, and confusion. Arch Pathol Lab Med 109: 1063-1066

Lafreniere R, Ketcham AS (1985) Primary squamous carcinoma of the rectum. Report of a case and review of the literature. Dis Colon Rectum 28: 967-972

Lafreniere R, Ketcham AS (1986) Squamous cell carcinoma of the rectum: a multimodality approach. J Surg Oncol 32: 106-109

Langevin JM, Nivatvongs S (1984) The true incidence of synchronous cancer in the large bowel. A prospective study. Am J Surg 147: 330-333

Laurence DJR, Neville AM (1983) The detection and evaluation of human tumor metastases. Cancer Metastasis Rev 2: 351-374

Lavery IC, Jones IT, Weakley FL, Saxton JP, Fazio VW, Jagelman DG (1987) Definitive management of rectal cancer by contact (endocavitary irradiation). Dis Colon Rectum 30: 835-838

Lee MYT (1986) Smooth muscle tumors. In: Fielding JWL, Priestman TJ (eds) Gastrointestinal oncology. Castle House, Tunbridge Wells, pp 293-305

Leyden MJ, Thompson CH, Lichtenstein M, Andrewes JT, Sullivan JR, Zalcberg JR, McKenzie IFC (1986) Visualization of metastases from colon carcinoma using on iodine-131 radiolabelled monoclonal antibody. Cancer 57: 1135-1137

Liebel SA, Order SE, Rominger CJ, Asbell SO (1981) Palliation of liver metastases with combined hepatic irradiation and misonidazole. Results of a Radiation Therapy Oncology Group phase I-II study. Cancer Clin Trials 4: 285-293

Lockhart-Mummery HE, Ritchie JK, Hawley PR (1976) The results of surgical treatment for carcinoma of the rectum at St. Mark's Hospital from 1948 to 1972. Br J Surg 63: 673-677

Lofgren EP, Waugh JM, Dockerty MB (1957) Local recurrence of carcinoma after anterior resection of the rectum and sigmoid. Arch Surg 74: 825-838

Logan WPD (1976) Cancers of the alimentary tract: international mortality trends. WHO Chron 30: 413-419

MacKeigan JM, Ferguson JA (1979) Prophylactic oophorectomy and colorectal cancer in premenopausal patients. Dis Colon Rectum 22: 401-405

Madden JL, Kandalaft S (1973) Electrocoagulation in the treatment of cancer of the rectum. RI Med J 56: 415-433

Manning MR, Cetas TC, Miller RC, Oleson JR, Connor WG, Gerner EW (1982) Clinical hyperthermia: Results of a phase I trial employing hyperthermia alone or in combination with external beam or interstitial radiotherapy. Cancer 49: 205-216

Marks G, Mohiuddin M (1983) The surgical management of the radiation-injured intestine. Surg Clin North Am 63: 81-96

Martin EW, Minton JP, Larey LC (1985) CEA-directed second-look surgery in the asymptomatic patient after primary resection of colorectal carcinoma. Ann Surg 202: 310-317

Mayer M, Papillon J, Bokin JY, Ardiet JM (1982) La lymphadenectomie mesenterique inferieure et perirectal dans la traitement conservateur des cancers de la partie inferieure de l'ampoule rectale. Chirurgie 108: 479-483

Mella O, Dahl O, Horn A, Morild I, Odland G (1984) Radiotherapy and resection for apparently inoperable rectal adenocarcinoma. Dis Colon Rectum 27: 663-668

Mendenhall WM, Million RR, Bland KI, Pfaff WW, Copeland EM (1985) Preoperative radiation therapy for clinically resectable carcinoma of the rectum. Ann Surg 202: 215-222

Michelassi F, Vannucci LE, Montag A, Chappell R, Rodgers J, Block GE (1988) Ras oncogene expression as a prognostic indicator in rectal adenocarcinoma. J Surg Res 45: 15-20

Moertel CG, Childs DS, Reitemeier RJ, Colby MY, Holbrook MA (1969) Combined 5-fluorouracil and supervoltage radiation therapy of locally unresectable gastrointestinal cancer. Lancet 2: 865-867

Mohiuddin M, Marks GJ (1987) High dose preoperative radiation and sphincter preservation in the treatment of rectal cancer. Int J Radiat Oncol Biol Phys 13: 839-842

Mohiuddin M, Marks G, Kramer S, Pajak T (1984) Adjuvant radiation therapy for rectal cancer. Int J Radiat Oncol Biol Phys 10: 977-980

Moossa AR, Ree PC, Marks JE, Levin B, Platz CE, Skinner DB (1975) Factors influencing local recurrence after abdominoperineal resection for cancer of the rectum and rectosigmoid. Br J Surg 62: 727-730

Morgan CN (1965) Carcinoma of the rectum. Ann R Coll Surg Engl 36: 73-97

Morgenstern L, Sanders G, Wahlstrom E, Yadegar J, Amodeo P (1984) Effect of preoperative irradiation on healing of low colorectal anastomoses. Am J Surg 147: 246-249

Morson BC (1985) Histological criteria for local excision. Br J Surg 72 [Suppl]: 53-54

Morson BC, Bussey HJR (1967) Surgical pathology of rectal cancer in relation to adjuvant radiotherapy. Br J Radiol 40: 161-165

Morson BC, Bussey HJR (1985) Magnitude of risk for cancer in patients with colorectal adenomas. Br J Surg 72 [Suppl]: 23-25

Morson BC, Bussey HJR, Samoorian S (1977) Policy of local excision for early cancer of the colorectum. Gut 18: 1045-1050

Morson BC, Bussey HJR, Day DW, Hill MJ (1983) Adenomas of large bowel. Cancer Surv 2: 451-477

Naunheim KS, Zeitels J, Kaplan EL, Sugimto J, Shen KL, Lee CH, Straus FH (1983) Rectal carcinoid tumors - treatment and prognosis. Surgery 94: 670-676

Netri G, Coco C, Valentini V, Fioravanti PM, Aronne D, Cellini N, Peglionisi A (1985) Clinical staging of rectal cancer. Results of a prospective continuing study. It J Surg Sci 15: 169-174

Newman HK, Stearns MW Jr (1975) Re-exploration for "unresectable" colonic cancer. Dis Colon Rectum 18: 576-580

Nias AHW (1967) Radiobiological aspects of preoperative irradiation. Br J Radiol 40: 166-169

Nicholls RJ (1982) Surgery. Recent Results Cancer Res 83: 101-112

Nicholls RJ, York Mason A, Morson BC, Dixon AK, Kelsey Fry I (1982) The clinical staging of rectal cancer. Br J Surg 69: 404-409

Nolan DJ (1982) Radiological assessment. Recent Results Cancer Res 83: 77-85

Northover JMA (1985) Carcinoembryonic antigen and recurrent colorectal cancer. Br J Surg 72 [Suppl]: 44-45

O'Connell MJ, Childs DS, Moertel CG, Holbrook MA, Schutt AJ, Rubin J, Ritts RE (1982) A prospective controlled evaluation of combined pelvic radiotherapy and methanol extraction residue of BCG (MER) for locally unresectable or recurrent rectal carcinoma. Int J Radiat Oncol Biol Phys 8: 1115-1119

Overgaard M, Overgaard J, Sell A (1984) Dose-response relationship for radiation therapy of recurrent, residual, and primarily inoperable colorectal cancer. Radiother Oncol 1: 217-225

Pacini P, Cionini L, Pirtoli L, Ciatto S, Tucci E, Sebaste L (1986) Symptomatic recurrences of carcinoma of the rectum and sigmoid: the influence of radiotherapy on the quality of life. Dis Colon Rectum 29: 865-868

Pahlman L, Glimelius B (1984) Local recurrences after surgical treatment for rectal carcinoma. Acta Chir Scand 150: 331-335

Pahlman L, Glimelius B, Ginman C, Graffman S, Adalsteinsson B (1985a) Preoperative irradiation of primarily nonresectable adenocarcinoma of the rectum and rectosigmoid. Acta Radiol Oncol 24: 35-39

Pahlman L, Glimelius B, Graffman S (1985b) Pre- versus postoperative radiotherapy in rectal carcinoma: an interim report from a randomized multicentre trial. Br J Surg 72: 961-966

Papillon J (1982) Rectal and anal cancers. Conservative treatment by irradiation - an alternative to radical surgery. Springer, Berlin Heidelberg New York

Papillon J (1984) New prospects in the conservative treatment of rectal cancer. Dis Colon Rectum 27: 695-700

Papillon J (1987) The future of external beam irradiation as initial treatment of rectal cancer. Br J Surg 74: 449-454

Parturier-Albot M (1981) Statistique de 2045 cas de cancers du rectum traites par la radiotherapie de contact avec un recul de 5 a 30 ans. Bull Acad Natl Med (Paris) 165: 967-974

Pereslegin IA, Zolotkov AG, Raifel BA, Kagan IL, Chushkin NA (1985) Twice daily irradiation of patients with rectal cancer. Med Radiol (Mosk) 30: 31-34

Perren TJ, Blackledge G (1986) Gastrointestinal lymphomas. In: Fielding JWL, Priestman TJ (eds) Gastrointestinal oncology. Castle House, Tunbridge Wells, pp 237-255

Pestana C, Reitemeier RJ, Moertel CG, Judd ES, Dockerty MB (1964) The natural history of carcinoma of the colon and rectum. Am J Surg 108: 826-829

Peters LJ, Brock WA, Johnson T, Meyn RE, Tofilon PJ, Milas L (1986) Potential methods for predicting tumor radiocurability. Int J Radiat Oncol Biol Phys 12: 459-467

Phillips RKS, Hittinger R, Blesovsky L, Fry JS, Fielding LP (1984) Local recurrence following "curative" surgery for large bowel cancer: I. The overall picture. Br J Surg 71: 12-16

Phillips TL, Wasserman TH, Stetz J, Brady LW (1982) Clinical trials of hypoxic cell sensitizers. Int J Radiat Oncol Biol Phys 8: 327-334

Pihl E, Malahy MA, Khankhanian N, Hersh EM, Mavligit GM (1977) Immunomorphological features of prognostic significance in Dukes' class B colorectal carcinoma. Cancer Res 37: 4145-4149

Pilipshen SJ, Heilweil M, Quan SHQ, Sternberg SS, Enker WE (1984) Patterns of pelvic recurrence following definitive resection of rectal cancer. Cancer 53: 1354-1362

Polk HC, Spratt JS (1971) Recurrent colorectal carcinoma. Detection, treatment and other considerations. Surgery 69: 9-23

Porter NH, Nicholls RJ (1985) Preoperative radiotherapy in operable rectal cancer: interim report of a trial carried out by the Rectal Cancer Group. Br J Surg 72 [Suppl]: 62-64

Postlethwait RW (1949) Malignant tumors of the colon and rectum. Ann Surg 129: 34-46

Powers WE, Tolmach LJ (1964) Preoperative radiation therapy: biological basis and experimental investigation. Nature 201: 272-273

Pujol H, Solassol C, Gary-Bobo J (1978) La radiotherapie preoperatoire des cancers du rectum, actualisation des resultats. A propos de 179 cas. Chirurgie 104: 606-611

Purdy JA, Prasad SC, Walz BJ, Cotter GW (1985) Radiation protection considerations for endocavitary X-ray units. Int J Radiat Oncol Biol Phys 11: 2177-2181

Puthawala AJ, Syed AMN, Gates TC, McNamara C (1982) Definitive treatment of extensive anorectal carcinoma by external and interstitial irradiation. Cancer 50: 1746-1750

Quirke P, Durdey P, Dixon MF, Williams NS (1986) Local recurrence of rectal adenocarcinoma due to inadequate surgical resection. Histopathological study of lateral tumor spread and surgical excision. Lancet 2: 996-999

Recio P, Bussey HJR (1965) The pathology and prognosis of carcinoma of the rectum in the young. Proc R Soc Med 58: 789-790

Reddy BS, Cohen LA, McCoy GD, Hill P, Weisburger JM, Wynder EL (1980) Nutrition and its relationship to cancer. Adv Cancer Res 32: 237-345

Reports on Public Health and Medical Subjects (1927) Cancer of the rectum. Report no 46. Ministry of Health, London, pp 1-70

Rich T, Gunderson LL, Lew R, Galdabini JJ, Cohen AM, Donaldson G (1983) Patterns of recurrence of rectal cancer after potentially curative surgery. Cancer 52: 1317-1329

Rich TA, Lokich JJ, Chaffey JA (1985a) A pilot study of protracted venous infusion of 5-fluorouracil and concomitant radiation therapy. J Clin Oncol 3: 402-406

Rich TA, Weiss DR, Mies C, Fitzgerald TJ, Chaffey JT (1985b) Definitive treatment of low rectal cancer with sphincter preservation by radiation therapy with or without local excision or fulguration. Radiology 156: 527-531

Rider WD (1975) Is the Miles operation really necessary for the treatment of rectal cancer? J Can Assoc Radiol 26: 167-175

Rider WD, Palmer JA, Mahoney LJ, Robertson CT (1977) Preoperative irradiation in operable cancer of the rectum: report of the Toronto trial. Can J Surg 20: 335-338

Roberson SH, Heron HC, Kerman HE, Bloom TS (1985) Is anterior resection of the rectosigmoid safe after preoperative radiation? Dis Colon Rectum 28: 254-259

Robinson E, Bartal A, Cohen Y, Hairn H, Mohilever J, Mekori T (1982) Combined adjuvant therapy of radically operated colorectal cancer patients (chemotherapy, radiotherapy, and MER-BCG). Cancer Chemother Pharmacol 8: 35-40

Rodriguez-Antunez A, Chernak ES, Jelden GL, Hunter TW (1973) Preoperative irradiation of carcinoma of the rectum. Radiology 108: 89-90

Rominger CJ, Gunderson LL, Gelber RD, Conner N (1985) Radiation therapy alone or in combination with chemotherapy in the treatment of residual or inoperable carcinoma of the rectum or rectosigmoid or pelvic recurrence following colorectal surgery. Radiation Therapy Oncology Group Study (76-16). Am J Clin Oncol 8: 118-127

Rosenberg IL, Russell CW, Giles GR (1978) Cell viability studies on the exfoliated colonic cancer cell. Br J Surg 65: 188-190

Rothenberger DA, Wong WD (1985) Rectal cancer - adequacy of surgical management. Surg Ann 17: 309-336

Rubin P, Casarett CW (1968) Alimentary tract; small and large intestine and rectum. In: Clinical radiation pathology, vol 1. Saunders, Philadelphia, pp 193-240

Russ JE, Smoron GL, Gagnon JD (1984) Omental transposition flap in colorectal carcinoma: adjunctive use in prevention and treatment of radiation complications. Int J Radiat Oncol Biol Phys 10: 55-62

Sapozink MD, Gibbs FA, Egger MJ, Stewart JR (1986) Regional hyperthermia for clinically advanced deep-seated pelvic malignancy. Am J Clin Oncol 9: 162-169

Scherer E (1972) Rectum. In: Zuppinger A, Krokowski E (eds) Radiation therapy of malignant tumors, part 1. Springer, Berlin Heidelberg New York, pp 635-658

Second Report of an MRC Working Party (1984) The evaluation of low dose preoperative X-ray therapy in the management of operable rectal cancer: results of a randomly controlled trial. Br J Surg 71: 21-25

Seefeld PH, Bargen JA (1943) The spread of carcinoma of the rectum: invasion of lymphatics, veins and nerves. Ann Surg 118: 76-90

Shamsuddin AKM, Weiss L, Phelps PC, Trump BF (1981) Colon epithelium. IV. Human colon carcinogenesis. Changes in human colon mucosa adjacent to and remote from carcinomas of the colon. JNCI 66: 413-419

Shank B, Enker W, Santana J, Morrissey K, Daly J, Quan S, Knapper W (1987) Local control with preoperative radiotherapy alone versus "sandwich" radiotherapy for rectal carcinoma. Int J Radiat Oncol Biol Phys 13: 111-115

Shindo K (1974) Recurrence of carcinoma of the large intestine: a statistical review. Am J Proctol 25: 80-90

Simon JB (1985) Occult blood screening for colorectal carcinoma: A critical review. Gastroenterology 88: 820-837

Sischy B (1985) The use of endocavitary irradiation for selected carcinomas of the rectum: Ten years experience. Radiother Oncol 4: 97-101

Sischy B, Graney MJ, Hinson EJ, Qazi R (1985) Preoperative radiation therapy with sensitizers in the management of carcinoma of the rectum. Dis Colon Rectum 28: 56-57

Sklaroff DM (1973) Radiation as primary therapy for rectal carcinoma. Am Fam Physician 8: 81-85

Slanetz CA, Herter FP, Grinnell RS (1972) Anterior resection versus abdominoperineal resection for cancer of the rectum and rectosigmoid. Am J Surg 123: 110-116

Slaney G (1971) Results of treatment of carcinoma of the colon and rectum. In: Irvine WT (ed) Modern trends in surgery, vol 3. Butterworths, London, pp 69-89

Smith TJ, Kemeny MM, Sugarbaker PH, Jones AE, Verness M, Shawker TM, Edwards BK (1982) A prospective study of hepatic imaging in the detection of metastatic disease. Ann Surg 195: 486-491

Sparso BH, von der Maase H, Kristensen D, Christiansen J, Nieksen SAD, Hebjorn M, Anderson B (1984) Complications following postoperative combined radiation and chemotherapy in adenocarcinoma of the rectum and rectosigmoid. A randomized trial that failed. Cancer 54: 2363-2366

Spiegelman D, Wegman DH (1985) Occupation-related risks of colorectal cancer. JNCI 75: 813-821

Spratt JS, Spjut HJ (1967) Prevalence and prognosis of individual clinical and pathologic variables associated with colorectal carcinomas. Cancer 20: 1976-1985

Stahle E, Glimelius B, Bergstrom R, Pahlman L (1988) Preoperative serum markers in carcinoma of the rectum and rectosigmoid. I. Prediction of tumour stage. Eur J Surg Oncol 14: 277-286

Stearns MW (1980) Pre- or postoperative radiation in resectable tumors. In: Welvaart K, Blumgart LH, Kreuning J (eds) Colorectal cancer. Leiden University Press, The Hague, pp 153-159

Stearns MW Jr, Deddish MK (1959) Five year results of abdominopelvic lymph node dissection for carcinoma of the rectum. Dis Colon Rectum 2: 169-172

Steel GG (1988) The search for therapeutic gain in the combination of radiotherapy and chemotherapy. Radiother Oncol 11: 31-53

Stevens KR, Allen CV, Fletcher WS (1976) Preoperative radiotherapy for adenocarcinoma of the rectosigmoid. Cancer 37: 2866-2874

Stevens KR, Fletcher WS, Allen CV (1978) Anterior resection and primary anastomosis following high dose preoperative irradiation for adenocarcinoma of the rectosigmoid. Cancer 41: 2065-2071

Stevens KR Jr, Fletcher WS, Allen CV (1979) A review of the value of radiation therapy for adenocarcinoma of the rectum and sigmoid. Front Gastrointest Res 5: 93-101

Stockholm Rectal Cancer Study Group (1987) Short-term preoperative radiotherapy for adenocarcinoma of the rectum: an interim analysis of a randomized multicenter trial. Am J Clin Oncol 10: 369-375

Streffer C, van Beuningen D, Gross E, Schabronath J, Eigler FW, Rebmann A (1986) Predictive assays for the therapy of rectum carcinoma. Radiother Oncol 5: 303-310

Stubbs RS (1983) The aetiology of colorectal cancer. Br J Surg 70: 313-316

Sugarbaker PH (1983) Intrapelvic prosthesis to prevent injury of the small intestine with high dosage pelvic irradiation. Surg Gynecol Obstet 157: 269-271

Sugarbaker PH, Corlew S (1982) Influence of surgical techniques on survival in patients with colorectal cancer. A review. Dis Colon Rectum 25: 545-557

Sugarbaker PH, Gunderson LL, Wittes RE (1985) Colorectal cancer. In: DeVita VT, Hellman S, Rosenberg SA (eds) Cancer. Principles and practice of oncology, vol 1, 2nd edn. Lippincott, Philadelphia, pp 759-884

Symonds DA, Vickery AL (1976) Mucinous carcinoma of the colon and rectum. Cancer 37: 1891-1900

Talbot IC (1982) Pathology and natural history. Recent Results Cancer Res 83: 59-66

Talbot IC, Ritchie S, Leighton MH (1980) The clinical significance of invasion of veins by rectal cancer. Br J Surg 67: 349-352

Taylor FW (1962) Cancer of the colon and rectum - a study of routes of metastases and death. Surgery 52: 305-308

Taylor RE, Kerr GR, Arnott SJ (1987) External beam radiotherapy for rectal adenocarcinoma. Br J Surg 74: 455-459

Tepper JE, Gunderson LL, Orlow E, Cohen AM, Hedberg SE, Shipley WU, Blitzer PH, Rich T (1984) Complications of intraoperative radiation therapy. Int J Radiat Oncol Biol Phys 10: 1831-1839

Tepper JE, Cohen AM, Wood WC, Hedberg SE, Orlow E (1986) Intraoperative electron beam radiotherapy in the treatment of unresectable rectal cancer. Arch Surg 121: 421-423

Tepper JE, Cohen AM, Wood WC, Orlow EL, Hedberg SE (1987) Postoperative radiation therapy of rectal cancer. Int J Radiat Oncol Biol Phys 13: 5-10

Thomas PRM, Lindblad AS, Stablein DM, Knowlton AH, Bruckner HW, Childs DS, Mittelman A (1986a) Toxicity associated with adjuvant postoperative therapy for adenocarcinoma of the rectum. Cancer 57: 1130-1134

Thomas PRM, Stablein DM, Kinzie JJ, Novak JW, Childs DS, Knowlton AH, Mittelman A, and the Gastrointestinal Tumor Study Group (GITSG) (1986b) Perineal effects of postoperative treatment for adenocarcinoma of the rectum. Int J Radiat Oncol Biol Phys12: 167-171

Tierie AH (1978) Radiotherapy in marginal resectable and non-resectable rectum cancer. Radiologia Clinica (Basel) 47: 222-227

Trott KR, Kummermehr J (1985) What is known about tumor proliferation rates to choose between accelerated fractionation or hyperfractionation. Radiother Oncol 3: 1-9

Trotter GA, Morgan GR, Cooper AJ, Kirkham N, Whitehouse JMA, Taylor I (1985) Cell kinetics and in vitro clonogenicity of primary colorectal cancer: clinicopathological relationships and the implications for chemotherapy. Gut 26: 267-273

Umpleby HC, Fermor B, Symes MO, Williamson RCN (1984) Viability of exfoliated colorectal carcinoma cells. Br J Surg 71: 859-863

Urdaneta LF, Hoyne RF, Loh PM, Fidler R, McGrady D (1983) A search for malignant cells in peritoneal fluid from patients with colorectal adenocarcinoma. Am Surg 46: 76-81

Van den Heule B, Taylor CR, Terry R, Lukes RJ (1982) Presentation of malignant lymphoma of the rectum. Cancer 49: 2602-2607

Varma JS, Smith AN, Busuttil A (1985) Correlation of clinical and manometric abnormalities of rectal function following chronic radiation injury. Br J Surg 72: 875-878

Veazey PR, McBride CM (1979) Pelvic recurrence of cancer after abdominoperineal resection of the rectum. South Med J 72: 1545-1547

Vigliotti A, Rich TA, Romsdahl MM, Withers HR, Oswald MJ (1987) Postoperative adjuvant radiotherapy for adenocarcinoma of the rectum and rectosigmoid. Int J Radiat Oncol Biol Phys 13: 999-1006

Villalon AH, Green D (1981) The use of radiotherapy for pelvic recurrence following abdominoperineal resection for carcinoma of the rectum: a 10-year experience. Aust NZ J Surg 51: 149-151

Vogelstein B, Fearon ER, Hamilton SR, Kern SE, Preisinger AC, Leppert M, Nakamura Y, White R, Smits AMM, Bos JL (1988) Genetic alterations during colorectal-tumor development. N Engl J Med 319: 525-532

Von der Maase H, Magnusson K, Sveinsson T (1982) Combined irradiation and three-drug chemotherapy in inoperable colorectal carcinoma. Acta Radiol Oncol 21: 369-372

Vongtama V, Douglass HO, Moore RH, Holyoke ED, Webster JH (1975) End results of radiation therapy alone and combined with 5-fluorouracil in colorectal cancers. Cancer 36: 2020-2025

Walsh TH, Mann CV (1984) Smooth muscle neoplasms of the rectum and anal canal. Br J Surg 71: 597-599

Wang CC, Schultz MD (1962) The role of radiation therapy in the management of carcinoma of the sigmoid, rectosigmoid and rectum. Radiology 79: 1-5

Wassink WF (1956) The curative treatment of carcinoma recti by means of electrocoagulation and radium. Arch Chir Neerland 8: 313-330

Welch JP, Donaldson CA (1979) The clinical correlation of an autopsy study of recurrent colorectal cancer. Ann Surg 189: 496-502

Welin S, Yanker J, Spratt JS (1963) The rates and patterns of growth of 375 tumors of the large intestine and rectum observed serially by double contrast enema study (Malmo technique). AJR 90: 673-687

Wendling P, Manz R, Thews G, Vaupel P (1984) Heterogeneous oxygenation of rectal carcinomas in humans: a critical parameter for preoperative irradiation? Adv Exp Med Biol 180: 293-299

Whiteley HW Jr (1967) Preoperative radiation therapy in simulated cancer of the colon in rabbits. Dis Colon Rectum 10: 100-102

Whiteley HW, Stearns MW, Leaming RH, Deddish MR (1970) Palliative radiation therapy in patients with cancer of the colon and rectum. Cancer 25: 343-346

Williams IG, Schulman IM, Todd IP (1957) The treatment of recurrent carcinoma of the rectum by supervoltage X-ray therapy. Br J Surg 44: 506-508

Williams MP, Husband JE (1987) CT scanning in carcinoma of the rectum: a review. J R Soc Med 80: 701-703

Williams NS (1984) The rationale for preservation of the anal sphincter in patients with low rectal cancer. Br J Surg 71: 575-581

Williams NS, Johnston D (1983) The quality of life after rectal excision for low rectal cancer. Br J Surg 70: 460-462

Williams NS, Price R, Johnston D (1980) The long term effect of sphincter preserving operations for rectal carcinoma on function of the anal sphincter in man. Br J Surg 67: 203-208

Winawer SJ, Miller DG, Sherlock P (1984) Risk and screening for colorectal cancer. In: Stollerman GH (ed) Advances in internal medicine. Year Book Medical, Chicago, pp 471-496

Winkler R (1985) Adjuvant radiotherapy in rectosigmoid cancers. Zentralbl Chir 110: 124-136

Withers HR, Mason KA (1974) The kinetics of recovery in irradiated colonic mucosa of the mouse. Cancer 34: 896-903

Withers HR, Cuasay L, Mason KA, Romsdahl MM, Saxton J (1981) Elective radiation therapy in the curative treatment of cancer of the rectum and rectosigmoid. In: Stroehlein JR, Romsdahl MM (eds) Gastrointestinal cancer. Raven, New York, pp 351-362

Wolmark N, Cruz I, Redmond CK, Fisher B, Fisher ER, Contributing NSABP Investigators (1983) Tumor size and regional lymph node metastasis in colorectal cancer. Cancer 51: 1315-1322

Woolam GL, Jackman RJ, Ramirez RJ (1965) Scirrhous carcinoma of the lower intestine. Surg Gynecol Obstet 121: 753-755

Working Party MRC (1982) A trial of preoperative radiotherapy in the management of operable rectal cancer. Br J Surg 69: 513-519

York Mason A (1976) Rectal cancer: the spectrum of selective surgery. Proc R Soc Med 69: 237-244

Young JL, Percy CL, Asire AJ (1981) Surveillance, epidemiology and end results (SEER): Incidence and mortality data, 1973-1977. Natl Cancer Inst Monogr 57

Zaridze DG (1983) Environmental etiology of large bowel cancer. JNCI 70: 389-400

Ziegler RG, Devesa SS, Fraumeni JF (1986) Epidemiologic patterns of colorectal cancer. Important Adv Oncol 3: 209-232

Zorzitto M, Germanson T, Cummings B, Boyd NF (1982) A method of clinical prognostic staging for patients with rectal cancer. Dis Colon Rectum 25: 759-765

11 Anal Cancer

MAHROO HAGHBIN, E. JOSEPHINE HINSON, and BEN SISCHY

CONTENTS

MAHROO HAGHBIN, M. D., Clinical Assistant Professor of Radiation Oncology, School of Medicine and Dentistry, University of Rochester Medical Center, Rochester, NY 14620, USA

E. JOSEPHINE HINSON, R. T. T., Research Assistant, Daisy Marquis Jones Radiation Oncology Center, Highland Hospital, 1000 South Avenue, Rochester, NY 14620, USA

BEN SISCHY, M. D., Director, Clinical Professor of Radiation Oncology, School of Medicine and Dentistry, University of Rochester Medical Center, Daisy Marquis Jones Radiation Oncology Center, Highland Hospital, 1000 South Avenue, Rochester, NY 14620, USA

Neoplasms of the anus are divided into two groups (TURELL 1962; BEAHRS 1979), carcinoma of the anal canal and malignancies of the perianal region or anal margin. The perianal area is defined as the space which lies outside of the anal verge within a 5 cm radius of the anus (BEAHRS and WILSON 1976).

11.1 Anatomy of the Anal Canal

The anal canal is the terminal portion of the large intestine. The anatomic anal canal is defined as the distance between the dentate line and the anal verge; the surgical anal canal is defined as the distance between the anorectal ring and the anal verge (GOLIGHER 1984a) (Fig. 1). The anorectal ring is a palpable fibromuscular structure which is composed of the upper fibers of the internal sphincter (GOLIGHER 1984a). The surgical anal canal on average measures 4.2 cm in length (range 3-5.3 cm) and the anatomic canal has an average length of 2.1 cm (range 1.0-3.8 cm) (NIVATVONGS et al. 1981). The canal is slightly longer in males as compared to females.

In the normal living subject the anal canal is completely collapsed owing to the tonic contraction of the anal sphincter. The anal orifice is represented by an anterior-posterior slit in the anal skin. Posteriorly, the canal is related to the coccyx, with a certain amount of fibrous, fatty, and muscular tissue intervening. Laterally, the ischiorectal fossae on either side contain fat, inferior hemorrhoidal vessels, and nerves. Anteriorly, in the male the canal is related to the perineum, the bulba of the urethra, and the posterior border of the urogenital diaphragm, which contains the membranous urethra. In the female, the canal is related in the front to the perineal body and to the lowest portion of the posterior vaginal wall. The canal is surrounded by external and internal sphincter muscles; the lower edges of these may be palpated as a distinct groove on digital rectal examination. As may be seen in the excised specimen of the rectum and anal canal and in living subjects by proctoscopy, a serrated fringe marks the junction between the rectum and the anus, which is called the dentate (*dentatus* = toothed) or pectinate (*pecten* = cock's comb) line (GOLIGHER 1984a). This serrated margin is formed by the anal valves (Fig. 1). These valves are a series of semilunar folds which surround the circumference of the canal and immediately superior, are 8-14 longitudinal folds known as the columns of Morgagni, each adjacent, two of which are connected below at the pectinate line by an anal valve.

The anal canal above the anal valves is lined by rectal mucosa except for a narrow zone of 0.5-1.0 cm immediately above the valves which is called the transitional or junctional zone (GRINVALSKY and HELWIG 1956; WALLS 1958). Here the structure of the epithelium is a composite of the glandular rectal mucosa above and the squamous mucous membrane below the valves (GRINVALSKY and HELWIG 1956; WALLS 1958). This supravalvular region contains a variety of epithelia. A transitional type of mucosa resembling urinary tract epithelium, stratified columnar epithelium, and squamous mucous membrane in addition to mixed epithelia, which may contain mucus-secreting goblet cells, can be seen in this region (MORSON 1960). In recent studies of the histology of this junctional zone it has been shown that squamous and transitional cells range as far as 20 mm above and 6 mm below the dentate line (FENGER 1979). Below the pectinate

Fig. 1. Anatomy of the anal canal

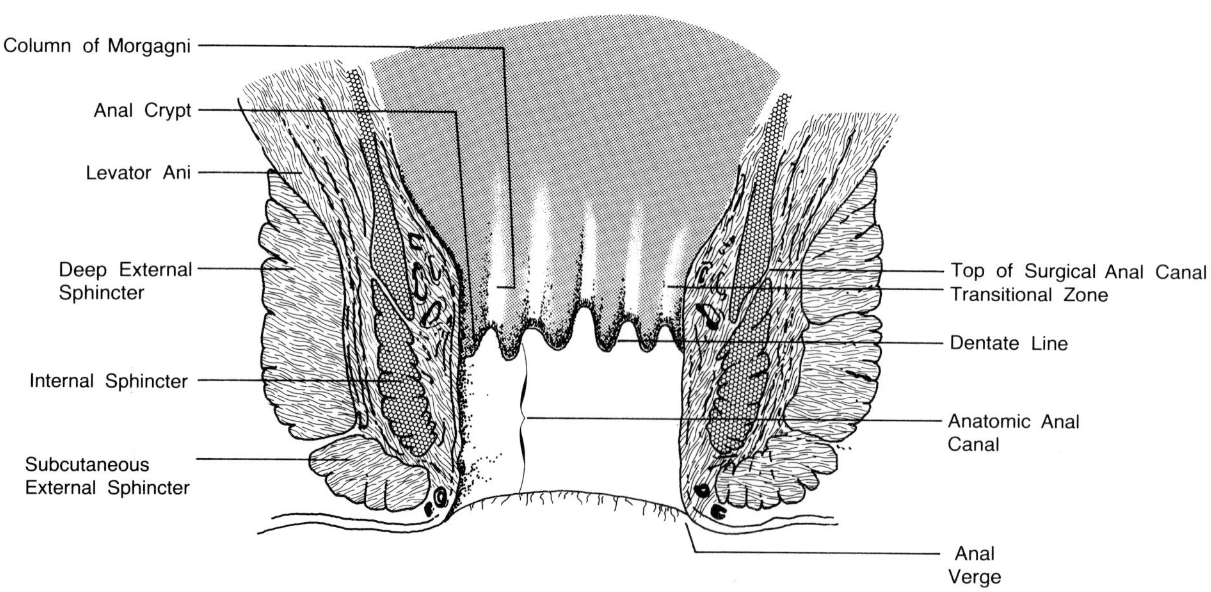

line the anal canal is covered by stratified squamous epithelium (WALLS 1958). This epithelium is a modified skin devoid of hair, sebaceous, and sweat glands which adheres closely to the underlying tissues. For approximately 1 cm distal to the anal valve the lining of the canal appears thin, smooth, and pale and this region is referred to as the pecten. Traced further inferiorly the lining becomes thicker and just outside the anal orifice acquires hair follicles, glands, and other biological features of normal skin (GOLIGHER 1984 a). Anal ducts arise in the anal crypts which extend both cephalad and caudad, penetrating the anal sphincters and sometimes extending to the perianal fat (GRINVALSKY and HELWIG 1956; WALLS 1958). These tubular tortuous glands with a variable epithelial lining of squamous, transitional, and columnar cells normally have a sparse mucus secretion (FENGER and FILIPE 1977). They may account for the presence of transitional neoplasms within the wall of the rectum and anal canal without mucosal involvement (KLOTZ et al. 1967).

11.1.1 Lymphatics

Three zones of lymphatics are described for the lower portion of the large bowel (RICHARDS et al. 1962):

1. The inferior group, which drains the perianal skin and the anal canal, terminating in the superficial inguinal nodes. The upper group of superficial inguinal nodes consists of five or six nodes and receives channels from the perianal region and the perineum. The lower group of four or five superficial inguinal nodes receives vessels from the anal canal (Fig. 2). Some efferent vessels from the superficial inguinal nodes drain to deep inguinal glands, but most channels end in the external iliac nodes (GRAY'S ANATOMY 1985).
2. The middle group of lymphatics are vessels that drain the bowel above the dentate line. These channels accompany the inferior and middle hemorrhoidal vessels to end in the internal pudendal and hypogastric nodes.
3. The superior group of collecting trunks drain the upper rectum and flow into the nodes along the superior hemorrhoidal and inferior mesenteric veins.

Fig. 2. Inguinal nodes

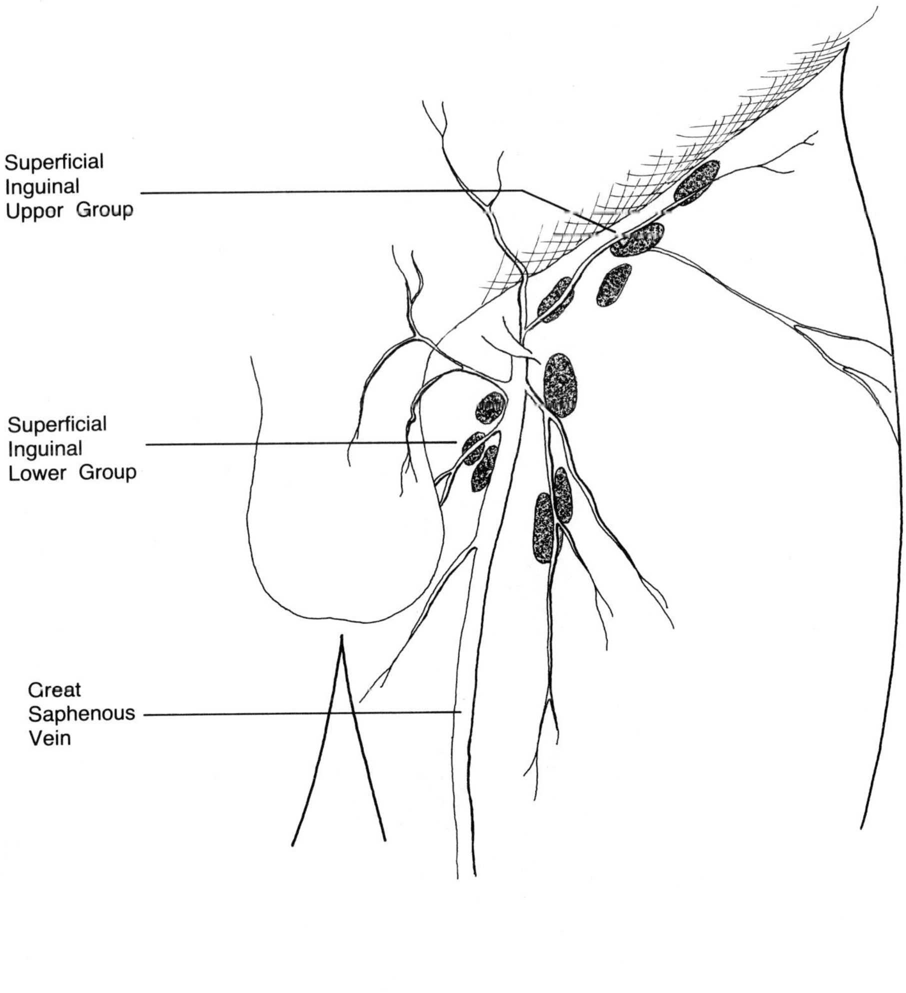

Many interconnections exist between these three zones (RICHARDS et al. 1962; STEARNS et al. 1980). Since many of the squamous cell carcinomas which originate in the anal canal cross the dentate line and extend to the rectal wall, they may metastasize to hypogastric nodes in addition to the superior hemorrhoidal chain (RICHARDS et al. 1962).

11.2 Gross Pathology

11.2.1 Location of Tumor

In 200 cases of anal malignancies reviewed by KUEHN et al. (1968), four lesions were located in the rectal region, 109 were in the mucocutaneous area, 26 were confined to the anal canal, and in 21 instances the tumor occupied the canal in addition to the perianal skin. In 29 cases the cancer was outside the canal and localized to the perineal area (Fig. 3). KLOTZ et al. (1967), in a survey of transitional carcinomas of the anal canal, found 13.3% of the tumors superior to the pectinate line (60% within 1 cm) and the most proximal lesions were at 6 cm from this anatomic site. Forty-four percent of the cancers originated inferior to the pectinate line. In an analysis of 113 squamous cell carcinomas of the anal canal at the Mayo Clinic (BEAHRS 1979), 58 were anorectal involving the dentate line, 24 lesions arose in the canal, and 31 were located in the perianal area. In the basaloid histology group, 53 were in the anorectal region and 11 were confined to the anal canal. STEARNS et al. (1980) have stated that cancers which are located in the upper canal are frequently nonkeratinizing while those in the lower part are more keratinizing.

11.2.2 Configuration and Size

Polypoid lesions account for a small number of anal cancers. In the series of KLOTZ et al. (1967) 13.8% of the patients presented with a polypoid mass and 11.7% of the tumors were observed to be circumferential. At the Mayo Clinic, 10 of 58 specimens examined were polypoid (RICHARDS et al. 1962). In most instances anal tumors are fungating, ulcerating, or excavating lesions. Occasionally no intraluminal mass is noted; however, obliteration of the normal mucosal pattern or minute ulceration may be seen in some cases (KLOTZ et al. 1967). The size of lesions may range from 0 cm for in situ carcinomas to 10 cm (RICHARDS et al. 1962; KLOTZ et al. 1967; WELCH and MALT 1977), with an average length of 4.2 cm.

11.3 Microscopic Pathology

The World Health Organization (WHO) classification of anal tumors recommends the following histologic typing (MORSON and SOBIN 1976):

1. Epithelial tumors consisting of squamous cell carcinoma, basaloid or cloacogenic carcinoma, adenocarcinoma, undifferentiated carcinoma, and unclassified carcinomas
2. Nonepithelial tumors
3. Malignant melanoma

11.3.1 Epithelial Tumors

11.3.1.1 Carcinoma In Situ

This type of carcinoma is found more frequently in the transitional epithelium than in the squamous membrane of the lower anal canal and the histolog-

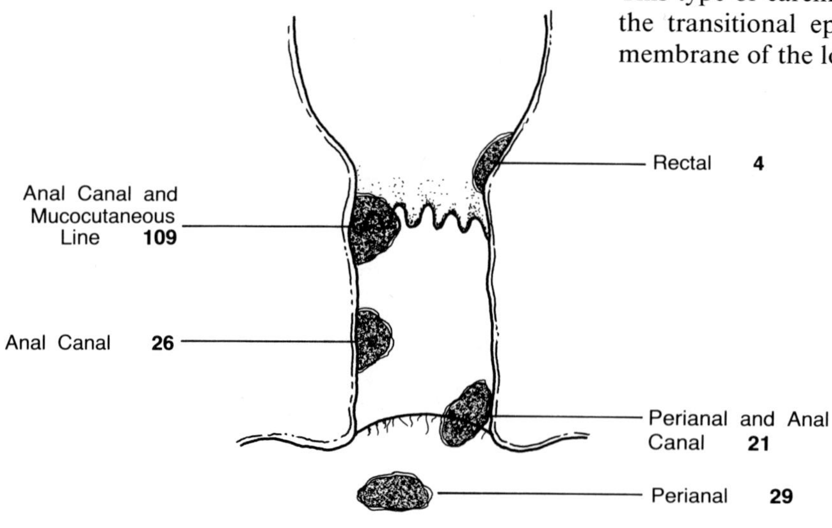

Fig. 3. Cancer of the anus: treatment regimen

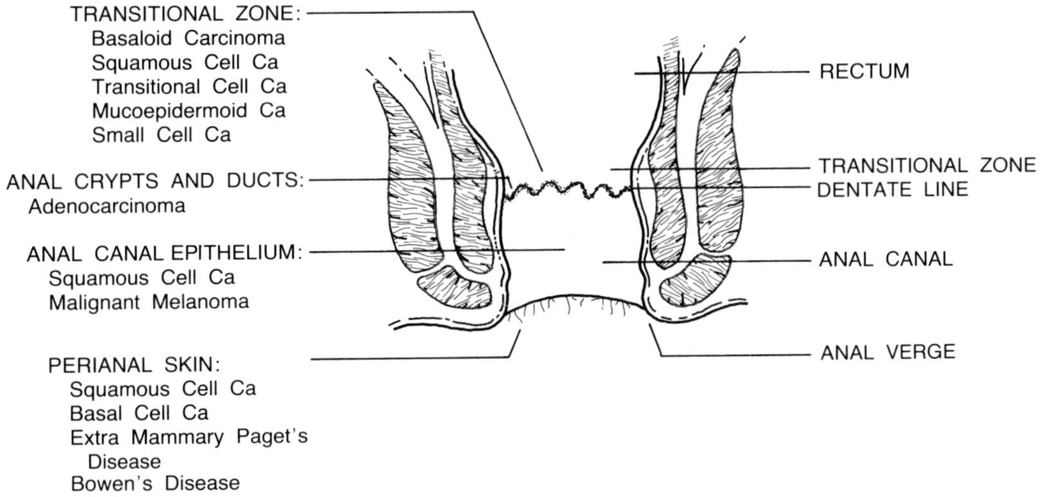

Fig. 4. Location of histologic subtypes of anal carcinoma

ic appearance has much in common with analogous lesions of the cervix (MORSON 1979). Carcinoma in situ may be discovered during the histologic examination of benign lesions (KLOTZ et al. 1967).

11.3.1.2 Invasive Tumors of the Anal Canal

Due to the diverse histologic structure of the anal canal, tumors arising in this area present a variety of histologic pictures and a broad terminology has been used to designate these varieties. Detailed microscopic examination of these lesions often reveals a mixed histologic pattern containing elements of squamous cells, transitional cells resembling bladder carcinoma, or even elements of mucoepidermoid differentiation (MORSON 1960; GRODSKY 1969). It has been suggested that most anal canal cancers originate from the unstable transitional zone above the anal valve (GRODSKY 1969; MORSON 1979). Figure 4 demonstrates the location of different histologic subtypes of anal carcinoma.

11.3.1.2.1 Squamous Cell Carcinoma. These tumors arise from the squamous epithelium of the anal canal and the supravalvular region, and demonstrate variable degrees of differentiation, closely resembling cancer of the cervix histologically (MORSON 1979). This type of tumor does not show intercellular cytoplasmic bridges or prickle cell formation and generally produces less keratin than is seen in typical skin cancers such as those in the anal margin (MORSON 1979).

11.3.1.2.2 Basaloid or Cloacogenic Carcinoma. The incidence of these tumors varies widely according to different reports. These tumors originate from the anorectal junction or transitional zone. This area is believed to represent a vestige of the cloacal membrane, an endodermal-ectodermal plate formed by the cloaca and surface ectoderm during early embryonic development (GRINVALSKY and HELWIG 1956; GILLESPIE and MACKAY 1978). Tumors in this area are therefore referred to as cloacogenic. The term basaloid stems from the fact that the histologic pattern of some of these tumors mimics that of basal cell carcinoma of the skin or rodent ulcer (MORSON 1960). The use of the term "basaloid" for the description of these lesions has been debated (GRODSKY 1969); however, there is some support from electron microscopic studies which demonstrate features of basal cell tumors of skin in cases of cloacogenic carcinoma (FISHER 1969). Three histologic subdivisions for cloacogenic neoplasms have been proposed: well differentiated, moderately differentiated, and undifferentiated or anaplastic. The well differentiated tumors are so designated by the presence of palisading at the periphery of the clumps of tumor cells (GRODSKY 1969; MORSON 1979). Other features which resemble basal cell carcinomas are a pseudoacinar pattern which is really a manifestation of palisading and the formation of small concentric whorls of squamoid cells undergoing incomplete differentiation toward keratin. One feature unlike basal cell carcinoma of the skin is the presence of masses of eosinophilic necrosis within the cell nests (MORSON 1979). In some cases the tumors resemble infiltrating transitional cell carcinoma of the bladder (MORSON 1960; HARRISON et al. 1966). Moderately differentiated carcinomas have less peripheral cell palisading and there is more cellular pleomorphism (Grodsky 1969; Morson 1979). Anaplastic or undifferentiated basaloid carcinomas, also called "small-cell basaloid" (Harrison et al. 1966; Grodsky 1969; Morson 1979), are rare and considered highly malignant. They may resemble the pleomorphic anaplastic "oat cell" of lung carcinoma. Ultrastructural features of neuroendocrine differentiation identical to the lung lesions have been observed in small cell carcinomas in other locations of the large bowel (Mills et al. 1983). There are no published reports of the presence of similar ultrastructural findings in anal small cell cancers.

11.3.1.2.3 Mucoepidermoid Carcinoma. Mucoepidermoid cancers of the anal canal are reported to represent one out of every 12 other histologic varieties at the Memorial Hospital in New York (Berg et al. 1960), while at St. Mark's Hospital in London they account for one of every eight anal canal tumors (Morson and Volkstadt 1963 a). These neoplasms resemble their counterparts in the salivary gland, lung, and esophagus and are defined as squamous or transitional cell carcinomas containing areas which secrete mucin. In addition to these lesions that represent classic mucoepidermoid cancers, anaplastic types have also been described (Berg et al. 1960). It has been suggested that anal canal mucoepidermoid carcinoma arises from the supravalvular region rather than the anal glands. The prognosis of these neoplasms appears to be comparable to that of other varieties of anal canal tumors (Berg et al. 1960; Morson and Volkstadt 1963 a).

The different morphologic types of anal cancer cited above do not influence prognosis (Morson 1960; Dougherty and Evans 1985). Considering the multiplicity of terms used in defining these lesions, Morson (1960) proposed classifying them in two categories, keratinizing and nonkeratinizing. Generally speaking, anal canal lesions produce little keratin.

11.3.1.2.4 Adenocarcinoma. True adenocarcinomas of the anal canal are very rare (Wellman 1962; Zaren et al. 1983; Goliger 1984 b, pp 780–793). Adenocarcinoma in this region can have different origins; it may be an extension of rectal carcinoma into the anal canal or an implantation metastasis from a colorectal cancer onto a raw area such as a hemorrhoidectomy wound (Harrison et al. 1966; Morson 1979; Goligher 1984 b, pp 780–793). True adenocarcinomas are submucosal growths with a tendency to spread in an annular fashion. The reported cases have been divided into three groups: (1) anal, (2) perianal, perirectal, and ischiorectal, (3) those developing in a fistulous tract. Such a classification has importance in regard to lymphatic spread according to the location of the tumor (Wellman 1962). Carcinoma in an anal fistula is very infrequent and in order to determine that a lesion arose primarily from a fistula, three criteria have been proposed (Kline et al. 1964). First, the fistula should have been present sufficiently long to exclude the reasonable possibility that cancer existed before the fistula. Second, no part of such a tumor should be present in the rectum or anal canal unless it be definitely secondary. Third, the opening of the fistula should be into the anal canal or crypt, not into the malignant tissue. Adenocarcinomas associated with fistula *in ano* are commonly of the colloid variety with abundant mucin production. The development of squamous cell carcinoma has also been described in an anal fistula (Kline et al. 1964). Apocrine glands which surround the anus and are vestigial remnants of scent glands found in animals may also be the source of anal adenocarcinoma (Thompson 1956).

The type of mucin secreted by anal glands differs histochemically from that produced by the goblet cells of the rectal mucosa. A strongly positive periodic acid-Schiff (PAS) reaction is shown in both but in the anal glands this is totally abolished after borohydride-KOH treatment (Fenger and Filipe 1977). This method might aid in the differential diagnosis of anal carcinomas (Morson 1979).

11.3.2 Nonepithelial Tumors

11.3.2.1 Sarcomas

A variety of rare anal and perianal sarcomas have been described. These include malignant lymphoma (McGraw and Bonefant 1960; Steele et al. 1985), rhabdomyosarcoma (McGregore and Jewett 1965), and liposarcoma (Harrison et al. 1966). Anorectal lymphomas appear as solitary, sharply circumscribed, submucous nodules or broad-based polypoid tumors. Their surfaces are smooth and mobile and ulceration or erosion is very uncommon. They may be firm in consistency, but never hard. Malignant lymphoma of the anus is a very rare condition, but it is not uncommon for leukemia to present with perianal involvement (Sehdev et al. 1973; Dubois et al. 1985). Benign lymphoid hyperplasia should not be confused with malignant lymphoma.

Very few cases of embryonal rhabdomyosarcoma of the anorectal region are reported in the medical literature. They may present as an anal or perianal mass and disseminate widely with hematogenous and lymphatic metastases. In children, this condition is closely related to botryoid sarcoma which is observed in the genitourinary tract (McGregore and Jewett 1965).

11.3.2.2 Miscellaneous Tumors

Very rarely a metastatic tumor originating from adenocarcinoma of the large bowel may be identified in a hemorrhoid. A case has been described where squamous cell carcinoma of the lung metastasized to the anal canal (Ger and Reuben 1968).

11.3.3 Malignant Melanoma

Primary malignant melanoma of the anal canal is a rare and highly malignant lesion (Morson and Volkstadt 1963 b; Chiu et al. 1980; Wanebo et al. 1981; Cooper et al. 1982). It accounts for approximately 0.25% of all anorectal malignant neoplasms (Harrison et al. 1966), and originates in the vicinity of the pectinate line, possibly involving the adjacent rectum. It is accepted that malignant melanoma of the anorectal region arises from squamous epithelium and the occasional appearance of this lesion above the anal valve is the result of contiguous upward spread of squamous epithelium or ectopia (Morson and Volkstadt 1963 b). These tumors are often large and polypoid and in some patients are mistaken for thrombosed hemorrhoids due to the dark pigmentation. Almost half of these lesions appear nonpigmented on gross inspection; however, only approximately 25% are amelanotic on microscopic examination (Harrison et al. 1966; Cooper et al. 1982). The neoplasm spreads rapidly with early involvement of the lymph nodes of the superior hemorrhoidal group, and approximately 60% of patients have nodal or visceral metastases at presentation (Cooper et al. 1982). The microscopic structure of anorectal malignant melanoma is variable (Morson 1979), and the most useful diagnostic criteria are considered to be melanin production, junctional change, and a nesting growth pattern (Cooper et al. 1982).

11.4 Incidence

The incidence of anal carcinoma varies from 1% to 4% of large bowel cancers in reported series. A report from the Presbyterian Hospital in New York cited an incidence of 1.8% in a series of 2341 patients with large bowel cancer (Grinnell 1954), while of 3200 patients seen between 1954 and 1958 at the Mayo Clinic with large bowel adenocarcinoma, 1% were afflicted with anal cancer (Richards et al. 1962). In an updated report from the latter institution, anal carcinoma constituted 2% of 9500 cases of rectal and anal malignancies (Boman et al. 1984). Data from the Liverpool Cancer Registry in England indicate a frequency of 2.8% for anal cancer among a group of 5189 patients with anal and rectal tumors (McConnell 1970). At St. Mark's Hospital in London the figure was 3.7% of 3672 cases of anal and rectal neoplasms (Morison and Volkstadt 1963 b), and at the Memorial Sloan-Kettering Cancer Center in New York anal carcinoma constituted less than 4% of the malignant tumors of the distal 15 cm of the large bowel (Quan 1986). The Swedish Cancer Registry records an incidence rate of 0.25 per 100000 men and 0.90 per 100000 women (Bohe et al. 1982).

11.5 Patient Population

11.5.1 Age

The reported age range is between 22 and the late 80s (Quan 1986). The median age in the Mayo Clinic series is 59 years (Boman et al. 1984). The average in three different series is recorded as 57.9 (Grinnell 1954), 58 (Quan 1986), and 60.4 years (Klotz et al. 1967). The average age is slightly younger for patients with anal cancer as compared to rectal adenocarcinoma (Quan 1986). A correlation between age and site of tumor was found in one report (McConnell 1970). Anal margin malignancies occurred more often in middle aged and younger individuals, while anal canal carcinomas were observed in the older age group. However, a similar correlation has not been found in other series (Stearns et al. 1980).

11.5.2 Gender

Women outnumber men in most series of patients. The ratio between female and male patients varies at different institutions. At the Mayo Clinic, the ratio was reported as 125:47 (Boman et al. 1984) and at Memorial Sloan-Kettering Cancer Center it was 76:41 (Stearns et al. 1980). In one study, two-thirds of the patients were women (Welch and Malt 1977) and reports from Sweden (Bohe et al. 1982) and Switzerland (Merlini and Eckert 1985) also show a female preponderance. In a series originating from Vanderbilt University in Tennessee, an equal distribution of females and males was noted (Sawyers et al. 1963). At St. Mark's Hospital in London (Morson 1960), the female to male ratio for anal canal tumors was 59:44, while for anal margin malignancies, the figure for males was higher than that for females (31:7). An epidemiologic survey of the Los Angeles County population demonstrated that anal canal carcinoma was more common among males under the age of 35 compared to females; beyond this age there was a substantial female preponderance (Peters and Mack 1983).

11.6 Epidemiology

An association between benign anal conditions and anal malignancy is often observed. In a study of 51 patients with anal carcinoma, 60% had a pre-existing history of benign lesions such as hemorrhoids, fistula, prolapse, or pruritus (Buckwalter and Jurayj 1957) and many patients harbored these conditions for over 5 years. This association was observed more often in men than women. A similar connection has been recorded in other reports (Grinnell 1954; McConnell 1970; O'Brien et al. 1982); however, Stearns et al. (1980), state that the pre-existing benign lesions are not causally related to anal cancer. A high incidence of anorectal tumors is reported from India and it is stated that these neoplasms constitute 30% of the large bowel malignancies seen in the All India Institute of Medical Sciences Hospital (Kapur et al. 1977). The pressure which the anorectal mucosa sustains during defecation in a squatting position has been cited as a contributing factor. In Northern Brazil anal carcinoma appears with a high frequency, with a female to male ratio of 4:1. Poor hygiene and the type of underclothing that the population wears have been implicated in the etiology of these tumors (De Oliveira 1976).

An increased incidence of anal cancer is reported in patients with Crohn's disease. The immunosuppressed state of many of these patients and the presence of prolonged perianal inflammatory disorders in the region have been considered as promoting factors (Slater et al. 1984).

Condyloma acuminatum, a nonmalignant condition of the anorectum caused by the papilloma virus, has been observed to be associated with intraepithelial neoplasm of the anus, as well as invasive lesions (Prasad and Abcarian 1980; Croxson et al. 1984). In situ carcinoma was diagnosed in nine cases of condyloma acuminatum within the condyloma tissue or adjacent anal mucosa. Five of these patients had a diagnosis of acquired immunodeficiency syndrome (Croxson et al. 1984).

Male homosexuals have been identified in a number of reports as individuals at high risk for the development of anal cancer (Cooper et al. 1979; Leach and Ellis 1981; Li and al. 1982).

Epidemiologic studies have indicated a greater incidence of anal carcinoma in single men, as compared to married individuals (Daling et al. 1982; Austin 1982). In a survey from Los Angeles County the marital status and other characteristics of 970 patients with anal and perianal cancers, diagnosed between 1972 and 1981, were determined (Peters and Mack 1983). As compared to all county residents and all other persons with a diagnosis of cancer, the incidence of anal cancer in single males was 6.1 times higher than in married males (Peters and Mack 1983). The female to male ratio for this neoplasm was 1.6. Single women were not at an increased risk, but divorced and separated individuals were compared to married persons. It was concluded that anal carcinoma was related to sexual activity involving the anus. In another epidemiologic study of Los Angeles County residents, several aspects of anogenital cancers were surveyed (Peters et al. 1984). Three parallels were found in the epidemiology of epidermoid carcinomas of the anus, cervix, vagina, vulva, and penis. The incidence of each site increased with decreasing social class of the patient. It was low among Jews and elevated among separated and divorced individuals of both sexes. It has been postulated that the anogenital epithelium constitutes a single organ at least as far as the etiology of carcinoma is concerned (Peters et al. 1984). Clinical evidence suggests that coital behavior and/or hygiene play an etiologic role at each site. While a higher incidence of penile cancer was observed among blacks and Hispanics compared to Caucasians, there was a deficit of anal neoplasia among these ethnic groups which could be the result of differences in patterns of anal coitus (Peters et al. 1984). The epithelium of the anus, cervix, vagina, vulva, and penis have a common embryologic origin in the urogenital sinus, which evolves from the cloacogenic membrane (Stern and KAPLAN 1969). Trauma, chronic irritation, and infection may be common etiologic factors in the genesis of carcinoma. Sexually transmitted viruses have been implicated in the evolution of these tumors (PETERS et al. 1984). The concomitant or sequential occurrence of female genital carcinoma with cloacogenic cancer of the anal canal has been observed (STERN and KAPLAN 1969).

11.7 Patterns of Spread

There are three pathways of tumor spread: (1) direct or local invasion, (2) lymphatic spread, (3) hematogenous dissemination.

11.7.1 Local Extension

The extension of anal canal neoplasms to the perineal tissue and sphincter muscles is commonly seen, and upward spread to the rectum is often observed, occasionally as high as 6–7 cm (GRINNELL 1954; MORSON 1960). In one series invasion of the neighboring structures was seen in 17.5% of cases and the vaginal wall was involved in 40.2% (KLOTZ et al. 1967). In another series the incidence of vaginal wall involvement was reported to be 12% (STEARNS et al. 1980). Adherence or extension of anal tumor to the prostate and seminal vesicles has also been described. While prostatic involvement was noted in 5% of the patients in the series of KLOTZ et al. (1967), only one of the nine patients had invasion of the seminal vesicles. The vagina is the most common site of extension when the anal tumor is located in the anterior portion of the canal. Involvement of the base of the bladder and/or urethra, cervix, sacrum, and coccyx is also reported (GRINELL 1954; KLOTZ et al. 1967). Vaginal and bladder extension was observed in 16% of cases in one series (KUEHN et al. 1968).

11.7.2 Lymphatic Spread

Lymphatic metastases occur frequently and the most common lymph nodes to be involved are the perirectal group, with an incidence of 75% in one series (KLOTZ et al. 1967). Metastases to the perirectal and mesocolic lymph nodes along the route of the inferior mesenteric vessels were reported in 7 of 25 surgical specimens, or at a rate of 28% (GRINNELL 1954). The frequency of inguinal nodal metastases at presentation is reported to be 14% (DILLARD et al. 1963) and 15% (STEARNS et al. 1980; FROST et al. 1984) in three different series. The overall incidence of inguinal disease has been reported to be 30% (STEARNS et al. 1980), 41% (KUEHN et al. 1968), and 63% (KLOTZ et al. 1967). The lesions that are located high in the anal canal at the mucocutaneous junction are said to be associated with a higher incidence of inguinal gland metastases. STEARNS et al. (1980) observe that delayed inguinal metastases are twice as common in keratinizing compared to nonkeratinizing anal tumors.

At the Memorial Sloan-Kettering Cancer Center a number of patients with anal canal carcinoma underwent extramesenteric node dissection at the time of abdominoperineal (AP) resection (STEARNS et al. 1980). Fourteen of 39 patients (36%) were found to have pelvic nodal involvement (external iliac, internal iliac, and obturator). Mesenteric nodal disease was noted in 22 of 86 cases (25%). Those with more distal lesions had a lower incidence of nodal metastases, while the rate approached 50% for patients with extensive carcinoma. Because of lymphatic interconnections even cancers that do not cross the mucocutaneous junction may have mesenteric nodal involvement and pelvic nodal disease may be present without superficial inguinal metastases (STEARNS et al. 1980). In one series, mesenteric nodal disease was detected in 18 of 50 patients (36%) who underwent an AP resection (DILLARD et al. 1963). A correlation has been found between the presence of lymph node metastases and the size of the primary lesion. Nodal disease is seen more frequently in patients with larger cancers; however, metastases have been noted with tumors as small as 5 mm (DILLARD et al. 1963). WELCH and MALT (1977), in a study of anal canal tumors, did not encounter any positive mesenteric nodes with negative inguinal glands. Para-aortic lymph node involvement has been reported in 9% of anal canal carcinomas (KLOTZ et al. 1967).

11.7.3 Hematogenous Spread

The frequency of hematogenous spread varies in different series. In three reports with large numbers of patients the rate of distant metastasis was cited to be 10% (KUEHN et al. 1968), 19% (KLOTZ et al. 1967), and 25% (FROST et al. 1984), and the liver was the most common site with a frequency of 6.5% (KUEHN et al. 1968) or 8.3% (KLOTZ et al. 1967). Lung and bone metastases occurred at a rate of 4% and 2%, respectively. Other organs such as the kidneys, adrenals, and brain may also harbor metastatic deposits (KLOTZ et al. 1967).

11.8 Natural History

Untreated anal carcinoma managed solely by a palliative colostomy leads to death within 8-36 months, with a median of 20 months. On the average, patients succumb to this disease in a period of 13 months (KLOTZ et al. 1967).

11.9 Patterns of Recurrence

In the Mayo Clinic's series of surgically treated patients, 84% of recurrences were locoregional (BOMAN et al. 1984). This figure takes into account those patients who in addition to distant metastases had a component of locoregional disease. The time to the development of recurrence (distant and local) ranged from 2.2 months to 17.5 years with a median of 15 months. In the report by WELCH and MALT (1977) the median time to recurrence was 7.5 months with all recurrences being observed within a period of 2 years from surgery. In the report of KLOTZ et al. (1967), which incorporated several institutions, local recurrence occurred in 26% of patients. They recorded an interval of 6-144 months with an average of 21 months to the development of recurrence. The perineum was the most common site followed by the pelvis, vagina, vulva, bladder, and sacrum. The time from primary treatment to the appearance of inguinal gland metastases may range from 3 to 88 months with a median of 12 months (FROST et al. 1984). The time to recurrence decreases as the stage advances and local recurrence develops sooner than distant metastases. The rate of local recurrence is dependent on the type of surgical procedure, being high with local excision (SINGH et al. 1981).

11.10 Staging

The design of a comprehensive staging system for carcinoma of the anal canal is hampered by the lack of sufficient numbers of patients with accurate detailed clinical information. Various staging systems have been proposed and some of these in ad-

Table 1. Different staging classifications for anal carcinoma

TNM UICC (1982)	Mayo Clinic BOMAN et al. (1984)	Roswell Park Memorial Institute SINGH et al. (1981)	Centre Léon Berard PAPILLON (1982)
Tis: Preinvasive carcinoma (in situ) T0: No evidence of primary tumor T1: Tumor occupying not more than ⅓ of the circumference or length of the canal and not infiltrating the external sphincter muscle T2: Tumor occupying more than ⅓ of the circumference or length of the anal canal or tumor infiltrating the external muscle T3: Tumor with extension to rectum or skin but not to other neighboring structures T4: Tumor with extension to neighboring structures N0: No evidence of regional lymph node involvement N1: Evidence of involvement of regional lymph nodes	A: Confined to anal epithelium and subepithelial connective tissue B: Tumor penetration into the muscle or adjacent pelvic tissue B1: invasion of internal sphincter B2: external sphincter B3: adjacent pelvic tissues C: Regional node involvement, inguinal or pelvic D: Unresectable regional tumor, or distant metastasis	0: Carcinoma in situ I: Invasion into submucosa IIA: Invasion into surrounding muscles only IIB: Invasion of contiguous soft tissue, ischiorectal fossa, vagina, bladder, urethra IIIA: Lymph nodes - retrorectal or mesocolic IIIB: Lymph nodes - inguinal, iliac, obturator IVA: Distant metastasis to high para-aortic nodes, sacrum IVB: Distant organ metastasis - liver, lung, bone, brain	T1: Tumor not exceeding 2 cm in diameter T2: Tumor between 2 and 4 cm T3: Tumor larger than 4 cm, mobile, infiltrating neither the vaginal mucosa nor the genitourinary tract T4a: Tumor invading the vaginal mucosa T4b: Tumor extending to neighboring structures other than skin, rectum, and vaginal mucosa

dition to the TNM classification of the UICC (1982) are shown in Table 1. The TNM staging system has been criticized for a variety of reasons (PAPILLON 1982). It is difficult to judge the involvement of the internal sphincter, and the presence of extension to the rectal or perianal skin does not necessarily imply a poor prognosis. It has been stated that T1-T2 tumors constitute only one-third of anal carcinomas, while two-thirds are T3 or T4 lesions at presentation (PAPILLON 1982).

11.11 Symptoms

The symptoms of anal canal carcinoma are similar to benign conditions such as hemorrhoids, fissures, and pruritus (STEARNS et al. 1980). The most common presenting symptom is bleeding and the second most frequent symptom is pain (KLOTZ et al. 1967; BENSAUDE and NORA 1968; MCCONNELL 1970; BEAHRS 1979). Bleeding is usually in small volume and anemia due to blood loss indicates a poor prognosis. It is estimated that 40% of patients will experience pain. Since the anal canal has a greater sensory nerve innervation than other parts of the large bowel, pain due to the presence of tumor is an early symptom (GRINNELL 1954; BEAHRS 1979). Other signs and symptoms, in decreasing order of frequency, are change in bowel habits, pruritus, and presence of an anal mass. Occasionally, symptoms are followed by a 2- to 3-month symptom-free interval, and a recurrence may not alarm the patient. When invasion of the vagina has taken place, the patient may experience vaginal bleeding or be aware of a mass (KLOTZ et al. 1967). The duration of symptoms may range from 2 weeks (SINGH et al. 1981) to a number of years (KLOTZ et al. 1967; STEARNS et al. 1980) and those patients with a long duration of symptoms invariably have associated anal pathology. Incontinence due to the involvement of the anal sphincter or the development of a rectovaginal fistula is seen only in advanced cases (PAPILLON 1982). Patients with anal cancer may be totally asymptomatic, their disease being discovered during a routine physical examination or a survey for cancer detection (KLOTZ et al. 1967; BEAHRS 1979).

11.12 Clinical Findings and Diagnosis

Anal malignancies are usually diagnosed after a patient presents with local symptoms and the suspecting physician undertakes an examination of the anus. Unsuspected anal malignancies have been detected in 1.9% of hemorrhoidectomy, fistulectomy, or fissurectomy surgical specimens (GRODSKY 1967). In a review of 117 patients with anal carcinoma the diagnosis was established in seven patients from the histologic examination of unsuspected hemorrhoidectomy specimens (STEARNS et al. 1980).

The diagnosis and assessment of anal canal tumors are based on palpation and proctoscopic evaluation carried out in one of the positions suitable for anorectal examination. It is important to determine the relationship of the tumor to the dentate line, as well as its upper and lower extent, and the number of involved quadrants. The depth of invasion, especially in regard to the rectovaginal septum in the female and the prostate in the male, has to be ascertained. If severe pain due to a deeply ulcerating lesion impedes the examination, or the circumferential growth of the tumor prohibits a digital assessment, some form of anesthesia is required for a thorough evaluation (QUAN 1983; GOLIGHER 1984b).

Anal canal lesions present with different configurations. An early lesion may be a nodular elevation of mucosa or have the feeling of a warty growth (STEARNS et al. 1980; PAPILLON 1982). The tumor may have the appearance of a thrombosed hemorrhoid or an anal fissure, but generally a cancer has a harder edge than might otherwise be expected. Induration is an important diagnostic sign, but sometimes it is difficult to differentiate tissue hardening due to malignancy from an inflammatory process and only a biopsy will clarify the issue. As was discussed earlier in Sect. 11.2, most anal tumors are ulcerated lesions and those with a polypoid configuration are uncommon. In the series of PAPILLON (1982), they constitute 20% of the cancers. Occasionally, a malignant lesion does not project into the anal canal and is covered with swollen mucosa with little ulceration to be seen (KLOTZ et al. 1967; PAPILLON 1982). A generous biopsy is needed in such a situation to avoid an erroneous diagnosis which might be made from a superficial sample. Advanced cancers may protrude outside the anal canal and surround the anus, becoming firmly adherent to adjacent tissues. A complete annular growth with narrowing of the lumen may be observed in some cases (GOLIGHER 1984b, pp 780-793). Palpation of the perianal region through the intact skin is helpful in assessing the presence of induration and the lateral extent of the tumor. A digital vaginal examination will assist in defining the upper limit of anterior wall lesions and will also demonstrate whether the posterior vaginal wall is infiltrated. If the initial examination was performed without an anesthetic, PAPILLON (1982) recommends a re-examination under general anesthesia. With relief of pain and relaxation of the sphincter, the true limits of the tumor are appreciated and the perirectal and pelvic regions can be explored. Metastatic perirectal nodes may be felt as indurated nodules, and pelvic nodes (hypogastric chain), if enlarged, can be appreciated on the lateral walls of the pelvis.

The inguinal glands should be examined carefully for evidence of metastatic disease. The superficial inguinal nodes in one or both groins may be enlarged due to sepsis originating from an infected anal carcinoma, but such glands are often soft as opposed to the hard metastatic nodes. A needle biopsy is a reliable technique in differential diagnosis (PAPILLON 1982; QUAN 1983; GOLIGHER 1984b, pp 780–793).

Primary adenocarcinomas of the anal canal are often manifest as so-called extramucosal tumors and spread in an annular extraluminal fashion, infiltrating widely in the perianal tissue. They may also break through the perianal skin or occasionally give rise to a scaling erythematous eczematoid lesion (Paget's disease) (GOLIGHER 1984b, pp 780–793). Because of the absence of a lesion in the anal mucosa, deep biopsies are necessary to establish a tissue diagnosis. Regional lymph nodes are often extensively involved at presentation (GOLIGHER 1984b, pp 780–793).

11.12.1 Diagnostic Procedures

A histologic examination is always required for confirmation of malignancy and for distinction between squamous carcinoma, adenocarcinoma, and other rare tumors. A needle biopsy of perianal or pelvic nodes by a transrectal or transvaginal approach can verify the presence of metastatic disease (PAPILLON 1982). A chest X-ray and routine laboratory evaluations are required in all cases. Pedal lymphangiography is not advised unless a patient has inguinal nodal involvement (PAPILLON 1982). Considering the pitfalls and inadequacy of lymphography in the evaluation of pelvic nodes, delineation of these nodes with radiocolloid lymphography has been proposed (EGE and CUMMINGS 1980). The procedure involves the injection of ^{99m}Tc antimony sulfide colloid into both ischiorectal fossae. In addition to external and common iliac nodes, perirectal, internal iliac, and obturator nodes are also visualized by the gamma camera with this method. Computed tomography (CT) has been used in the preoperative evaluation of colorectal carcinoma. The sensitivity of this technique for the detection of nodal metastases is low, but it has a high sensitivity as well as specificity for hepatic involvement (FREENY et al. 1986). The local tumor extension can be correctly assessed in fewer than half of the patients (FREENY et al. 1986). There are no reported studies concerning the utility of CT scanning in the staging of anal carcinoma. However, a result similar to that of colorectal tumors can be expected for the assessment of liver and lymph nodes.

Unlike adenocarcinoma of the colorectum, no tumor markers of clinical utility have been explored for anal cancer. The level of squamous cell carcinoma antigen is now being determined in anal carcinoma. The preliminary results indicate that if might be a useful marker for monitoring patients with this neoplasm (SHAW et al. 1986).

11.13 Prognostic Factors

11.13.1 Age

Most studies indicate that age has no significance on survival.

11.13.2 Gender

In a study of 192 patients with carcinoma of the anorectum at the M. D. Anderson Hospital and Tumor Institute, females had a higher survival rate than males at 5 years (62% vs 50%) (FROST et al. 1984). A similar advantage for women has not been observed in other large series (KLOTZ et al. 1967; BOMAN et al. 1984).

11.13.3 Duration of Symptoms

A short duration of symptoms is associated with a better prognosis. Two studies have shown that those patients who present with symptoms of less than 3-month duration fare better compared to patients with a longer symptomatic period (KLOTZ et al. 1967; STEARNS et al. 1980).

11.13.4 Size

All studies agree that the size of the primary lesion influences the prognosis (SCHRAUT et al. 1983; BOMAN et al. 1984; FROST et al. 1984; QUAN 1986).

Small lesions 2 cm or less have a significantly better prognosis than larger tumors. In the Mayo Clinic series (Boman et al. 1984), patients with lesions measuring ≤ 2 cm had a 5-year survival rate of 83% compared to 55% for those patients with tumors >5 cm in size. The report from the M. D. Anderson Hospital and Tumor Institute (Frost et al. 1984) records a 5-year survival rate of 78% for patients with cancers 1–2 cm in diameter compared to 40% for those patients with lesions measuring 6 cm. It appears that small tumors 2 cm or less seldom metastasize to nodes (Schraut et al. 1983; Boman et al. 1984).

11.13.5 Location

The location of a lesion in relation to the pectinate line as a prognostic indicator has been studied (Klotz et al. 1967), and no difference in survival has been observed between tumors above, at, or below the pectinate line. Quan (1986) states that more proximal lesions of the anal canal have a worse prognosis and this has been confirmed elsewhere (Frost et al. 1984).

11.13.6 Histology

Most studies indicate that various histologic subgroups have no influence on survival (Welch and Malt 1977; Singh et al. 1981; Schraut et al. 1981; Boman et al. 1984). Quan (1986), in a survey of cases seen at the memorial Sloan-Kettering Cancer Center, indicates a better prognosis for purely basaloid tumors compared to epidermoid tumors (75% vs 50% survival rate at 5 years). No difference in survival rate was observed for patients with keratinizing compared to nonkeratinizing tumors (Stearns et al. 1980). Klotz et al. (1967) observed a relationship between grade and survival rate, with high grade tumors having a poor prognosis. In the Mayo Clinic series (Boman et al. 1984), the patients with low grade tumors had a better survival rate, but this was closely related to the fact that they were in the earlier stage groups. When corrected for stage, no difference in survival rate with regard to different grades was observed. Other studies have found that the grade of tumor is of borderline value (Frost et al. 1984), or no value (Morson 1960; Stearns et al. 1980). It has been stated that an increase in the desmoplastic activity of the tumor adversely affects the prognosis. Patients with lesions with maximum fibroblastic reaction have a 30.5% survival rate compared to 60% in those without such a reaction (Klotz et al. 1967). The depth of infiltration, which is a function of stage, correlated with the prognosis in all studies (Klotz et al. 1967; Schraut et al. 1983; Greenall et al. 1985a) and has a significant impact on survival. The 5-year survival rate for patients with superficially invasive cancer is 78%–87% compared with 43% for patients with deeply penetrating tumors (Boman et al. 1984; Frost et al. 1984). The existence of vascular and lymphatic invasion has an unfavorable influence on survival (Klotz et al. 1967; Stearns et al. 1980).

The presence of inguinal nodal metastases at presentation carries a poor prognosis (Dillard et al. 1963; Klotz et al. 1967; Wolfe and Bussey 1968; Quan 1986). The 5-year survival rate for patients with synchronous groin node metastases ranges from 21% (Klotz et al. 1967; Wolfe and Bussey 1968) to 9% (Quan 1986). While patients with positive inguinal nodes carried the same prognosis as mesenteric or perirectal nodal metastases in the M. D. Anderson Hospital and Tumor Institute series (Frost et al. 1984), data from Memorial Sloan-Kettering Cancer Center indicate a better outcome for patients with mesenteric and pelvic lymph node involvement (Quan 1986). A high rate of nodal metastases was observed for nonkeratinizing basaloid tumors in one report (Boman et al. 1984), while no difference in relation to nodal disease was seen between keratinizing and nonkeratinizing lesions in another series (Stearns et al. 1980), except that delayed inguinal nodal metastases were twice as frequent with keratinizing cancers.

11.14 Uncommon Anal Tumors

11.14.1 Malignant Melanoma

Malignant melanoma of the anal canal is a rare disease. For every eight squamous cell carcinomas of the anus seen at St. Mark's Hospital in London, one case of malignant melanoma was encountered (Morson and Volkstadt 1963b). The largest reported series from a single institution (Memorial Sloan-Kettering Cancer Center) consisted of 51 patients (Wanebo et al. 1981). It is observed mainly in the 5th and 6th decades of life but the age range is wide (20–90 years). Bleeding, pain, and the presence of a mass are the common presenting symptoms (Chiu et al. 1980; Wanebo et al. 1981; Cooper et al. 1982). Most patients have large advanced tumors at the time of diagnosis and the prognosis is poor, the 5-year survival rate being approximately

12% (CHIU et al. 1980; WANEBO et al. 1981). The survival rate is related to the thickness of the tumor and the depth of penetration. Only patients with lesions 2 mm or less in thickness have been reported to survive beyond 5 years (WANEBO et al. 1981). The disease disseminates widely, through lymphatic channels and hematologic routes to involve the liver, lung, brain, bony skeleton, and skin. The nodal status at diagnosis has a significant bearing on survival. The mean survival time for node negative patients is reported to be 16 months compared to 6 months for those with positive nodes (WANEBO et al. 1981). In the Memorial Sloan-Kettering Cancer Center series, only one of the six node positive patients survived beyond 5 years.

11.14.2 Small Cell Anal Carcinoma

These tumors are aggressive neoplasms with a rapid fatal course. Among the 188 patients with anal carcinoma seen at the Mayo Clinic (BOMAN et al. 1984), 13 had a small cell carcinoma histology. Five of the 13 patients had evidence of distant metastases at diagnosis and lived for a median time of 2.2 months. Only one patient in the group survived more than 5 years without recurrence. Postoperative recurrence developed in six patients following an AP resection in a median interval of 4 months, with death following shortly (BOMAN et al. 1984).

11.15 Perianal Tumors

In the study of KUEHN et al. (1968), of 189 cases of anal canal and perianal tumors only 29 cases (15%) had tumors which were confined to the perianal region. BEAHRS (1979) reports that among 113 squamous cell carcinomas of the anus, 31 were located in the perianal area. At the Memorial Sloan-Kettering Cancer Center, anal margin cancers accounted for 25% of all epidermoid cancers of the anal region (GREENALL et al. 1985 b). The percentage of the cases in any series will depend on the anatomic boundaries that are used. MORSON (1960) considers anal verge tumors to be those lying inferior to the dentate line. In his series, 38 of 141 (27%) cases are designated as anal margin tumors. It is difficult sometimes to determine whether a tumor arises in the anal canal or the anal margin. In the series of KUEHN et al. (1968), 21 of the 189 lesions had such a position. Anal margin tumors are biologically distinct from anal canal cancers and tend to be less invasive (GABRIEL 1960; TURELL 1962; DILLARD et al. 1963; SAWYERS 1972; BEAHRS 1979).

11.15.1 Pathology

The anal margin or perianal tumors are classified by the WHO into the following pathologic groups (MORSON and SOBIN 1976):

A. Malignant lesions:
   1. Squamous cell carcinoma
   2. Basal cell carcinoma
   3. Others
B. Bowen's disease
C. Paget's disease

11.15.1.1 Squamous Cell Carcinoma

These are mostly well differentiated cancers similar to those occurring elsewhere in the skin. Perianal squamous cell carcinomas are keratin producing and may be compared to squamous cell carcinomas of the lip, which grow slowly and metastasize to lymph nodes at a late stage (MORSON 1979). Verrucous carcinoma, a variant of squamous cell carcinoma, also occurs at this site (MORSON and SOBIN 1976), and can appear as a pale, cauliflower-like mass in the perianal skin. Histologically, it is characterized by voluminous rete pegs extending into deeper tissues, with keratin filling the crevices between papillary structures (GINGRASS et al. 1978).

11.15.1.2 Basal Cell Carcinoma

Basal cell carcinomas of the perianal region are very rare (QUAN 1978; KRAUS 1978; NIELSEN and JENSEN 1981; WHITE et al. 1984). They have a similar histologic appearance to that of basal cell carcinoma of the skin in other sites (MORSON and SOBIN 1976). A number of microscopic features have been suggested which distinguish these lesions from basaloid or cloacogenic carcinomas. Basaloid carcinomas have a greater variation in nuclear size and shape, an increase in mitoses, absence of characteristic induced stroma, and prominent peripheral palisading (WHITE et al. 1984).

11.15.1.3 Bowen's Disease

Bowen's disease, which is rarely found in the perianal region (GRODSKY 1965; SCOMA and LEVY 1975; QUAN 1978; STRAUSS and FAIZO 1979), is a chronic intraepithelial squamous cell carcinoma in situ which invades and metastasizes in less than 5% of cases. Bowenoid cutaneous lesions may be associated with simultaneous or subsequent development of other systemic or cutaneous cancers. Early disease appears as a dull red, spreading, irregular, plaque-like eczematoid lesion. Microscopic findings are diagnostic and there is a striking loss of normal cell progression with characteristic giant, hyperchromatic, multinucleated, vacuolated cells in the malpighian cell layer (GRODSKY 1965). Unlike Paget's disease, a bowenoid cell does not pick up aldehyde-fuchsin stain (QUAN 1978).

11.15.1.4 Perianal Paget's Disease

Paget's disease of the anus is seen very infrequently (GRODSKY 1960; HUTCHESON et al. 1960; WILLIAMS et al. 1976; QUAN 1978) and is surrounded by a great deal of controversy. The histogenesis of Paget's cell and whether there must be a concomitant underlying invasive malignancy are debated. Grossly, it appears as a chronic perianal inflammation with eczematoid changes. The lesion is not unlike that which has been described for mammary disease. Histologically, characteristic Paget's cells are seen in the epidermis and sometimes in the ducts of underlying apocrine glands (MORSON 1979). It has been suggested that these cells originate from the epithelium of apocrine gland ducts and migrate into the epidermis (MORSON 1979).

In an analysis of 48 patients with epidermoid carcinoma of the anal margin which excluded Bowen's disease, Paget's disease, and simple basal cell carcinoma, GREENALL et al. (1985b) found that 67% of the lesions had evidence of keratin formation. Twenty-two percent of the cancers were well differentiated and only 4% were poorly differentiated or anaplastic, the remaining being classified as moderately differentiated. In 46 patients the size of the tumor was recorded; 14 had a lesion of 2 cm or less, in 21 patients it measured 2-5 cm, and for the remaining 11 patients, the mass exceeded 5 cm. In this study, cancers below the dentate line were considered in the anal margin group.

11.15.2 Clinical Features

Carcinoma of the anal margin has a higher frequency in males (GABRIEL 1960; WOLFE and BUSSEY 1968; GREENALL et al. 1985b). It has been stated that patients with anal margin carcinoma have a younger median age than those with anal canal tumors (MCCONNELL 1970). However, the median age of 56 years in the report of GREENALL et al. (1985b) is similar to that of patients with anal canal cancers. Coexisting conditions such as condyloma, fistula, leukoplakia, and chronic pruritus have been seen more frequently with anal margin neoplasms compared to anal canal tumors (GREENALL et al. 1985b). A history of previous irradiation to this area has been noted in a number of patients with anal verge cancers (GOLIGHER 1984b, pp 780-793; GREENALL et al. 1985b). In the early stages, neoplasms of the anal margin may be indolent and not give rise to pain until later stages are reached. Due to the absence of pain, the growth is often detected by accident during the cleaning of the region (BENSAUDE and NORA 1968). In the case of Bowen's disease, the patient may be without symptoms until recognition of the lesion during tissue examination from an anorectal procedure such as a hemorrhoidectomy (STRAUSS and FAIZO 1979). In symptomatic patients, the sensation of a lump or an ulcer, bleeding, pain, perianal itching, mucoid discharge, or change in the bowel habits lead to the diagnosis (NIELSEN and JENSEN 1981). In the early stages, anal margin neoplasms may look similar to a simple erosion of the skin, fissure, or an unusual type of hemorrhoid; the lesion may also present as a small hard nodule which grows slowly and eventually ulcerates (BENSAUDE and NORA 1968; PAPILLON 1982). Basal cell carcinoma of the perianal region appears as a chronic indurated growth with raised pearly borders and a central depression (SAWYERS 1972; NIELSEN and JENSEN 1981). Bowen's disease with reddish eczematoid features can slowly extend laterally, eventually covering a wide area. The dermatologic manifestations of Paget's disease with a crusty plaque-like lesion can resemble Bowen's disease and those with an underlying cancer often present with a mass and even metastatic inguinal nodes (QUAN 1978).

11.15.3 Patterns of Spread

These tumors grow mainly on the surface of the perianal region and may extend in the direction of the buttocks and vulva, possibly invading the anal canal (PAPILLON 1982). The perianal region is drained by efferent collecting lymphatics that end in the superficial inguinal nodes, and lymphatic spread to inguinal and iliac nodes is seen in 15%-20% of cases (PAPILLON 1982). Lesions of the anal margin, however, unlike anal canal cancers, rarely metastasize to the hemorrhoidal glands (AL-JURF et al. 1979). After local excision, recurrent disease may be observed in 42% of cases, locally in the perineal area or inguinal glands (9.5%) with the most common site being the perianal skin (GREENALL et al. 1985 b). Recurrences are usually observed within 5 years of primary surgery (GREENALL et al. 1985 b). Anal margin tumors rarely metastasize distantly (PAPILLON 1982).

11.15.4 Staging

The clinical staging system of UICC (1982) for anal margin or orifice carcinomas is not widely used. Staging is based on clinical examination, endoscopy, and radiography.

Primary tumor

- Tis: Preinvasive carcinoma (in situ)
- T0: No evidence of primary tumor
- T1: Tumor 2 cm or less in its greatest dimension, strictly superficial or exophytic
- T2: Tumor more than 2 cm but not more than 5 cm in its greatest dimension or tumor with minimal infiltration of the dermis
- T3: Tumor more than 5 cm in its greatest dimension or tumor with deep infiltration of dermis
- T4: Tumor with extension to muscle, bone, etc.
- Tx: The minimum requirements to assess the primary tumor cannot be met

Regional nodes

- N0: No evidence of regional lymph node involvement
- N1: Evidence of involvement of movable unilateral regional lymph nodes
- N2: Evidence of involvement of movable bilateral regional lymph nodes
- N3: Evidence of involvement of fixed regional lymph nodes
- Nx: The minimum requirements to assess the regional lymph nodes cannot be met

11.15.5 Diagnosis

The extent of the lesion in the perianal area and anal canal should be determined and the limit of the tumor in the anal canal should be verified by digital examination and endoscopy. The great variation in the appearance of anal margin cancers is such that every suspicious lesion should be biopsied (BENSAUDE and NORA 1968). When in doubt, several biopsies should be taken, especially at the edge of the lesion (PAPILLON 1982). The inguinal region should be thoroughly evaluated and if groin nodes are involved, lymphangiography is recommended (PAPILLON 1982).

11.15.6 Prognosis

There are very few reports of sizable numbers of patients with anal margin tumors. In addition, differences in opinion as to the boundaries of anal margin cancers add to the problem of assessing prognosis. In the series of BEAHRS (1979) from the Mayo Clinic, consisting of 31 patients, a survival rate of 74.2% at 5 years and 70.7% at 10 years was recorded. At the Memorial Sloan-Kettering Cancer Center, the absolute 5-year survival rate for 31 patients with anal margin carcinomas who underwent local excision was 68% with an adjusted survival rate of 88% (GREENALL et al. 1985 b). Of the ten patients who underwent excision for primary tumors at the Cleveland Clinic (AL-JURF et al. 1979), five remained without evidence of recurrence up to 20 years. At the same institution, seven patients were seen for recurrent disease after local excision elsewhere. Recurrences were observed within 3 months to 14 years following the treatment of the primary lesion.

11.15.7 General Management

Management decisions are often difficult as relatively few patients are seen by any one physician over a period of time. However, during the last decade many articles on the subject have appeared in the literature. A number of these reports discuss moderate sized series of patients and have therefore been instrumental in defining management strategies more clearly.

11.15.7.1 Squamous Cell Carcinoma

Squamous cell carcinoma of the anal margin should be managed as any epidermoid carcinoma elsewhere in the skin. Treatment may be complicated by involvement of the lower part of the anal canal and lymphatic spread to the inguinal glands is a common occurrence. This disease should be treated conservatively and in many cases a wide local excision produces a good local result but skin grafting may occasionally be required. Care must be taken that the resected margins are free of tumor. A report from the Mayo Clinic shows a 74.2% 5-year survival rate following surgery alone (BEAHRS 1979).

Squamous cell carcinomas at this site and elsewhere in the skin are responsive to radiation with or without the use of chemotherapeutic agents for radiation potentiation. Reports on the use of interstitial irradiation for tumors in this area have shown a high incidence of failure and complications (McConnell 1970). Occasionally, an anorectal fistula can become associated with squamous cell carcinoma and an AP resection may be required for successful management. A patient with this pathology arising in a fistula has a slight survival advantage over a patient with an adenocarcinoma arising in a fistula.

11.15.7.2 Basal Cell Carcinoma

This is a form of rodent ulcer involving the skin in the area of the anal margin. These neoplasms never invade deeply or metastasize and an adequate surgical excision virtually guarantees a cure. It is very rare not to be able to achieve primary surgical closure as these tumors are usually small.

11.15.7.3 Bowen's Disease

This disease is seen only rarely in the anal area. The treatment of choice is a wide local full thickness excision. It may be possible to produce a response by applying 5-fluorouracil (5-FU) ointment.

11.15.7.4 Perianal Paget's Disease

Very often this disease is the presenting symptom of an underlying adenocarcinoma in the rectum and inguinal nodal metastases are not uncommon. The early symptoms of this neoplasm are not severe but eventually become sufficient for the patient to seek medical care. If an underlying adenocarcinoma does exist, complete surgical eradication by an AP resection may be required.

11.15.7.5 Malignant Melanoma

This is the most discouraging lesion in the anorectal area to treat. The lesion is often advanced at presentation and accompanied by the metastatic involvement of inguinal nodes. The prognosis is resultantly poor. Radical surgery is the only form of treatment which has produced any successful results and the administration of chemotherapy and immunotherapy has not proven beneficial.

11.16 Anal Canal Tumors

As stated above, cells in the region of the dentate line are highly variable and this accounts for the histological subgroup of cancer that is seen arising in this region. Even though pathologists use such terms as cloacogenic, basaloid, transitional, epidermoid, and squamous cell carcinoma, it is advisable to consider all these to be variants of squamous cell carcinoma when considering management strategies as these cell subtypes do not appear to influence the therapeutic outcome.

11.16.1 Clinical Aspects

Unfortunately, at the beginning of this disease symptoms may be minimal and there is a reluctance among certain patients, for whatever reason, to come forward with complaints involving the anorectal area. There is also a lack of awareness by the medical profession and these two factors are often combined to cause considerable delay in diagnosis. As the size of the growth increases, however, the pain becomes sufficiently severe for the patient to seek medical advice. The symptoms of anal canal tumors mimic the more common inflammatory and nonmalignant conditions and malignancy is often discovered following an examination for "hemorrhoids" or following a so-called hemorrhoidectomy. When bleeding occurs from anal canal tumors, the loss is usually very slight and the outstanding clinical features of anal carcinoma which should be recognized are hardness and induration on digital examination.

A thorough assessment must be made of the lesion at presentation and a careful digital examination must be carried out. In the female patient, a vaginal examination should be performed. Proctoscopy should be undertaken to assess the extent of the tumor around the circumference of the anus, and its upper and lower limits. This examination is often painful due to narrowing of the lumen of the anus and there should be no hesitation about subjecting the patient to a thorough examination under either general or spinal anesthesia. A careful search should be made for perirectal nodes as metastatic nodules in most cases are low lying in the rectum and may be palpated. If involvement is suspected, needle biopsies of these nodes may be taken. A thourough examination of the inguinal area must also be carried out, bearing in mind that inguinal nodes may be enlarged due to inflammation. If involved, these nodes are usually hard on palpation, but, should any doubt remain as to their nature, a needle biopsy should be performed.

11.6.2 Treatment

Both surgery and radiation therapy may be offered as primary treatment for carcinoma of the anal canal but disagreement exists concerning the respective roles of these modalities between surgeons and radiotherapists. There is a current interest in multimodality management, including the use of certain chemotherapeutic agents to potentiate the effectiveness of irradiation, and new therapeutic approaches may result.

11.6.2.1 Surgery

A few, very small, superficial anal canal tumors may be amenable to local resection with anal sphincter preservation. On the whole, however, this procedure has very little place in the management of anal canal carcinoma. Some reports which claim good results probably concern carcinomas of the anal margin as the distinction between these and anal canal tumors is often not fully appreciated. NIGRO (1984) states that squamous cell carcinoma of the anal canal 2 cm or less in diameter may be excised locally with a fairly wide margin of normal tissue. It is recognized that local excision is an adequate operation only for well differentiated tumors that are sufficiently small to allow for wide removal. These patients should be followed carefully at regular intervals. Five-year survival rates between 77% and 86% are reported but many of these tumors received adjuvant irradiation (HOLM and JACKMAN 1964; BEAHRS 1979). STEARNS and QUAN (1970) treated 30 patients by local excision for lesions of the anal canal or perineal area and 63% of the tumors recurred locally. CLARK et al. (1986) reported on 67 patients, nine of whom had a complete local excision. In this group, seven patients (78%) experienced recurrence with a median disease-free interval of 13 months. In these seven patients the tumor recurred locally in four patients, in the inguinal lymph nodes in one patient, and in two patients extensive disseminated disease developed.

It is debatable whether invasive tumors, however small, can be excised with an adequate margin. If the size of the tumor dictates the removal of a large portion of the sphincter muscles, the patient may become incontinent. Postoperative irradiation following local excision is probably advisable for all patients to reduce the possibility of locoregional recurrence, as irradiation of the pelvis is able to eradicate micrometastasis in any involved nodes. It would appear that all patients suitable for a local excision would also be suitable for definitive radiation therapy with an excellent chance for local cure.

The most widely acceptable therapy for invasive squamous cell carcinoma of the anal canal has been an AP resection. This should include a wide perianal excision to include the levator muscles in addition to the contents of the ischiorectal fossa. In females, a posterior vaginectomy should be performed. The procedure may be extended to include a pelvic lymphadenectomy but increasing the extensiveness of the resection does not appear to improve the 5-year survival rate (PARADIS et al. 1975). Because of the profuse network of both blood and lymphatic vessels, local spread can be anticipated but it is difficult to remove the lateral and downward extension of the lymphatic vessels and their nodes, and extended surgery is of questionable benefit as the morbidity is increased (CLARK et al. 1986). A pelvic exenteration may be indicated in males with involvement of the prostate, bladder, or seminal vesicles, and when the uterus is involved the AP resection may have to be extended into a hysterectomy and vaginectomy.

There are significant disadvantages to surgery, i.e., the loss of bowel function and the fairly high mortality (Table 2), especially in the elderly, and there is also considerable morbidity with frequent urinary disturbances and sexual dysfunction, especially in males.

Generally speaking, the incidence of local recurrence following an AP resection is higher in patients with anal cancer compared to adenocarcinoma of the rectum. GOLDEN and HORSLEY (1976) reported a cumulative 5-year survival rate of 47.6% following an AP resection in a large series of 487 patients with anal canal carcinoma. Other reviews from single institutions with relatively large numbers of patients treated in the last few decades report survival rates of approximately 60%, but reports with survival rates as low as approximately 30% are not unusual (STEARNS 1979) (Table 3). Surgical results appear to be affected by the depth of invasion of the primary tumor. In one group of 89 patients with invasion of tumor into greater than half of the sphincter muscle, the absolute 5-year survival rate following an AP resection was 50%. For those patients with superficial invasion only, the absolute 5-year survival rate was 79%.

Table 2. Anal cancer: mortality following radical surgery

Author	No. of cases	Operative mortality
PARADIS et al. (1975)	28	4%
MACCONNELL (1970)	22	8%
KUEHN et al. (1968)	83	5%

Table 3. Anal cancer: 5-year survival following AP resection

Author/institution	No. of cases	5-year survival rate
GOLDEN and HORSLEY (1976)	26	54%
Mayo Clinic (1976, 1979)	80	60.7%
LOYGUE et al. (1981)	33	53%
FROST et al. (1984)	109	62%
BRENNEN and STEWART (1972)	16	37%
PARADIS et al. (1975)	28	50%

11.16.2.1.1 Surgical Management of Lymph Node Metastases. It is recognized that patients who present with lymph node involvement usually have a poorer prognosis than those who present with tumor confined to the anal canal. It appears that the 5-year survival rate of patients with epidermoid carcinoma of the anal canal with synchronous inguinal node metastases is approximately 16%. When patients develop metachronous inguinal metastases following therapy, the survival rate is over 50% (GOLDEN and HORSLEY 1976).

Until recently the treatment of choice for involved inguinal nodes was surgery. GOLIGHER (1984b, pp 785–789) describes two approaches: (a) a radical dissection which involves the division of the inguinal ligament and the removal of the superficial inguinal nodes in addition to glands along the external and common iliac vessels; (b) a superficial inguinal gland dissection during which the inguinal ligament is not cut and the pelvis is not entered. Both these procedures are accompanied by considerable morbidity. The lesser procedure is often carried out on patients who are in poor general condition and adjuvant postoperative irradiation may be administered. Published reports show that the 5-year surgical cure rate for patients treated by nodal dissection for snychronous inguinal metastases averages 20%, indicating that this treatment is not particularly successful (PARADIS et al. 1975).

Prior to 1970, bilateral groin dissections were carried out at the time of the resection of the primary tumor, or as a second procedure a few weeks later. STEARNS and QUAN (1970), however, demonstrated that prophylactic node dissection produced no therapeutic gain. It was pointed out that at least 53 bilateral dissections would have to be considered in order to salvage a few patients. GOLIGHER (1984b, pp 785–789) states that the "operative complications outweigh the possible gain of prophylactic dissection."

On reviewing the literature there does not appear to be a role for prophylactic pelvic lymphadenectomy. Patients with large, inoperable inguinal node masses are usually referred for radiation therapy.

11.16.2.2 Radiation Therapy

The effectiveness of alternative modalities of therapy for epidermoid carcinoma of the anal canal should be measured against that of radical surgery. Radiation therapy alone or in combination with other forms of therapy is becoming the primary modality for the treatment of anal canal carcinoma. It is now apparent that most anal canal tumors are radiosensitive when moderately high, tumoricidal doses of radiation are delivered. Local control may be achieved together with the preservation of normal sphincter function.

The use of radiation therapy in the management of anal canal carcinomas can be traced back for 70 years. The early experience, however, was limited and only small series of patients are reported. The characteristics of the X-ray equipment dictated suboptimal treatment, and complications, undesirable side-effects, and failures were often encountered. For this reason for many years surgery was the treatment of choice. Due to the dosimetric factors of orthovoltage therapy, interstitial irradiation was commonly used for distally situated tumors. The morbidity rate, however, remained high due to the limited experience of each operator and the lack of understanding at that time of accurate dosimetry. Most of the patients who were referred for radiation therapy in those days were medically inoperable.

In 1947 SWEET asserted a belief that anal canal carcinoma responds favorably to radiation therapy. ROUX-BERGER and ENNUYER (1948) from the Institute Curie reported on a series of patients who had been treated by various combinations of radiation therapy and this encouraged the use of the modality. COURTAIL and COLMEIRO (1960) reported on 183 patients treated between 1941 and 1953 by orthovoltage irradiation and the overall survival in this group was 37%. It is interesting to note that the slight improvements in survival over the years bore a direct relationship to the increase in the kilovoltage used. The dose limiting factor at that time was the radiotolerance of the skin. In particular, perineal reactions were very poorly tolerated, especially by elderly patients.

The problem of the inadequate depth dose was resolved in the 1950s with the advent of cobalt 60 and supervoltage equipment. As high energy radiation beams, 1 MeV and higher, became available, more patients with anal carcinomas were treated by irradiation, less morbidity was experienced by the patient, and the treatment results improved. Surgeons nevertheless remained reluctant to refer their patients for radiotherapy and the modality remained in use for large inoperable lesions only. Over recent years, however, greater interest has arisen in sphincter preserving techniques and more and more patients are now referred for definitive radiation therapy. The role of surgery for residual or persistent carcinoma has not always been clear and certain controversy exists between the various techniques. When used as a definitive modality, optimal radiation therapeutic techniques should include the use of radiation fields to the pelvis in addition to direct fields to the anal area. The use of an electron beam therapy directly on the anus as a major component of the treatment is not to be advocated as fibrosis and sphincter incompetence may result. Electron beams may be employed to deliver boost doses to the perineum or inguinal nodes, as described later.

11.16.2.2.1 Technique. The radiation therapy technique that will be used for any particular patient will be governed by the stage of the disease and the philosophy of the radiation therapist. The status of the inguinal nodes at presentation will also dictate what technique is appropriate. Because many patients present with a fairly long history and due to the fact that the pattern of behavior of anal canal carcinoma is such that local and regional nodes may be involved when the primary disease is fairly advanced, it is wise to treat the inguinal and pelvic nodes on all patients. External beam irradiation should be delivered with supervoltage equipment, including cobalt 60, and the minimum source-skin distance should be 80 cm. Treatment should be given by two opposed anterior and posterior fields. The anterior field may be approximately 15 cm high by 17 cm wide and the posterior field approximately 15 cm high by 14 cm wide. The anterior field should have approximately 1.5 cm shielded from each side in such a manner that the inguinal nodes remain in the field (inverted T) while keeping the amount of bowel exposed to the irradiation to a minimum. Great care must be taken during simulation of the pelvic fields and lead markers may be used to ensure that the perineum is included in the field. The position of the anal orifice and the soft tissue of the perineum in relation to the bony landmarks is often deceptive in obese patients. The superior border of the pelvic field should extend to the L5/S1 interspace and the lateral borders of the field should extend 1 cm onto the bony pelvis.

The optimal dose to the pelvis is approximately 45–50 Gy at a daily dose of approximately 1.8 Gy, 5 days a week over a period of beween 5 and 8 weeks. The dose should be calculated at the midseparation of the patient in the central axis of the field. The dose to the inguinal nodes should be calculated at depth of 3 cm from the anterior surface of the patient and a total dose of between 45 and 50 Gy should be delivered to the inguinal area. Following completion of the pelvic field, small fields may be utilized over the inguinal areas to boost the dose to the appropriate level and some contribution may be delivered by an electron beam.

Due to the fact that the perineum is a potentially moist area, particularly in obese patients, severe local skin reactions may be experienced during radiation therapy. This may necessitate a rest period for the patient and interruptions from treatment of up to 3 weeks do not appear to have a negative effect on the outcome of therapy.

Following the delivery of 45–50 Gy to the pelvis, the treatment plan may be reviewed and, depending on the preference of the radiotherapist, a direct perineal field approximately 10 cm × 10 cm in size may be used to deliver a boost dose to the anal region. With the patient in the knee-chest position, an additional dose of 10–20 Gy with either photons or electrons may be delivered. Some radiotherapists (Holm and Jackman 1964; Rousseau et al. 1973) describe techniques of definitive radiation in which a direct perineal beam is used to treat the anal area alone. Papillon (1982) uses a 240° arc technique for which the patient lies in the prone position and the photon beam is directed to the anorectal area. The target volume in this technique includes the anal canal and the lymphatic drainage areas of the posterior portion of the pelvis. Alternatively, a four field box technique may be used which must be carefully defined to avoid uninvolved structures such as the bowel and bladder.

A report by Cantril et al. (1983) of a selected group of 32 patients treated by primary radiation therapy demonstrated a local control rate of 84.3%. Three patients in this group received an iridium implant to boost the dose to the anus following initial external beam pelvic irradiation therapy and although the control rate was encouraging in this report, the complication rate was high. Other results are shown in Table 4.

11.16.2.2.2 Interstitial Radiation Therapy. Interstitial radiation therapy delivers a high dose of radiation to a localized area in a relatively short time. It appears that the anal canal is most suited to single plane implants due to the dosimetric distribution so produced. There is a fairly high rate of complications such as necrosis and the disturbance of sphincter function due to excessive local fibrosis which may occur, particularly when multiple implants are used. Initially, radium needles were used for implantation but as these have a short active length, the proximal extensions of larger tumors were inaccessible to this type of source. More recently, the scope of interstitial therapy has been increased by the use of reactor produced radionuclides, e.g., iridium 192, which are available in longer, flexible source strands and permit the use of afterloading techniques. During the early years of interstitial radiation therapy, a temporary diverting colostomy was recommended prior to the procedure but this is no longer required. The use of a low residue diet, optimal preoperative preparation, bed rest, and opiates enable the procedure to be carried out satisfactorily.

Interstitial implants are performed under spinal or general anesthesia with the patient in either the lithotomy position or in a modified genupectoral position. The target volume must be carefully defined and a surgical clip is usually inserted to delineate the distal end of this area. Active radium needles may be inserted or, alternatively, other techniques may be used which involve the use of an anal template and the insertion of hollow cannulae which are afterloaded with iridium wires at an appropriate time. Doses up to 60 Gy are delivered and the dosimetry is calculated according to Paterson Parker rules. Following the procedure appropriate simulation films are taken, prior to the insertion of active sources in situations where afterloading techniques are used. Following a review of these films the final dose calculations are made, often with the aid of a computer to ensure accurate dosimetry. There are some reports on the use of ^{125}I seeds in elderly patients as this type of implant does not require removal (Gofinet et al. 1978).

The major criticism of interstitial radiation therapy is that it is a purely local form of treatment and ignores the possibility of nodal involvement. Some authors (Devois and Decker 1960) felt that the effectiveness of this modality was limited due to the characteristic upward and lymphatic spread of this disease. There was also some concern about the lack of stability of implants but this latter problem is not confirmed elsewhere. Dalby and Pointon (1961) reported on a series of 171 patients with anal tumors of various sizes (Table 5). In a subgroup of 59 patients with anal canal cancer followed over a 5-year period, the 5-year survival rate was 39%. The authors concluded that interstitial radium implantation was a satisfactory method of treatment for early and moderately sized lesions. The results, however, were poor in patients with more advanced tumors.

Papillon (1982) reported from the Centre Léon Berard that between 1949 and 1970 interstitial implantation was the usual method by which radiation therapy was delivered. Eighty-eight patients were thus treated, 58 with limited tumors of less than 4 cm in size of 30 patients with tumors equal to or larger than 4 cm. Patients with larger and more invasive tumors were treated by surgery. Between 1949 and 1962 iridium implants were performed in a one-stage procedure. A dose of between 40 and 50 Gy was delivered over 3–4 days. Since 1963 fractionated interstitial implants have been used. A dose of 40 Gy is delivered during the first implant and a dose of between 25 and 30 Gy is delivered during a second implant which is carried

Table 4. Anal cancer: results with the use of external beam irradiation

	No. of patients	Energy	Dose (Gy)	Overall 5-year survival (%)
Cummings et al. (1982a)	34	60 Co	45–55	70
Green et al. (1980)	16	25 MeV +1 MeV	60–75	81
Eschwege et al. (1973)	87	60 Co	60	31
Salmon et al. (1984)		60 Co	60–65	58

Table 5. Anal cancer: results with the use of interstitial therapy alone

Author	No. of cases	5-year survival rate
Dalby and Pointon (1961)	59	39%
Papillon (1982) (fractionated in 63%)	88	68%

out approximately 8 weeks following the first. The 5-year survival rate in this group of patients is 68%.

At this time it is unusual for patients to be treated solely by means of an interstitial implant. External beam radiation therapy either with or without chemotherapeutic agents for radiosensitization is usually administered, and 4–6 weeks following the completion of this part of the treatment the implant is inserted. Interstitial radiation therapy may be used either in addition to or as an alternative to the external boost therapy to the perineum.

11.16.2.2.3 Prognostic Considerations. a) *Tumor size;* the size of the anal tumor at presentation appears to influence survival (Salmon et al. 1984). Patients with tumors less than 4 cm in height demonstrated a 5-year survival rate of 70% while patients with tumors larger than 6 cm showed a survival rate of 33%. A survival rate of 70% was obtained following radiation therapy for those patients whose tumors occupied less than one-quarter of the circumference, but this figure dropped rapidly to 19% at 5 years when more than 75% of the circumference of the canal was affected at the time of disease presentation. It seems evident that in most radiation therapy series of patients, the highly favorable results are confined to those patients with smaller tumors (Salmon et al. 1986). Rousseau et al. (1973) record a 5-year survival rate of 73% in a group of 26 patients presenting with small primary lesions (UICC stage T1 and T2) but only a 28% 5-year survival rate among 50 patients with more extensive primary tumors. The probability of keeping a functional anus following radiation therapy is also related significantly to the initial size of the tumor (Salmon et al. 1986).

b) *Inguinal nodal involvement;* synchronous inguinal node metastases are associated with a poor prognosis, but occasionally a patient may be cured by radiation therapy, either alone or in conjunction with an iliofemoral nodal dissection. Metachronous inguinal node metastases in the absence of distant spread can be controlled by radiation therapy if the area has not been previously treated. Patients who have received a maximum radiation dose to the inguinal area may undergo surgery if metachronous inguinal node metastases develop subsequent to treatment.

11.16.2.3 Surgery vs Radiation Therapy: Comparison of Results

It is a very difficult task, and one which may be misleading, to compare the relative merits of primary radiation therapy vs. an AP resection for cancer of the anal canal by reviewing the literature. The overall number of cases reported is relatively small and many series include patients accrued over a long time period. Radiation therapy technology is changing constantly and this not only may account for differing results and complication rates but may have had an influence on the selection of patients. For many years, only patients who were medically inoperable or who had unresectable tumors were offered radiation therapy and patients presenting with large lesions are most likely to have nodal involvement and occult distant metastases. A randomized clinical trial to compare results of radiation therapy vs an AP resection would be inappropriate due to the relatively high morbidity subsequent to surgery, i.e., a colostomy and potential sexual dysfunction.

In the series of Papillon (1982), 64 patients treated by definitive radiation therapy and followed over a 5-year period showed a 68% disease-free survival rate. Dalby and Pointon (1961) reported on 31 patients who presented with early lesions and demonstrated a 72% 5-year survival rate following irradiation. Green et al. (1980) reported a 75% 3-year disease-free survival rate for patients treated with radiation therapy. Boman et al. (1984), in a recent report, showed a 5-year survival rate of 71% for patients who had undergone an AP resection, although not all surgical series report results that are this favorable. It should be noted that the use of radiation therapy does not preclude salvage surgery if this should become necessary following treatment.

11.16.2.4 Chemotherapy

An extensive review of the world literature indicates that chemotherapy does not have a major role to play in the management of primary carcinoma of the anus. It has only been occasionally of palliative benefit and few reports are available which confirm its use.

There appears to be no single agent that has been used extensively in the treatment of anal carcinoma but there are a number of reports on the use of multiagent chemotherapy for advanced, metastastic, or recurrent disease. To date, most of the drug combinations include bleomycin, cis-platinum, vincristine, and high dose methotrexate with Leucovorin rescue (Wilking et al. 1985). Other possible drug combinations include 5-FU, mitomycin-C, Adriamycin, and methyl-CCNU. A small percentage of patients respond to a variety of combinations of these drugs although the response is usually short-lived.

As will be discussed below, certain chemotherapeutic agents are being used together with irradiation in the management of anal cancer. When used in such a multimodality situation, these agents function as radiation potentiators.

11.16.2.5 Multimodality Treatment

The objective in the use of a multimodality approach is to increase the survival rate of patients while reducing the complications of treatment. Over the past decade an increased cooperation between surgeons, radiotherapists, and medical oncologists has aided the development of a number of multimodality treatment regimens for malignant diseases at a variety of sites. Considerable interest has arisen in the use of radiation combined with chemotherapeutic agents with or without surgery in the management of anal canal carcinoma. It would be particularly helpful if the effectiveness of irradiation in the perineal area could be increased while reducing the required total dose.

In a recent report discussing the combination of radiation therapy with surgery for patients with large infiltrating tumors it appeared that there was a therapeutic advantage when these two modalities were combined (SVENSON and MONTAGUE 1980). Most series describe postoperative radiation treatment following incomplete excisions when persistent disease was apparent.

11.16.2.5.1 Radiation Therapy Combined with Chemotherapeutic Agents for Radiation Potentiation. Although it is possible to achieve local control of primary anal canal carcinoma with radiation therapy, moderately high doses in the range of 55–65 Gy are required for total tumor eradication. Significant local morbidity may be anticipated during and following the delivery of doses of this magnitude to the perineal area. It is unlikely that further improvements can be made in the physical dosimetry of either interstitial or external beam radiation therapy even though sophisticated computerization is now available and precise treatment techniques have evolved. It has become important therefore to attempt to increase the effectiveness of radiation therapy biologically, particularly in such areas as the perineum where a reduction in the optimal total dosage could lower the morbidity considerably. The therapeutic gain could thus be enhanced such that patients could be offered definitive treatment with a reasonable chance of cure while still maintaining sphincter function and sexual potency.

Preliminary work in 1958 by HEIDELBERGER et al. (1958) gave rise to an ever-increasing interest in the interaction of 5-FU and irradiation. In the 1960s a number of clinical reports were published which suggested that radiation therapy for gastrointestinal malignancies could be potentiated by the addition of infusional 5-FU. VIETTI et al. (1971) were able to show synergy between the two treatment modalities in AKR mice and a few years later, in 1974, NIGRO et al. (1974) first reported the use of combined radiation therapy and intravenous 5-FU and mitomycin-C prior to AP resection in patients with anal canal carcinoma. Their experience was confirmed by BUROKER et al. (1979), QUAN (1979), CUMMINGS et al. (1982b), and SISCHY et al. (1980), and it became apparent that a significant diminution in the size of the tumor was seen consistently following radiation doses which varied between 30 and 35 Gy. An AP resection was performed within 6 weeks of the conclusion of external beam irradiation and, in a large percentage of patients, no tumor was identified in the resected specimen. This led to the suggestion that these patients could be observed carefully following treatment, surgery being reserved until recurrence occurred.

The mechanism of the interaction between 5-FU, mitomycin-C, and X-irradiation is not fully understood at this time, yet BYFIELD et al. (1980) were able to show that enhanced cytotoxicity is dependent upon both the 5-FU concentration and the duration of exposure, radiation synergy requiring exposure durations of at least 48 h. If 5-FU is administered before the radiation, an additive effect only is demonstrated. Other investigators have also reported on the interaction of 5-FU and radiation (NAKAJIMA et al., 1979; LOONEY et al. 1979). Much work is also being done on the interaction between mitomycin-C and X-ray therapy (SPREMULLI 1983; ROCKWELL 1982).

External beam irradiation combiend with chemotherapeutic agents for radiation potentiation is now widely used in the management of patients with carcinoma of the anal canal. The treatment regimen varies from institution to institution but modifications are usually fairly minor (NIGRO et al. 1974). Patients receive between 40 and 45 Gy to the whole pelvis at a dose rate of between 9 and 10 Gy per week. On the 2nd day of the radiation therapy, a bolus injection of mitomycin-C at a dose which varies between 10 and 15 mg/m^2 is administered. This is the only time that mitomycin-C is given. A

96-h infusion of 5-FU at a dose of approximately 1000 mg/m^2 is also commenced on day 2 of radiation therapy. A second 96-h infusion of 5-FU is commenced on day 28 of the radiation therapy at the same dose rate (see Fig. 5). Following this a perineal boost of approximately 15 Gy incident dose may be delivered. Between 4 and 6 weeks later patients either receive an interstitial implant if induration is still present or undergo a full thickness biopsy. If this is positive an AP resection is performed, if negative careful follow-up is undertaken.

In an attempt to avoid hematological toxicity it is recommended that the doses of both 5-FU and mitomycin-C should be based on the patient's ideal weight and that the dose of mitomycin-C should not exceed 18 mg and that of 5-FU, 1800 mg over a 24-h period. If the white cell and platelet counts fall unduly the second course of chemotherapy should be either modified, temporarily postponed, or even omitted. Patients with large lesions (T3 and T4) may require the upper dose levels of irradiation in an attempt to avoid the necessity of surgery. These patients may require an interruption in their course of therapy in order to avoid an excessive local skin reaction. If the patient is not receiving treatment on day 28, the second course of 5-FU should be postponed until the recommencement of radiation therapy.

The follow-up period of patients treated by combined chemotherapy and radiation therapy is still relatively short but local control rates appear to be excellent (85%–100%) (Table 6). In some series the numbers of patients were relatively small but the local control and survival rates appear consistent. As the follow-up of these patients is extended it will be of interest to see if patients, particularly those who have not undergone surgery, remain free of disease. It appears that those patients who do recur following the combined treatment show evidence of disease within 2 years of completion of therapy (MICHAELSON et al. 1983; SISCHY 1985; CUMMINGS et al. 1982b).

Fig. 5. Location of anal malignancies (data from KUEHN et al. 1968)

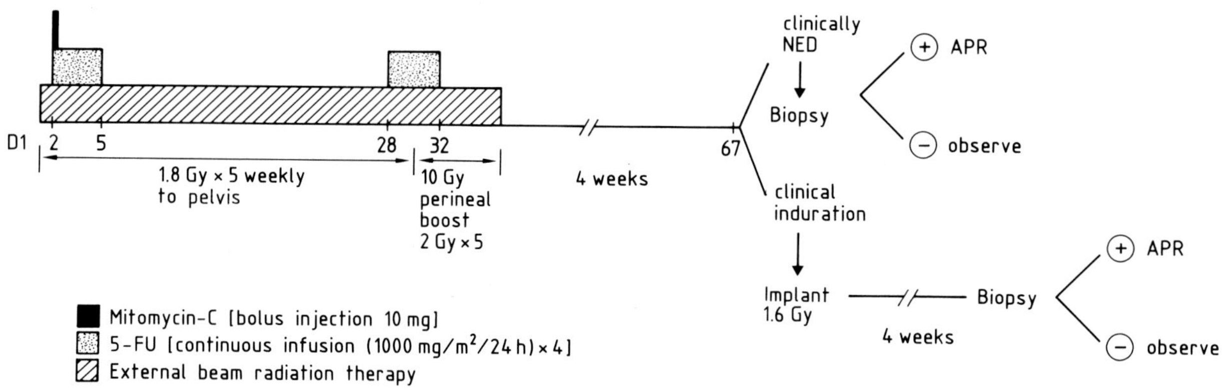

Table 6. Anal cancer: results with the use of combined therapy: 5-FU, mitomycin-C, and radiation therapy

Author	No. of cases	Dose schedule		External RT Dose/Fx (Gy)	Survival NED	Follow-up
		5-FU (mg m^{-2}/day)	Mitomycin-C (mg m^{-2})			
NIGRO et al. (1974)	28	1000 × 4q4W × 2	15 Day 1	30/15	12-APR 50% 16-Local excision 100%	1–8 years
MICHAELSON et al. (1983)	37	750 × 5	15 Day 1	30/15	20 APR 85% 17 Local excision 76%	Median 28 months
SISCHY et al. (1985)	32	1000 × 4q4W × 2	10 Day 1	55–65 Some implants	4 APR 29 No surgery 86.2%	1–4 years
CUMMINGS et al. (1982)	13	1000 × 4	10 Day 1	50/20	No surgery 100%	Median 1 year

11.6.2.6 Complications

In order to deliver optimal radiation therapy to the anus and node bearing areas, i.e., pelvis and inguinal fossae, moderate to high doses in the range of 40–60 Gy are required. The ratio between the therapeutic effect and the complication rate is narrow, particularly as the tissues of the perineum have a low radiotolerance. Tumoricidal doses of radiation to this area may give rise to acute and possibly late sequelae. There appears to be a correlation between the incidence of post-treatment complications and the fractionation scheme used and the total dose delivered. Modern radiation therapy, however, requires the acceptance of a calculated risk of complications in order to achieve an optimal cure rate. Locoregional recurrence and distant metastases may also be regarded as severe complications of primary disease.

There is a subset of patients who may be at risk for radiation injury and this group may be identified prior to the initiation of therapy (MORGENSTERN et al. 1985). Steps may therefore be taken to minimize the risk by formulating a less aggressive treatment plan than might otherwise be offered and lower doses may be used for these patients. Elderly patients, females, and patients with diabetes or hypertension are at particularly high risk for injury, as are those with a history of inflammatory bowel disease, diverticulitis, or prior surgery. If injudicious dose rates and radiation techniques are used, the small intestine may be injured without these factors being present.

11.16.2.6.1 Short-Term Complications. During a course of radiation therapy to the pelvis, approximately 70% of patients will experience mild diarrhea and approximately 27% will experience dysuria (HOSKINS et al. 1985). These toxicities may be managed successfully by conservative means such as diphenoxylate hydrochloride with atropine sulfate (Lomotil) and phenazopyridine hydrochloride (Pyridium). Skin reactions varying between dry desquamation and moist mucositis will develop in the perineum and inguinal areas of the majority of patients. A bacteriostatic preparation may be applied and an interruption of treatment may be necessary. Strict attention should be given to skin care and hygiene during treatment. All acute side-effects normally settle within a few weeks of the completion of therapy.

If external beam irradiation is delivered in combination with radiation potentiators such as 5-FU and mitomycin-C, the short-term complications may be exacerbated. It is extremely important to restrict the daily dose of radiation to approximately 1.8 Gy, as higher doses may produce severe skin reactions in the perineum and groin. The hematologic profile of such patients should be carefully monitored and twice weekly blood counts may be advisable as the nadir of the white cell and platelet counts may occur between the 16th and 20th day and modification of the second course of 5-FU may be required. Mucositis and stomatitis may occur following the infusion of 5-FU but do not normally require interruption of treatment. A few patients experience a mild temporary hair loss following the delivery of 5-FU and mitomycin-C. An early menopause is produced in premenopausal female patients who undergo a course of pelvic irradiation.

11.16.2.6.2 Long-Term Complications. A small number of patients experience late mild adverse reactions such as ischial tenderness, balanitis, proctitis, and cystitis. A frequent reaction in this category is bowel dysfunction and it appears that further investigation is required into the multiple factors that are involved in the etiology of radiation induced diarrhea. Ileal dysfunction may play a role and this, when caused by radiation therapy, may be less rapid to recover than the morphologic changes which occur in the bowel during the course of therapy (STRYKER et al. 1977). Investigations have shown that the bile acid concentraiton in the blood decreases during a course of irradiation, suggesting malabsorption of bile acids leading to diarrhea (MERRWALDT 1984). Following therapy, the bile acid concentration in the serum appears to return to normal limits, and diarrhea in the follow-up period, therefore, may be due to an impaired capacity of the rectal ampulla to retain feces as a direct result of irradiation. This causes frequent evacuation without an increase in fecal weight or transit time and this may be accompanied by an urgency to defecate.

Severe long-term complications of radiation include contraction of the bladder and chronic radiation enteritis, both of which require surgical intervention for correction. Radiation enteritis may cause either obstruction, fistula formation, or ulceration of the mucosa. Symptoms may present from 3 months to several years after treatment but commonly arise within the 1st year. Perforation and stricture of the bowel may be the result of local ischemia due to small vessel disease. Microscopic vascular changes may extend into bowel which appears macroscopically normal and for this reason, radical excision may be necessary in order to secure sound anastomotic healing (SCHOFIELD 1983).

An element of rectal fibrosis or telangiectasia and some mild vulvar edema is not unusual following therapy for cancer of the anus. Severe and painful necrosis of the anal canal may sometimes occur, requiring either a temporary colostomy or an AP resection. If the sphincter muscles are involved at presentation by extensive tumor, varying degrees of incontinence may be experienced by the patient following therapy due to reactional fibrosis.

Complications such as anal necrosis or fibrosis may be observed in patients with advanced, ulcerative and infiltrating tumors (T3 and T4) who have undergone a two-plane or volume radionuclide implant. The risk of radionecrosis is greater for patients with poor circulation or suboptimal nutritional status. Infrequently a single-plane implant may produce similar complications. The risk of producing radionecrosis is dependent upon the modality of treatment, the total dose, the time-dose relationship, and the volume treated and it is extremely important to select the appropriate treatment and technique for each patient.

There are reports in the literature which discuss serious late toxicities following radiation therapy for anal canal carcinoma, some of which required surgery (Cantril et al. 1983; Cummings et al. 1984). In one report (Cantril et al. 1983), 28% of the patients required an interruption of their therapy due to acute reactions but these were all managed on an outpatient basis without untoward chronic sequelae. Chronic complications were noted in 13 of 47 patients, including two patients who required a colostomy for severe anal stenosis and two who required an AP resection for large painful ulcers. Two of these four patients had received a portion of their treatment by radionuclide implant and one received an excessively high dose of perineal irradiation (80 Gy). Some of the patients in this series received 2.5 Gy per treatment, four times a week. Those patients who were treated preoperatively received approximately 45 Gy while those treated by radiation alone received approximately 65 Gy.

Cummings et al. (1984) report on a group of 51 patients treated by radiation alone or radiation combined with 5-FU and mitomycin-C either with a continuous or split course of therapy. Patients treated by radiation therapy alone received a total dose of between 45 and 60 Gy in 4–6 weeks. Patients receiving radiation combined with chemotherapy received 50 Gy in 4 weeks (2.5 Gy per treatment). Patients in this latter group received the first 25 Gy to the pelvis and the final 25 Gy was delivered by a reduced field to the anal region only. After 16 of 30 patients had been treated in this manner, the radiation treatment course was split in an effort to reduce acute hematologic and intestinal toxicities. The remaining 14 patients were treated in an identical manner to those receiving a continuous course of therapy but there was a 4-week interval between the large volume and the reduced volume treatment. Chemotherapy was given in a similar fashion and was started on day 1 of each part of the split course radiation therapy. Virtually all patients in each of the three treatment groups developed a moist desquamation of the perineum, even when the radiation dose was only 25 Gy in 2 weeks. Five of the 16 patients who received continuous radiation treatment combined with 5-FU and mitomycin-C developed an ileus and peritonism or diarrhea and tenesmus which did not respond readily to antidiarrheal agents. An interruption in treatment of more than 7 days was necessary due to acute toxicities in four patients treated by radiation therapy alone and in three patients treated by continuous radiation combined with 5-FU and mitomycin-C.

Late complications which presented 3 months following the completion of therapy included minor asymptomatic perineal fibrosis and telangiectasia. Four patients, two who had received radiation therapy alone and two who had received radiation therapy combined with 5-FU and mitomycin-C, one as continuous treatment and one with a split course, experienced minor intermittent bleeding from the anal canal, rectum, or bladder. Five patients subsequently developed anal ulcers, four requiring surgical management. Three of these patients had been treated by combined therapy and the ulcers were attributed to the rapid regression of primary carcinoma more than 7 cm in diameter and the subsequent lack of epithelialization of the residual defect. Late developing anal necrosis has not been seen at this time. There was one serious late intestinal complication which required surgical management in each of the groups treated by irradiation combined with chemotherapy. One patient suffered a perforated sigmoid colon almost 4 years after treatment.

Preoperative doses in the range of 40–45 Gy delivered to the pelvis do not typically produce severe acute or late toxicities and any sequelae may be managed usually by conservative methods. In the last few years it has become apparent that it is possible to cure squamous cell carcinoma of the anal canal without surgery when doses in the range of 50–55 Gy are delivered. If 5-FU and mitomycin-C or other chemotherapeutic agents are used for radiopotentiation it may be possible to cure these patients with a slightly lower dose of radiation in the range of 40 Gy, but complications at these dose levels and following the use of radiopotentiators pose a potentially serious problem. Strict attention must be given to the daily dose, the diet, and skin care of the patient and every effort must be made to limit the amount of small bowel exposed to radiation. It is of little gain to the patient if, following the eradication of disease by the use of radiotherapy alone, surgery is required for the treatment of late complications.

11.16.2.7 Clinical Trials

For many years the accepted treatment for carcinoma of the anal canal was an AP resection. This procedure is accompanied by inevitable morbidity, i.e., a colostomy and in males, the likelihood of sexual impotency. A local excision, if performed with a sufficiently wide margin to eradicate all disease, may produce an incompetent sphincter.

The eradication of squamous cell carcinoma of the anal canal by the use of radiation therapy requires high doses to be delivered to an area which has a low radiotolerance. Injudicious use of high daily doses and wide fields to the perineum may cause unacceptable skin reactions. The sole use of an electron beam to treat the anal area may produce fibrosis of the sphincter muscles which may lead to an incompetence of the anus. When chemotherapeutic agents are combined with radiation therapy for potentiation, the total dose of radiation may be reduced but the cytotoxic agents may produce an exacerbated radiation reaction in the groin and perineum.

Five-year survival rates following a curative resection for patients with anal canal carcinoma are approximately 50%. During the last decade it has been observed that the survival rates of patients treated by a combination of radiation therapy, 5-FU, and mitomycin-C equal, or in many situation, surpass those of surgery. It is imperative, however, that this treatment be administered in an optimal fashion in order to avoid unnecessary complications. The radiation therapy fields should be tailored to avoid as much bowel as possible and the daily dose should be such that morbidity is reduced to a minimum. The patient should be observed carefully on a regular basis during the course of treatment and rested from therapy should the need arise. An electron beam may be used to boost the total dose to the anus but should not be used for the entire treatment. When the inguinal nodes are involved at presentation, the inguinal area should be included in the radiation field.

The Radiation Therapy Oncology Group and the Eastern Cooperative Oncology Group are currently sponsoring phase II trials to assess the effect of radiation therapy combined with 5-FU and mitomycin-C for anal canal squamous cell carcinoma. According to these trials, 2 months after completion of radiation therapy, if residual disease is present the patient undergoes and AP resection. If there is no clinical evidence of disease in the anus the patient undergoes a full thickness biopsy of the anal wall. If this is negative for tumor, the patient is observed, but if the biopsy is positive, an AP resection is performed immediately. The total dose of radiation delivered, even when combined with 5-FU and mitomycin-C for sensitization, does not preclude surgery should this be necessary and no untoward complications have been observed in those patients who have undergone surgery.

The Eastern Cooperation Oncology Group is currently accruing patients to a trial which uses combination chemoptherapy for metastatic or recurrent cancer of the anal canal. Patients receive mitomycin-C, 10 mg/m^2 IV every 4 weeks for two courses then every 10 weeks; Adriamycin, 30 mg/m^2 IV every 4 weeks for two courses then every 5 weeks; and cis-platinum, 60 mg/m^2 IV given every 4 weeks $\times$ 2 (with mannitol diuresis), then every 5 weeks. Patients with stable disease or an objective response continue on treatment; patients showing objective evidence of disease progression are eligible for crossover to bleomycin, 10 mg/m^2 IV given daily to a maximum cumulative dose of 280 mg or toxicity. CCNU, 130 mg/m^2 is given orally on day 1, 4 weeks after completion of bleomycin, and every 6 weeks thereafter. The purpose of this study is to determine the antitumor activity of the combination regimen of mitomycin-C, Adriamycin, and cis-platinum (MAP) for patients with metastatic anal canal cancer and to determine the antitumor activity of the combination of bleomycin and CCNU as second line treatment in patients with anal canal cancer who have failed primary MAP therapy. The results of this trial have not been analyzed at this time. The Southwest Oncology Group and the Southest Group have both initiated trials using either combination chemotherapy or methyl-GAG for patients with squamous cell carcinoma of the anal canal; these studies are now closed for analysis.

The physical aspects of radiation therapy have probably peaked at this time and much work is now underway to evaluate the use of biologic response modifiers and the use of hypoxic cell sensitizers and radioprotectors in an effort to increase the biologic effectiveness of radiation therapy, particularly for patients with large necrotic tumors. Patients with neoplasms of the anal margin may be successfully treated by either local excision or in some situations an AP resection, and adjuvant therapy is not usually required.

References

Al-Jurf AS, Turnbull RB, Fazio VW (1979) Local treatment of squamous cell carcinoma of the anus. Surg Gynecol Obstet 148: 576–578

Austin DF (1982) Etiological clues from descriptive epidemiology: Squamous carcinoma of the rectum or anus. NCI Monogr 62: 89–90

Beahrs OH (1979) Management of cancer of the anus. AJR 133: 790–795

Beahrs OH, Wilson SM (1976) Carcinoma of the anus. Ann Surg 1984: 422–428

Bensaude A, Nora J (1968) Differential diagnosis of carcinoma of the anal margin. Proc R Soc Med 61: 624–626

Berg JW, Lone F, Stearns MW Jr (1960) Mucoepidermoid anal cancer. Cancer 13: 914–916

Bohe M, Lindstrom G, Ekelund G, Leandoer L (1982) Carcinoma of the anal canal. J Gasteroenterol 17: 795–800

Boman BM, Moertel CG, O'Connell MJ, Scott M, Weiland LH, Beart RW, Gunderson LL, Spencer RJ (1984) Carcinoma of the anal canal. A clinical and pathologic study of 188 cases. Cancer 54: 114–125

Brennan JT, Stewart CF (1972) Epidermoid carcinoma of the anus. Ann Surg 176: 787–790

Buckwalter JA, Jurayj MN (1957) Relationship of chronic anorectal disease to carcinoma. Arch Surg 75: 352–361

Buroker T, Nigro ND, Considine B et al. (1979) Mitomycin C, 5 fluorouracil and radiation therapy in squamous (epidermoid) cell carcinoma of the anal canal. In: Carter SK, Crooke SK (eds) Mitomycin C. Academic, New York, pp 183–188

Byfield JE, Barone R, Mendelson J et al. (1980) Infusional 5-fluorouracil and X-ray therapy for non-resectable esophageal cancer. Cancer 45: 703–708

Cantril ST, Green JP, Schall GL et al. (1983) Primary radiation therapy in the treatment of anal carcinoma. Int J Radiat Oncol Biol Phys 9: 1271–1278

Chiu YS, Unni KK, Beart RW (1980) Malignant melanoma of the anorectum. Dis Colon Rectum 23: 122–124

Clark T, Petrelli N, Herrera L et al. (1986) Epidermoid carcinoma of the anal canal. Cancer 57: 400–406

Cooper HS, Patchefsky AS, Marks G (1979) Cloacogenic carcinoma of the anorectum in homosexual men. Dis Colon Rectum 22: 557–558

Cooper PH, Mills SE, Allen MS (1982) Malignant melanoma of the anus. Report of 12 patients and analysis of 255 additional cases. Dis Colon Rectum 25: 693–703

Courtail J, Fenadez Colmeiro JM (1960) Les indications et les resultats de la roentgentherapie et de la curie therapie dans le cancer du canal anal. Arch Mal Appar Dig [Suppl] 49: 43–54

Croxson T, Chabon AB, Rorat E, Barash IM (1984) Intraepithelial carcinoma of the anus in homosexual men. Dis Colon Rectum 27: 325–330

Cummings BJ, Thomas GM, Keane TJ et al. (1982a) Primary radiation therapy in the treatment of anal canal carcinoma. Dis Colon Rectum 25: 778–728

Cummings BJ, Rider WD, Harwood AR et al. (1982b) Combined radical radiation therapy and chemotherapy for primary squamous cell carcinoma of the anal canal. Cancer Treat Rep 66: 489–492

Cummings B, Keane T, Thomas G et al. (1984) Results and toxicity of the treatment of anal canal carcinoma by radiation therapy or radiation therapy and chemotherapy. Cancer 54: 2062–2068

Dalby JE, Pointon RS (1961) The treatment of anal carcinoma by interstitial irradiation. AJR 85: 515–520

Daling JR, Weiss NS, Klopfenstein LL et al. (1982) Correlation of homosexual behavior and the incidence of anal cancer. JAMA 247: 1988–1990

DeOliveira D (1976) Anal carcinoma in northern Brazil. Dis Colon Rectum 19: 18–19

Devois A, Decker R (1960) La curie puncture du cancer de l'anus. Arch Fr Mal Appar Dig 49: 54–67

Dillard BM, Spratt JS, Ackerman LV et al. (1963) Epidermoid cancer of anal margin and canal. Arch Surg 86: 772–777

Dougherty BG, Evans HL (1985) Carcinoma of the anal canal: A study of 79 cases. Am J Clin Pathol 83: 159–164

Dubois JD, Dilly SA, Gazet JC (1985) Leukaemic infiltration of the anus. Eur J Surg Oncol 11: 365–367

Ege GN, Cummings BJ (1980) Interstitial radiocolloid iliopelvic lymphoscintigraphy: Technique, anatomy and clinical application. Int J Radiat Oncol Biol Phys 6: 1483–1490

Eschwege F, Fajbisowicz S, Otmezguina Y et al. (1973) Radiotherapie des cancers malpighiens de l'anus. J Radiol Electol Med Nuc 54: 636

Fenger C (1979) The anal transitional zone. Acta Pathol Microbiol Immunol Scand [A] 87: 379–386

Fenger C, Filipe MI (1977) Pathology of the anal glands with special reference to their mucin histochemistry. Acta Pathol Microbiol Immunol Scand [A] 85: 273–285

Fisher ER (1969) The basal cell nature of the so-called transitional cloacogenic carcinoma of anus as revealed by electron microscopy. Cancer 24: 312–322

Freeny PC, Marks WM, Ryan IA et al. (1986) Colorectal carcinoma evaluation with CT: Preoperative staging and detection of postoperative recurrence. Radiology 158: 347–353

Frost DB, Richards PC, Montague ED et al. (1984) Epidermoid cancer of the anorectum. Cancer 53: 1285–1293

Gabriel WB (1960) Discussion on squamous cell carcinoma of the anus and anal canal. Proc R Soc Med 53: 403–409

Ger R, Reuben J (1968) Squamous cell carcinoma of the anal canal: A metastatic lesion. Dis Colon Rectum 11: 213–219

Gillespie JJ, MacKay B (1978) Histogenesis of cloacogenic carcinoma. Fine structures of anal transitional epithelium and cloacogenic carcinoma. Hum Pathol 9: 579–587

Gingrass PJ, Rubrick MP, Hickcock CR et al. (1978) Anorectal verrucose squamous carcinoma. Dis Colon Rectum 21: 120–122

Glass RE, Ritchie TK, Thompson HR et al. (1985) The results of surgical treatment of the rectum by radical resection and extended abdominoiliac lymphadenectomy. Br J Surg 72: 599–601

Gofinet DR, Mastinez A, Poller D et al. (1978) Perineal brachytherapy. Front Radiat Ther Oncol 12: 72–81

Golden GT, Horsley JS (1976) Surgical management of epidermoid carcinoma of the anus. Am J Surg 131: 275–280

Goligher J (1984a) Surgical anatomy and physiology of the anus, rectum and colon. In: Goligher J (ed) Surgery of the anus, rectum and colon, 5th edn. Bailliere Tindall, London, pp 7–18

Goligher J (1984b) Carcinoma of the anal canal and anus. In: Goligher J (ed) Surgery of the anus, rectum and colon, 5th edn. Bailliere Tindall, London

Gray's Anatomy (1985) The lymphatic system, 13th edn. Lea and Febiger, Philadelphia, p 916

Green JP, Schaupp WC, Cantril ST et al. (1980) Anal carcinoma: therapeutic concepts. Am J Surg 140: 151–155

Greenall MJ, Quan SHQ, Urmacher C et al. (1985a) Treatment of epidermoid carcinoma of the anal canal. Surg Gynecol Obstet 161: 509–517

Greenall MJ, Quan SHQ, Stearns MW et al. (1985b) Epidermoid cancer of the anal margin. Pathologic features, treatment, and clinical results. Am J Surg 149: 95–101

Grinnell RS (1954) An analysis of forty-nine cases of squamous cell carcinoma of the anus. Surg Gynecol Obstet 98: 29–39

Grinvalsky HT, Helwig EB (1956) Carcinoma of the anorectal junction. 1. Histological considerations. Cancer 9: 480–488

Grodsky L (1960) Extramammary Paget's disease of the perianal region. Dis Colon Rectum 3: 502–510

Grodsky L (1965) Rare nonkeratinizing malignancies of anal region. Arch Surg 90: 216–221

Grodsky L (1967) Unsuspected anal cancer discovered after minor anorectal surgery. Dis Colon Rectum 10: 471–478

Grodsky L (1969) Current concepts on cloacogenic transitional cell anorectal cancers. JAMA 207: 2057–2061

Harrison EG, Beahrs OH, Hill JR (1966) Anal and perianal malignant neoplasms. Dis Colon Rectum 9: 255–267

Heidelberger C, Griesbach L, Montag BJ et al. (1958) Studies on fluorinated pyrinidines II. Effects on transplanted tumors. Cancer Res 18: 305–317

Holm WH, Jackman RJ (1964) Anorectal squamous cell carcinoma. Conservative or radical treatment. JAMA 188: 162–172

Hoskins RB, Gunderson LL, Dosoretz DE (1985) Adjuvant postoperative radiotherapy in carcinoma of the rectum and rectosigmoid. Cancer 55: 61–71

Hutcheson JB, Gordon JB, Fuqua WN Jr (1960) "Extramammary Paget's disease" of the anorectal junction. Arch Pathol 69: 728–732

Kapur BML, Dhawan IK, Singhal KK (1977) Epidermoid carcinoma of the anorectum. Dis Colon Rectum 20: 252–254

Kline RJ, Spencer RJ, Harrison EG Jr (1964) Carcinoma associated with fistula-in-ano. Arch Surg 89: 989–994

Klotz RG, Pamukcoglu T, Souilliard DH (1967) Transitional cloacogenic carcinoma of the anal canal. Cancer 20: 1727–1745

Kraus EW (1978) Perianal basal cell carcinoma. Arch Dermatol 114: 460–461

Kuehn PG, Eisenberg H, Reed JF (1968) Epidermoid carcinoma of the perianal skin and anal canal. Cancer 22: 932–938

Leach RD, Ellis H (1981) Carcinoma of rectum in male homosexuals. J R Soc Med 74: 490–491

Li FP, Osborn D, Cornin CM (1982) Anorectal squamous carcinoma in two homosexual men. Lancet 2: 391

Looney WB, Hopkins HA, MacLeod MS et al. (1979) Solid tumor models for the assessment of different treatment modalities. XII Combination chemotherapy-radiotherapy: Variation of time interval between time of administration of 5-fluorouracil and radiation and its effect on the control of tumor growth. Cancer 44: 437–445

Loygue J, Laugier A, Parc et al. (1981) Carcinoma epidermoide de l'anus, a propos de 149 observations. Chirurgie 109: 710

McConnell EM (1970) Squamous carcinoma of the anus – a review of 96 cases. Br J Surg 57: 89–92

McGraw JY, Bonefant JL (1960) Anorectal lymphomas. Can J Surg 3: 225–228

McGregor JK, Jewett TC Jr (1965) Perianal rhabdomyosarcoma in an infant. Dis Colon Rectum 8: 52–55

Merrwaldt JH (1984) Post-irratiation diarhea. Erasmus University of Rotterdam, DecorDavids, Alblasserdam, Netherlands

Merlini M, Eckert P (1985) Malignant tumors of the anus. A Study of 106 cases. Am J Surg 150: 370–372

Michaelson RA, Magill GB, Quan SHQ et al. (1983) Preoperative chemotherapy and radiation therapy in the management of anal epidermoid carcinoma. Cancer 51: 390–395

Mills SE, Allen MS Jr, Cohen AR (1983) Small-cell undifferentiated carcinoma of the colon. Am J Surg Pathol 7: 643–651

Morgenstern L, Host M, Lugo D et al. (1985) Changing aspects of radiation enteropathy. Arch Surg 120: 1225–1228

Morson BC (1960) The pathology and results of treatment of squamous cell carcinoma of the anal canal and anal margin. Proc R Soc Med 53: 22–26

Morson BC (1979) Tumors of anorectal region. In: Morson BC, Dawson (eds) Gastrointestinal pathology, 2nd edn. Blackwell, Oxford, pp 735–756

Morson BC, Sobin LH (1976) Histological typing of intestinal tumors. International histological classification of tumors no 15. World Health Organization, Geneva, Switzerland, pp 62–65

Morson BC, Volkstadt H (1963a) Muco-epidermoid tumors of the anal canal. J Clin Pathol 16: 200–205

Morson BC, Volkstadt H (1963b) Malignant melanoma of the anal canal. J Clin Pathol 16: 126–132

Nakajima Y, Miyamoto T, Tanake M et al. (1979) Enhancement of mammalian cell killing by 5-fluorouracil in combination with X-rays. Cancer Res 39: 3763–3767

Nielsen OV, Jensen SL (1981) Basal cell carcinoma of the anus – a clinical study of 34 cases. Br J Surg 68: 856–857

Nigro ND (1984) Treatment of squamous cell cancer of the anus. In: DeCosse JT, Sherlock P (eds) Clinical management of gastro-intestinal cancer. Nijhoff, Boston, pp 221–242

Nigro ND, Vaitkevicius VK, Considine BJ (1974) Combined therapy for cancer of the anal canal: A preliminary work. Dis Colon Rectum 17: 354–356

Nivatvongs S, Stern HS, Fryd DS (1981) The length of the anal canal. Dis Colon Rectum 24: 600–601

O'Brien PH, Jenrette JM, Wallace KM et al. (1982) Epidermoid carcinoma of the anus. Surg Gynecol Obstet 155: 745–751

Papillon J (1974) Radiation therapy in the management of epidermoid carcinoma of the anal region. Dis Colon Rectum 17: 181–187

Papillon J (1982) Rectal and anal cancers. Springer, Berlin Heidelberg New York, pp 109–188

Paradis P, Douglass HO Jr, Holyoke ED (1975) The clinical implications of a staging system for carcinoma of the anus. Surg Gynecol Obstet 141: 411–416

Peters RK, Mack TM (1983) Patterns of anal carcinoma by gender and martial status in Los Angeles County. Br J Cancer 48: 629–636

Peters RK, Mack TM, Bernstein L (1984) Parallels in the epidemiology of selected anogenital carcinomas. JNCI 72: 609–615

Petrelli NJ, Shaw N, Bhargava A et al. (1988) Squamous cell carcinoma antigen as a marker for squamous cell carcinoma of the anal canal. J Clin Onc 6: 782–785

Prasad ML, Abcarian H (1980) Malignant potential of perianal condyloma acuminatum. Dis Colon Rectum 23: 191–197

Quan SHQ (1978) Anal and para-anal tumors. Surg Clin North Am 58: 591–602

Quan SHQ (1979) Squamous cancer of the anorectum. Int J Radiat Oncol Biol Phys 5 [Suppl]: 63 (Abstract 55)

Quan SHQ (1983) Carcinoma of the anus. Int Adv Surg Oncol 6: 323–335

Quan SHQ (1986) Cancer of the anus. In: Beahrs OH, Higgins GA, Weinstein JJ (eds) Colorectal tumors. Lippincott, Philadelphia, pp 215–222

Richards JC, Beahrs PH, Woolner LB (1962) Squamous cell carcinoma of the anus, anal canal and rectum in 109 patients. Surg Gynecol Obstet 114: 475–482

Rockwell S (1982) Cytotoxicities of mitomycin C and X-rays to aerobic and hypoxic cells in vitro. Int J Radiat Oncol Biol Phys 8: 1035–1039

Rousseau J, Mathieu G, Fenton J et al. (1973) La telecobaltherapie des cancers du canal anal. J Radiol Electrol 54: 2–20

Roux-Berger JL, Ennuyer A (1948) Carcinoma of the anal canal. Statistic Fondation Curie. AJR 60: 807–815

Salmon RJ, Fenton J, Asselain B (1984) Treatment of epidermoid anal canal cancer. Am J Surg 147: 43–48

Salmon RJ, Zafrani B, Labib A et al. (1986) Prognosis of cloacogenic and squamous cell cancers of the anal canal. Dis Colon Rectum 29: 336–340

Sawyers JL (1972) Squamous cell cancer of the perianus and anus. Surg Clin North Am 52: 935–941

Sawyers JL, Herrington JL Jr, Main FB (1963) Surgical considerations in the treatment of epidermoid carcinoma of the anus. Ann Surg 157: 817–823

Schofield PF, Holden D, Carr ND (1983) Bowel disease after radiotherapy. J R Soc Med 76: 463–466

Schraut WH, Wang CH, Dawson PJ et al. (1983) Depth of invasion, location, and size of cancer of the anus dictate operative treatment. Cancer 51: 1291–1296

Scoma JA, Levy EI (1975) Bowen's disease of the anus. Dis Colon Rectum 18: 137–140

Sehdev MK, Dowling MD, Seal HS et al. (1973) Perianal and anorectal complications in leukemia. Cancer 31: 149–152

Singh R, Nime F, Mittleman A (1981) Malignant epithelial tumors of the anal canal. Cancer 48: 411–415

Sischy B (1985) The use of radiation therapy combined with chemotherapy in the management of squamous cell carcinoma of the anus and marginally resectable adenocarcinoma of the rectum. Int J Radiat Oncol Biol Phys 11: 1587–1593

Sischy B, Remington JH, Sobel SH et al. (1980) Treatment of carcinoma of the rectum and squamous cell carcinoma of the anus by combination chemotherapy, radiotherapy and operation. Surg Gynecol Obstet 151: 369–371

Slater G, Greenstein A, Aufses AH JR (1984) Anal carcinoma in patients with Crohn's disease. Ann Surg 199: 348–350

Spremulli EN, Leith JT, Bliven SF et al. (1983) Responses of human colon adenocarcinoma (DLD-1) to X-irradiation and mitomycin C in vivo. Int J Radiat Oncol Biol Phys 9: 1209–1212

Stearns MW (1979) Anus and anal canal cancer. In: Digestive cancer. Pergamon, Oxford, pp 51–57 (Advances in medical oncology, research and education, vol 9)

Stearns MW, Quan SH (1970) Epidermoid carcinoma of the anorectum. Surg Gynecol Obstet 131: 953–957

Stearns MW, Urmacher C, Stenberg SS et al. (1980) Cancer of the anal canal. Curr Probl Cancer 4: 1–44

Steele RJC, Eremin O, Krajewski AS et al. (1985) Primary lymphoma of the anal canal presenting as perianal suppuration. Br Med J 291: 311

Stern BD, Kaplan L (1969) Multicentric foci of carcinomas arising in the structures of cloacal origin. Am J Obstet Gynecol 104: 255–266

Strauss RJ, Faizo VW (1979) Bowen's disease of the anal and perianal area. Am J Surg 137: 231–234

Stryker JA, Hepner GW, Mortel R (1977) The effect of pelvic irradiation on ileal function. Radiology 124: 213–216

Sugarbaker PH, Gunderson LL, Wittes RE (1985) Cancer of the anal region. In: DeVita VT Jr, Hellman S, Rosenberg SA (eds) Cancer principles and practice of oncology, 2nd edn. Lippincott, Philadelphia, pp 885–894

Svenson EW, Montague ED (1980) Results of treatment in transitional cloacogenic carcinoma. Cancer 46: 828–830

Sweet RH (1947) Results of treatment of epidermoid carcinoma of the anus and rectum. Surg Gynecol Obstet 84: 967–972

Thompson HR (1956) Carcinoma of the anorectal region arising from the intramuscular and apocrine glands. Proc R Soc Med 49: 469–472

Turell R (1962) Epidermoid squamous cell cancer of the perianus and anal canal. Surg Clin North Am 42: 1235–1241

UICC (1982) TNM atlas Spiessl B, Hermanek P, Scheibe O, Wagner G (eds). Springer, Berlin Heidelberg New York, pp 132–144

Vietti T, Eggerding R, Valeriole F (1971) Combined effect of X-radiation and 5-fluorouracil on survival of transplanted leukemic cells. JNCI 47: 865–870

Walls EW (1958) Observations on the microscopic anatomy of the human anal canal. Br J Surg 45: 504–512

Wanebo HJ, Woodruff JM, Farr GH et al. (1981) Anorectal melanoma. Cancer 47: 1891–1900

Welch JP, Malt RA (1977) Appraisal of the treatment of carcinoma of the anus and anal canal. Surg Gynecol Obstet 145: 837–841

Wellman KF (1962) Adenocarcinoma of anal duct origin. Can J Surg 5: 311–318

White WB, Schneiderman H, Sayre JT (1984) Basal cell carcinoma of the anus: clinical and pathological distinction from cloacogenic carcinoma. J Clin Gastroenterol 6: 441–446

Wilking N, Petrelli N, Herrera L et al. (1985) Phase II study of combination bleomycin, vincristine and high dose methotrexate (BOM) with Lencovorin rescue in advanced squamous cell carcinoma of the anal canal. Cancer Chemother Pharmacol 15: 300–302

Williams SL, Rogers LW, Quan SHQ (1976) Perianal Paget's disease: Report of seven cases. Dis Colon Rectum 19: 30–40

Wolfe HRI, Bussey HJR (1968) Squamous cell carcinoma of the anus. Br J Surg 55: 295–301

Zaren HA, Delone FX, Lerner HJ (1983) Carcinoma of the anal gland: Case report and review of literature. J Surg Oncol 23: 250–254

12 Medical Management of Gastrointestinal Cancer

Guillermo Ramirez and Paul P. Carbone

CONTENTS

Guillermo Ramirez, M. D., Professor of Oncology, Paul P. Carbone, M. D., Professor, University of Wisconsin, Department of Human Oncology, Wisconsin Clinical Cancer Center, 600 Highland Avenue, Madison, WI 53792, USA

12.1 Introduction

Gastrointestinal (GI) cancer includes a wide variety of malignancies, starting in the esophagus and ending at the other end of the GI tract, the anus. They are derived from the ectoderm, the mesoderm, or the entoderm and the histologies encompass many subtypes including squamous cell carcinoma, adenocarcinoma, neuroendocrine tumors, small cell tumors, sarcomas, and lymphomas. As a group they make up more than 25% of all new cancers and are responsible for 24.7% of all cancer deaths (Silverberg and Lubera 1988). The various histologies in the several anatomic sites have distinct clinical features and responsiveness to systemic therapy. Surgery, which plays a major role in management of some GI cancers such as colon, has a lesser role in the management of hepatocellular or pancreatic cancers. These latter cancers tend to present with multicentric or far advanced disease that does not lend itself to surgical resection. Likewise, sensitivity to radiation therapy is quite varied.

12.2 Medical Management

12.2.1 Noncytotoxic Therapy

In addition to the problems associated with the cytotoxic therapy of these cancers, other important management features need to be emphasized. First, the rich lymphatic system of the GI tract leads to early local spread and dissemination. In other anatomic areas such as the rectum and esophagus the lack of a serosal surface is associated with local spread and regional lymph node involvement and accounts for increased locoregional recurrences (Dukes 1932; Grinnell 1950). Vascular dissemination occurs through the portal system as well as the arterial circulation, leading to a high incidence of hepatic metastases from intra-abdominal malignancies. Intra-abdominal spread is also likely to occur, presenting as deposits on the serosal surfaces of the bowel, mechanical obstruction, and/or ascites.

12.2.2 Tumor Markers

Another feature of some of these cancers involves the production of specific markers. These include carcinoembryonic antigen (CEA), alpha-fetoprotein, 5-hydroxyindoleacetic acid, etc. (Table 1). These markers can be quantitatively measured and used to diagnose the disease, look for recurrences, or measure the impact of treatment. However, their lack of specificity and the absence of a quantitative correlation between the levels and the volume of cancer has led to confusion about their utility. For instance the measurement of CEA is done routinely after colon resection, yet the value of reoperation after detecting elevation of the CEA has not been accepted routinely or its use definitively has not made an impact on outcome (Rittgers et al. 1978).

Table 1. Serum markers in GI cancer

Marker	Tumor site or type
ACTH	Pancreatic islet cell
PTH	Liver
Insulin-like substances	Liver
Growth hormone	Carcinoid islet cell

12.2.3 Nutritional Care

Another feature of these tumors is the rather common negative effects of the cancer on the nutrition of the patient. Mechanical blockage of the GI tract leads to obvious interference with normal alimentation. Lack of pancreatic juice egress into the intestine may interfere with absorption of nutrients. Tumors of the upper GI tract often lead to loss of appetite, possibly by production of cachectin. The loss of normal GI integrity also interferes with the administration of narcotics for pain. Special diets, supplemental feedings, and hyperalimentation may all need to be used to maintain nutrition. Appropriate surgical intervention may be required to alleviate obstruction and insure oral nutrition integrity.

12.2.4 Pain Control

These intra-abdominal tumors invade the splanchnic nerves in the celiac plexus, leading to deep-seated discomfort and pain. The pain is diffuse and not well localized. The combined pain and discomfort from mechanical obstruction of the bowel or liver capsule encroachment can be emotionally disturbing to the patient and his family, adding to the physical problems. Appropriate pain control requires recognition of the cause, judicious use of medications, and sometimes surgical intervention. Radiation therapy also may be effective in reducing pain, particularly from bone metastases. Surgical approaches to pain control need to be considered early in the course of the illness since splanchnic blocks may be useful to alleviate the abdominal pain.

12.2.5 Supportive Care

An important aspect of the medical management of the cancer patient involves providing the patient with support and care that allows the patient and his family to participate and create a feeling of control by the patient of his/her body and the disease. The patient and the family must be fully informed about the current status of the clinical condition and therapy plans. Many of these patients will ultimately develop metastatic disease. While the cytotoxic chemicals may play a role in the treatment of some patients, the physician must be fully aware of the limitations of therapy. He must offer to refer the patient for more experimental therapy. He must not allow the patient to feel abandoned when the conventional, available therapies run out. Symptomatic control is a major role of the oncologist even if he cannot effectively treat the primary cancer.

12.3 Cytotoxic Therapies: General Principles

12.3.1 Major Drug Actions

The classification systems for the cytotoxic drugs usually relate to the presumed mechanism of action, class of drugs, or their cell cycle specificity. These systems are not mutually exclusive. For instance, alkylating agents work by interacting with DNA to cause crosslinking or strand breaks. The alkylating agents are also non-cycle-specific, that is the affect cells that are in all phases of the cell division cycle. Antibiotics such as Adriamycin act by intercalation of the DNA, interfering with strand duplication, a lesion also produced by alkylating agents. Natural products such as the vinca alkyloids seem to affect the metaphase spindle and are definitely cycle specific. Since the mechanisms and the cell cycle specificity are not always clear, the system of describing drugs by their chemical class makes the most sense. These classes are alkylating agents, natural products, hormones, and miscellaneous agents.

12.3.2 Alkylating Agents

This class of agents has been a major tool in the therapy of several malignancies but except for methyl-CCNU (semustine) and streptozotocin, both nitrosoureas, and mitomycin, an antibiotic, this class of drugs plays a relatively minor role in GI cancers. The agents act by forming covalent bonds with nucleophiles such as DNA. The bifunctional alkylating agents crosslink the DNA or cause the binding of the DNA to other macromolecules. Cells throughout the cell cycle are affected by these agents with some special sensitivity occurring in late G1 or early S phases.

Semustine or methyl-CCNU is one of the class of nitrosureas, characterized by lipophilicity and the ability to alkylate DNA. The drug is administered orally, with rapid absorption, and there is pharmacologic evidence for CNS penetration (SPONZO et al. 1973). The major acute side-effects include nausea, vomiting, and bone marrow suppression. More chronic toxicity has been reported to be renal failure and induction of acute leukemias (HARMON et al. 1979). This drug has not been released for general use and even in GI cancer has only a limited utility, despite some evidence for synergism with 5-fluorouracil (5-FU).

Streptozotocin, a glycosylated nitrosourea that occurs naturally, has the rather unique feature of being diabetogenic in animals. In man it has mainly renal toxicity and causes relatively little marrow suppression. The renal problems are both tubular and glomerular (WEISS RB 1982). Nausea and vomiting occur after parenteral administration regularly.

Mitomycin C is an antibiotic that is activated in vivo to an alkylating agent that can act as a bifunctional or trifunctional alkylating agent (REICH 1979). It must be administered intravenously and is accompanied by nausea and vomiting. The drug's dose limiting toxicities include effects on the platelets, leukocyte precursors, renal parenchyma, and lung tissues. The myelosuppression may be cumulative and profound, particularly on the platelets, reaching a nadir in 4–6 weeks after administration. The pulmonary effects do not seem to be dose dependent and can result in interstitial fibrosis (OSWOLL et al. 1978). In patients with pulmonary toxicity the death rate is high and some evidence exists to indicate that the pulmonary toxicity is accentuated by hyperoxygenation (CROOKE and BRADNER 1976). Renal failure may be accompanied by microangiopathic hemolytic anemia (MAHA). The kidney damage is both glomerular and cortical. The treatment of the MAHA syndrome includes steroids, plasmapheresis, and more recently, vincristine. This latter treatment seems to increase the platelets and decrease the intravascular destruction (ANTMAN et al. 1979). While mitomycin has been used in combination with 5-FU and Adriamycin, the value of this combination routinely in stomach or pancreatic cancers has been questioned (CULLINAN et al. 1985). While the response rate may be slightly higher, no impact was seen on survival over other single agents such as 5-FU or Adriamycin.

12.3.3 Antimetabolites

The antimetabolites were among the original antineoplastic drugs. However, one stands out as the most effective single agent against a variety of GI cancers, namely, 5-FU. Methotrexate, an antifolic acid, has some activity against colon cancer but its largest role is in the treatment of squamous cancer of the oral cavity, not to be discussed in this book. Hydroxyurea has had a mixed clinical usage in GI cancer. Several early papers suggested synergism with 5-FU; however, no well controlled trial has proven the clinical enhancement.

5-Fluorouracil was first synthesized by HEIDELBERGER in 1957, based on the observation that uracil uptake was enhanced in rat hepatomas as compared to normal liver cells (HEIDELBERGER et al.

1957). Despite many elegant biochemical studies with 5-FU the exact mechanism of action is not clear. It affects both the DNA and RNA of tumor cells. The RNA effect leads to messenger disruption and faulty protein synthesis (MANDEL 1969). Alternatively, fluorodeoxyuridine produced in vivo after 5-FU administration interferes with thymidine synthesis and thus DNA (MAJOR et al. 1982). 5-FU toxicity and/or effect can be enhanced by adding folic acid, thymidine, allopurinol, hypoxanthine, methotrexate, and physiologic pyrimidines and purines. In addition, the sequence of 5-FU and MTX administration has been suggested to affect the action of 5-FU (CADMAN et al. 1979). More recently 5-FU administration by continuous infusion has been recommended to enhance the effect or overcome apparent resistance (CABALLERO et al. 1985).

5-Fluorouracil is erratically absorbed orally and it should not be administered in this fashion. When given as an intravenous bolus, 5-FU achieves a relatively high blood level but its half-life is less than 10 min in the plasma. 5-FU has been administered by prolonged infusion intravenously, intra-arterially, and intracavitarially. The toxicity is determined by the rate and route of administration. Mucositis, diarrhea, and myelosuppression are the major toxicities associated with IV bolus administration. These effects usually occur sequentially, with the oral or GI toxicity preceding the marrow effects. Since the marrow effect may occur later, the usual schedule has been worked out to be 12–15 mg/kg daily for 4–5 days with a rest period of 1 day followed by 6–7.5 mg/kg every other day for up to six doses (ANSFIELD et al. 1977). The toxicity is also affected by prior radiation therapy, nutritional status, and perhaps patient age. A relatively rare toxic manifestation may be cerebellar ataxia, somnolence, and slurred speech (WEISS HD et al. 1974). This is reversible by stopping the drug. Other less serious but bothersome side-effects include hyperpigmentation and photosensitization of the skin. Long-term 5-FU administration may result in severe dryness of the skin.

Fluorodeoxyuridine (FUDR) is a metabolic product of 5-FU and is a substrate for thymidine kinase (CLARKSON et al. 1964). FUDR seems equally as effective as 5-FU when given intravenously as a bolus. However, it is preferred as a drug for intra-arterial infusion by implantable pump (ENSMINGER et al. 1978), the reason being the volume of infusion and the clearance from the liver on first pass as compared to 5-FU.

12.3.4 Natural Products

The natural products include the vinca alkyloids, antibiotics, and epiphylotoxins. These agents unfortunately have little activity in GI cancer except for Adriamycin.

Adriamycin is one of the anthracycline antibiotics, which are among the most active of all cytotoxic drugs. They come from the fungus *Streptomyces peutcetiud*. Adriamycin is the most common analogue used in GI cancer. It was isolated in 1963 (DIMARCO et al. 1963). The exact mechanism of action is unkown. There are several known effects on cancer cells, including DNA binding, chelation of divalent cations, and binding to the cell surface membranes (DIMARCO 1975; BENJAMIN 1975). The anthracyclines bind to the DNA rapidly by intercalating between base pairs. The cascade of effects of this as well as free radical formation results in inhibition of DNA and RNA, membrane alteration, as well as binding to cardiolipin. In addition to the usual bone marrow suppression, Adriamycin has the potential to cause cardiac toxicity. This is related to the dose administered and rises to clinically significant proportions at doses of 500 mg/m^2 or greater. The treatment or prevention of cardiac damage is beyond the scope of this chapter. However, it is rare that patients with GI cancer will have treatment limited by cardiotoxicity. Alopecia is a frequent accompaniment of Adriamycin toxicity. The schedule of Adriamycin administration is usually by bolus every 3 weeks, or weekly. The use of this agent alone has not been a part of the armamentarium against GI cancer. The usual administration schedule includes a combination of 5-FU, Adriamycin, and mitomycin (MACDONALD et al. 1980).

12.3.5 Miscellaneous Agents

Cis-dichlorodiammineplatinum (Cisplat) was synthesized by Rosenberg when he noted inhibition of bacteria in a salt solution occurred by passing electric current through two platinum electrodes (ROSENBERG et al. 1965). He isolated the active agent and tested the agent against a wide variety of animal tumors. The compound is an inorganic complex, soluble in water, with the biologic effects clearly related to the geometry of the molecule. In water the compound's chloride dissociates and the transformed product resembles a bifunctional alkylating agent. Crosslinking of DNA occurs and damages the DNA (ZWELLING and KOHN 1979). It is not schedule dependent and seems to be non-cycle-dependent. Cisplat is given with saline hydration and diuretics since the excretion product damages the renal tubule and that is clearly the dose limiting toxicity. Careful monitoring of the serum creatinine and the creatinine clearance is important since the toxicity may be accumulative. Another common side-effect is severe nausea and vomiting requiring intensive sedation and major antiemetic regimens. Ototoxicity, neurotoxicity, and marrow suppression complete the spectrum of adverse effects (VON HOFF et al. 1979 a). Despite these problems the drug has major utility in upper GI neoplasms that have squamous cell histologies. It may also play a role in colon cancer.

12.4 Chemotherapy Strategies

12.4.1 Adjuvant Therapy

Chemotherapy appears to have the best chance of being curative when the tumor burden is small. Presumably at that time not only are the numbers of cells small, but the growth rate and the proportion of cells in DNA synthesis are highest. These three factors usually herald increased sensitivity to drugs. The smallest volume of cells occurs after primary surgery when all gross tumor is removed. Based on pathologic or clinical staging the risk of recurrence can be estimated. These parameters usually include: size of the tumor, depth of penetration, lymph node status, and liver involvement. In general there is some agreement that patients with recurrence rates of 25% or more at 10 years would constitute appropriate target populations for postoperative chemotherapy approaches.

The other variable that must be determined in devising strategies for adjuvant therapy is the degree of clinical activity of the drug or drugs against advanced disease. It is felt that this predicts for effectiveness when used after surgery. This has not been well defined. In general one hopes to have drugs that will have higher response rates (75%) and significant numbers of complete regressions (more that 25%). Unfortunately, few combinations of drugs in GI cancer meet these characteristics (LOKICH and SKARIN 1972; KOVACH et al. 1974; MOERTEL CG 1976). This did not inhibit clinical trials in the early 1980s but the results were marginal (NAKAJIMA et al. 1980; ENGSTROM et al. 1981). More recently adjuvant trials employing primarily biologic agents that affect immune responses in the host have been proposed to be used in the postoperative situation. In these trials clinicians have not demanded activity in late disease but have suggested that the low tumor burden may be exactly the right situation for biologic modifiers that enhance immune stimulation. Several studies are underway now with tumor vaccines, LAK cells, and levamisole (TERRY and ROSENBERG 1982) (SONDEL et al. 1987).

The combination of drugs, surgery, and radiotherapy has attracted some interest in the oral cancers as well as the esophagus and anus. These cancers are squamous cell type and seem to be sensitive to combinations of cisplat and 5-FU. Another combination with some promise includes 5-FU and mitomycin as proposed by SISCHY et al. (1980). This set of drugs combined with X-ray therapy is being employed by the Radiation Therapy Oncology Group (RTOG) and the Eastern Clinical Oncology Group (ECOG) in anal and esophageal cancers in a presurgery approach. The surgery is done when the chemotherapy is finished and a biopsy is done, (confirming, in some patients, no obvious cancer). This use of drugs before surgery has been termed "neoadjuvant".

12.4.2 Metastatic Cancer

12.4.2.1 Single Agents

The use of chemotherapy proceeds logically from the discovery of activity against advanced cancers with a single drug with the progression to development of combinations. The dismal record of single agents in GI cancers has been a major disappointment. Because of this poor track record for new agents in GI cancer it was felt that somehow prior treatment with any drug may enhance resistance to other drugs. This form of resistance has been termed pleiotropic because the resistance occurs to a wide spectrum of drugs, even those not utilized in the original program (SHOEMAKER et al. 1983). This concept, well known in the laboratory, has yet to be shown to be a reason why drugs have significantly less activity in GI cancer. Except for 5-FU in the major epithelial tumors, few if any other agents have been shown to cause tumor regressions. Recent trials therefore have sought to have patients with no prior therapy entered on activity seeking trials (phase II) (BEER et al. 1983; MOERTEL C et al. 1984). Despite this design new active agents continue to elude discovery in most GI cancers.

12.4.2.2 Drug Combinations

The concept of drug combinations is based on the principles that adding non-cross-resistant drugs with nonoverlapping toxicities at full doses will result in synergistic or additive effects. This has been shown to be so in the hematologic malignancies, such as Hodgkins's disease (Frei et al. 1973), large cell lymphomas (Schein et al. 1976), and acute leukemias (Frei and Freireich 1965). However, the frustration in GI cancer is that relatively few classes of drugs have activity and most agents have similar toxicity, namely, bone marrow. Thus, relatively little evidence exists for effective combinations except those with cisplatin and 5-FU in squamous cell cancers.

12.4.2.3 Combined Modality

The use of different modalities sequentially or simultaneously again has been advocated to increase the chance of cure. The principles basically involve the use of surgery to remove bulky disease, radiotherapy to decrease local-regional recurrences, and chemotherapy to diminish the impact of occult metastases. This principle again has found limited utility in GI cancer except in the upper tract and possibly in the pancreas, rectum, and anus (Taylor et al. 1985; Leichman et al. 1984 b).

12.4.2.4 Regional Infusion

The infusion of drug into an anatomic regional arterial supply has the potential of achieving a high concentration of drug close to or in the tumor with lesser amounts outside the systemic circulation. This concept has been most effectively applied in the hepatic artery infusion protocols (Cady and Oberfield 1974; Sullivan and Zurek 1965). In addition, because FUDR has a rapid clearance from the liver there is lessened systemic toxicity. There is no question that in some patients the infusion of FUDR or 5-FU into the liver filled with metastases has caused significant shrinkage of tumor (Sullivan et al. 1960). However, there is still controversy as to whether this approach has been adequately tested in a proper controlled clinical trial to prove efficacy of the intra-arterial route over the systemic intravenous route. The recent use of implantable pumps has made the infusion of drugs much easier, but really has only added to the problems of doing a good trial. Many intra-arterial infusions have been done in the past 20 years, and there are some data as to how effective the infusions are in improving survival as compared to intravenous administration (Kemeny et al. 1987).

12.5 Immunotherapy

The role of immune modulators in the GI tract neoplasms has yet to be defined. The basic approach has been to use nonspecific immune stimulants such as bacillus Calmette-Guerin (BCG) or levamisole. In general, these trials have led to equivocal results except for some intriguing data with levamisole in colon cancer. Investigators from the University of Wisconsin conducted a pilot study with 5-FU and levamisole in Dukes D colon cancer and observed some survival advantage (Borden et al. 1980). Other trials done at the Mayo Clinic and by the NCCTG have indicated some early beneficial results with levamisole alone or 5-FU plus levamisole in patients with Dukes B and C cancers (Laurie et al. 1986). This trial is being replicated now as an intergroup effort in the NCI/CTEP trials program.

12.5.1 Vaccines

Tumor vaccines have long been advocated, based on animal data, as a potentially effective tool. The idea is to make a vaccine that is intimately associated with an immune stimulant. The best example of the effort is in the tumor cells of the guinea pig model by Hanna and Peters (1981). Here BCG mixed with live tumor cells was able to improve the survival of animals as compared to tumor cells alone, or BCG and tumor cells injected at separate sites. This led to a small clinical trial by Hoover et al. (1985) and subsequently to a multi-institutional study in the ECOG.

12.5.2 LAK Cells/IL-2

Rosenberg has shown in laboratory animals that IL-2 alone or IL-2 with LAK cells can cause significant regressions and prolongations of survival. The IL-2 is a lymphokine that enlarges a specific population of cytotoxic lymphocytes and also activates the cells to attack cancer cells without affecting normal cells (Rosenberg et al. 1985). With the advent of genetic engineering techniques large amounts of IL-2 have been produced and have been administered to patients. The early results indicate a definite role for this material in certain cancers such as renal cell and melanoma (ROSENBERG 1986). The role in GI cancers needs to be defined.

12.5.3 Interferons

These substances are natural materials that have a variety of biologic effects. In particular they stimulate immune reactions, have a direct effect against tumor cells, and also enhance certain HLA antigens that may play a role in the immune system (BORDEN 1984). To date the interferons have not been tested adequately against the GI cancers except possibly some of the neuroendocrine tumors such as islet cell or carcinoid tumors (OLDHAM and SMALLEY 1984).

12.5.4 Tumor Necrosis Factor

An interesting new agent is a biologic product of a human gene that experimetally causes regressions of cancers by lethal nonimmune direct action (CARSWELL et al. 1975 a). This material is now being used in early trials in a wide variety of cancers. It is interesting that this material is also biochemically similar to another product isolated from genes that is called cachectin (BEUTLER et al. 1985). Cachectin has been shown to be a cause of weight loss in animals with cancer.

12.5.5 Monoclonal Antibodies

An antibody specific and lethal to cancer cells has been a goal for many years. Recently with new laboratory techniques specific high activity antibodies have been produced that will bind to cancer cells with some specificity. These techniques also have the potential to produce antibodies in purified form in large quantities. The utility of these anitbodies can be expanded by adding radioisotopes or toxins that will kill the cancer cells even at low binding concentrations (OLDHAM and SMALLEY 1984). These antibodies can also be used to localize cancers and to diagnose them at early stages (HERLYN et al. 1983; MURANO et al. 1985; MAGNANI et al. 1983). The major efforts in GI cancers have been in hepatomas and pancreatic/colon cancers. Interestingly the data from Johns Hopkins indicate that the use of polyclonal antibodies has the best effect in hepatic cancers (SITZMANN et al. 1987). The obvious limitation of allogeneic polyclonal antibodies is that they will cause sensitization and antibodies that will prove to be limiting. However, the monoclonal studies are still in the early stages with only rare reports of activity that will need to be confirmed.

12.6 Esophageal Cancer

12.6.1 Incidence and Etiologic Factors

Esophageal cancers represent about 7% of all GI cancers in the United States with about 9000 new cases per year. The disease occurs predominantly in men, in patients over 60. The disease is also three times as common in blacks as whites. In the past 25 years there has been a remarkable increase in mortality in blacks of over 100%. This selective increase in blacks is not understood.

There are major geographic differences in incidence throughout the world. These national differences cannot be explained simply on geography since marked variations in incidence occur within the same region. For example France is a high incidence country while Denmark is relatively low. The well known high incidence areas are located around the Caspian Sea, in northern China, Transkei, S. Africa, Kenya, Puerto Rico, and certain parts of India (LEVIN et al. 1974). Specific predisposing factors have been associated with esophageal cancer. These are alcohol, smoking, benign esophageal strictures, Barrett's esophagus, bracken fern ingestion, dietary nitrosamines, and Plummer-Vinson syndrome (POLEYNARD et al. 1977, JOSKE and BENEDICT 1959; JACOBSON 1961).

12.6.2 Pathology

Most esophageal cancers are squamous cell histopathologically. Adenocarcinomas tend to occur in the Barrett's esophagus and at the gastroesophageal junction, where they may actually be fundal stomach cancers that have spread upwards. Because of the rather rich lymphatics, blood supply, and lack of serosal surface, the esophageal tumors tend to spread extensively to nodes in the mediastinum, neck, and abdomen (MCCORT 1952). These tumors also invade into the adjacent tissues, particularly the tracheobronchial tree, causing fistulas. The intraluminal extension often extends submucosally beyond the original superficial mass. The intraluminal lesions cause the pain and dysphagia characteristic of these cancers.

Table 2. Single agents in esophageal cancer: bleomycin (modified from KELSEN 1984 a)

Authors	Route	Dose	No. of patients evaluable	Response (CR/PR)	Duration
EORTC (1970)	IV/IM	10–20 mg/m^2/d	5	1	
BONADONNA et al. (1972)	IV	10–20 mg/m^2 daily × 5 15–30 mg/m^2 2 × /wk	10	2	NS
STEPHENS (1973)	IM	15 mg daily	3	1	NS
RAVRY et al. (1973)	IM	20 mg/m^2 daily to total 280 mg	14	0	NS
TANCINI et al. (1974)	IV	10–20 mg/m^2 daily × 5 15–30 mg/m^2 2 × /wk	29	4	1–2 mo
KOLARIC et al. (1976)	IV	15/m^2 2 × /wk	15	4	1–4 mo
YAGODA et al. (1972)	IV	0.25 mg/kg/d	4	0	
Total			80		

NS, not significant

12.6.3 Natural History

The problem of treating esophageal cancer can be well illustrated by looking at the outcome in 100 typical patients who present to American centers. On diagnostic evaluation 42% will be defined as inoperable. Of the 58% who are subjected to surgery 19% more will be found to be unresectable. The postoperative mortality will be about 13% in most series, primarily due to the effects of extensive operative resections, the preoperative nutritional deprivation of most patients, and the associated medical conditions in these elderly patients who are very likely to have been heavy smokers and alcohol imbibers (ELLIS and GIBB 1979). Only 18 patients will be alive in 1 year. At 2 years only 9% will be alive. After 5 years a modest 4% will be still alive. The use of preoperative X-ray therapy seems to increase the number of resectable cases but not the 5-year survival rate (PEARSON 1977). This dismal record is a challenge for all oncologists.

12.6.4 Chemotherapy

The use of single agents has been reviewed recently by KELSEN (1984) (Table 2). He reiterated the difficulty of assessing response and the need to document the completeness of response pathologically by endoscopic biopsy or surgical resection. In his review he stated that only ten drugs have sufficient published information with a minimum of 19 patients per drug to be assessed accurately for antitumor activity.

Bleomycin: In the review by Kelsen, a total of 80 patients on eight different studies were evaluable for response. The doses of bleomycin ranged from 10 to 30 mg/m^2/day using different schedules. The overall response rate was 15% with a median duration of 2–3 months.

Mitomycin C: The major evaluable study was by the ECOG and involved 32 patients treated with 20/mg/m^2 on an every 4-week schedule for two doses and then every 6 weeks (ENGSTROM et al. 1983). Severe toxicity occurred in 19 patients (50%). The response rate was 10/32 or about 27%. Other authors report a much lower response rate but also less toxicity.

Adriamycin: A total of 33 patients have received this drug as a single agent with an average response rate of 15%. Of interest is one of the largest studies by ECOG reported that only 1 of 20 patients responded to Adriamycin.

5-FU: The use of 5-FU alone has had limited evaluation. Again the ECOG has reported a 15% response rate with durations of remissions of 5–26 weeks.

Methotrexate: The only good study was an ECOG trial with 26 evaluable agents and a response in 3/24 patients (EZDINLI et al. 1980).

CCNU: In a single trial 19 patients were treated with three responses (16%) (MOERTEL et al. 1976 b).

Methyl-GAG: Methyl-glyoxal bis(guanylhydrazone) (methyl-GAG) is a polyamine systhesis inhibitor with little or no myelosuppression. In two trials, utilizing either a daily or weekly schedule, approximately 20% of the patients had some response.

Cisplatin: In four separate trials a total of 73 patients were treated. The overall response rate was 22% with relatively short responses and rare complete ones.

Vindesine: This new vinca alkaloid has been given to 83 patients in four separate trials with a surprising 33% partial response rate (KELSEN 1984). However, the responses were short in duration.

VP-16: This new epipodophyllotoxin derivative

Table 3. Combination chemotherapy for esophageal carcinoma (modified from Kelsen 1984)

Regimen	Evaluable	Response rate (%)	Median response duration (mo)	Authors
DDP-Bleo	61	15	6	Popkin et al. (1983)
DDP-Bleo	30	17	–	Kelsen et al. (1978)
Bleo-Adria	16	19	4	Wittes (1975)
DDP-DVA-Bleo	68	53	7	Bosset (1983)
DDP-MTX-Bleo	10	50	7	Kolaric (1980)
DDP-Adria-5FU	21	33	–	Kelsen et al. (1983 b)
DDP-Bleo-VP16	16	31	–	Vogel et al. (1981)
DDP-5FU	10	80	8	Gisselbrecht et al. (1983)
DDP-DVA-MGBG	13	46	–	Forestiere et al. (1983)
DDP-Bleo-MTX-MGBG	14	64	–	Hellerstein (1983)
DDP-VCR-Bleo-5FU	10	60	–	Kelsen et al. (1983 c)

DDP = cisplatin; Bleo = bleomycin; Adria = Adriamycin; DVA = vindesine; MTX = methotrexate; 5FU = 5-fluorouracil; MGBG = methyl-glyocall bis (quanhydrannonel); VCR = vincristine

has been administered to 30 patients with a 7% response rate.

12.6.4.1 Drug Combinations (Table 3)

Cisplatin/Bleomycin: The approach with drug therapy for these patients has taken on new enthusiasm with the report of Kelsen et al. (1984), who described a 15% overall response rate to a two drug combination of cisplatin and bleomycin in both extensive disease and locoregional disease, with 10% (four patients) alive at 41 months. In a review Kelsen reports an overall 15% CR/PR response rate with durations of 5–9.5 months (Bosset et al. 1983).

Cisplatin/Vindesine/Bleomycin: This approach has been explored extensively at Memorial Sloan Kettering by Kelsen and co-workers using cisplatin 3 mg/kg and vindesine 3 mg/m^2 on day 1, bleomycin by bolus and continuous infusion from days, 3–6 at 10 mg/m^2, followed again by vindesine on days 8, 15, and 22 at 3 mg/m^2 (Kelsen et al. 1983). The whole regimen is recycled at 28-day intervals and then given as maintenance involving vindesine and cisplatin. For both advanced and surgically resected disease 36 of 68 neoadjuvant patients responded. While several patients had excellent clinical responses, complete responses were not confirmed at the time of surgical resection. The marrow toxicity was severe, requiring hospitalization for sepsis in eight. There were two drug-related deaths, one related to kidney failure (Kelsen et al. 1983).

Cisplatin/5-FU: This regimen was used in head and neck squamous cell carcinoma, achieving a 70% response rate. In esophageal cancer effectiveness with cisplatin and 5-FU alone is also well known. This combination's activity actually has been demonstrated in patients undergoing preoperative therapy with two cycles of drug followed by surgery. In 19 cancer patients undergoing surgery after two cycles of chemotherapy and simultaneous radiotherapy, 15 tumors were resected for cure and in five of the specimens there was no cancer. This was reported in 1981 by Steiger and colleagues with four of the five patients alive and disease free at 5 years (Steiger et al. 1981). More will be discussed in the combined modality section. DeBessi and colleagues reported good initial results in advanced esophageal cancer with cisplatin, bleomycin, and methotrexate (DeBessi et al. 1984). Their patients had a 51% rate of shrinkage of tumor with a 26% objective response rate.

Other combinations have been reported and reviewed by Kelsen et al. (1984). In general the above combinations seem to have the greatest clinical utility and have been used in combined modality trials. The common determinants of response include the enrollment primarily of good performance status patients and the use of cisplatin. Toxicity has been significant in terms of marrow suppression. The duration of remission appers to be around 6 months. The trend now seems to be towards utilizing these therapies in combined modality approaches.

12.6.5 Combined Modality Therapy

Preoperative radiotherapy has been the most widely applied form of adjunctive radiotherapy, often reserved for those patients who may be marginally resectable to decrease the bulk of local cancer. In a recent review of combined modality therapy, Kelsen reported over 1280 patients treated with XRT (Kelsen et al. 1984). Three-quarters of these patients went to surgery and 58% of the tumors were resectable. There was a 21% mortality with an 8% 5-year survival rate. His conclusion was that while the proportion of patients with resectable disease was higher with preoperative RT, the overall 5-year survival was not improved.

At Memorial Sloan-Ketterin Cancer Center, a combined chemotherapy program with cisplatin, bleomycin, and vindesine and XRT has been given to 34 patients mainly with stage II disease. Surgery was done on day 56. Downstaging occurred in ten patients after the two cycles of chemotherapy as demonstrated by significant shrinkage or disappearance of the tumor by esophagogram. Three patients had no pathologically demonstrable tumor in the specimen after surgery. XRT was given to patients with T_1 or N_1 disease postoperatively. The overall therapy related mortality was 9%. With a 3-year median follow-up 26% of all patients remain free of disease (Kelsen et al. 1983).

A large experience in preoperative chemo-radiotherapy has been gained by the Wayne State Group and a follow-up study from the RTOG/SWOG. This has been referred to above in part as the experience with cisplatin and 5-FU. After an initial study with 5-FU, mitomycin, and preoperative XRT, the Wayne State investigators piloted a trial with cisplatin/5-FU in 21 patients (Leichman et al. 1984a). The chemotherapy was given as a bolus of cisplatin, 100 mg/m^2 on days 1 and 29. The 5-FU was administered as an infusion, 1000 mg/m^2 continuously daily for 4 days, on days 1–4 and 29–32. Radiation therapy was given preoperatively (30 Gy in 15 fractions over 3 weeks). If no metastatic disease was detected, surgery was planned at 8 weeks. Of the 21 patients completing the preoperative therapy, 19 went to surgery with 71% of the tumors resectable for cure. Five patients had no tumor in the specimen. The treatment fatality rate was 27%. The median survival of the patients undergoing resection was 24 month. Three of the five without evidence of disease remain alive and free of disease. The authors state that the absence of cancer in the resectable specimen is a favorable predictor of long-term disease-free survival. Based on these results a larger trial was done in the SWOG and the RTOG. Over 138 patients have been entered at last report of accruals (Leichman et al. 1984b).

Thus, the current trend in investigative therapy is to treat patients with good performance status only and to give preoperative chemotherapy combined with radiation. While the results of two combined modality programs are good, no controlled trial is available comparing these complex treatments to surgery and XRT alone or to a single drug, namely 5-FU. When 5-FU was used alone in a group of patients with head and neck cancer so treated because they had poor renal function, 8 of 11 had objective responses, with four of four patients without prior therapy responding. A comparison of the two regimens in esophageal cancer appears to be appropriate since 5-FU alone carries much less toxicity. Parker and his colleagues reported a series of sequential results with the latest being combined modality therapy with mitomycin and 5-FU, preoperative XRT of 30 Gy, and surgery (Parker et al. 1985). They caution that the 2-year survival rate is still not significantly better than their experience with XRT alone. Moreover, in their hand the absence of tumor in the resected specimen did not confer a good outcome. As in all of medicine, we must be prepared to test our hypotheses with good, well-designed prospective clinical trials.

12.7 Gastric Cancer

12.7.1 Incidence

Two important features dominate the geographic occurrence of gastric cancer. First, the incidence differs by a factor of 8 around the globe. Japan and China have the highest rates, about 60/100000, as compared to the US rate of 8/100000. Secondly, there has been a continuing reduction in the rate in many countries over the past 50 years. This is best illustrated in the US male population, where the rate has dropped from 29 to 8/100000 in the past 50 years. Diet appears to play a major role in the etiology of this cancer. Japanese who move to the United States achieve the lowered US rate of this disease within two generations. Major contenders for a contributing exogenous factor are the nitroso compounds that can be directly ingested or produced from dietary nitrates. In addition the increased use of refrigeration and the higher intake of vitamin C have also been used to explain the lowered occurrence of gastric cancer. Predisposing factors such as pernicious anemia, atrophic gastritis, and postgastrectomy states all have been implicated to increase the gastric cancer rate in patients.

12.7.2 Pathology

Adenocarcinomas make up more than 95% of all gastric neoplasms, with lymphomas making up 2.5% more. The other tumors are rare. Most gastric cancers arise from the antral or pyloric regions of the stomach. They may take on ulcerated, polypoid, or diffuse appearances. The sites of metastases include adjacent organs or regional lymph nodes, liver, and supraclavicular nodes. The prognosis relates to the depth of invasion of the primary tumor, the presence or absence of lymph nodes, and metastases. Lesions that are superficial are highly curable, while those that have lymph node involvement or are deeply invasive carry a poor prognosis.

12.7.3 Natural History

Gastric cancer tends to be clinically silent for long periods of time. Eventually, pain, anorexia, anemia, or dysphagia causes the patient to seek medical attention. The diagnosis is established by upper GI radiologic examination, fiberoptic endoscopy, and/or exfoliative cytology. Eventually a tissue diagnosis must be obtained and the patient evaluated for disease extent. The use of CT scanning has been of major value to assess regional lymph nodes, liver metastasis, and spread to other organs in the abdomen or lung. Screening is not considered cost-effective in the United States; however, in Japan the procedure has been almost automated and is finding large numbers of early cancers. The technique begins with a double contrast upper GI study followed by appropriate endoscopy in the patients with radiologic abnormalities (YAMADA 1977). The result is a 5-year survival rate of 80%–90% for those cases picked up by screening.

In patients with extensive disease Lavin and his colleagues have examined pretreatment information to estimate response and survival (LAVIN et al. 1982). Performance status was a major determinant of response and survival. Other factors included more than 5% weight loss in 6 months, and normal liver chemistries. Response to therapy was associated with an improvement in survival, 36 versus 14 weeks, that was above that predicted based on prognostic factors.

Table 4. Single-agent chemotherapy for advanced gastric cancer (O'CONNELL 1985)

Drug	No. of patients treated	Objective response rate (%)
Mitomycin C	189	29
Cisplatin	50	26
5-FU	457	20
Hydroxyurea	31	19
BCNU (carmustine)	33	18
Adriamycin (doxorubicin)	180	17
Chlorambucil	18	17
Triazinate	26	15
Dacarbazine	15	13
Methotrexate	28	11
Methyl-CCNU	37	8

12.7.4 Chemotherapy

The therapy of gastric cancer with single agents has resulted in demonstration of objective tumor regressions with several drugs (Table 4).

5-Fluorouracil has had a 20-year history of use given in a variety of schedules with the best results reported by using the loading course. The response rate averages about 21% with a duration of 4 months (O'CONNELL 1985). Other single agents include mitomycin C and BCNU (carmustine). These reportedly effect a 29% and 16% response, respectively (MOERTEL CG 1973). Cisplatin (BEER et al. 1983), hydroxyurea (COMIS and CARTER 1974), Adriamycin (EARL et al. 1984), methotrexate, and triazinate (BRUCKNER 1982) all have been reported to effect a response rate of 11% or more in at least 25 patients. These phase 2 results have led to attempts at developing combinations hoping to develop a higher complete response rate. It is important to note that almost all the single agent activities are manifested as incomplete responses.

12.7.4.1 *Drug Combinations* (Table 5)

The many combination studies began with 5-FU and a nitrosourea. In 1974 KOVACH et al. reported a randomized study involving 5-FU, BCNU, and the combination there of. The combination was superior, producing a 41% PR/CR rate with 29% for 5-FU and 19% for BCNU. No survival advantage was seen for the combination. Subsequent studies done by the ECOG utilized 5-FU with methyl-CCNU, resulting in a 30% response rate, while that with methyl-CCNU alone was 8% (KOVACH et al.

Table 5. Prospectively randomized clinic trials of chemotherapy in advanced gastric cancer (O'CONNELL 1985)

Regimen	No. of patients treated	No. of patients with measurable disease	Objective response (%)	Survival (median wk)
5-FU	28	28	29	30
BCNU	23	23	17	14
5-FU + BCNU	34	34	41	31
Methyl-CCNU	67	67	7	13
5-FU + methyl-CCNU	60	60	30	20
5-FU	10	10	20	18
5-FU + methyl-CCNU	29	29	21	18
5-FU + methyl-CCNU	87	54	9	18
5-FU + mitomycin C	80	43	14	25
5-FU	95	0	–	36
5-FU + methyl-CCNU	84	0	–	25
Adriamycin	36	17	24	8
5-FU + mitomycin C + cytosine arabinoside	36	18	17	9
5-FU + Adriamycin + methyl-CCNU (FAMe)	38	15	47	13
Adriamycin	37	37	22	17
5-FU + mitomycin C	53	53	34	18
5-FU + methyl-CCNU	49	49	24	16
FAMe	34	10	30	34
5-FU + Adriamycin + mitomycin C	43	12	25	30
5-FU + ICRF + methyl-CCNU	46	19	21	17
5-FU + methyl-CCNU	58	18	6	23
FAMe	76	16	25	28
5-FU + Adriamycin + mitomycin C	78	18	17	26
5-FU + Adriamycin	78	19	5	25

1974). Another trial done by the SWOG failed to show benefit of the combination over 5-FU alone (20% vs 21%) (PANETTIERE et al. 1984). O'CONNELL, in his review of gastric cancer chemotherapy, illustrated the great variability of 5-FU/methyl-CCNU combinations, with only one trial showing any benefit from the combination (O'CONNELL et al. 1985).

A large number of studies utilizing Adriamycin, 5-FU, and mitomycin C (FAM) have been reported. Response rates with FAM in nonrandomized trials ranged from 18% to 55%; however, median survivals were closely arranged around 25 weeks. The SWOG tested FAM given as a simultaneous combination versus the same drugs given sequentially. Their conclusions were that the two regimens were similar in response rates and survival. However, a critical comparison between 5-FU alone and FAM has been done recently by the NCCTG with no demonstrable survival benefit for the combination (CULLINAN et al. 1985). The general conclusions are that differences in responses are seen but little evidence has been uncovered that this translates into longer term survivals for the various regimens.

In a single phase 2 trial the combination of 5-FU, Adriamycin, and BCNU was given by LOPEZ and colleagues, producing a response rate of 50% but a dismal 28-week median survival, illustrating again a lack of therapeutic gain (LOPEZ et al. 1984). In three randomized trials with 5-FU, Adriamycin, and methyl-CCNU (FAMe) by the GASTROINTESTINAL TUMOR STUDY GROUP (GITSG), there was a significant short-term survival gain over single agents or two drug combinations (GITSG 1984a). The FAM regimen FAM has not been compared to FAMe combinations. However, the failure to achieve significant long-term survival gain has been disappointing. The overall conclusion of these studies is that it is difficult to point to a major benefit of any specific regimen in terms of survival. Depending on the mix of patients, extent of disease, measurability, prior therapy, and performance status, a spectrum of response rates can be obtained. The survival gain for all the trials seems to be about 28 weeks as a median. Certainly additional trials looking for effective agents, including "biologics," need to be done.

12.7.5 Combined Modality Studies

Locally unresectable stomach cancer was the basis of one trial attempting to evaluate combined radiation and chemotherapy at the Mayo Clinic (MOERTEL CG et al. 1969). The trial employed short-term 5-FU for 3 days given at the initiation of 35 Gy over 4 weeks. A second arm consisted of radiation therapy alone. The survival gain for the 5-FU/radiotherapy arm was substantial, 13 months versus 5.9 months for XRT alone. A follow-up trial by ECOG used the same schedule of 5-FU and radiation or weekly 5-FU with both groups going on to receive maintenance 5-FU. Patients with both gastric and pancreatic cancer were included. The combined arm was significantly more toxic without any evidence for therapeutic benefit (KLAASSEN et al. 1985). Another attempt to develop an intensive regimen utilizing 5-FU with Adriamycin was started at the Mayo Clinic with 18 patients in a pilot study. The conclusion was that the combination was too toxic for general use (KLAASSEN et al. 1985).

The early trials of adjuvant therapy in resected gastric cancer were done in the late 1950s by the VASAG (V. A. COOPERATIVE SURGICAL ADJUVANT STUDY GROUP 1965) and the University Surgeons (SERLIN et al. 1969). These studies employed thio-TEPA and FUDR, both indicating no benefit. With the rebirth of interest and enthusiasm for adjunctive chemotherapy in the early 1970s, several studies were initiated attempting to show that combination chemotherapy in patients after complete resection of the primary tumor might produce beneficial results. One of the first trials reported by the GITSG indicated that 5-FU with methyl-CCNU given for 2 years after surgery significantly decreased recurrence rates when compared to surgery alone (GASTROINTESTINAL TUMOR STUDY GROUP 1982). Not only was the disease-free survival better but a survival advantage was reported for the chemotherapy group. A second trial with the same drugs and control reported 3 years later by the ECOG failed to confirm a survival advantage. Moreover, ANLL (Acute non-lymphocytic leukemia) was seen in two patients. A later survey by Boice uncovered at least 17 cases with the same problem, further detracting from the continued study of this combination (ENGSTROM et al. 1985). The VAHSOG did a similar study with 12 month of therapy with 5-FU and methyl-CCNU and noted no statistical improvement in survival rate, or disease-free survival (HIGGINS et al. 1983).

The Japanese have done several large studies looking at mitomycin C alone or in combination with Futraful, a derivative of 5-FU. Because of the lack of a surgery only control group, little value comes out of all this work involving over 1000 patients. The authors can only conclude that there are no differences in survival between the two arms (INOKUCHI et al. 1984). In another study, 5-FU in combination with mitomycin and ara-C was compared to ftorafur in the same combination and a surgery alone arm (NAKAJIMA et al. 1984). The two drug arms were superior to surgery alone but because of the low numbers of patients the differences were not statistically significant except in subgroups. A single trial reported in 1983 from Spain showed a benefit to using mitomycin C postoperatively in what appears to be a controlled trial; however, only a total of 70 patients entered the study (ALCOBENDAS et al. 1983).

In summary, despite enthusiasm and clinical leads indicating a response rate of 30% with combinations, the current trials of adjuvant therapy done in the United States fail to show a consistent advantage for chemotherapy. In addition, all the large trials in the United States utilized methyl-CCNU, an agent suspected of inducing acute leukemia. In Japan and Spain mitomycin C seems to have induced a survival gain, although the data suffer from low patient numbers, or lack of controls. It is fair to say that no evidence exists that would preclude, on the basis of clinical ethics, a trial that has a control arm of surgery alone. The lack of benefit can be explained in part by the absence of effective agents that cause high rates of complete responses in advanced disease. The lesson learned based on the trials experience in the 1970s is that the adjuvant trials, while designed to deal with a statistically average lower tumor burden, in fact involve two patient populations, those who are cured with surgery alone and those who have a residual volume that will recur. These latter patients do not seem to benefit with sufficient frequency to be seen in the reported studies because of too few patients or too short a follow-up period. Recently it was shown that when many separate studies that fail to show individual benefit are subjected to analyses that combine the trends in these studies, overall benefits are revealed. These analyses, when applied in breast cancer adjuvant trials, indicate a small but significant benefit for chemotherapy (CARBONE 1986). To the best of the authors' knowledge this has not been done in gastric cancers.

12.8 Pancreatic Cancer

Carcinoma of the pancreas accounts for about 12% of all cancers of the gastrointestinal tract and is the fifth most common cause of cancer deaths. About 27 000 cases will be diagnosed this year (SILVERBERG and LUBERA 1988). They rarely occur in persons under 25 years of age and are most commonly seen between the ages of 30 and 70 (BERG an CONNELLY 1979). The incidence is higher in blacks (BUCHER 1980). No definitive association has been encountered between pancreatic cancer and various socioeconomic indicators.

12.8.1 Etiology

Although the causes of pancreatic carcinoma have yet not been determined, epidemiologic studies have provided some clues as to its causes and related factors. Smoking, chronic alcoholism, and exposure to certain industrial carcinogens such as coal tar, β-naphthylamine, and benzidine, have been associated with a higher incidence of this tumor (KRAIN 1970). It is believed to have greater frequency in patients with pre-existing chronic pancreatitis (PAULINO-NETTO et al. 1960), a fact disputed by others (BOWDEN 1972). The association with diabetes mellitus has not been clearly established either, for it is known that diabetes and the pathologic changes of chronic pancreatitis may be produced by carcinoma of the pancreas.

Histologic types vary widely in carcinoma of the pancreas, but adenocarcinomas account for over 95% of all tumors of the exocrine pancreas. They may be subdivided into ductal cell, acinar cell, cystadenocarcinomas, or anaplastic. Other types include adenoacanthomas, squamous cell carcinomas, and sarcomas. Neoplasms of the islets of Langerhans can be both benign and malignant, but in the absence of metastasis the histologic distinction may be impossible. The clinical course differs in many aspects from the exocrine tumors, but is of great clinical importance because of the systemic effects associated with the production of endocrine secretions (VERNER 1969; WILDER et al. 1927).

12.8.2 Diagnosis

Symptoms of exocrine pancreatic cancer, predominantly pain, jaundice, and weight loss, depend on whether the tumor is located in the head, body, or tail of the organ. Jaundice is not always painless, as previously thought, and there is an association with either abdominal discomfort or severe pain. Emotional disturbances, personality changes, and depression are commonly present in patients with pancreatic carcinoma. Except for signs of obstructive jaundice and encroachment on the major vessels, early physical signs of carcinoma of the pancreas are rarely observed. Hepatomegaly, palpable masses, ascites, and signs of metastatic involvement are frequently the presenting clinical findings.

Abnormal laboratory findings are mostly observed in tumors of the head of the pancreas, with elevation of both direct and indirect bilirubin levels, marked increase of alkaline phosphatase levels, but minimal alterations of other hepatic enzymes. Elevation of the carcinoembryonic antigen (CEA) is unreliable for detecting early disease. α-fetoprotein (AFP) is produced by some pancreatic cancer cells. A number of other studies such as serum leucine aminopeptidase, arginine exopeptidase, trypsin inhibitor, and plasma antithrombin activity have not been established as reliable tests for definitive diagnosis (KLAVINS 1981).

Conventional radiologic studies, like X-rays of the upper GI tract series, may suggest the diagnosis. The most common abnormalities observed are duodenal or pyloric abstruction, narrowing of the first or second portion of the duodenum, widening of the duodenal loop, the so-called reverse figure-three sign from local indentation of the concave surface of the duodenum, and distortion of the mucosal pattern from direct tumor invasion (EATON et al. 1968). However, angiography, percutaneous transhepatic cholangiography, computerized axial tomography, and transduodenal pancreatography are more helpful in making the diagnosis (FREENY and BALL 1981). Only biopsy can prove the presence of cancer and this can be accomplished by retrograde pancreatocholangiography or percutaneous needle biopsy, guided by CT scanning (BOURBEAU et al. 1979). The advantage of this method is that it allows definitive tissue diagnosis without subjecting the patient to laparotomy, but also precludes accurate surgical staging or an attempt at surgical cure. To some, these considerations are of no significance since surgical cure can only rarely be accomplished and CT scan provides enough information for adequate staging. Others feel that a small chance of cure is preferable to none and that the information provided by the CT scan might not be sufficiently accurate to determine unresectability (HOLYOKE 1981).

Monoclonal antibodies have been developed to pancreatic carcinoma, using the human cell line

HPAF as immunogen and collagenase-treated adult human pancreas. Although they recognize some pancreatic carcinomas, they also react with other normal and malignant neoplasias (SCHLOM and WEEKS 1985).

12.8.3 Treatment

Surgical approaches to the treatment of carcinoma of the pancreas vary widely. Some surgeons propose radical pancreatoduodenectomy (WHIPPLE et al. 1935; MONGUE et al. 1964), while others totally oppose it (CRILE 1970).

Biliary bypass procedures for immediate relief of obstructive jaundice, followed by pancreatic resection when feasible, is advocated by others (BROOKS 1979; FORREST and LONGMIRE 1978). However, surgical results remain poor when the overall problem is considered. Most of the tumors are diagnosed after they have spread to contiguous structures, and in many cases metastases to the liver or other organs are present at the time of the exploratory surgery.

Quite frequently, complete surgical resection of locally advanced disease, where there is involvement of the regional lymph nodes and encroachment upon the blood vessels, is not feasible and definitive radiation therapy, using external beam, interstitial, or intraoperative radiation techniques, has been used (DOBELBOWER 1979).

Intraoperative radiation therapy was employed as early as 1915, when Finsterer treated a patient with gastric carcinoma. In 1937 Eloesser reported his results in some patients with intra-abdominal malignancies (ELOESSER 1937). For many years the technique was not used as widely, but in the last two decades the interest in its use has revived (ABE 1975; DOBELBOWER and ABE, in press) and the experiences in various centers in this country have been encouraging, although the technique requires careful monitoring to prevent complications. The subject is discussed in Chap. 7.

12.8.3.1 Chemotherapy

Systemic chemotherapy of pancreatic carcinoma has not been successful, for the clinical antitumor activity observed with cytotoxic therapy has not translated into a significant prolongation in survival. The majority of the patients with advanced disease died within 6 months of diagnosis. Single agent therapy employing 5-FU, mitomycin C, CCNU, streptozotocin, and Adriamycin produced responses in less than 25% of the patients (CARTER 1975; MOERTEL CG 1976). More recent reports showed some activity by isofosfamide (LOEHRER et al. 1985).

Drug combinations, most of them including 5-FU as one of the agents, were reported as producing a higher response rate and some improvement in survival. The combination of 5-FU, Adriamycin, and mitomycin C (FAM) yielded partial responses in 37% of the patients treated, with a median survival of 12 months (SMITH et al. 1980). Unfortunately, studies conducted by the North Central Cancer Treatment Group (NCCTG) and the Cancer and Acute Leukemia Group B (CALGB) did not confirm the high response rate obtained by the Georgetown University group, using the FAM regimen (CULLIMAN et al. 1985; OSTER et al. 1982). The combination of 5-FU + testolactone or spironolactone was found capable of increasing survival in patients with pancreatic carcinoma (WADDELL 1973), but the small number of patients treated did not allow definitive conclusions.

12.8.3.2 Combination Therapy

In locally advanced, unresectable carcinoma of the pancreas, where there is involvement of the lymph nodes and encasement of the blood vessels, but no evidence of hepatic involvement or distant metastasis, many have attempted to control the disease by giving a combination of radiation and chemotherapy. In a double-blind study investigators at the Mayo Clinic reported that the combination of 5-FU (15 mg/kg/day × 3) and radiation therapy (35 Gy given over 4 weeks) offered an advantage over the treatment with radiation alone. There was an improvement in survival to 10.4 months in the combined group, as compared with 6.3 months in the patients treated only with radiation (MOERTEL CG et al. 1969). Similar results were reported by the Gastrointestinal Study Group from their randomized study where they found the median survival for patients treated with radiation therapy alone (60 Gy) to be 4.3 months, lower than the 8.5 months in the group treated with 40 Gy + 5-FU and the 11.4 months for those treated with 60 Gy + 5-FU (MOERTEL CG et al. and the GASTROINTESTINAL TUMOR STUDY GROUP 1981).

12.9 Pancreatic Endocrine Tumors

This cluster of diseases comprises two main groups of neoplasms: (1) APUDomas, and (2) carcinoids. APUDomas are derived from cells that are known by their ability to produce polypeptide hormones and to synthesize biogenic amines from amine precursors [amine precursor uptake and decarboxylation (=APUD) system] (FRIESEN 1982).

Tumors of the islets of Langerhans are interesting clinically because of the striking systemic effects associated with them, as a result of excessive endocrine secretion, which occurs in about 20% of these tumors. Commonly they are associated with hyperplastic or tumoral changes of two or more endocrine glands and the syndromes of multiple endocrine neoplasia (MEN) have been described as clinically defined entities. The main endocrine tumors of the pancreas are: insulinoma, glucagonoma, SRIFoma, VIPoma, PPoma, gastrinoma, and ACTHoma.

Glucagonomas are α-cell tumors, commonly located in the tail of the pancreas; over 60% are malignant and the metastases frequently occur in the liver and peripancreatic lymph nodes.

Somastimomas (SRIFomas) were first reported in 1977 (GANDA et al. 1977); they are slow growing and are diagnosed when a tumor mass is detected, since they produce few metabolic disturbances and clinical symptoms.

VIPomas, so designated for the production of vasoactive intestinal peptide (VIP), were first described in 1958, when a syndrome of refractory diarrhea and severe hypokalemia was observed in two patients with non-insulin-secreting islet cell tumors (VERNER and MORRISON 1958). The syndrome was given the authors' name, and later described as "pancreatic cholera" (MATSUMOTO et al. 1966). When another component of achlorhydria was added, the condition was named „WHDA syndrome" (MARKS et al. 1967). The diagnosis is commonly made when there are symptoms of abdominal colic and pronounced weight loss, not totally explained by diarrhea alone. Potassium levels were below 2.5 mmol/l, and bicarbonate levels were below 15 mmol/l in the majority of the patients. Fasting levels of VIP were elevated in all patients, with a range between 48 and 760 pmol/l (n = 0.5–16 pmol/l) (WEIL 1985).

Gastrinomas, which constitute 20%–25% of endocrine pancreatic tumors are mainly located in the head or the tail of the pancreas. Their association with recurrent peptic ulcers and gastric hypersecretion is known as the Zollinger-Ellison syndrome (ZOLLINGER and ELLISON 1955). Although they grow slowly, 50%–75% have metastases at the time of diagnosis. They are small and difficult to localize; their resection is not possible in 80%–90% of the cases, because of their location or the presence of metastases.

Insulinomas, or β-cell tumors, are the most commonly diagnosed functioning islet cell tumors for their classic symptom is hypoglycemia, which frequently occurs in the morning or after vigorous exercise. Early associated symptoms include irritability, flushing, sweating, tachycardia, and nausea, most of them related to sympathetic nervous system irritation. The first islet cell tumor was described in a patient with episodic hypoglycemia and β-cell carcinoma (WILDER et al. 1927). However, about 90% of these tumors are benign. If the hypoglycemia is not treated, central nervous system symptoms appear, including erratic psychiatric behavior, convulsions, and coma. Repeated untreated attacks may cause irreversible brain damage. Diagnosis is made by keeping the patient fasting for at least 24 h, while kept active. Another diagnostic method is the tolbutamide tolerance test.

Carcinoid tumors are a group of neoplasms derived from the disseminated endocrine cells of the primitive gut and characterized by the presence of argentaffin granules in their cells, accounting for the term "argentaffinomas," by which they are also known. They are found in the trachea, bronchial tree, or the gastrointestinal tract, pancreas, and biliary ducts. They produce 5-hydroxytryptophan, which is broken down and excreted mainly as 5-hydroxyindoleacetic acid (5-HIAA) in the urine, a useful test in making the diagnosis of carcinoid syndrome. They are commonly located in the appendix and often found in incidental appendectomies. However, they may be present in any segment of the intestine.

Their malignancy, as judged by the presence of metastasis, is highest in tumors of the cecum; the tendency to metastasize correlates with their size, and pancreatic carcinoids, since they are discovered late, are almost always considered malignant.

Carcinoid syndrome is frequently associated with carcinoid tumors metastatic to the liver. It is commonly seen in carcinoid tumors of the small intestine, but may be seen in tumors originating in other locations. Clinically, it is characterized by cutaneous flushing, diarrhea, and asthma-like attacks. Not infrequently, palpitations, tachycardia, hypotension, headaches, nausea, and vomiting are accompanying symptoms. In addition, cardiac valvular damage may occur as a result of the endocardium being covered with a fibrotic layer that may compromise the function of the valves. The flushing can occur spontaneously or be triggered by efforts, a meal, or palpation of the tumor (TILSON 1974). The mechanisms by which the syndrome occurs are varied, but is likely associated with the release of histamine, serotonin, bradykinin, or 5-hydroxytryptidine, accounting for the elevated urinary excretion of 5-HIAA, useful in making the diagnosis.

12.9.1 Treatment

Surgery is the treatment of choice for eradicating endocrine tumors of the pancreas. However, recurrences are frequent and on occasion, because of their location or the presence of metastases, total resection is not possible. When hepatic metastases are present, if they cannot be completely resected, debulking of the lesions may provide symptomatic relief. Embolization of the hepatic artery may be quite effective in controlling the carcinoid syndrome.

Chemotherapy with streptozotocin, alone or combined with 5-FU, is effective and produces long-term remissions. Other drugs used include carmustine, dacarbazine (DTIC), mitomycin C, cyclophosphamide, and Adriamycin. The treatment of Zollinger-Ellison syndrome has been made easier with the use of H^2-receptor antagonists. Symptomatic relief of the symptoms in the carcinoid syndrome can be accomplished by using different drugs: flushing may be relieved by phenothiazines, pentholamine, or methyldopa. Diarrhea may be treated with opiates and 5-HT antagonists, like cyproheptadine.

Somatostatin analogs, by their ability to inhibit the secretion of various pituitary and gastropancreatic regulatory peptides, provide significant symptomatic and objective improvement in many cases (KVOLS et al. 1986).

12.10 Colorectal Carcinoma

12.10.1 Incidence and Natural History

Carcinoma of the large bowel (colon and rectum) is the third most common malignancy in the United States, after skin and lung cancers. It is estimated that approximately 5% of all men and women will develop this malignancy; over 147 000 new cases are diagnosed annually and 61 000 patients will die from this disease every year (SILVERBERG and LUBERA 1988). Although many new screening techniques have become available and new therapies are constantly appearing, the incidence of and mortality from colorectal carcinoma have not changed significantly over the past 50 years.

There is a slight difference in the incidence of these cancers by sex, with a small predominance of colonic cancers in females and of rectal carcinomas in males. The highest incidence occurs in the sixth and seventh decades, but 8% of colorectal carcinomas are diagnosed in people less than 40 years old. Colorectal carcinomas occur with different frequencies at various locations in the colon. Over two-thirds of these lesions occur in the distal colon, rectum, and anus. About 30% are reached by digital examination and 60% are within reach of a sigmoidoscope. The use of a flexible sigmoidoscope allows visualization up to 50–60 cm. It appears that right colon lesions are becoming relatively more frequent and rectosigmoid lesions less frequent.

The gross appearance may vary from polypoid to ulcerating, with the former carrying a better prognosis if the tumor has not invaded the stalk. Microscopically, most carcinomas of the colon and rectum resemble the mucosa of the organ of origin, but the degree of differentiation may vary from well differentiated to poorly differentiated, although this does not correlate as well with the prognosis for the patient as does the Dukes' classification.

The spread of colorectal carcinoma occurs by several means, including direct extension and hematogenous spread, but most commonly it occurs by lymphatic dissemination. From 30% to 68% of reported large bowel cancers encountered surgically have already metastasized to the regional lymph nodes (DEPEYSTER and GILCHRIST 1969). Vascular invasion occurs in 15%–61% of cases and there is a strong correlation between this finding and the later appearance of metastases. The prognosis of colorectal tumors corresponds quite well to the degree of penetration, with the best prognosis for superficial tumors and the worst for those penetrating the entire intestinal wall and invading the lymph nodes.

Clinical and laboratory research has proven that despite efforts to adequately stage the tumors a great number of them are disseminated at the time of surgery and the unrecognizable minimal residual tumor remaining after surgery is responsible for recurrences and death. About one-half of the patients die from distant metastases and the other half from local recurrences.

12.10.2 Classification

Dukes classified the cancers of the rectum in three categories: *A*, tumors limited to the rectal wall, with no extension into the extrarectal tissues and no metastases in lymph nodes; *B*, those tumors in which the carcinoma had spread by direct continuity to the extrarectal tissues but had not yet invaded the regional nodes; and *C*, those cases in which metastases were present in the regional lymph nodes (DUKES 1932). KIRKLIN et al. (1949) modified Dukes' staging, applying it to the entire colon, and introduced the concept of prognosis of subgroups related to depth of infiltration of the bowel wall. Further modification of previous classifications (ASTLER and COLLER 1954) labeled type A as limited to the mucosa; type B_1 involved the muscularis propria but did not extend through the wall, while B_2 involved the serosa. Furthermore, type C was separated into C_1, defined as disease not through the wall but associated with positive lymph nodes, and C_2, defined as disease extending through all layers of the intestinal wall plus positive nodes.

12.10.3 Patterns of Relapse

Survival rates have changed very little over recent decades after patients have undergone curative resections, particularly if the lymph nodes were positive at the time of surgery. The most common sites of metastases are the lungs and the liver. Less commonly they appear in the brain, bone, adrenal glands, and kidneys. Local recurrences are observed frequently as well as peritoneal implants and abdominal wall recurrences. Autopsy series of patients dying of colorectal cancer confirm that abdominal failure is common. Nearly 75% of patients with these tumors die from intra-abdominal disease. Although lung and liver metastases are common, they account for only 25% of the deaths (OLSON et al. 1980). Patients with apparently isolated liver metastases form a small but interesting and extensively studied group, for those patients with solitary resectable lesions are the only ones who have achieved 5- and 10-year survival (FOSTER 1978).

12.10.4 Diagnosis

Diagnosis of recurrence may be made by a number of techniques, including radiographic and scanning methods, but quite commonly the patients present with vague, nonspecific symptoms. The use of CEA as a tumor marker has been extensively evaluated and reports by Putzki and others suggest that elevation of CEA is frequently an early sign of recurrence, more sensitive than other markers like tissue polypeptide (TPA) or carbohydrate antigen (CA 19-9) (PUTZKI et al. 1987). MINTON et al. have used the elevation of CEA as a criterion to direct second-look surgery. The goal of such surgery is to determine whether early detection, leading to early intervention, would lead to improved survival in patients with recurrent disease. It is noteworthy that in their report they emphasize the need to perform the surgery soon after confirmation of the CEA elevation, for delays, in operation have resulted in the lesions being unresectable. Their results suggest a 30% 5-year survival for those patients undergoing second-look resections with curative intent (MINTON et al. 1985). However, this approach has not been universally accepted.

The levels of secretory immunoglobulins A and M in the serum, as well as the presence of TPA and CA 19-9, have been suggested as a way to detect recurrences from colorectal carcinomas, but they appeared elevated only in advanced stages of the disease and thus they are of no value in detecting early recurrences (KVALA et al. 1987).

12.10.5 Therapy

For purposes of better planning their therapy, patients with colorectal carcinoma may be divided into three groups: those with local or regional recurrences, those with hepatic metastases exclusively, and those with disseminated disease.

Radical surgery is the only curative treatment of invasive carcinoma of the large bowel. However, despite some improvements in the surgical techniques, the 5- and 10-year survival rates have not been altered significantly, except in the published series of patients operated on utilizing the "no touch" technique, in which a favorable change in the survival rates of those patients was reported (TURNBULL 1970).

Surgical resection is the treatment of choice for patients with solitary metastasis or with the lesions confined to one lobe of the liver; this has produced 5- and 10-year survival in a small number of patients. Since it was reported that the primary blood supply to metastatic lesions in the liver came from the hepatic artery (BREEDIS and YOUNG 1954), patients without the possibility of hepatic resection have been treated by ligation of the hepatic artery, with or without postligation therapy with 5-FU. Thrombosis of the hepatic artery and multiple embolizations of the hepatic artery and its branches have also been successfully employed (WALLACE et al. 1984). All of these methods produce a temporary regression in the size of the tumor masses, but have not yet proven to have any impact on patient survival.

12.10.5.1 Regional Intrahepatic Infusion

Intrahepatic infusion has been tried both through the portal system and through the hepatic artery, by placing the catheters surgically or percutaneously. The results obtained showed that intrahepatic infusion, by delivering a higher concentration of the drug directly into the tumor bed, was more effective in controlling hepatic metastases than when the drug was administered systemically by the intravenous route. The fluorinated pyrimidines 5-FU and 5-FUDR are well suited pharmacokinetically for intrahepatic infusion because of their rapid clearance from the systemic circulation, and the large quantities of the drug that can be extracted by the liver, thus causing minimal systemic toxicity. Many of the patients seek treatment when they are severely jaundiced and thus do not tolerate systemic administration of 5-FU; in many instances a dramatic clearing of jaundice and marked improvement in the patient's condition occur after intra-arterial administration of the drug (RAMIREZ and ANSFIELD 1982). However, dislodgement of the catheter may deliver some of the drug to the surrounding organs and patients might develop severe gastritis, duodenitis, or, in a few instances, peptic ulcers which may perforate (NARSETTE et al. 1977). To improve the regional selectivity of hepatic intra-arterially administered drugs, degradable starch microspheres have been used (GYVES et al. 1983). The superiority of the intra-arterial approach over the systemic therapy has not been demonstrated unequivocally; the results of the Central Oncology Group (COG) studies showed no difference between the two methods of administration, but those patients randomized to receive intra-arterial infusion for a short period of time were not allowed to receive a second course, despite progression of the hepatic metastases (GRAGE et al. 1979). The reports of OBERFIELD et al. (1979), ANSFIELD et al. (1975), and others favored the intrahepatic administration of the drug, for it produced better control of hepatic metastases.

The recent resurgence of hepatic infusion therapy with the introduction of totally implantable pumps

Table 6. Results of hepatic artery infusion for liver metastases

Authors	Drugs	Response (%)
WEISS (1983)	FUDR 0.2–0.3 mg/kg/d	29
PETTAVEL (1978)	5-FU 2–3 g/d × 10 + FUDR 0.1–0.4 mg/kg/d	81
NIDERHUBER (1984)	FUDR 0.2–0.3 mg/kg/d 14 of 28 days	79
SHEPARD (1985)	FUDR 0.2–0.3 mg/kg/d + mitomycin 15 mg/m^2 or dichloromethotrexate 2 mg/m^2 on day 15	32
KEMENY (1984)	FUDR 0.2–0.3 mg/kg/d 14 of 28 days	52

has rendered more promising results. Several investigators have reported responses of over 70% with median survival ranging between 13 and 23 months (WEISS GR et al. 1983); NIEDERHUBER et al. 1984; KEMENY et al. 1984). Most investigators report their results using FUDR (Table 6).

Some complications have occurred, including biliary sclerosis, cholecystitis, and chemical hepatitis (HOHN et al. 1985). To avert some of the problems, the surgical technique was modified to include cholecystectomy at the time of the insertion of the catheter.

Although it is clear that the intrahepatic infusion is capable of controlling the hepatic disease, many patients died from extrahepatic metastases (WEESE and RAMIREZ 1986).

12.10.5.2 Systemic Chemotherapy

Systemic chemotherapy of metastatic colorectal carcinoma has been extensively studied and the fluopyrimidines have been the drugs most widely used. The development of 5-FU provided the foundation stone for the study of chemotherapy in gastrointestinal cancer and a wide range of approaches have been devised in an attempt to improve its effectiveness. The response rate is about 20% (range 8%–85%) but the optimal dose route and schedule have not been determined. Since the report by ANSFIELD et al. (1962), who proposed that the maximum benefit from the drug might be obtained only if some degree of toxicity is reached, multitudes of dosage schedules and routes of administration, including 2-, 8-, and 24-h infusions, administration by

Table 7. Results of 5-FU therapy in colorectal carcinoma

Authors	Response (%)
ROCHLIN et al. (1962)	55
WEISS and JACKSON (1961)	35
HURLEY (1960)	31
LAVIN et al. (1980)	27
ANSFIELD et al. (1962)	17
MOERTEL et al. (1976 a)	17
KENNEDY et al. (1960)	9

Table 8. Results of combination chemotherapy in colorectal carcinoma

Drugs	Response (%)	Authors
5-FU + methyl-CCNU	20	MOERTEL (1976)
5-FU + methyl-CCNU + vincristine	23	KEMENY et al. (1979)
5-FU + methyl-CCNU + DTIC	14	ENGSTROM et al. (1978)
5-FU + methyl-CCNU + vincristine + streptozotocin	27	KEMENY et al. (1980)

weekly injections or by an intensive course followed by weekly injections, and oral use in capsule or liquid form have been proposed. The comparison of four different dose schedules by the COG favored the intensive course over the other routes of administration (ANSFIELD et al. 1977) (Table 7).

More recently, the development of ambulatory infusion pumps permited the low-dose, continuous administration of the fluorinated pyrimidines 5-FU and 5-FUDR for prolonged periods (more than 30 days), via central venous catheters (LOKICH et al. 1981; CABALLERO et al. 1985). Responses using this technique range between 30% and 45%. The continuous administration of the drug allows the exposure of a greater number of malignant cells to a cytotoxic agent during a sensitive phase of the growth cycle and this might improve the response rate. Although most of the studies report milder toxicities using this approach, side-effects like stomatitis, dermatitis, chemical phlebitis, diarrhea, and bacteremia related to the central venous catheter have occurred (FAINTUCH et al. 1985).

Besides the fluopyrimidines, numerous cytotoxic drugs have been evaluated for therapeutic efficacy against advanced colorectal cancer, including nitrosoureas (methyl-CCNU, CCNU, PCNU, chlorozotozin), anthracyclines (Adriamycin, mitozantrone), antibiotics (mitomycin C), cisplatinum, methotrexate, and triazinate. Tumor response to any of these agents is typically partial rather than complete, and is frequently obtained at the expense of considerable toxicity, if the drugs are given near their maximum tolerated dose.

12.10.5.3 Combination Chemotherapy

Attempts to improve the results obtained with single agents have primarily focused on the addition of other drugs to 5-FU, in many combinations and dosage schedules. Although preliminary trials with small numbers of patients suggested improved response rates with various drug combinations, larger series invariably showed these regimens to be no better than 5-FU alone and/or frequently more toxic and unable to enhance the overall group survival (Table 8).

A current clinical research approach in the scientific development of combination chemotherapy regimens involves the use of biochemical modulators (such as thymidine and *N*-phosphoacetyl-L-aspartate) to selectively enhance the cytotoxic effects of 5-FU on malignant cells by altering tumor cell metabolism. In an effort to bypass cellular mechanisms of drug resistance and to maximize the effectiveness of 5-FU, other attempts to biochemically modulate the fluopyrimidines have employed sequential methotrexate and 5-FU (BERTINO et al. 1984), cisplatinum + 5-FU, and 5-FU and folinic acid (BUROKER et al. 1985).

To appreciate the rationale for these approaches we have to understand some of the pathways of 5-FU metabolism. Upon entry into the cell, the reaction with phosphoribosyl pyrophosphate yields 5-fluoridine diphosphate (FUDP). From this point FUDP can follow one of two different metabolic pathways, which explains the two proposed mechanisms of antitumor effects of 5-FU: (a) further phosphorylation to produce 5-fluoridine triphosphate that can be incorporated into RNA and subsequently inhibit RNA processing; and (b) activation to deoxyuridine monophosphate which binds to thymidylate synthase by forming a complex with the cofactor 5,10-methylene-tetrahydrofolate and inhibiting the formation of nucleotides, which are necessary for production of DNA.

Modulation of the first mechanism should be achieved by the administration of thymidine, which blocks catabolism of 5-FU. However, in clinical trials, thymidine increased the toxicity of 5-FU but did not affect its therapeutic activity (STERNBERG et al. 1984). Although 5,10-methylene-tetrahydrofolate appears to be the major folate participating in the formation of the ternary complex, the calcium salt of folinic acid (also called citrovorum factor or calcium leucovorin) is a biologically active, exogenous source of reduced folate. Available in oral and parenteral forms it is the folate most tested in clinical trials as a means of modulating the second mechanism described above, blocking the formation of DNA (MADAJEWICZ et al. 1984; MACHOVER et al. 1986).

The optimal dosage and schedule of administration of folinic acid and 5-FU have not yet been defined. Ongoing studies comparing this combination to 5-FU alone should allow for an accurate assessment of the role of this form of biochemical modulation in the treatment of colorectal carcinoma.

12.10.5.4 Biologic Response Modifiers

In an attempt to enhance immunoreactivity and possibly attain antitumor effect, the methanol extraction residue fraction of BCG was employed in advanced colorectal cancer with disappointing results (MOERTEL et al. 1976a).

Clinical trials with interferon α-2 and recombinant clone α-interferon have shown minimal activity in patients with advanced disease (SILGALS et al. 1984; NEEFE et al. 1984). Ongoing studies employing IFN$\beta^{s}{}_{er}$, γ-IFN, and different interferon combinations are trying to establish the optimal dosage and schedules of administration, but it is too early to judge their clinical usefulness.

Levamisole, a drug widely used as an anthelminthic, has been shown to be an immune stimulant capable of increasing antibody formation to a variety of antigens, of increasing delayed hypersensitivity, of stimulating blastogenic response, and of increasing phagocytic macrophages (CHIRIGOS and AMERY 1978). Clinical studies in colorectal carcinoma have proven its effectiveness (VERHAEGEN et al. 1982). A randomized study conducted at the Wisconsin Clinical Cancer Center in patients with colorectal cancer demonstrated that the addition of levamisole to 5-FU produced a significant improvement in survivorship when compared to 5-FU alone (BORDER et al. 1980).

Monoclonal antibodies are emerging as an effective tool in the treatment of colorectal carcinoma. The monoclonal antibody 17-1A IgG2A (17-1A) is a mouse immunoglobulin with a high specificity for colorectal and pancreatic cancers. Some clinical responses have been reported (SEARS et al. 1984) and there are some current studies attempting to determine the best dose and schedules of administration.

12.10.5.5 Adjuvant Therapy

The prognosis of patients with colon cancer is clearly related to the extent of the tumor spread at the time of the diagnosis. If the patients are treated only with surgery and the tumor is confined to the mucosa or submucosa, probably over 80% will survive 5 years. Deeper penetration into the bowel wall will lower the percentage to about 60%, and growth through the serosa into the pericolic fat will be associated with a 45% 5-year survival rate. Regional node involvement further reduces the cure rate to the range of 25%. Since about 50% of all patients with colon carcinoma have serosal penetration, nodal involvement, or both, at the timie of surgical resection, there is an obvious need for an effective adjuvant program which would improve the cure rate.

During the last 20 years there have been several adjuvant trials, most of them using 5-FU, in patients at high risk for developing recurrence after surgical resection. Some of the trials involved the intraluminal administration of the drug at the time of the surgery (ROUSSELOT et al. 1972) or systemic intravenous 5-FU (LI and ROSS 1976). However, the patients were compared with historical controls, degrading the validity of the results. Prospective randomized studies using untreated controls were conducted by the Veterans Administration Surgical Group (HIGGINS et al. 1976), and the COG (GRAGE et al. 1978). In both studies there was a slight advantage for the treated group, but the difference did not reach statistical significance (Table 9).

Drug combinations of 5-FU and nitrosoureas, with or without vincristine, effective in advanced disease, led the GITSG to conduct a randomized study using 5-FU, nitrosoureas, and immunotherapy, comparing it with a control group, after "curative" surgical resection in patients with colon carcinoma stage B_2, C_1, or C_2. Their reported results after a median of 5½ years of follow-up (GASTROINTESTINAL TUMOR STUDY GROUP 1984b) failed to

Table 9. Adjuvant chemotherapy in colorectal cancer

Drugs	Authors	Prolonged DFS
5-FU	GROSSI et al. (1972)	+
5-FU	LI et al. (1976)	+
5-FU + BCG	MAVLIGIT et al. (1976)	+
5-FU	HIGGINS, VASG (1976)	?
5-FU	GRAGE, COG (1981)	+

DFS, disease-free survival

demonstrate any superiority in the survival or the recurrence rate of the treated patients over the controls. However, results of the NSABP trial showed a significant advantage in disease-free survival and in survival for young males with rectal carcinomas, Dukes' B or C, treated with a combination of 5-FU, methyl-CCNU, and vincristine (MOF) following surgery, over those treated by surgery alone (FISHER et al. 1988).

To assess the effects of postoperative radiation therapy and chemotherapy on tumor recurrence and patient survival, the GITSG conducted a study in patients who underwent surgical curative resection for rectal adenocarcinoma and prospectively randomized them to receive one of four treatments: adjuvant therapy (control group), postoperative radiotherapy with 40 or 48 Gy, postoperative chemotherapy (5-FU and methyl-CCNU), or a combination of radiation therapy and chemotherapy. Their results (GASTROINTESTINAL TUMOR STUDY GROUP 1985), with a median follow-up of all survivors for 80 months, showed that the recurrence rate was highest among the control patients and lowest among the patients receiving the combination of adjuvant radiation and chemotherapy. The time to tumor recurrence was significantly prolonged by the combination of radiation and chemotherapy, but comparison of the survival in the four treatment arms did not establish the superiority of any form of treatment.

The report of the NASBP randomized study comparing surgery alone with surgery followed by radiation showed that postoperative radiation therapy reduces the local-regional recurrences, but produces no significant benefit in overall disease-free survival (FISHER et al. 1988). The ECOG is conducting a study (EST 5283) for surgically resected lesions with positive nodes (Dukes' C), or with penetration through the muscularis propria (B_2) or serosa (B_3). The patients receive: (a) surgical resection alone, or (b) postoperative immunotherapy (autologous irradiated cells plus BCG). This study will attempt to confirm in a larger multi-institutional clinical trial the encouraging results reported by HOOVER et al. (1985).

The suggested usefulness of levamisole in prolonging interval to progression and survival time when used as an adjuvant to surgery in colorectal cancer was reported by the NCCTG (LAURIE et al. 1986). An intergroup study by the ECOG, NCCTG, and SWOG is evaluating levamisole alone or in combination with 5-FU as an adjuvant surgical treatment for resectable adenocarcinoma of the colon.

Based on the relatively high risk of liver recurrence after curative surgery, some studies have attempted liver-directed adjuvant chemotherapy. TAYLOR et al. (1985) reported their results of a study administering 5-FU into the portal venous circulation for 1 week after curative surgery for colon and rectal cancer, to prevent liver and distant metastasis. Their experience suggested that the incidence of hepatic metastases decreased and the disease-free period increased in patients receiving postoperative portal vein infusion chemotherapy, compared with patients treated by surgery alone. The question of whether these results can be confirmed in a larger, randomized trial will be answered by the several studies currently in progress.

The use of intrahepatic arterial mitomycin C and floxuridine as an adjuvant perioperative therapy combined with surgical resection of metastatic colorectal cancer in the liver has been reported by PATT et al. (1987). Their results in 20 patients compare favorably with previously reported results and suggest that this approach may increase the survival of patients undergoing this aggressive adjuvant treatment.

Despite some gains with the use of chemotherapy in colorectal cancer, there is an urgent need for the development of more effective systemic chemotherapeutic agents for the treatment of colorectal carcinoma.

12.11 Carcinoma of the Anal Canal

Squamous cell carcinoma of the anal canal, often called epidermoid carcinoma, is a relatively rare tumor. In the past it was commonly treated by surgery, usually an abdominoperineal resection (APR) with a permanent colostomy. Because the cure rate ranges between 50% and 60% and the morbidity associated with it is also high, with impotence occurring in more than 50% of the patients, alternative modes of treatment have been sought. Because a single group of physicians is likely to encounter relatively few patients with this disease, and because of the variety of cell types and stages of the disease, changes in the management of these patients may be slow to evolve.

Alternatives to surgery include treatment with radiation therapy, alone or in combination with chemotherapy. External radiation therapy alone, brachytherapy alone, or combinations of the two approaches have been employed. The results published by PAPILLON et al. (1983) showed that with a consistent technique a 67% 5-year survival rate and 81% local control can be obtained.

Combined chemotherapy, radiation, and surgery has been tried by several investigators. Since the studies by the Wayne State University Group (NIGRO et al. 1981), the drug combination most commonly used is 5-FU plus mitomycin C. In some studies the combination of chemotherapy and radiation precedes surgery (MICHAELSON et al. 1983) with actuarial 5-year survival of 77%; in others, APR is only performed in patients whose disease was incompletely controlled or who developed recurrences (MEEKER et al. 1986). A recently completed ECOG study (EST 7283) showed the benefit of using a combined approach giving the patients chemotherapy with 5-FU + mitomycin C and radiation therapy to the pelvis (40 Gy), with a boost to the perineum of 10–13 Gy. A full thickness biopsy was performed after the combined therapy and those patients with microscopic disease were candidates for APR.

One problem still remains in dealing with patients with known pelvic and inguinal node metastases. The survival rate is only 9% for patients initially diagnosed with inguinal node metastases. The results of PAPILLON et al. (1983) have shown better results with their approach in this group.

The question of whether cure can be achieved without surgical resection and which sphincter-conserving method is the best may be answered by current studies, but it is possible that with more refined radiotherapy techniques and better chemotherapeutic combinations the future of patients with anal canal carcinoma will appear brighter.

12.12 Hepatocellular Carcinoma

Hepatocellular carcinoma is relatively rare in the United States, but it is estimated that about 1 000 000 cases occur annually around the world. It is more common in males than in females (ratio 4:1) and in China, Korea, Taiwan, and Africa ranks as the fourth most common tumor in adults (LONDON 1981).

Primary tumors of the liver originate either in the parenchymal cells or in the cells of the bile ducts. Those arising from the parenchymal cells are denominated *hepatomas* whereas those of ductular origin are called *cholangiomas*, with the hepatomas accounting for about 90% of the primary hepatic tumors. the distinction between the two types is of little use for both types may be found in different parts of the same tumor and their clinical course is essentially the same.

12.12.1 Etiology

Etiologic agents implicated in the genesis of hepatoma include cirrhosis, parasites (liver fluke), fungal contamination of food with carcinogenic metabolites, aflatoxin, congenital rests, inflammation, and irritation. Postnecrotic cirrhosis sometimes is complicated by primary hepatic carcinoma. There is strong evidence of the correlation between hepatocarcinoma and hepatitis B virus infection. Carriers of the virus may be as much as 94 times more at risk of developing hepatocellular carcinoma than noncarriers (DI BISCEGLIE et al. 1988). The incidence of hepatocellular carcinoma is highest in sub-Sahara Africa, China, and Southeast Asia, where the prevalence of hepatitis B virus carriers is also very high. Conversely, areas with a low incidence of hepatoma have a low prevalence of hepatitis B virus carriers. Liver cancer is uncommon in the United States, Europe, and Australia, where no significant sex or racial differences are noted. In Asia and Africa there is a higher frequency in males (MUÑOZ and LINSELL 1982).

Pathologically, three major categories of hepatocellular carcinomas can be identified. Two-thirds of the cases are of the nodular form, in which the liver appears studded with multiple nodules. About 5% of cases are of the diffuse form and almost invariably associated with cirrhosis. In 30% of cases the presentation is the large, dominant cancer mass, on occasion associated with satellite lesions. An encapsulated form of hepatocellular carcinoma also has been described and seems to be associated with a better survival.

The clinical presentation is dependent on the presence of cirrhosis because its symptoms can mask the tumor, which may go undetected. Rapid development of cachexia, wasting, persistent fever, jaundice, and ascites is common, as well as rapid deterioration of hepatic functions, leading to hepatic failure and death. Hemorrhage occurs frequently and has been implicated in the cause of death in 50% of the cases. Some patients develop Budd-Chiari syndrome and present with tender hepatomegaly, ascites, jaundice, and a rapid downhill course. In others, the first symptoms are related to the metastases or associated neuroendocrine and hematologic syndromes.

12.12.2 Diagnosis

The diagnosis of hepatocellular carcinoma carries a grave prognosis. Several reports from the United

States (ECKARDT 1979) and abroad (KEW et al. 1981) have shown median survivals of 18 weeks from the time of diagnosis. Physical findings may lead to a diagnosis only in advanced cases and those easier to detect, like ascites, jaundice, and hemorrhage, carry the worst prognosis. Laboratory tests may be of some help, but are frequently nonspecific. AFP can be useful in diagnosis and follow-up of the patients; persistent elevation or increase in the serum levels, following any therapeutic approach, is invariably associated with poor prognosis. Elevation of other liver function tests is not necessarily indicative of hepatoma and their relative value rests on the fact that when their high elevation represents deterioration of hepatic function, this commonly precedes hepatic failure, a frequent cause of death. Radiologic and nuclear medicine techniques are useful in making the diagnosis, but in patients with cirrhosis the interpretation of the liver scan is somewhat more difficult. Angiography is useful mostly in preoperative assessment to determine resectability. Arteriovenous anastomoses are frequently seen, associated with increased vascularity. The difficulty in making a histologic diagnosis of hepatocellular carcinoma results from the fear of provoking severe hemorrhage due to the marked vascularity of these tumors. However, some series (CHLEBOWSKI et al. 1984) report that needle biopsy, either blind percutaneous or peritoneoscopy directed, provided diagnostic material in over 85% of cases with little morbidity. Although the possibility of bleeding after liver biopsy cannot be ignored, it appears that needle biopsy affords a relatively safe means of rapidly making the diagnosis of hepatocellular carcinoma.

12.12.3 Treatment

Surgical resection is the most successful therapy for hepatocellular carcinoma and some series (LIN 1976) showed survival rates of 65% at 3 years, 36% at 5 years, and 33% at 10 years. Unfortunately, hepatoma is rarely diagnosed at a time when resection is possible. As the majority of these tumors are highly vascular, many therapeutic approaches have been directed toward decreasing the blood supply of the tumors by either arterial embolization (SATO et al. 1985) or arterial ligation followed by chemotherapy (NAGASUE et al. 1977). Patient survival in those studies was markedly decreased by the presence of large tumors occupying more than 20% of the liver, occlusion of the first- or second-order portal branches, and hypovascular tumors.

Intra-arterial hepatic infusion is particularly appealing in hepatocellular carcinoma because the majority of these tumors are highly vascular and the blood supply to the liver tumors is almost exclusively by the hepatic artery. Thus, hepatic artery infusion affords a marked concentration of the drug in the tumor, with an estimated magnitude of 5–20 times, compared with the surrounding hepatic tissue. Many drugs have been given by this route: 5-FU, 5-FUDR, methotrexate, mitomycin C, dichloromethotrexate, and others. The results vary greatly from one investigator to another and there is little uniformity in what constitutes an objective response. However, some reports show a 42% response rate with 5-FU or 5-FUDR (AL-SARRAF et al. 1974), clearly superior to the results reported with the same drugs given systemically, and the survival is also longer in patients receiving the drug intra-arterially (RAMMING 1983).

Systemic chemotherapy has employed several drugs, including fluopyrimidines, antimetabolites, and Adriamycin. Since the results obtained with Adriamycin published by OLWENY et al. (1975), many studies have reported some benefit from the use of this drug and some investigators (IHDE et al. 1985) have identified some prognostic factors which predict a more favorable outcome in some patients. They include a good performance status, absence of jaundice, lack of cirrhosis, and the presence of fibrous stroma within the tumors, constituting a separate histologic variant denominated fibrolamellar carcinoma (CRAIG et al. 1980) or polygonal cell carcinoma with fibrous stroma (BERMAN et al. 1980).

The use of specific immunoglobulins coupled with isotopes for therapeutic irradiation has been the basis of a study by the RTOG. The work of RICHTER (1965) indicated that ferritin, a tumor associated protein synthesized and secreted by diverse malignancies, was present in high concentration surrounding both the tumor cells and the stroma of hepatoma.

The high concentration of ferritin offered an ideal target for iodine 131 antiferritin antibodies. Ferritin was isolated and purified from Hodgkin's splenic infiltrate and several animals were immunized to obtain the antibody and antiferritin immunoglobulin G (IgG). After adequate preparation and testing for sterility and pyrogenicity, the IgG samples were radiolabeled with ^{131}I. The protocol involved the use of external beam irradiation, chemotherapy with 5-FU and Adramycin, in addition to the ^{131}I ferritin. The encouraging results (ORDER et al. 1985) showed responses in approximately 50% of cases with several complete responders and long-term survivors. More studies are underway with different isotopes and they may elucidate the future role of radiolabeled antibodies in the management of GI cancer patients.

12.13 Carcinomas of the Gallbladder and Biliary Tract

Cholangiocarcinomas and gallbladder carcinomas are uncommon in the United States, with an annual incidence estimated between 6000 and 10 000. Patients' symptoms usually are the same as those of cholelithiasis and seldom is the diagnosis made before surgery. In the majority of the cases the tumor is discovered at the time of surgical intervention for cholecystectomy. There are some reports of the use of radiation alone (KOPELSON et al. 1977) or in combination with 5-FU (DAVIS et al. 1974), but the results were disappointing, mostly because the tumor frequently involves the liver, which is not usually included in the radiation field.

Chemotherapy trials have included 5-FU, Adriamycin, or mitomycin C alone with reported response rates between 10% and 20%. The combination of the three drugs (FAM), as reported by HARVEY et al. (1984), produced a response rate of 31% in 14 evaluable patients. Hepatic artery infusion employing 5-FU and mitomycin C was tried by SMITH et al. (1984) in 11 patients with one complete response, but the median survival was only 3 months. Being an infrequent tumor, most of the reported studies include only a few patients, making it impossible to draw definitive conclusions.

References

Abe M, Takahashi M, Yabumoto E, Onoyama Y, Torizuka K, Tobi T, Mori K (1975) Techniques, indications and results of intraoperative radiotherapy of advanced cancers. Radiology 116: 693-702

Alcobendas F, Milla A, Estape J et al. (1983) Mitomycin C as an adjuvant in resected gastric cancer. Ann Surg 198: 13-17

Al-Sarraf et al. (1974) Primary liver cancer: A review of the clinical features, blood groups, serum enzymes, therapy, and survival of 65 cases. Cancer 33: 574

Ansfield FJ, Schroeder JM, Curreri AR (1962) Five years clinical experience with 5-fluorouracil. JAMA 181: 295-299

Ansfield FJ, Ramirez G, Davis HL et al. (1975) Further clinical studies with intrahepatic arterial infusion. Cancer 36: 2413-2417

Ansfield FJ, Klotz J, Nealon T, Ramirez G et al. (1977) A phase III study comparing the clinical utility of four regimens of 5-fluorouracil. Cancer 39: 34-40

Antman KH, Skarin AT, Mayer RJ, Hargreaves HK, Canellos GP (1979) Microangiopatic hemolytic anemia and cancer: A review. Medicine 58: 377-384

Astler VB, Coller FA (1954) The prognostic significance of direct extension of carcinoma of the colon and rectum. Ann Surg 139: 846-852

Beer M, Cocconi G, Ceci G et al. (1983) A phase II study of cisplatin in advanced gastric cancer. Eur J Cancer Clin Oncol 19: 717-720

Benjamin RS (1975) Clinical pharmacology of Adriamycin. Cancer Chemother Rep 6: 183-185

Berg JW, Connelly RR (1979) Updating the epidemiologic data on pancreatic cancer. Semin Oncol 6: 275-284

Berman MM, Libbey NP, Foster JH (1980) Hepatocellular carcinoma: polygonal cell type with fibrous stroma - an atypical variant with a favorable prognosis. Cancer 46: 1448-1455

Bertino JR, Mini E, Sobrero A (1984) Sequential methotrexate and 5-fluorouracil in the treatment of solid tumors. In: Kimura K, Fujii S, Ogawa M (eds) Fluopyrimidines in cancer therapy. Excerpta Medica, New York, pp 251-260

Beutler B, Milsark IW, Cerami AC (1985) Passive immunization against cachectin/tumor necrosis factor protects mice from lethal effects of endotoxin. Science 220: 869-871

Bonadonna G, deLena M, Monfardini S et al. (1972) Clinical trial with bleomycin in lymphomas and in solid tumors. Eur J Cancer 8: 205-215

Borden EC et al. (1980) Levamisole effects in primary and recurrent colorectal carcinoma. Second International Conference on Immunotherapy of Cancer: Present status of trials in man. p 16

Borden EC (1984) Progress toward therapeutic application of interferons, 1979-1983. Cancer 54: 2770-2776

Bosset J, Hurteloup P, Bontemas P et al. (1983) A phase II trial of bleomycin and cisplatin in advanced oesophagus carcinoma. Proceedings of the 13th International Cancer Congress, p 41 (Abstract)

Bourbeau D, Sylvestre J, Levesque HP, Dussault RG, Bivin Y, Dobe S (1979) Computerized axial tomography and fine needle biopsy in surgery of the pancreas. Am J Surg 22: 29-37

Bowden L (1972) Cancer of the pancreas. CA 22: 275-283

Breedis C, Young G (1954) Blood supply of neoplasms in the liver. Am J Pathol 30: 969-977

Brooks JR (1979) Operative approach to pancreatic cancer. Semin Oncol 6: 357-367

Bruckner HW, Lokich JJ, Stablein DM (1982) Studies of Baker's antifol, methotrexate, and razoxane in advanced gastric cancer: A Gastrointestinal Tumor Study Group report. Cancer Treat Rep 66: 1713-1717

Buncher CR (1980) Epidemiology of pancreatic cancer. In: Moossa AR (ed) Tumors of the pancreas. William and Wilkins, Baltimore, pp 415-427

Buroker T, Moertel CG, Fleming TR et al. (1985) A controlled evaluation of recent approaches to biochemical modulation for enhancement of 5-fluorouracil therapy in colorectal carcinoma. J Clin Oncol 3: 1624-1631

Caballero GA, Ausman RK, Quebbeman EJ (1985) Long-term, ambulatory, continuous intravenous infusion of 5-fluorouracil for treatment of advanced adenocarcinoma. Cancer Treat Rep 69: 13-15

Cadman EC, Heimer R, Davis L (1979) Enhanced 5-fluorouracil nucleotide formation following methotrexate: biochemical explanation for drug synergism. Science 205: 1135-1137

Cady B, Oberfield RA (1974) Regional infusion chemotherapy of hepatic metastases from carcinoma of the colon. Am J Surg 127: 220-227

Carbone PP (1986) Advanced and adjuvant breast cancer therapy. In: Jordan VC (ed) Estrogen/antiestrogen action and breast cancer therapy. University of Wisconsin Press, Madison, pp 484-497

Carswell EA et al. (1975 a) Effect of tumor necrosis factor on cultured human melanoma cells. Nature 258: 31-32

Carswell EA, Old LJ, Kassel RL et al. (1975 b) An endotoxin-induced serum factor that causes necrosis of tumors. Proc Nat Acad Sci USA 72: 366-367

Carter SK (1975) The integration of chemotherapy into a combined modality approach for cancer treatment. VI. Pancreatic adenocarcinoma. Cancer Treat Rev 2: 193-214

Ch'ng JLC, Anderson JV, Williams SJ, Carr DH, Bloom SR (1986) Remission of symptoms during long term treatment of metastatic pancreatic endocrine tumours with long acting somatostatin analogue. British Med J 292: 982-983

Chirigos MA, Amery WK (1978) Combined levamisole therapy: An overview of its protective effects. In: Terry WD, Rosenberg SA (eds) Immunotherapy of human cancer. Raven Press, New York, pp 181-196

Chlebowski RW, Tong M, Weissman J et al. (1984) Hepatocellular carcinoma - diagnostic and prognostic features in North American patients. Cancer 53: 2701-2706

Clarkson B, O'Connor A, Winston L, Hutchinson D (1964) The physiological disposition of 5-fluorouracil and 5-fluoro-2-deoxyuridine in man. Clin Pharmacol Ther 5: 581-610

Comis RL, Carter SK (1974) A review of chemotherapy in gastric cancer. Cancer 34: 1576-1586

Craig JR, Peters RL et al. (1980) Fibrolamellar carcinoma of the liver: a tumor of adolescents and young adults with distinctive clinicopathologic features. Cancer 46: 372-379

Crile G Jr (1970) The advantages of bypass procedures over radical pancreatoduodenectomy in the treatment of pancreatic carcinoma. Surg Gynecol Obstet 130: 1049-1053

Crooke ST, Bradner WT (1976) Mitomycin C: a review. Cancer Treat Rev 3: 121-139

Cullinan SA, Moertel CG, Fleming TR et al. (1985) A comparison of chemotherapeutic regimens in the treatment of advanced pancreatic and gastric carcinoma. Fluorouracil vs. fluorouracil and doxorubicin vs. fluorouracil, doxorubicin and mitomycin. JAMA 253: 2061-2067

Davis HL, Ramirez G, Ansfield FJ (1974) Adenocarcinomas of stomach, pancreas, liver and biliary ducts. Cancer 33: 193-197

DeBessi P, Salvagno L, Endrizzi L et al. (1984) Cisplatin, bleomycin and methotrexate in the treatment of advanced oesophageal cancer. Eur J Cancer Clin Oncol 20: 743-747

Depeyster FA, Gilchrist RK (1969) Pathology and manifestations of cancer of the colon and rectum. In: Turrel R (ed) Diseases of the colon and anorectum. Saunders, Philadelphia, pp 428-452

Di Bisceglie AM, Rustgi VK, Hoofnagle JH et al. (1988) Hepatocellular carcinoma-NIH conference. Ann Intern Med 108: 390-401

DiMarco A (1975) Adriamycin (NSC-123127): Mode and mechanism of action. Cancer Chemother Rep 6: 91-106

DiMarco A, Gaetani M, Orezzi P et al. (1963) Antitumor activity of a new antibiotic: daunomycin. Communications 3rd international congress of chemistry, 22-27 July 1963

Dobelbower RR (1979) The radiotherapy of pancreatic cancer. Semin Oncol 6: 378-389

Dobelbower RR, Abe M (1989) Intraoperative radiation therapy, CRC, Boca Raton, in press

Douglass HO, Lavin PT, Goudsmit A, Klaassen DJ, Paul AR (1984) An Eastern Cooperative Oncology Group evaluation of combinations of methyl-CCNU, mitomycin C, adriamycin, and 5-fluorouracil in advanced measurable gastric cancer (EST 2277). J Clin Oncol 2: 1372-1381

Dukes CE (1932) The classification of cancer of the rectum. J Pathol 35: 323-332

Earl HM, Coombes RC, Schein PS (1984) Cytotoxic chemotherapy for cancer of the stomach. Clin Oncol 3: 351-369

Eaton SB, Fleischli DJ, Polard JJ, Nebesar RA, Potsaid MS (1968) Comparison of current radiologic approaches to the diagnosis of pancreatic disease. N Engl J Med 279: 389-396

Eckhardt S (1979) Chemotherapy of primary liver cancer. J Toxicol Environ Health 5: 359-400

Ellis FH, Gibb SP (1979) Esophagectomy for carcinoma: Current hospital mortality and morbidity rates. Ann Surg 190: 699

Eloesser I (1937) The treatment of some abdominal cancers by irradiation through the open abdomen combined with cautery excision. Ann Surg 106: 645-652

Engsminger W, Rosowsky A, Raso V et al. (1978) A clinical-pharmacologic evaluation of hepatic arterial infusions of 5-fluoro-2'deoxuridine and 5-fluorouracil. Cancer Res 38: 3784-3792

Engstrom P, MacIntyre J et al. (1978) Combination chemotherapy of advanced bowel cancer. Proc Am Assoc Cancer Res and Am Soc Clin Oncol 19: 384

Engstrom PF, Lavin P, Douglass HO Jr (1981) Adjuvant therapy of gastric carcinoma using 5-fluorouracil (5FU) plus semustine (me-CCNU): a preliminary report. In: Gerard A (ed) Progress and Perspectives in the treatment of gastrointestinal tumors. Pergamon, Oxford, pp 31-35

Engstrom P, Lavin P, Kalssen D (1983) Phase II evaluation of mitomycin and cisplatin in advanced esophageal carcinoma. Cancer Treat Rep 67: 713-715

Engstrom PF, Lavin PT, Douglass HO Jr, Brunner KW (1985) Postoperative adjuvant 5-fluorouracil plus methyl-CCNU therapy for gastric cancer patients. Eastern Cooperative Oncology Group Study (EST 3275). Cancer 55: 1868-1873

ETORC/Clinical Screening Group (1970) Study of the clinical efficacy of bleomycin in human cancer. B Med J 2: 643-645

Ezdinli EZ, Gelber R, Desai D, Falkson G et al. (1980) Chemotherapy of advanced esophageal carcinoma: Eastern Cooperative Oncology Group experience. Cancer 46: 2149-2153

Faintuch K, Shepard KV, O'Laughlin K et al. (1985) Toxicity of continuous infusion of 5-FU. Proc Am Soc Clin Oncol 4: 92

Fisher B, Wollmark N, Rockette H, Redmond C et al. (1988) Postoperative adjuvant chemotherapy or radiation therapy for rectal cancer: results from NSABP protocol R-01. JNCI 80: 21-29

Forestiere A, Patel H, Hankins J et al. (1983) Cisplatin, bleomycin and VP-16-213 in combination for epidermoid carcinoma of the esophagus. Proceed ASCO 2: 123

Forrest JF, Longmire WP (1978) Carcinoma of the pancreas and periampullary region. Ann Surg 189: 129-138

Foster JH (1978) Survival after liver resection for secondary tumors. Am J Surg 135: 389-394

Freeny PC, Ball TJ (1981) Endoscopic retrograde cholangiopancreatography (ERCP) and percutaneous transhepatic cholangiography (PTC) in the evaluation of suspected pancreatic carcinoma: Diagnostic limitations and contemporary roles. Cancer 47: 166-1678

Frei E III, Freireich EJ (1965) Progress and perspectives in the chemotherapy of acute leukemia. Adv Chemother 2: 269-298

Frei E III, DeVitae BT, Moxley JH III, Carbone PP (1966) Approaches to improving the chemotherapy of Hodgkin's disease. Cancer Res 26: 1284

Frei E III, Luce JK, Gamble JE et al. (1973) Combination chemotherapy in advanced Hodgkin's disease: Induction and maintenance of remission. Ann Intern Med 79: 376-382

Friesen SR (1982) The APUD syndromes. Prog Clin Cancer 8: 75-87

Ganda OP, Weir GC, Soelner JS, Legg MA et al. (1977) "Somastimoma": a somatostatin-containing tumor of the endocrine pancreas. N Engl J Med 296: 963-967

Gastrointestinal Tumor Study Group (1982) Controlled trial of adjuvant chemotherapy following curative resection for gastric cancer. Cancer 49: 1116-1122

Gastrointestinal Tumor Study Group (1984a) Randomized study of combination chemotherapy in unresectable gastric cancer. Cancer 53: 13-17

Gastrointestinal Tumor Study Group (1984b) Adjuvant therapy of colon cancer. Results of a prospectively randomized trial. N Engl J Med 310 (12): 737-743

Gastrointestinal Tumor Study Group (1985) Prolongation of the disease-free interval in surgically treated rectal carcinoma. Cancer 53: 13-17

Gisselbrecht C, Calvo F, Mignot L et al. (1983) Fluorouracil, Adriamycin and cisplatin combination chemotherapy of advanced esophageal carcinoma. Cancer 52: 974-977

Grage TB, Moss SE et al. (1981) Adjuvant chemotherapy in cancer of the colon and rectum: demonstration of effectiveness of prolonged 5-FU chemotherapy in a prospectively controlled, randomized trial. Surg Clin North Am 61: 1321-1329

Grage TB, Vassilopoulos PP, Shingelton WW et al. (1979) Result of a prospective randomized study of hepatic artery infusion with 5-fluorouracil versus intravenous 5-fluorouracil in patients with hepatic metastases from colorectal cancer: A Central Oncology Group study. Surgery 86: 550-555

Grinnell RS (1950) The spread of carcinoma of the colon and rectum. Cancer 3: 641-652

Grossi CE, Nealon TF, Rouselot AM (1972) Adjuvant chemotherapy in resectable cancer of the colon and rectum. Surg Clin North Am 52: 925-933

Gyves JW, Ensminger WD, van Harken D et al. (1983) Improved regional selectivity of hepatic arterial mitomycin by starch microspheres. Clin Pharmacol Ther 34: 259-263

Hanna MG, Peters LC (1981) Morphologic and functional aspects of active specific immunotherapy of established pulmonary metastases in guinea pigs. Cancer Res 41: 4001-4009

Harmon WE, Cohen HJ, Schneeberger EE, Grupe WE (1979) Chronic renal failure in children treated with methyl CCNU. N Engl J Med 300: 1200-1203

Harvey JH, Smith FP, Schein PS (1984) 5-Fluorouracil, mitomycin, and doxorubicin (FAM) in carcinoma of the biliary tract. J Clin Oncol 2 (11): 1245-1248

Heidelberger C, Chandhari NK et al. (1957) Fluorinated pyrimidines; a new class of tumor inhibitory compounds. Nature 179: 663-666

Hellerstein S, Rosen S, Kies M et al. (1983) Diamminedichloroplatinum and 5-FU combined chemotherapy of epidermoid esophageal cancer. Proceed ASCO 2: 127

Herlyn M, Blaszczyk M, Sears HF, Verrill H et al. (1983) Detection of a carcinoembryonic antigen and related antigens in sera of patients with gastrointestinal tumors using monoclonal antibodies in double determinant radioimmunoassays. Hybridoma 2: 329-339

Higgins GA Jr, Humphrey E, Juler GL et al. (1976) Adjuvant chemotherapy in the surgical treatment of large bowel cancer. Cancer 38: 1461-1467

Higgins GA, Amadeo JH, Smith DE, Humphrey EW, Keehn RJ (1983) Efficacy of prolonged intermittent therapy with combined 5-FU and methyl-CCNU following resection for gastric carcinoma. A Veterans Administration Surgical Oncology Group report. Cancer 52: 1105-1112

Hohn D, Melnick J, Stagg R et al. (1985) Biliary sclerosis in patients receiving hepatic arterial infusions of floxuridine. J Clin Oncol 3: 98-102

Holyoke ED (1981) New surgical approaches to pancreatic cancer. Cancer 47: 1719-1723

Hoover HC Jr, Peters LC, Brandhorst JS, Hanna MG Jr (1981) Therapy of spontaneous metastases with an autologous tumor vaccine in a guinea pig model. J Surg Res 30: 409-415

Hoover HC Jr, Surdyke MG, Dangel RB et al. (1985) Prospective randomized trial of adjuvant active-specific immunotherapy for human colorectal cancer. Cancer 55: 1236-1243

Hurley JD, Ellison EH et al. (1960) Chemotherapy of solid carcinomas - Indications, agents and results. JAMA 174: 1696-1701

Ihde DC, Matthews MJ, Makuch RW et al. (1985) Prognostic factors in patients with hepatocellular carcinoma receiving systemic chemotherapy - Identification of two groups of patients with prospects for prolonged survival. Am J Med 78: 399-405

Inokuchi K, Hattori T, Taguchi T, Abe O, Ogawa N (1984) Postoperative adjuvant chemotherapy for gastric carcinoma. Analysis of data on 1805 patients followed for 5 years. Cancer 53: 2393-2397

Jacobsson F (1961) The Paterson-Kelly (Plummer-Vinson) syndrome and carcinoma of the esophagus. In: Tanner NC, Smithers DW (eds) Neoplastic diseases at various sites, vol 4. Tumors of the oesophagus. Livingston, Edinburgh

Joske RA, Benedict EB (1959) The role of benign esophageal obstruction in the development of carcinoma of the esophagus. Gastroenterology 36: 749

Kaplan RS, Wiernik P (1982) Neurotoxicity of antineoplastic drugs. Semin Oncol 9: 103-130

Kelsen D (1984) Chemotherapy of esophageal cancer. Semi Oncol 11: 159-168

Kelsen KP, Cvitkovic E, Bains M et al. (1978) Cis-diamminedichlros platinum (11) and bleomycin in the treatment of esophageal carcinoma. Cancer Treat Rep 62: 1041-1046

Kelsen D, Hilaris B, Coonley C, Chapman R, Lesser M et al. (1983a) Cisplatin, vindesine, and bleomycin chemotherapy of local-regional and advanced esophageal carcinoma. Amer J Med 75: 645-652

Kelsen DP, Coonley C, Hilaris B et al. (1983b) Cisplatin, vindesine and bleomycin combination chemotherapy of local-regional and advanced esophageal carcinoma. Amer J M 75: 639-652

Kelsen DP, Coonley C, Bains M et al. (1983c) Cisplatin, vindesine and methyl-glyoxal bis (quanylhydrazone) combination chemotherapy of esophageal cancer. Proceed ASCO 2: 128

Kelsen D, Bains M, Hilaris B, Martini N (1984) Combined-modality therapy of esophageal cancer. Seminars in Oncology 11: 169–177

Kemeny N, Yagoda A et al. (1979) A randomized study of two different schedules of methyl-CCNU, 5-FU + vincristine for metastatic colorectal carcinoma. Cancer 43: 78–82

Kemeny N, Yagoda A et al. (1980) Therapy for metastatic colorectal carcinoma with a combination of methyl-CCNU, 5-fluorouracil, vincristine and streptozotocin. Cancer 45: 876–881

Kemeny N, Daly J, Oderman P et al. (1984) Hepatic artery pump infusion: Toxicity and results in patients with metastatic colorectal carcinoma. J Clin Oncol 2: 595–600

Kemeny N, Daly J, Reichman B, Geller N et al. (1987) Intrahepatic or systemic infusion of fluorodeoxyuridine in patients with liver metastases from colorectal cancer – a randomized trial. Ann Intern Med 107: 459–465

Kennedy BJ, Theologides A (1961) The role of 5-fluorouracil in malignant disease. Ann Int Med 55: 719–730

Kew MC (1981) Clinical, pathological, and etiological heterogeneity in hepatocellular carcinoma. Evidence from Southern Africa. Hepatology 1: 366–369

Kirklin JW, Dockerty MB, Waugh JM (1949) The role of the peritoneal reflection in the prognosis of carcinoma of the rectum and sigmoid colon. Surg Gynecol & Obstet 88: 326–331

Klaassen DJ, MacIntyre JM, Catton GE et al. (1985) Treatment of locally unresectable cancer of the stomach and pancreas: A randomized comparison of 5-fluorouracil alone with radiation plus concurrent and maintenance 5-fluorouracil – An Eastern Cooperative Group study. J Clin Oncol 3: 373–378

Klavins JV (1981) Tumor markers of pancreatic carcinoma. Cancer 47: 1597–1601

Kolaric K, Moricic Z, Dujmovic I et al. (1976) Therapy of advanced esophageal cancer with bleomycin, irradiation and combination bleomycin and irradiation. Tumor 62: 255–262

Kolaric K, Maricic Z, Roth A et al. (1980) Combination of bleomycin and Adriamycin with and without radiation in the treatment of inoperable esophageal cancer. Cancer 45: 2265–2273

Kopelson G, Harisiadis L, Tretter P et al. (1977) The role of radiation therapy in cancer of the extra-hepatic biliary system. Int J Radiat Oncol Biol Phys 2: 883–894

Kovach JS, Moertel CG, Schutt AJ et al. (1974) A controlled study of combined 1,3-bis-(2-chloroethyl)-1-nitrosourea and 5-fluorouracil therapy for advanced gastric and pancreatic cancer. Cancer 33: 563–567

Krain LS (1970) The rising incidence of the pancreas – real or apparent? J Surg Oncol 2: 115–124

Kvale D, Rognum TO, Brandtzaeg P (1987) Elevated levels of secretory immunoglobulins A and M in serum of patients with large bowel carcinoma indicate liver metastasis. Cancer 59: 203–207

Kvols LK, Moertel CG, O'Connel MJ et al. (1986) Treatment of the malignant carcinoid syndrome – Evaluation of a long-acting somatostatin analogue. N Engl J Med 315: 663–666

Laurie J, Moertel C, Fleming T et al. (1986) Surgical adjuvant therapy of poor prognosis colorectal cancer with levamisole alone or combined levamisole and 5-fluorouracil (5-FU). Proc Am Soc Clin Oncol 5: 81

Lavin P, Mittelman A et al. (1980) Survival and response to chemotherapy for advanced colorectal adenocarcinoma: An Eastern Cooperative Oncology Group report. Cancer 46: 1536–1543

Lavin PT, Bruckner HW, Plaxe SC (1982) Studies in prognostic factors relating to chemotherapy for advanced gastric cancer. Cancer 50: 2016–2023

Leichman L, Steiger Z, Seydel HG, Dindogru A, Kinzie J et al. (1984a) Preoperative chemotherapy and radiation therapy for patients with cancer of the esophagus: A potentially curative approach. J Clin Oncol 2: 75–79

Leichman L, Steiger Z, Seydel HG, Vaitkevicius VK (1984b) Combined preoperative chemotherapy and radiation therapy for cancer of the esophagus: The Wayne State University, Southwest Oncology Group and Radiation Therapy Oncology Group experience. Semin Oncol 11: 178–185

Levin DL, Devesa SS, Goodwin JD II, Silverman DT (1974) Cancer rates risks, 2nd edn. National Institutes of Health, Washington, DC

Li MC, Ross ST (1976) Chemoprophylaxis for patients with colorectal cancer: prospective study with five-year follow-up. JAMA 235: 2825–2828

Lin TY (1976) Recent advances in technique of hepatic lobectomy and results of surgical treatment for primary carcinoma of the liver. Prog Liv Dis 5: 668–682

Loehrer PJ Sr, Williams SD, Einhorn LH, Ansari R (1985) Ifosfamide: an active drug in the treatment of adenocarcinoma of the pancreas. J Clin Oncol 3: 367–372

Lokich JJ, Skarin AT (1972) Combination therapy with 5-fluorouracil (5FU; NSC'19893) and 1,3-bis(2-chloroethyl)-1-nitrosourea (BCNU; NSC'409962) for disseminated gastrointestinal cancer. Cancer Chemother Rep 56: 653–657

Lokich J, Bothe A, Fine N et al. (1981) Phase I study of protracted venous infusion of 5-fluorouracil. Cancer 48: 2565–2568

London WT (1981) Primary hepatocellular carcinoma – etiology, pathogenesis, and prevention. Hum Pathol 12: 1085–1097

Lopez M, Perno CF, DiLauro L, Papaldo P (1984) 5-Fluorouracil, adriamycin, and BCNU (FAB) combination chemotherapy for advanced gastric cancer. Cancer Chemother Pharmacol 12: 194–197

MacDonald JS, Schein PS, Woolley PV, Smythe T, Ueno W et al. (1980) 5-Fluorouracil, doxorubicin, and mitomycin (FAM) combination chemotherapy for advanced gastric cancer. Ann Intern Med 93: 533–536

Machover D, Goldscmidt E et al. (1986) Treatment of advanced colorectal and gastric adenocarcinomas with 5-FU and high dose folinic acid. J Clin Oncol 4: 685–696

Madajewicz S, Petrelli N, Rustum Y et al. (1984) Phase I–II trial of high-dose calcium leucovorin and 5-fluorouracil in advanced colorectal cancer. Cancer Res 44: 4667–4669

Magnani JL, Steplewski Z, Koprowski H, Ginsberg V (1983) Identification of the gastrointestinal and pancreatic cancer-associated antigen detected by monoclonal antibody 19-9 in the sera of patients as mucin. Cancer Res 43: 5489–5492

Major PP, Egan E, Herrick D, Kufe DW (1982) 5-Fluorouracil incorporation in DNA of human breast carcinoma cells. Cancer Res 42: 3005–3009

Mandel HG (1969) Incorporation of 5-fluorouracil into RNA and its molecular consequences. Prog Mol Subcell Biol 1: 82–135

Marks IN, Bank S, Louw JH (1967) Islet cell tumor of the pancreas with reversible water diarrhea and achlorhydria. Gastroenterology 52: 695–708

Matsumoto KK, Peter JB, Schultze RJ, Hakim AA, Franck PT (1966) Watery diarrhea and hypokalemia associated with pancreatic islet cell adenoma. Gastroenterology 50: 231–242

Mavligit GM, Burgess MA, Gutterman JU et al. (1976) Prolongation of post-operative disease-free interval and survival in human colorectal cancer by BCG or BCG plus 5-fluorouracil. Lancet 1: 871–876

McCort JJ (1952) Radiographic identification of lymph node metastases from carcinoma of the esophagus. Radiology 59: 694

Meeker WR, Sickle-Santanello BJ, Philpott G et al. (1986) Combined chemotherapy, radiation and surgery for epithelial cancer of the anal canal. Cancer 57: 522–529

Michaelson RA, Magill GB, Quan SH et al. (1983) Preoperative chemotherapy and radiation therapy in the management of anal epidermoid carcinoma. Cancer 51: 390–395

Minton JP, Hoehn JL, Gerber DM et al. (1985) Results of a 400-patient carcinoembryonic antigen second-look colorectal cancer study. Cancer 55: 1284–1290

Moertel C, Schutt A, Reitemeier R et al. (1976 a) Therapy for gastrointestinal cancer with the nitrosoureas alone in drug combination. Cancer Treat Rep 60: 729–732

Moertel C, Fleming T, O'Connell M et al. (1984) A phase II trial of combined intensive course 5-FU, Adriamycin and cis-platinum in advanced gastric and pancreatic carcinoma. Am Soc Clin Oncol 3: 137 (Abstract)

Moertel CG (1973) Therapy of advanced gastrointestinal cancer with the nitrosoureas. Cancer Chemother Rep 4: 27–34 [Suppl]

Moertel CG (1976) Chemotherapy of gastrointestinal cancer. Clin Gastroenterol 5: 777–793

Moertel CG, Childs DS, Reitemeier RJ, Colby MY, Holbrook MA (1969) Combined 5-fluorouracil and super-voltage radiation therapy of locally unresectable gastrointestinal cancer. Lancet 2: 865–869

Moertel CG, Mittelman A, Bakemeier RF, Engstrom P, Hanley J (1976 b) Sequential and combination chemotherapy of advanced gastric cancer. Cancer 38: 678–682

Moertel CG, Frytak S, Hahn RG, the Gastrointestinal Tumor Study Group (1981) Therapy of locally unresectable pancreatic carcinoma: A randomized comparison of high dose (6000 rads) radiation alone, moderate dose radiation (4000 rads) + 5-fluorouracil, and high dose radiation + 5-fluorouracil. Cancer 48: 1705–1710

Mongue JJ, Jud SS, Gage RP (1964) Radical pancreatoduodenectomy: A 22-year experience with the complications, mortality rate, and survival rate. Ann Surg 160: 711–722

Muñoz N, Linsell A (1982) Epidemiology of primary liver cancer. In: Correa P, Haenszel W (eds) Epidemiology of cancer of the digestive tract. Nijhoff, The Hague, pp 161–195

Murano R, Wunderlich D, Thor A, Lundy J et al. (1985) Definition by monoclonal antibodies of a repertoire of epitopes on carcinoembryonic antigen differentially expressed in human colon carcinomas versus normal adult tissues. Cancer Res 45: 5769–5780

Myers CE, Diasio R, Eliot HM, Chabner BA (1976) Pharmacokinetics of the fluoropyrimidines: implications for their clinical use. Cancer Treat Rep 3: 175–183

Naef AP, Savary M, Ozzello L (1975) Columnar-lined lower esophagus: An acquired lesion with malignant predisposition. J Thorac Cardiovasc Surg 70: 826–835

Nagasue N, Inokuchi K et al. (1977) Serum alpha-protein levels after hepatic artery ligation and postoperative chemotherapy. Correlation with clinical status in patients with hepatocellular carcinoma. Cancer 40: 615–618

Nakajima T, Fukami A, Takagi K, Kajitani T (1980) Adjuvant chemotherapy with mitomycin C and with a multi-drug combination of mitomycin C, 5-fluorouracil and cytosine arabinoside after curative resection of gastric cancer. Jpn J Clin Oncol 10: 187–194

Nakajima T, Takahashi T, Takagi K, Kuno K, Kajitani T (1984) Comparison of 5-fluorouracil with ftorafur in adjuvant chemotherapies with combined inductive and maintenance therapies for gastric cancer. J Clin Oncol 2: 1366–1371

Narsete T, Ansfield FJ, Wirtanen G et al. (1977) Gastric ulceration in patients receiving intrahepatic infusion of 5-fluorouracil. Ann Surg 186 (6): 734–736

Neefe JR, Silgals R, Schein PS (1984) Minimal activity of recombinant clone. A interferon in metastatic colon cancer. J Biol Response Mod 3: 366–370

Niederhuber J, Ensminger W, Gyves J et al. (1984) Regional hepatic chemotherapy of colorectal cancer metastatic to the liver. Cancer 53: 1336–1343

Nigro ND, Vaitkevicius VK, Buroker T, Bradley GT, Considine B (1981) Combined therapy for cancer of the anal canal. Dis Colon Rectum 24: 73–75

Oberfield RA, McCaffrey JA, Polio BS et al. (1979) Prolonged and continuous percutaneous intra-arterial hepatic infusion chemotherapy in advanced metastatic liver adenocarcinoma from colorectal primary. Cancer 44: 414–23

O'Connell MJ (1985) Current status of chemotherapy for advanced pancreatic and gastric cancer. J Clin Oncol 3: 1032–1039

O'Connell MJ, Gunderson LL, Moertel CG, Kvols LK (1985) A pilot study to determine clinical tolerability of intensive combined modality therapy for locally unresectable gastric cancer. Int J Radiat Oncol Biol Phys 11: 1827–1831

Oldham RK (1983) Monoclonal antibodies in cancer therapy. J Clin Oncol 1: 582–590

Oldham RK, Smalley RV (1984) The role of interferon in treatment of cancer. In: Zoon KC, Noguchi PO, Liu TY (eds) Interferon research, clinical application and regulatory consideration. Elsevier, New York, pp 191–206

Oldham RK, Morgan AC, Woodhouse CS et al. (1984) Monoclonal antibodies in the treatment of cancer: Preliminary observations and future prospects. Med Oncol Tumor Pharmacotherapy 1: 51–62

Olson RM, Perencevich NP, Malcolm AW et al. (1980) Patterns of recurrence following curative resection of adenocarcinoma of the colon and rectum. Cancer 45: 2969–2974

Olweny CLM, Toya T et al. (1975) Treatment of hepatocellular carcinoma with adriamycin – Preliminary communication. Cancer 36: 1250–1257

Order SE, Stillwagon GB, Klein JL et al. (1985) Iodine 131 antiferritin, a new treatment modality in hepatoma: A Radiation Therapy Oncology Group study. J Clin Oncol 3 (12): 1573–1582

Oster MW, Theologides A, Cooper MR et al. (1982) Fluorouracil (F) + Adriamycin (A) + mitomycin (M) (FAM) versus fluorouracil (F) + streptozotocin (S) + mitomycin (M) (FSM) in advanced pancreatic cancer. Proc Am Soc Clin Oncol 1: 90 (Abstract)

Oswoll ES, Kiessling PJ, Patterson JR (1978) Interstitial pneumonia from mitomycin. Ann Intern Med 89: 352–355

Panettiere FJ, Haas C, McDonald B, Costanzi JJ, Talley RW et al. (1984) Drug combinations in the treatment of gastric adenocarcinoma: a randomized Southwest Oncology Group study. J Clin Oncol 2: 420–424

Papillon J, Mayer M, Montbarbon JF et al. (1983) A new approach to the management of epidermoid carcinoma of the anal canal. Cancer 51: 1830–1837

Parker EF, Marks RD Jr, Kratz JM, Chaikhouni A et al. (1985) Chemoradiation therapy and resection for carcinoma of the esophagus: short-term results. Ann Thorac Surg 40: 121–125

Patt Y, McBride CM, Ames FC et al. (1987) Adjuvant perioperative hepatic arterial mitomycin C and floxuridine combined with surgical resection of metastatic colorectal cancer in the liver. Cancer 59: 867-873

Paulino-Netto A, Dreiling DA, Baronofsky ID (1960) The relationship between pancreatic calcification and cancer of the pancreas. Ann Surg 5: 530-537

Pearson JG (1977) The present status and feature potential of radiotherapy in the management of esophageal cancer. Cancer 39: 882

Pettaval J, Mortgenthaler F (1978) Protracted arterial chemotherapy of liver tumors. An experience of 107 cases over a twelve year period. In: Ariel IM (ed) Progress in clinical cancer, vol 7. Grune and Stratton, New York, pp 217-233

Poleynard GD, Marty AT, Birnbaum WB et al. (1977) Adenocarcinoma in the columnar-lines (Barrett) esophagus. Arch Surg 112: 997

Putzki H, Student A, Jablonski M, Heymann H (1987) Comparison of the tumor markers CEA, TPA, and CA 19-9 in colorectal carcinoma. Cancer 59: 223-226

Popkin J, Bromer R, Byrne R et al. (1983) Continuous 48-hour infusion of vindesine in squamous cell carcinoma of the upper aerodigestive tract. Proceed 13th Internat'l Cancer Congress, p 40

Radigan LR, Glover JL, Shipley FF, Shoemaker RE (1977) Barrett esophagus. Arch Surg 112: 486-491

Ramirez G, Ansfield FJ (1982) Chemotherapy of liver metastases. In: Weiss L, Gilbert HA (eds) Liver metastasis. Hall, Boston, pp 348-359

Ramming K (1983) The effectiveness of hepatic artery infusion in treatment of primary hepatobiliary tumors. Semin Oncol 10 (2): 199-205

Ravry M, Moertel CG, Schutt AJ et al. (1973) Treatment of advanced squamous cell carcinoma of the gastrointestinal tract with bleomycin (NSC 125066). Cancer Chemother Rep 57: 493-495

Reich SD (1979) Clinical pharmacology of mitomycin C. In: Carter SK, Crooke ST (eds) Mitomycin C: current status and new developments. Academic Press, New York, pp 243-250

Richter GW (1965) Comparison of ferritins from neoplastic and nonneoplastic human cells. Nature 207: 616-618

Rittgers RA, Steele G, Zamcheck N et al. (1978) Transient carcinoembryonic antigen (CEA) elevations following resection for colorectal cancer: A limitation in the use of serial CEA levels as an indicator for second look surgery. JNCI 61: 315-318

Rochlin DB, Shiner J, Langdon E et al. (1962) Use of 5-FU in disseminated solid neoplasms. Ann Surg 156: 105-113

Rosenberg B, Van Camp L, Krigas T (1965) Inhibition of cell division in *Escherichia coli* by electrolysis products from a platinum electrode. Nature 205: 698-699

Rosenberg SA, Lotze MT, Muul LM et al. (1985) A new approach to the therapy of cancer based on the systemic administration of autologous lymphokine activated killer cells and recombinant interleukin-2. N Engl J Med 313: 1485-1492

Rosenberg SA (1986) Adoptive immunotherapy of cancer using lymphokine activated killer cells and recombinant interleukin-2. In: DeVita VT, Hellman S, Rosenberg SA (eds) Important advances in oncology. Lippincott, Philadelphia

Rousselot LM, Cole DR, Grossi CE et al. (1972) Adjuvant chemotherapy with 5-fluorouracil in surgery for colorectal cancer: eight-year progress report. Dis Colon Rectum 15: 169-174

Sato Y, Fujiwara K, Ogata I et al. (1985) Transcatheter arterial embolization for hepatocellular carcinoma - benefits and limitations for unresectable cases with liver cirrhosis evaluated by comparison with other conservative treatments. Cancer 55: 2822-2825

Schein PS, DeVita VT Jr, Hubbard S et al. (1976) Bleomycin, Adriamycin, cyclophosphamide, vincristine, and prednisone (BACOP) combination chemotherapy in the treatment of advanced diffuse histiocytic lymphoma. Ann Intern Med 85: 417-422

Schlom J, Weeks M (1985) Potential clinical utility of monoclonal antibodies in the management of human carcinomas. In: DeVita VT Jr, Hellman S, Rosenberg SA (eds) Important advances in oncology. Lippincott, Philadelphia, pp 170-192

Sears H, Mattis J, Herlyn D et al. (1984) Effects of monoclonal antibody immunotherapy on patients with gastrointestinal adenocarcinoma. J Biol Response Mod 3: 138-150

Serlin O, Wolkoff JS, Amaded JM, Keehn RJ (1969) Use of 5-fluoro-deoxyuridine (FUDR) as an adjuvant to the surgical management of carcinoma of the stomach. Cancer 24: 223-228

Shepard KV, Levin B, Karl RC et al. (1985) Therapy for metastatic colorectal cancer with hepatic artery infusion chemotherapy using a subcutaneous implanted pump. J Clin Onc 3: 161-169

Shoemaker RH, Curt GA, Carney DN (1983) Evidence for multidrug-resistant cells in human tumor cells populations. Cancer Treat Rep 67: 883

Silgals RM, Ahlgren JD, Neefe JR et al. (1984) A phase II trial of high-dose intravenous interferon alpha-2 in advanced colorectal cancer. Cancer 54: 2257-2261

Silverberg E, Lubera J (1986) Cancer Statistics, 1986. CA 36: 9-25

Silverberg E, Lubera JA (1988) Cancer Statistics, 1988. CA 38: 5-22

Sischy B, Remington JH, Sobel SH, Savlov ED (1980) Treatment of carcinoma of the rectum and squamous carcinoma of the anus by combination chemotherapy, radiotherapy and operation. Surg Gynecol Obstet 151: 369-371

Sitzmann JV, Order SE, Klein JL, Leichner PK et al. (1987) Conversion by new treatment modalities of nonresectable to resectable hepatocellular cancer. J Clin Oncol 5: 1566-1573

Smith FP, Hoth DF, Levin B, Karlin DA, MacDonald JS, Woolley PV, Schein PS (1980) 5-Fluorouracil, Adriamycin, and mitomycin-C (FAM) chemotherapy for advanced adenocarcinoma of the pancreas. Cancer 46: 2014-2018

Smith GW, Bukowski RM, Hewlett JS, Groppe CW (1984) Hepatic artery infusion of 5-fluorouracil and mitomycin C in cholangiocarcinoma and gallbladder carcinoma. Cancer 54: 1513-1516

Sondel P, Hank J, Kohler PC (1987) The clinical potential of interleukin-2 (IL-2) for the treatment of neoplastic disease. Oncology 1: 41-49

Sponzo RW, DeVita VT, Oliverio VT (1973) Physiologic disposition of 1-(2-chloroethyl)-3-cyclohexyl-1-nitrosourea (CCNU) and 1-(2-chloroethyl)-3-(4-methyl cyclohexyl)-1-nitrosourea (MeCCNU) in man. Cancer 31: 1154-1159

Steiger Z, Franklin R, Wilson RF, Leichman L, Seydel H et al. (1981) Eradication and palliation of squamous cell carcinoma of the esophagus with chemotherapy, radiotherapy, and surgical therapy. J Thorac Cardiovasc Surg 82: 713-719

Sternberg A, Petrelli N, Au J et al. (1984) A combination of 5-Fluorouracil and thymidine in advanced colorectal carcinoma. Cancer Chemother Pharmacol 13: 218-222

Stephens F (1973) Bleomycin - A new approach in cancer chemotherapy. Med J Aust 1: 1277-1283

Sullivan RD, Zurek WZ (1965) Chemotherapy for liver cancer by protracted ambulatory infusion. JAMA 194: 481-490

Sullivan RD, Young CW, Miller E, Glatstein N et al. (1960) The clinical effects of the continuous administration of fluorinated pyrimidines (5-FU and 5-FUdR). Cancer Chemother Rep 8: 77-83

Tancini G, Bajetta E, Bonadonna G (1974) Terapia con bleomycin da sola o in associazione con methodtrexate nel carcinoma epidermoide dell' esofago. Tumor 60: 65-71

Taylor I, Rowling J, West C (1979) Adjuvant cytotoxic liver perfusion for colorectal cancer. Br J Surg 66: 833-837

Taylor SG, Murthy AK, Showel JL, Caldarelli DD, Hutchinson JC Jr, Holinger LD, Kramer T, Kiel K (1985) Improved control in advanced head and neck cancer with simultaneous radiation and cisplatin/5-FU chemotherapy. Cancer Treat Rep 69: 93-939

Terry WD, Rosenberg SA (Eds) (1982) Immunotherapy of human cancer. Elsevier, North Holland

Turnbull RB (1970) Cancer of the colon - The five-and-ten-year survival rates following resection utilizing the isolation technique. Ann R Coll Surg Engl 46: 243-250

Tilson D (1974) Carcinoid syndrome. Surg Clin North Am 54: 409-423

V. A. Cooperative Surgical Adjuvant Study Group (1965) Use of Thio TEPA as an adjuvant to the surgical management of cancer of the stomach. Cancer 18: 291-297

Verhaegen H, DeCree J, DeCook W et al. (1982) Levamisole therapy in patients with colorectal cancer. In: Terry WD, Rosenberg SA (eds) Immunotherapy of Human Cancer. Elsevier, New York

Verner JV, Morrison AB (1958) Islet cell tumor and a syndrome of refractory watery diarrhea and hypokalemia. Am J Med 25: 374-380

Verner JA (1969) Clinical syndromes associated with non-insulin producing tumors of the pancreatic islets. In: Demlin L, Ottenjann R (eds) Non-insulin Producing Tumors of the Pancreas. Thieme, Stuttgart, pp 165-183

Vogel S, Greenwald E, Kaplan B (1981) Effective chemotherapy for esophageal cancer with methotrexate, bleomycin and cis-diamminedichloroplatinum (11). Cancer 48: 2555-2558

Von Hoff DD, Reichert CM, Cuneo R et al. (1979 a) Demyelination of peripheral nerves associated with cis-diamminedichloroplatinum (II) (DDP) therapy. Proc Am Assoc Cancer Res 20: 91

Von Hoff DD, Schilsky R, Reichert CM et al. (1979 b) Toxic effects of cis-dichlorodiammineplatinum (II) in man. Cancer Treat Rep 63: 1527-1531

Waddell WR (1973) Chemotherapy for carcinoma of the pancreas. Surgery 74: 420-429

Wallace S, Charnsangavej C, Carrasco CH, Bechtel W (1984) Infusion-embolization. Cancer 54: 2751-2765

Weese JL, Ramirez G (1986) An implantable pump for regional chemotherapy of hepatic metastases. Wisc Med J 85: 33-37

Weil C (1985) Gastroenteropancreatic endocrine tumors. Klinische Wochenschrift 63: 433-459

Weiss AJ, Jackson LG, Carabasi R (1961) An evaluation of 5-fluorouracilin malignant disease. Ann Int Med 55: 731-741

Weiss HD, Walker MD, Wiernik PH (1974) Neurotoxicity of commonly used antineoplastic agents. N Engl J Med 291: 75-81, 127-133

Weiss GR, Garnick MB, Osteen RT et al. (1983) Long-term hepatic arterial infusion of 5-fluorodeoxyuridine for liver metastases using an implantable infusion pump. J Clin Oncol 1: 337-344

Weiss RB (1982) Streptozotocin: A review of its pharmacology, efficacy, and toxicity. Cancer Treat Rep 66: 427-438

Whipple AO, Parsons WB, Mullins CR (1935) Treatment of carcinoma of the ampulla of Vater. Ann Surg 102: 763-779

Wilder RJ, Allan FN, Power MH, Robertson HE (1927) Carcinoma of the islands of the pancreas. Hyperinsulinism and hypoglycemia. JAMA 89: 348

Wittes R, Brescia F, Young CW (1975) Combination chemotherapy with disciamminedichloro platinum (11) and bleomycin in tumors of the head and neck. Oncology 32: 202-207

Yagoda A, Mukherji B, Young C et al. (1972) Bleomycin, an antitumor antibiotic: clinical experience in 274 patients. Ann Int Med 77: 861-870

Yamada T (1977) Early detection of gastric cancer. In: Hirayman T (ed) Epidemiology of stomach cancer: key questions and answers. WHO Collaborating Center for Evaluation of Methods of Diagnosis and Treatment of Stomach Cancer, Tokyo, Japan, pp 99-104

Zollinger RM, Ellison EH (1955) Primary peptic ulcerations of the jejunum associated with islet cell tumors of the pancreas. Ann Surg 142: 709-728

Zwelling LA, Kohn KW (1979) Mechanism of action of cis-dichlorodiammineplatinum (II). Cancer Treat Rep 63: 1439-1444

13 Follow-up of Patients Treated for Gastrointestinal Cancer

JOSEF KORINEK and BERNARD LEVIN

CONTENTS

13.1 Esophageal Carcinoma

Squamous cell carcinoma is the most common malignant tumor of the esophagus, accounting for almost 98% of all esophageal cancers (APPELQVIST 1972). Primary adenocarcinoma of the esophagus is relatively uncommon, the reported worldwide incidence being from less than 1% to 15% of all esophageal malignancies (RAPHAEL 1966); however its age incidence, sex ratio, and clinical course are similar to those of squamous carcinoma.

In general, the prognosis of patients with esophageal carcinoma remains poor. The long-term survival and cure depend, in part, on the grade of tumor differentiation, but they depend even more dramatically on the depth of invasion and presence of lymph node metastases (SKINNER et al. 1982). These prognostic factors are often difficult to establish preoperatively. Fifty to sixty percent of diagnosed patients are able to undergo surgical resection. However, 41% of patients with surgically resected tumors already have lymph node metastases and 38% have extension of the tumor into the periesophageal tissue (MORSON and DAWSON 1979; MING 1973). Patients whose lymph nodes are found at the time of surgery to be involved have a 5-year survival rate half that of patients without spread to regional lymph nodes. Nonetheless, even with involvement of regional lymph nodes, 10%–15% of these patients can be cured by surgical resection. These survival and postsurgical follow-up of these patients is further complicated by the lack of clinical symptoms, which even some patients with widespread lymphatic dissemination do not have. One report showed celiac node metastases in 40% of such asymptomatic patients (GUERNSEY and KNUDSEN 1970).

Achieving 5-year survival, an accomplishment arbitrarily accepted as indicating cure, may not be a valid measure for patients who have been treated for esophageal carcinoma. After 5 years as many as 78% surviving patients still may die because of recurrent disease (CEDARQUIST et al. 1978). Autopsy findings revealed lymph node metastases and residual tumor in approximately 75% of patients. In 50% of these patients visceral metastases were most frequently seen in lungs, liver, bones, kidneys, and adrenals (in that order) (MANDARD et al. 1981).

The goal of surgical treatment of esophageal cancer is a curative resection with a reconstruction of the gastrointestinal tract (by esophagogastrostomy or colon interposition) that allows normal swallowing and prevents the reflux of gastric secretions. In order to improve the actual cure rate of these patients and establish criteria for their follow-up, it is important to analyze the reasons for failure in the postoperative period, the immediate and long-term effects of alterations in physiology and nutrition caused by the loss of the esophagus and stomach, and the recurrence of local and distant disease. The information collected from 22 different surgical departments in Europe over a 9-year period on 2400 surgical patients has helped to improve the operative approach to this malignancy (GIULI 1985). The continual monitoring of pulmonary wedge pressure, arterial blood gases, and intensive therapy decreased the operative mortality in these 22 centers to 13.5%. The major cause of operative mortality is the development of anastomotic fistulas; those as-

JOSEF KORINEK, M.D., Ph.D., BERNARD LEVIN, M.D., Section of Gastrointestinal Oncology and Digestive Diseases, Division of Medicine, The University of Texas, M.D. Anderson Cancer Center, Houston, P.O. Box 78, 1515 Holcombe Blvd., Houston, TX 77030, USA sociated with mediastinitis particularly have very high mortality rates (over 90%). Respiratory complications, atelectasis, and pneumonia are the most common complications in the immediate postoperative period. Stenosis at the anastomotic site, seen in approximately 15% of patients, is usually diagnosed early and managed by endoscopic dilatation. This technique of therapeutic endoscopy is now a safe and routine procedure in the hands of experienced endoscopists who can choose from a variety of rubber, metal olive, and synthetic tapered or balloon dilators. These can be safely passed over an endoscopically placed guide wire without fluoroscopic assistance.

Postgastrectomy syndromes are seen in a significant number of patients and include early dumping syndrome, late dumping (reactive hypoglycemia) syndrome, reflux gastritis and reflux esophagitis, chronic afferent loop syndrome, chronic efferent loop obstruction, small gastric remnant syndrome, postvagotomy diarrhea, and malabsorption and nutritional deficiencies. These complications are diagnosed in clinically symptomatic patients by a combination of endoscopic and radiologic techniques with support from laboratory tests. They are manages by symptomatic therapy in most patients and only a minority require reoperation.

The early detection of recurrent and metastatic disease can only be accomplished by frequent follow-up of all surgically treated patients regardless of whether or not they are symptomatic. There is great variation in the frequency and types of tests done as follow-up procedures of these patients in different institutions. The biology of this malignancy certainly justifies periodic evaluations extending beyond the 5-year period. An arbitrary but accepted rule in some institutions, including ours, is to examine patients 3 months after resection surgery and then to follow them at 6-month intervals (Table 1). Unless patients' symptoms require a specific test, the follow-up examination includes blood count, determination of liver enzymes, chest roentgenogram, CT scan of the abdomen, and esophagogastroscopy. The CT scan of the chest may also be included in the evaluation of those patients whose carcinoma was located in the upper esophagus. The endoscopic examination should always include multiple biopsies from the area of anastomosis in order to rule out recurrent disease that may be present only as microfoci of malignant cells in the early stage of recurrence. Most experienced endoscopists recommend obtaining six or more biopsy specimens with central bayonet-type biopsy forceps (Witzel et al. 1976). Brushing cytology samples

Table 1. Recommended follow-up of patients after surgery for esophageal carcinoma

Interval	Test or procedure
3 months	Blood count, liver enzymes, chest X-ray examination
6 months	Endoscopy, blood count, liver enzymes, chest X-ray examination, scan of the abdomen (CT or ultrasound)
12 months; then 6 months for 3 years	Endoscopy, blood count, liver enzymes, chest X-ray examination, scan of the abdomen
Every year for 2 years.	Endoscopy, blood count, liver enzymes

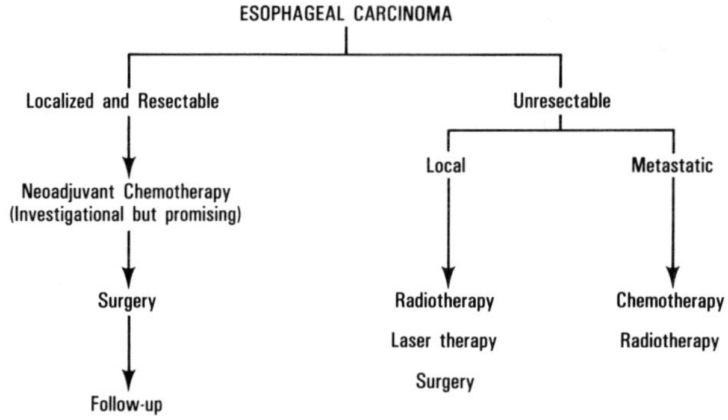

Fig. 1. Classification and treatment of Esophageal carcinoma

(Hanson et al. 1980) obtained at the end of endoscopic biopsies may increase the yield of positive findings, particularly in submucosal recurrent disease.

The success of the present treatment of patients with esophageal carcinomas depends primarily on early detection and preoperative assessment with accurate staging, overall evaluation before major abdominal and thoracic surgery, and nutritional preparation for surgical treatment rather than on the extent and frequency of follow-up examinations. Despite some progress in nonsurgical therapy of this malignancy in the past 5 years, attributed to combination chemotherapy and availability of new drugs as well as the use of chemotherapy in conjunction with radiation therapy and surgery (Shields et al. 1982), the cure of metastatic esophageal carcinoma still remains elusive (Fig. 1). Preoperative chemotherapy may produce an improvement in long-term survival but confirmatory trials are still incomplete.

13.2 Gastric Carcinoma

The number of cases of stomach cancer has declined significantly over the past five decades in the United States and in parts of Europe, but in Japan, Chile, Iceland, and Finland it continues to be a leading cause of death. This geographic variability of gastric cancer strongly suggests that environmental factors play an important role in its etiology. In addition, the results of some studies also support the existence of a higher risk for people with a familial aggregation of stomach cancer or long-standing history of atrophic gastritis, those who have pernicious anemia or adenomatous gastric polyps, and those who have undergone subtotal gastrectomy (MADDOCK 1966).

The term *gastric carcinoma* usually refers to adenocarcinoma, which makes up 95% of stomach malignancies; the remaining histologic types include leiomyosarcoma, squamous cell carcinoma, and carcinoid. Most gastric cancers, which are commonly ulcerated, begin in the distal portion of the stomach. The less common lesions appearing in the proximal stomach are usually polypoid. Recent observations, however, suggest an increased incidence of gastroesophageal adenocarcinoma. Because of the abundant lymphatic supply of the stomach, lymph node metastases are common in symptomatic patients who presumably have more advanced stages of cancer. The nodes involved are situated along the lesser and the greater curvature, near the porta hepatis, and in the subpyloric region. Additionally, gastric cancer may spread directly to adjacent organs – the esophagus, spleen, pancreas, liver – and through the gastrocolonic ligament to the transverse colon. Lymphatic spread may occur to supraclavicular nodes (Virchow's node) and hematogenous metastasis to lungs, bones, and ovaries.

Most patients with gastric cancer present at an advanced stage with lymph node or distant metastases or both. Common symptoms in these patients include weight loss with anorexia and pain that may be relieved initially by antacids and H_2 receptor antagonists. Cancer that involve the antral or pyloric region frequently cause symptoms of gastric outlet obstruction, whereas carcinomas of the cardia often grow into the distal esophagus and produce dysphagia. Bleeding from gastric carcinoma is usually chronic, but occasionally patients may present with massive upper gastrointestinal tract bleeding. The differences in presenting symptoms of early and late gastric cancer greatly overlap, precluding the use of symptom patterns to aid in gastric cancer staging (GREEN et al. 1981). However, the presence of ascites, great loss of muscle mass, jaundice from intrahepatic metastases, back pain, and occasional neurologic changes from brain metastases are all signs of advanced disease.

The diagnostic tests for gastric carcinomas are based on barium air-contrast radiologic and endoscopic techniques, both perfected by Japanese investigators (KREEL et al. 1973; MORRISSEY 1976). Recent technical advances in upper gastrointestinal endoscopy have made the procedure very effective and less difficult for the patients. The reported overall accuracy for diagnostic endoscopy and biopsy is 98.8% (LLANOS et al. 1982). The general belief is that all patients with gastric ulcers should be evaluated by early endoscopy and biopsy. The unresolved question is how the Japanese experience with early detection of gastric cancer can be applied to other countries.

In spite of improvements in diagnostic techniques, there has been no demonstrated impact of early detection on mortality rates outside of Japan. Surgery remains the only treatment that can cure a patient with gastric cancer. Therefore, a careful preoperative evaluation enables identification of candidates for curative surgery and determines what operative procedure will follow. The best results seem to be achieved by radical total or subtotal gastrectomy (DUPONT et al. 1978). Computed tomography and magnetic resonance imaging may be used to assess the extent of spread of the tumor and potential for a curative resection (MOSS et al. 1981).

The most important factors influencing the cure rate of gastric cancer are the size of the primary lesions and the number of involved lymph nodes (CADY et al. 1976). Patients with advanced disease do poorly no matter what type of surgical procedure is used (SHIU et al. 1980). The overall 5-year survival rates of surgically treated patients are dismal and in most reported series are below 30% (ALFONSO et al. 1977; DUPONT et al. 1978; CADY et al. 1976 and SHIU et al. 1980). As would be expected, metastatic and recurrent disease are the major causes of failure after surgery, particularly within the first postoperative year. The follow-up of surgically treated patients is aimed at early detection of either recurrent disease in the anastomosis or the remaining part of the stomach and the finding of metastatic disease. There are great variations among different institutions in the extent and frequency of these follow-up examinations. At the University of Texas M.D. Anderson Hospital and Tumor Institute at Houston patients on various protocols are evaluated 3 months after the initial resection and then every 6 months for a period of 5 years

Table 2. Suggested follow-up of patients after curative surgery for gastric cancer

Time after surgery	Tests and procedures
3 months	Blood count, liver enzymes
6 months	Endoscopy, chest X-ray, CT scan, blood count, liver enzymes
12 months; then every 6 months for 3 years	Endoscopy, chest X-ray, CT scan, blood count, liver enzymes
1- to 2-year intervals	Endoscopy, blood count, liver enzymes

(Table 2). Those patients surviving 5 years after surgical resection without any evidence of disease and any other complications of surgery can be evaluated yearly or every other year. The routine follow-up procedures within the 5-year postoperative period indclude an endoscopic examination with biopsies and cytology studies of the anastomosis. Abdominal CT scanning helps detect disease in the intraabdominal lymph nodes as well as the liver. The chest X-ray film and routine laboratory tests (blood count, liver enzymes), including evaluation of CEA, are also integral parts of these follow-up examinations. Some symptomatic patients may require such additional tests as bone scan or brain scan. The nature and behavior of recurrent and metastatic gastric carcinoma usually preclude a second resection with curative intent.

Combination chemotherapy of unresectable and metastatic gastric carcinoma is under investigation (Gastrointestinal Tumor Study Group 1982), and it should only be used in those patients who are of good performance status. These patients also require limited follow-up checks of measurable disease in order to assess the effects of chemotherapy.

13.3 Carcinoma of the Liver and Bile Ducts

The two major malignant tumors of the liver are hepatocellular carcinoma and cholangiocarcinoma. Other less common primary malignant tumors of the liver include hepatoblastoma, bile duct cystadenocarcinoma, carcinoid, and squamous cell carcinoma. Hepatocellular carcinoma is by far the most frequent malignant tumor worldwide.

In most countries the hepatitis B virus has an important etiologic role in hepatocarcinogenesis, perhaps through the integration of its viral genome into the DNA of hepatocytes (Beasley and Hwang 1984). Hepatocellular carcinoma may be preventable through the wide use of hepatitis B virus vaccine (Zuckerman 1982). The prognosis of patients with liver cancer has been greatly improved in some instances because of early detection and better surgical treatment. Newer imaging methods and screening for tumor antigens, such as alpha-fetoprotein, in serum of high-risk populations allow early detection of primary liver carcinomas (Kubo et al. 1978). At the same time, surgical resection, which remains the only cure for this cancer, is carried out in increasing numbers of patients in selected institutions.

Hepatocellular carcinoma and cirrhosis frequently coexist in the same liver. In countries with a high hepatitis B surface antigen (HBsAg) carrier rate, up to 80% of patients with hepatocellular carcinomas have positive test results (Tong et al. 1971). Even in those countries in which HBsAg carriers are uncommon, primary liver carcinoma in patients with nonalcoholic cirrhosis is frequently associated with hepatitis B virus infection (Omata et al. 1979). Other clinical evidence for the role of hepatitis B virus in hepatocarcinogenesis is frequent family clustering of HBsAg carriers and chronic liver disease with hepatocellular carcinomas because of vertical transmission of virus from mother to child (Ohbayashi et al. 1972). Although hepatitis B virus is associated with hepatocarcinogenesis, it is not the sole etiologic factor. In Greenland, for example, the carrier rate among the Eskimo population is very high, yet hepatocellular carcinoma is rare (Skinho et al. 1978). Only some of the nonviral carcinogens in the environment are known, including mycotoxins (aflatoxins), plant carcinogens (cycasin), a variety of synthetic carcinogens, and steroid hormones (androgenic-anabolic) (Newberne 1984; Shar and Kew 1982). Ethanol probably acts as a cocarcinogen (Vitale and Gottlieb 1979), but no clear experimental evidence indicates that it is carcinogenic in animals; however, clinical and pathologic reports support the association of alcoholic liver disease and hepatocellular carcinoma (Norredam 1979). Among the other liver diseases associated with a high incidence of hepatocellular carcinoma are α-antitrypsin deficiency (ZZ and MZ phenotypes) and hemochromatosis.

In countries without screening programs for early detection of hepatocellular carcinoma, most patients present with symptoms of weight loss, upper abdominal fullness and pain, a palpable mass or an enlarged liver, and sometimes ascites. Patients with coexisting cirrhosis may show typical signs of chronic liver disease, having only small lesions that may defy early scintigraphic detection. Rarely, primary liver cancer presents as an acute abdomen with hemoperitoneum due to the rupture of the tumor. A few patients may also present with signs of extrahepatic metastases to lungs and bones. The manifestations of obstructive jaundice are more often seen in patients with bile duct carcinomas rather than primary hepatocellular cancer.

The diagnosis of primary liver cancer is based on specific laboratory findings in conjunction with imaging studies. Alpha-fetoprotein level is the most important diagnostic marker for hepatocellular carcinoma, and its serum level is roughly proportional to the number of carcinoma cells present (Kojiro et al. 1982; Bellet et al. 1984). Computed tomography of the liver with intravenous iodinated contrast reveals early opacification of arteries supplying the mass followed by enhanced contrast within the tumor. Real-time ultrasonography is another useful diagnostic technique for localized lesions of the liver. If diagnosis is still equivocal, then arteriography may help, particularly in finding small lesions that may escape detection by other diagnostic procedures. Biopsy of the lesions, either blindly or guided by scan, is essential for a definitive tissue diagnosis.

Surgical resection is the only treatment for primary cancer of the liver that can result in cure or extended survival; therefore, early detection enhances the prospects for curative surgery (Adson 1981). Newer surgical procedures, such as the left hepatic trisegmentectomy (Starzl et al. 1982), have improved significantly the surgical cure in patients who in the past were deemed inoperable. Patients with cirrhosis and limited liver function with impaired hemostasis, however, continue to have increased operative risks and postoperative morbidity. Patients with advanced unresectable disease should be considered for chemotherapy under carefully designed protocols.

A variety of trials have been conducted in the treatment of hepatocellular carcinoma, including doxorubicin alone or with combinations of 5-fluorouracil and semustine (methyl-CCNU) or 5-fluorouracil and streptozocin (Falkson et al. 1978). Doxorubicin was the most active single agent for treatment of hepatocellular carcinoma. For cholangiocarcinoma the combination of 5-fluorouracil, doxorubicin, and mitomycin C has been reported to produce partial response in 31% of the patients (Haskell et al. 1980). Presently there are multiple ongoing studies with hepatic arterial chemoinfusion in primary liver carcinoma. Some reports already show a 50% partial response rate with floxuridine (FUDR) and an occasional patient experiences long-term survival (Haskell et al. 1980). Other experimental treatments of hepatocellular carcinoma include use of radiolabeled immunoglobulins such as antiferritin and antibodies to alpha-fetoprotein. In the future, liver transplantation may offer cure to highly selected patient with primary liver cancer, especially those with the fibrolamellar type (Iwatsuki et al. 1985).

Table 3. Recommended follow-up of patients after surgery for hepatocellular carcinoma

Interval	Tests
3 and 6 months	Alpha-fetoprotein and liver enzyme levels, blood counts, liver scan (CT or ultrasonography), chest X-ray examination
9 months	Alpha-fetoprotein and liver enzyme levels, blood count
12 months; then every 6 months for 4 years, followed by yearly intervals	Alpha-fetoprotein and liver enzyme levels, blood count, liver scan, chest X-ray

Patients with primary hepatocellular carcinoma represent a very heterogeneous group whose disease may be linked to such variable factors as geographic location and ethnic origin. They may be characterized by a multicentric origin instead of a single focus in the liver associated diseases (cirrhosis, hepatitis B), and variations in hepatic function. All these factors greatly influence the treatment and subsequent follow-up. Those patients treated by curative surgery should be frequently monitored by following the alpha-fetoprotein levels, liver enzymes, imaging studies of the abdomen (ultrasonography or computerized tomography), in addition to chest X-ray examination and routine blood counts. The frequency of these follow-up evaluations varies, but 3-month intervals during the first postoperative year and then 6-month intervals up to 5 years after surgery would be a reasonable recommendation (Table 3).

13.4 Exocrine Tumors of the Pancreas

Approximately 23000 individuals develop pancreatic cancer each year in the United States. Unfortunately the majority will be dead within 1 year. The symptoms are often vague (Tables 4, 5). In greater than two-thirds of cases, the tumor involves the head of the pancreas. The most common symptoms of cancer of the head of the pancreas are weight loss, jaundice, and pain. Carcinoma of the body and tail does not usually cause jaundice but weight

Table 4. Clinical presentation of pancreatic cancer: symptoms (modified from HOWARD and JORDAN 1977)

Head		Body and tail	
Major symptoms	Percent of patients	Major symptoms	Percent of patients
Weight loss	92	Weight loss	100
Jaundice	82	Pain	87
Pain	72	Weakness	43
Anorexia	64	Nausea	43
Dark urine/ light stools	63	Vomiting	37
		Anorexia	33
Nausea/vomiting	45	Constipation	27
Weakness	35		
Pruritus	24		

Table 5. Clinical presentation of pancreatic cancer: physical signs (modified from HOWARD and JORDAN 1977)

Head		Body and tail	
Major signs	Percent of patients	Major signs	Percent of patients
Jaundice	87	Palpable liver	33
Palpable liver	83	Abdominal mass	23
Palpable gallbladder	29	Ascites	20
		Jaundice	13
Ascites	14		
Abdominal mass	13		

loss and pain are more evident than in carcinomas of the body and tail. Radiation of pain to the back is common (HOWARD and JORDAN 1977).

Physical examination of patients with carcinoma of the head of the pancreas reveals jaundice and hepatomegaly (HOWARD and JORDAN 1977). A palpable gallbladder (Courvoisier's sign) can be found in 30% of patients.

Computed tomography and ultrasonography are the two most commonly noninvasive tests used in the initial evaluation of patients with the suspicion of pancreatic cancer. Ultrasound is most helpful in the jaundiced patient to differentiate between nonobstructive and obstructive jaundice. CT is most helpful in determining the cause and site of the obstruction. (MOOSSA and LEVIN 1981).

A histologic diagnosis of pancreatic cancer can be made by ERCP and aspiration cytology or percutaneous needle aspiration cytology. The latter technique is best used for clearly unresectable lesions of the body and tail.

Resectability of the pancreatic lesion can often be determined by using CT; this technique can demonstrate vascular invasion or peripancreatic spread.

Table 6. Suggested follow-up of patients after curative surgery for pancreatic adenocarcinoma

Time after resection to initiation of therapy	Tests and procedures
3 months	Liver enzymes, CT scan, CEA
6 months	CT scan, CEA, liver enzymes
12 months and then every 6 months for 3 years	CT scan

The approach to therapy of patients with pancreatic cancer will vary considerably depending on the stage at which the diagnosis is made. Neoplasms of the head of the pancreas may be resected or the patient may undergo bypass of the biliary tract and/or duodenum. Neoplasms of the body and tail are usually unresectable and bypass procedures may sometimes be indicated for palliation of luminal obstruction.

For patients with locally unresectable cancers, radiation therapy administered intraoperatively or external beam techniques may be employed (DOBELBOWER 1986). For patients with overt metastatic disease, phase II chemotherapy or combination therapy may be used (LITKA and SCHEIN 1986).

For purposes of follow-up evaluation the most useful radiologic technique is the CT scan in that it permits evaluation of draining lymph nodes as well as other sites of involvement such as the liver. Serological markers such as CEA and CA 19-9 (FAINTUCH and LEVIN 1986) may also be occasionally used to monitor the results of therapy (Table 6).

13.5 Large Bowel Cancer

In the United States approximately 140000 individuals had the diagnosis of large bowel cancer made in 1987 and it is estimated that 60000 individuals died from this disorder (AMERICAN CANCER SOCIETY 1987).

Symptoms will vary with the location of the cancer (Table 7) (POSTLETHWAIT 1949).

The initial evaluation of the patient with suspected colorectal cancer includes detailed history, physical examination, blood count, and endoscopy (flexible sigmoidoscopy or colonoscopy) or a double contrast barium enema. Assessment of

Table 7. Comparison of the five most frequent symptoms in rectal, left colon, and right colon cancer (modified from Postlethwait 1949)

Rectum and rectosigmoid (258 persons)	Left colon (99 patients)	Right colon (984 patients)
Hematochezia (85%)	Abdominal pain (72%)	Abdominal pain (74%)
Constipation (46%)	Melena (53%)	Weakness (29%)
Tenesmus (30%)	Constipation (42%)	Melena (27%)
Diarrhea (30%)	Nausea (25%)	Nausea (24%)
Abdominal pain (26%)	Vomiting (23%)	Abdominal mass (23%)

Table 8. Evaluation of patients treated for large bowel cancer

Preoperative:	Colonoscopy or flexible sigmoidoscopy and double contrast barium enema, chest X-ray
Postoperative:	History, physical examination, biochemical profile at 3-month intervals for 3 years and then every 6 months for 2 years
CEA:	After 4 weeks and then every 3 months for 3 years and then at 6-month intervals for 2 additional years
Colonoscopy:	After 6–9 months and then every 24 months for 4 additional years
Chest X-ray:	After 6 months and then every 12 months for 5 years

CT scans are used if hepatomegaly is present or if liver enzymes are abnormal; bone and brain scans are employed in symptomatic patients

distant spread in the abdomen or pelvis will employ CT and/or ultrasonography, and chest X-ray will be useful in determining the presence of pulmonary metastases.

The measurement of CEA preceding primary colon and rectal cancer resection has been correlated with the stage of the disease (Wanebo 1978) and can be used as a rough guide to prognosis independent of the histopathologic staging (Herrera et al. 1976).

There is still considerable controversy about the most cost-effective manner by which to follow patients after curative resection of large bowel neoplasms. The approaches used vary from center to center but will usually include methods to detect metachronous neoplasms in the colon as well as distant spread or recurrence at the anastomosis. In addition to routinely performed blood chemistry tests and endoscopic and imaging studies, the physician should be sensitive to changes in the patient's symptoms.

13.5.1 Synchronous Lesions

Colonoscopy or double contrast barium enema should be performed prior to resection to exclude synchronous tumors. If this is not possible because of the presence of obstruction, colonoscopy should be carried out 2–3 months after resection.

13.5.2 Metachronous Lesions

Postoperatively, colonoscopy is the preferred method of surveillance. In a series of 240 patients who had undergone resection of colorectal cancer, Nava (1983) compared the value of different methods of colonic surveillance at 6 and 12 months and annually for the next 4 years. Out of 304 colonoscopies performed, 17 recurrent cancers, 11 new cancers, 66 tubular adenomas, and 9 villous adenomas were discovered. Barium enema (predominantly double contrast) detected approximately 40% of the lesions seen on colonoscopy. Occult blood testing was extremely insensitive, particularly in the detection of adenomas, and cannot be recommended. We perform colonoscopy within the first 6–9 months after resection and then at 2-yearly intervals for the next 4 years and then at 3-year intervals for the next 6 years (Table 8).

13.5.3 Other Methods of Follow-up

Laboratory studies, including blood count and liver enzymes, are obtained at 6-month intervals for the first 3 years and then at yearly intervals. Chest roentgenogram is obtained at yearly intervals for the first 4 years and then at 2- to 3-year intervals thereafter (Table 8).

13.5.4 Carcinoembryonic Antigen and Second-Look Procedures

Considerable controversy exists concerning the use of serial CEA monitoring after "curative" resection of large bowel cancer. Recent studies have suggested that approximately 20% of patients reoperated for a rising CEA will be rendered disease-free 5 years after resection of recurrent tumors (Steele 1986). Preoperative diagnosis of sites of recurrence using improved imaging methods such as CT and

MRI scans may help to exclude patients with clearly unresectable metastases. Unfortunately, the majority of patients will not benefit from these reoperative procedures and their ultimate curability will depend upon the future development of successful systemic therapy. Nevertheless, we do follow the CEA in an attempt to define those few patients who will benefit from aggressive surgical intervention.

The patient enrolled in experimental adjuvant therapy programs will require a variety of tests to follow the response to therapy. The recommendations outlined in Table 8 are those which are used in our center in the routine follow-up of our patients.

References

Adson MA (1981) Diagnosis and surgical treatment of primary and secondary solid hepatic tumors in the adult. Surg Clin North Am 61: 181-196

Alfonso A, Rosen P, Guerra O, Fortuer J (1977) Adenocarcinoma of the proximal third of the stomach: pitfalls in surgical management. Am J Surg 134: 325-330

American Cancer Society (1987) Cancer facts and figures. American Cancer Society, New York City

Appelqvist P (1972) Carcinoma of the oesohagus and gastric cardia: a retrospective study based on statistical and clinical material from Finland. Acta Chir Scand 430: 1-92

Beasley RP, Hwang LY (1984) Hepatocellular carcinoma and hepatitis B virus. Semin Liver Dis 4: 113-121

Bellet DH, Wands JR, Bohuon C (1984) Serum alpha-fetoprotein levels in human disease: perspective from a highly specific monoclonal radioimmunoassay. Proc Natl Acad Sci USA 81: 3869-3873

Cady B et al. (1976) Treatment of gastric cancer. Surg Clin North Am 56: 599-605

Cedarquist C et al. (1978) Cancer of the esophagus II. Therapy and outcome. Acta Clin Scand 144: 233

Dobelbower RR (1986) Therapy by irradiation. In: Go VLW (ed) The exocrine pancreas, biology, pathology and disease. Raven, New York, pp 699-711

Dupont BJ et al. (1978) Adenocarcinoma of the stomach. Review of 1497 cases. Cancer 42: 941-947

Falkson G et al. (1978) Chemotherapy studies in primary liver cancer: a prospective randomized clinical trial. Cancer 42: 2149-2156

Faintuch J, Levin B (1986) Clinical presentation and diagnosis of exocrine tumors of the pancreas. In: Go VLW (ed) The exocrine pancreas, biology, pathology and disease. Raven, New York, pp 675-678

Gastrointestinal Tumor Study Group (1982) A comparative clinical assessment of combination chemotherapy in the management of advanced gastric carcinoma. Cancer 49: 1362-1366

Giuli R (1985) Surgical complications and reasons for failure. In: DeMeester TR, Levin B (eds) Cancer of the esophagus. Grune and Stratton, Orlando, pp 199-208

Green PHR et al. (1981) Early gastric cancer. Gastroenterology 81: 247-256

Guernsey JM, Knudsen DF (1970) Abdominal exploration in the evaluation of patients with carcinoma of the thoracic esophagus. J Thorac Cardiovasc Surg 59: 62-66

Hanson JT et al. (1980) Brush cytology in the diagnosis of upper gastrointestinal malignancy. Gastrointest Endosc 26: 33-35

Haskell CM (1980) Cancer of the liver. In: Haskell CM (ed) Cancer treatment. Saunders, Philadelphia, pp 319-357

Herrera MA et al. (1976) Carcinoembryonic antigen as a prognostic and monitoring test in clinically complete resection of colorectal carcinoma. Ann Surg 183: 5-17

Howard JM, Jordan JL (1977) Cancer of the pancreas. Curr Probl Cancer 2: 43-47

Iwatsuki S et al. (1985) Role of liver transplantation in cancer therapy. Ann Surg 202: 401-407

Kojiro M et al. (1982) Hepatocellular carcinoma presenting as intra-bile duct tumor growth: A clinicopathological study f 24 cases. Cancer 99: 2144-2147

Kreel L et al. (1973) Techniques of double-contrast barium meal with examples of correlation with endoscopy. Clin Radiol 24: 307-314

Kubo Y et al. (1978) Detection of hepatocellular carcinoma during a clinical follow-up of chronic liver disease: observations in 31 patients. Gastroenterology 74: 578-582

Litka PA, Schein PS (1986) Chemotherapy of pancreatic cancer. In: Go VL (ed) The exocrine pancreas, biology, pathology and disease. Raven, New York, pp 689-697

Llanos O et al. (1982) Accuracy of the first endoscopic procedure in the differential diagnosis of gastric lesions. Ann Surg 195: 224-226

Maddock CR (1966) Environment and heredity factors in carcinoma of the stomach. Br J Cancer 20: 660-669

Mandard AM et al. (1981) Autopsy findings in 111 cases of esophageal cancer. Cancer 48: 329-335

Ming SC (1973) Tumors of the esophagus and stomach. Atlas Tumor Pathology 7: 27-79

Moossa AR, Levin B (1981) The diagnosis of "early" pancreatic cancer: The University of Chicago experience. Cancer 47: 1688-1697

Morrissey JF (1976) The diagnosis of early gastric cancer. Gastrointest Endosc 23: 14-15

Morson BC, Dawson IMP (1979) Gastrointestinal pathology, 2nd edn. Blackwell Scientific, Oxford, pp 48-49

Moss AA et al. (1981) Gastric adenocarcinoma: A comparison of the accuracy and economics of staging by computed tomography and surgery. Gastroenterology 80: 45-50

Nava H (1983) Postoperative surveillance of colorectal cancer. Cancer 49: 1043-1049

Newberne PM (1984) Chemical carcinogenesis: mycotoxins and other chemicals to which humans are exposed. Semin Liver Dis 4: 122-35

Norredam J (1979) Primary carcinoma of the liver. Acta Pathol Microbiol Immunol Scand 87: 227-236

Ohbayashi A et al. (1972) Familial clustering of asymptomatic carriers of Australia antigen and patients with chronic liver disease or primary liver cancer. Gastroenterology 62: 618-625

Omata M et al. (1979) Hepatocellular carcinoma in the U.S.A.: Etiologic considerations. Localization of hepatitis B antigens. Gastroenterology 76: 279-287

Postlethwait RW (1949) Malignant tumors of the colon and rectum. Ann Surg 129: 33-46

Raphael HA (1966) Primary adenocarcinoma of the esophagus: 18-year review of the literature. Ann Surg 164: 785-796

Shar SR, Kew MC (1982) Oral contraceptives and heatocellular carcinoma. Cancer 49: 407-410

Shin MH et al. (1980) Selection of operative procedure for adenocarcinoma of the midstomach. Ann Surg 192: 730-737

Skinho JP et al. (1978) Occurrence of cirrhosis and primary liver cancer in an Eskimo population hyperendemically infected with hepatitis B virus. Am J Epidemiol 108: 121-125

Skinner DB et al. (1982) Potentially curable cancer of the esophagus. Cancer 50: 2571

Steele G (1986) Follow-up plans after "curative" resection of primary colon or rectum cancer in colorectal cancer: current concepts in diagnosis and treatment. In: Steele G, Osteen RT (eds) Marcel Dekker, New York, chap 10, pp 247

Starzl TE et al. (1982) Left hepatic trisegmentectomy. Surg Gynecol Obstet 155: 21-27

Tong MJ et al. (1971) Hepatitis-associated antigen and hepatocellular carcinoma in Taiwan. Ann Intern Med 75: 687-691

Vitale JJ, Gottlieb LS (1979) Alcohol and alcohol-related deficiencies as carcinogens. Cancer Res 35: 1116-1138

Wanebo HJ (1978) The use of the preoperative carcinoembryonic antigen level as a prognostic indicator to complement pathologic staging. N Engl J Med 299: 448-460

Witzel L et al. (1976) Evaluation of specific value of endoscopic biopsies and brush cytology for malignancies of the esophagus and stomach. Gut 17: 375-377

Zuckerman AJ (1982) Virological approach to the prevention or primary liver cancer. Hepatology 2: 67-71

Subject Index